Advanced Underground Space Technology

Advanced Underground Space Technology

Editors

Chenjie Gong
Mingfeng Lei
Xianda Shen

MDPI • Basel • Beijing • Wuhan • Barcelona • Belgrade • Manchester • Tokyo • Cluj • Tianjin

Editors

Chenjie Gong
Central South University
China

Mingfeng Lei
Central South University
China

Xianda Shen
Clarkson University
USA

Editorial Office
MDPI
St. Alban-Anlage 66
4052 Basel, Switzerland

This is a reprint of articles from the Special Issue published online in the open access journal *Applied Sciences* (ISSN 2076-3417) (available at: https://www.mdpi.com/journal/applsci/special_issues/Advanced_Underground_Space_Technology).

For citation purposes, cite each article independently as indicated on the article page online and as indicated below:

LastName, A.A.; LastName, B.B.; LastName, C.C. Article Title. *Journal Name* **Year**, *Volume Number*, Page Range.

ISBN 978-3-0365-5533-1 (Hbk)
ISBN 978-3-0365-5534-8 (PDF)

Contents

About the Editors

Chenjie Gong

Chenjie Gong obtained his Ph.D. in Civil Engineering from Tongji University in 2018 and is an Associate Professor at Central South University. He was a visiting student researcher at University of California, Berkeley, from 2016 to 2018. Before that, he graduated with a degree in Civil Engineering from Tongji University in 2012. He is a member of the China Civil Engineering Society (CCES), China Railway Society (CRS), China Highway & Transportation Society (CHTS), and Chinese Society for Rock Mechanics & Engineering (CSRME). He has authored more than 50 scientific and technical papers and is a frequent reviewer of more than 10 computer technology and civil and infrastructure journals, such as *Computer-Aided Civil and Infrastructure Engineering, IEEE Transactions on Neural Networks and Learning Systems, Tunnelling and Underground Space Technology, International Journal for Numerical and Analytical Methods in Geomechanics*, among others. His research activity is focused on tunnel and underground infrastructures, with a special interest in the performance and health monitoring of tunnels using machine and deep learning.

Mingfeng Lei

Mingfeng Lei graduated in Civil Engineering at Central South University in 2013 and is a Full Professor at Central South University. His teaching career has covered degree, post-graduate, Master's and doctorate study courses. Additionally, he has supervised more than 30 Master's degrees and PhD dissertations. In terms of the transference of knowledge, he has performed numerous laboratory tests of tunnel mechanics and many technical studies concerning problems with tunnels and underground infrastructure. At present, he is the vice chair of the Committee for Tunnel Construction and Operation in China Civil Engineering Society (CCES), a member of the 4th Expert Committee for China Highway Construction Association (CHCA), and the vice secretary-general of the Hunan Railway Association. He has authored more than 100 scientific and technical papers, more than 50 patents and 10 books. He has supervised various research grants in the field of tunnel and underground structures. His research interests include the evolution mechanism of underground structure lifetime performance and construction technology of tunnel/underground engineering in complex conditions.

Xianda Shen

Xianda Shen earned his Ph.D. in Civil Engineering from Georgia Institute of Technology in 2019 and is an Assistant Professor at Clarkson University. At present, he is a member of the American Society of Civil Engineering (ASCE) and American Rock Mechanics Association (ARMA). He has authored more than 20 scientific and technical papers and is a frequent reviewer of more than 10 geotechnical journals. A major focal point of his work over the coming years is to gain a further understanding of the microscopic properties and the THCM coupling processes that characterize the macroscopic properties of geomaterials. The overarching goal of these activities is to develop a rigorous and mechanics-based framework to predict the evolution of intrinsic properties and the mechanical behavior of materials. His research aims to shed light on the resilient and sustainable geotechnical and geological engineering systems.

Preface to "Advanced Underground Space Technology"

The world is constantly changing, and so is tunnelling and underground space. The challenges posed by society are increasingly demanding in terms of the safety and stability of underground works, as well as their sustainability. The tunnelling and underground space community has been facing up to these challenges by presenting innovative solutions. Advances in high-performance materials and structures, information modeling, and aritificial intelligence, among other topics, are examples of the most recent advances, trends, and prospects in tunnelling and underground engineering. All types of underground works, such as mechanical tunnels, drill-and-blast tunnels, and subway stations, benefit from such advances.

The aim of this Special Issue book is not to provide an exhaustive summary of the latest advances in the field of tunnelling and underground space but, rather, to present some works that are considered innovative and may have practical applications in the near future. Therefore, this Special Issue, "Advanced Underground Space Technology", brings together 18 papers (from among 27 submitted papers) that, in addition to their scientific merit, address some of the current and future challenges in tunnelling and underground space. The published papers cover a wide range of these emerging topics with a specific focus on the research, design, construction, and performance of underground works. These works are expected to inspire the development of tunnelling and contribute to the future construction of more resilient and sustainable tunnel/underground structures.

In the end, we would like to take this opportunity to acknowledge all the authors who contributed to this Special Issue, "Advanced Underground Space Technology". Special thanks are also due to all reviewers for their valuable contributions and to the *Applied Sciences* MDPI editorial team for the invitation and for support given during the production of this Special Issue book, especially Mr. Enoch Li and Ms. Kimi Wang.

Chenjie Gong, Mingfeng Lei, and Xianda Shen
Editors

applied sciences

MDPI

Editorial

Advanced Underground Space Technology

Chenjie Gong [1,*], Mingfeng Lei [1] and Xianda Shen [2]

[1] School of Civil Engineering, Central South University, Changsha 410075, China
[2] Department of Civil and Environmental Engineering, Clarkson University, Potsdam, NY 13699, USA
* Correspondence: gongcj@csu.edu.cn

The Special Issue titled "Advanced Underground Space Technology" was launched with an invitation to authors from all over the world to address state-of-the-art challenging topics in tunnelling and underground space technology. Advances in high-performance materials and structures, information modeling, artificial intelligence, among other topics, are examples of the most recent advances, trends, and prospects in the fields of tunnel and underground engineering. Twenty-seventeen manuscripts were submitted to this Special Issue, and eighteen were accepted for publication. These contributions addressed some emerging topics, with a specific focus on the research, design, construction and operation of tunnelling and underground infrastructures.

Gong et al. [1] summarize the research progress on waterproofing of sealing gaskets to guide follow-up research and practical referencing for engineering design. Combined with the domestic and foreign scholars' research on the joint sealing performance of segmental tunnel lining and related analysis methods related topics, in this article, the waterproof performance, material characteristics, numerical simulation, and thermal–mechanical coupling analysis of the gasketed joint are discussed.

Lin et al. [2] evaluate the effect of closed-cell aluminum foam boards in tunnel noise reduction under two different working conditions, and obtain a noise reduction effect of fixed pure sound sources and moving vehicle noise in the tunnel, which can provide theoretical basis for the promotion of closed-cell aluminum foam in tunnel applications.

Huang et al. [3] propose a new idea of cross-section design for a minimum bending moment shield tunnel. The pioneering work performed here shows that a rational cross-section can greatly decrease the bending moment for underground structures. For shield tunnels or buried pipes, the minimum bending moment cross-section is similar to a vertical ellipse. The research results can be used as a reference for the design of a shield tunnel or other underground structures.

Tong et al. [4] introduce the rough set theory and conditional entropy and establish a suitability evaluation model of urban underground space (UUS). Rough set theory is used to construct a decision information table, preprocess sample data and classify the knowledge base, while conditional entropy is employed to calculate the attributes' own and relative importance. The results show that the overall suitability of underground space development and utilization in the starting area of Wuhan Changjiang New Town are good.

Lee et al. [5] propose the direct calculation method (DCM) to investigate the interaction behavior of the support and ground due to the excavation of a circular tunnel in the isotropic stress field. The feasibility of a direct algorithmic process for the analytical solution is examined by numerical analysis steps; in particular, the finite element method (FEM). The purpose of this paper is to provide computational concepts for tunnel analysis and to comprehensively discuss solutions for the support–ground interaction behavior in tunneling.

Nguyen et al. [6] present an improved HRM method that is developed to estimate the internal forces induced in square and rectangular tunnel linings. The developed HRM method is validated by comparing it with results obtained by FEM analysis. The HRM

Citation: Gong, C.; Lei, M.; Shen, X. Advanced Underground Space Technology. *Appl. Sci.* **2022**, *12*, 9613. https://doi.org/10.3390/app12199613

Received: 8 September 2022
Accepted: 22 September 2022
Published: 25 September 2022

method is then used to investigate the effects of different parameters of the tunnel lining and the surrounding soil, including the flexibility ratio of the tunnel lining F, and the coefficient of lateral earth pressure on the internal forces induced in the tunnel lining.

Zhao et al. [7] introduce 10 machine learning algorithms to predict the quality of surrounding rock using measure-while-drilling (MWD) data (drilling parameters) obtained from five tunnels of the Zhengzhou–Wanzhou line of the high-speed railway project in China. Through comparative analysis, three machine learning models with better comprehensive performance among them are selected to establish the tunnel surrounding rock intelligent classification system by the drill-and-blast method. The results of this study create a solid foundation for the dynamic design and intelligent construction of tunnels.

Dai et al. [8] study the structural response of segments in different assembling positions and under the condition of asymmetrical propulsion exerted by jacks. By establishing a finite element model of segment assembly, they explore the interaction and stress concentration of unlooped segments under asymmetric stress in the synchronous propulsion and assembly mode. This study determines the assembly error control range of the synchronous propulsion and assembly mode applied in practical projects.

Zeng et al. [9] carry out indoor direct shear tests to deeply study the influence of the compound bentonite slurry on the friction characteristics of the pipe–soil interface from two aspects: the slurry concentration and the slurry standing time. By analyzing the test data and phenomena, combined with the physical properties of the slurry, the influencing mechanisms of the concentration and standing time on the friction characteristics of the interface are revealed.

Wu et al. [10] establish a design optimization method for feet-lock steel pipes under two working conditions, including predicted large deformation and observed large deformation before the primary lining construction. The method is applied to the Yulinzi Tunnel, and the results show that the original design of the feet-lock steel pipes could not meet the requirements of the controlling settlement. The active bearing design can significantly reduce the amount of steel required and the engineering cost.

Huang et al. [11] adopt the Loganathan formula based on a virtual image principle to calculate a surface settlement trough without isolation piles. On this basis, Melan a solution in the semi-infinite plane based on elasticity is introduced to derive the ultimate surface settlement trough from the perspective of negative friction of pile edge. The applicability and reliability of the analytical and numerical solutions are verified by comparing field measured data, which have significant engineering guiding reference in analyzing real projects.

He et al. [12] propose a refined three-dimensional model considering bolts and other construction loads to study the effect of jack thrust angle change on the mechanical characteristics of shield tunnel segmental linings. Discussions are carried out on the displacement, deformation, and stress of the tunnel segments and the connecting bolts. According to the obtained results, final recommendations will be provided to control the jack thrust angle. The results in this paper provide novel insights into the effect of jack thrust on shield tunnels.

Chen et al. [13] analyze the impact of improvements in tunnel sidewall brightness on the lighting environment and visual characteristics of human eyes based on the lighting experiments when the tunnel sidewalls are decorated with two different types of materials and illuminated by LED lamps with five different color temperatures. The test results show that the tunnel sidewall luminance will increase if the energy storage and luminescent coating with high reflectance is decorated on tunnel sidewalls, and the pavement luminance increases with the increasing sidewall luminance.

Zhou et al. [14] consider the influence of internal pore structure characteristics and matrix material properties of foamed concrete on the overall elastic modulus. Combined with the two-layer embedded model, an inclusion theory model of the elastic modulus of foamed concrete is established. The influence law of the elastic parameters of foamed concrete is thoroughly explored in accordance with the laboratory tests and the existing theoretical models to verify the theory.

In the work of Wu et el. [15], the buried rubber waterstop commonly used in the project is the research object, and the waterproofing principle of the waterstop, the bonding of the waterstop and concrete, the extrusion effect of the concrete pouring, etc., are taken into account to carry out a finite element analysis, study the failure characteristics, and propose a relevant optimization scheme to provide a reference for the optimization of the waterproof measures of a mountain tunnel joint.

Zhang et al. [16] introduce an improved method to achieve this target, in which a 3D finite element model for the soil–tunnel system should be established to obtain the overall performance response under excitations of different input waves. Based on the joint opening and offset obtained from the integral model, a refined model of the anchor joint is then set up to further examine its mechanical behavior under the seismic action. The method developed in this paper can also be accessed for use in seismic response analysis of other types of new joints with complicated structures in shield tunnels.

Huang et al. [17] design a high-strength and high-rigidity initial tunnel support structure based on the current situation of the construction of shallow-buried large-section tunnels in complex urban environments. The bearing characteristics and failure mechanism are analyzed through indoor model tests, and industrial tests are carried out, relying on the Guangzhou subway tunnel project to provide a reasonable new support type for urban shallow-buried large-section tunnels.

Radovanović et al. [18] present a methodology for efficient and accurate modeling of water losses in hydraulic tunnels under inside internal water pressure, based on multiple linear regression (MLR). The methodology presented has been validated in modeling water losses in the hydraulic tunnel under the pressure of PSHPP "Bajina Bašta" in the Republic of Serbia. The obtained results have shown significantly better accuracy compared to the results published by other authors, proving that the developed model can be used as a powerful tool in future analyses of tunnel losses and remediation planning.

Overall, this Special Issue, together with many other papers published in *Applied Sciences* and other platforms, will provide useful sources of knowledge to researchers, engineers, designers, and other colleagues in the society of tunnelling and underground space. As the guest editors, we would like to thank all authors for their contributions and the reviewers for their valuable comments. Last but not least, we are grateful to Mr. Enoch Li and Ms. Kimi Wang, Section Managing Editors, and all the Academic Editors of this Special Issue for their great support during the entire editorial process.

Author Contributions: Writing—original draft preparation and review, C.G.; writing—editing, C.G., M.L. and X.S. All authors have read and agreed to the published version of the manuscript.

Funding: This Special Issue was supported by the National Natural Science Foundation of China (Grant No. 51908557, 52278421 and 51978669) and China Railway Corporation Science and Technology Research and Development Program (K2021G022).

Acknowledgments: As the guest editors, we would like to thank all the authors who submitted papers to this Special Issue. The papers published in this edition have been peer-reviewed by experts in the field, whose comments greatly improved the quality of the publication. Moreover, we also would like to thank the Editorial Board of *Applied Sciences* for their assistance in managing this Special Issue.

Conflicts of Interest: The authors declare no conflict of interest.

References

1. Gong, C.; Wang, Y.; Ding, W.; Lei, M.; Shi, C. Waterproof Performance of Sealing Gasket in Shield Tunnel: A Review. *Appl. Sci.* **2022**, *12*, 4556. [CrossRef]
2. Lin, Z.; Gong, W.; Wan, L.; Shen, J.; Zhang, H.; Huang, J.; Zhu, B. Field Measurements for Traffic Noise Reduction in Highway Tunnels Using Closed-Cell Aluminum Foam Board. *Appl. Sci.* **2022**, *12*, 538. [CrossRef]

3. Huang, D.; Jiang, H.; Xu, C.; Tu, W.; Li, X.; Wang, W. A New Design Method of Shield Tunnel Based on the Concept of Minimum Bending Moment. *Appl. Sci.* **2022**, *12*, 1082. [CrossRef]
4. Tong, D.; Tan, F.; Ma, B.; Jiao, Y.-Y.; Wang, J. A Suitability Evaluation Method of Urban Underground Space Based on Rough Set Theory and Conditional Entropy: A Case Study in Wuhan Changjiang New Town. *Appl. Sci.* **2022**, *12*, 1347. [CrossRef]
5. Lee, Y.-L.; Hsu, W.-K.; Chou, P.-Y.; Hsieh, P.-W.; Ma, C.-H.; Kao, W.-C. Verification and Comparison of Direct Calculation Method for the Analysis of Support–Ground Interaction of a Circular Tunnel Excavation. *Appl. Sci.* **2022**, *12*, 1929. [CrossRef]
6. Nguyen, C.T.; Do, N.A.; Dias, D.; Pham, V.V.; Alexandr, G. Behaviour of Square and Rectangular Tunnels Using an Improved Finite Element Method. *Appl. Sci.* **2022**, *12*, 2050. [CrossRef]
7. Zhao, S.; Wang, M.; Yi, W.; Yang, D.; Tong, J. Intelligent Classification of Surrounding Rock of Tunnel Based on 10 Machine Learning Algorithms. *Appl. Sci.* **2022**, *12*, 2656. [CrossRef]
8. Dai, Z.; Li, P.; Wang, X.; Liu, J.; Fan, J.; Kou, X. Asymmetric Force Effect and Damage Analysis of Unlooped Segment of Large-Diameter Shield under Synchronous Propulsion and Assembly Mode. *Appl. Sci.* **2022**, *12*, 2850. [CrossRef]
9. Zeng, C.; Xiao, A.; Liu, K.; Ai, H.; Chen, Z.; Zhang, P. Experimental Study on the Influence of Slurry Concentration and Standing Time on the Friction Characteristics of a Steel Pipe-Soil Interface. *Appl. Sci.* **2022**, *12*, 3576. [CrossRef]
10. Wu, Y.; Tian, C.; Xu, P.; Zhao, Z.; Zhang, J.; Wang, S. Design Optimization Method of Feet-Lock Steel Pipe for Soft-Rock Tunnel Based on Load-Deformation Coordination. *Appl. Sci.* **2022**, *12*, 3866. [CrossRef]
11. Huang, K.; Sun, Y.; Kuang, X.; Huang, X.; Liu, R.; Wu, Q. Study on the Restraint Effect of Isolation Pile on Surface Settlement Trough Induced by Shield Tunnelling. *Appl. Sci.* **2022**, *12*, 4845. [CrossRef]
12. He, L.; Jiang, Y.; Zhang, W. Effect of Jack Thrust Angle Change on Mechanical Characteristics of Shield Tunnel Segmental Linings Considering Additional Constrained Boundaries. *Appl. Sci.* **2022**, *12*, 4855. [CrossRef]
13. Chen, X.; Feng, S.; Li, J.; Zhou, L.; Mao, W.; Zhu, H. Effects of Sidewall Brightness on LED Lighting Environment and Visual Performance in Road Tunnels. *Appl. Sci.* **2022**, *12*, 4919. [CrossRef]
14. Zhou, Z.; Hu, J.; Li, F.; Zhang, J.; Lei, M. Elastic Modulus Prediction Model of Foamed Concrete Based on the Walsh Formula. *Appl. Sci.* **2022**, *12*, 5142. [CrossRef]
15. Wu, Y.; Wu, H.; Chu, D.; Feng, S.; Zhang, J.; Wu, H. Failure Mechanism Analysis and Optimization Analysis of Tunnel Joint Waterstop Considering Bonding and Extrusion. *Appl. Sci.* **2022**, *12*, 5737. [CrossRef]
16. Zhang, G.; Zhang, W.; Qi, J.; Niu, R.; Zhang, C. Seismic Response Analysis of Anchor Joint in Shield–Driven Tunnel Considering Soil–Structure Interaction. *Appl. Sci.* **2022**, *12*, 6362. [CrossRef]
17. Huang, M.; Song, Y.; Zhang, X.; Sun, T. Experimental Study and Engineering Application of the Spatial Reticulated Grid Bolt-Shotcrete Support Structure for Excavation Tunnels. *Appl. Sci.* **2022**, *12*, 8506. [CrossRef]
18. Radovanović, S.; Milivojević, M.; Stojanović, B.; Obradović, S.; Divac, D.; Milivojević, N. Modeling of Water Losses in Hydraulic Tunnels under Pressure Based on Stepwise Regression Method. *Appl. Sci.* **2022**, *12*, 9019. [CrossRef]

applied sciences

MDPI

Article

Field Measurements for Traffic Noise Reduction in Highway Tunnels Using Closed-Cell Aluminum Foam Board

Zanquan Lin [1], Weipeng Gong [2], Li Wan [3], Jiajia Shen [3], Hu Zhang [1], Juan Huang [1,*] and Binbin Zhu [1,*]

1 School of Civil Engineering, Central South University, Changsha 410075, China; lzq1205062393@csu.edu.cn (Z.L.); zhang-h@csu.edu.cn (H.Z.)
2 Research and Development Department, Shandong High-Speed High-Tech Materials Technology Co., Ltd., Jinan 250013, China; weipenggong@163.com
3 Department of Engineering Design, Shandong Provincial Communications Planning and Design Institute Group Co., Ltd., Jinan 250031, China; foxwl0901@163.com (L.W.); shenjiajia0720@163.com (J.S.)
* Correspondence: huangjuan77@csu.edu.cn (J.H.); zhubinbin@csu.edu.cn (B.Z.)

Abstract: In order to explore the sound absorption and noise reduction performance of closed-cell aluminum foam in the tunnel, the field test of the sound absorption performance of aluminum foam board was carried out based on the installation of aluminum foam board in the whole line of Haoshanyu Tunnel on Qinglan Expressway. Combined with the existing loudspeaker test and typical tunnel measurements, a new field test method for the noise reduction performance of closed-cell aluminum foam board was proposed for two different working conditions including fixed-point pure tone sound source condition and mobile vehicle sound source condition. The testing results of the new methods were analyzed, and it showed that the closed-cell aluminum foam has good sound absorption property at the frequency spectra between 250 Hz and 1000 Hz, and the farther away from the sound source, the better the sound absorption effect. In the research on the noise reduction effect of actual vehicle, it was found that the insertion loss of the closed-cell foam aluminum board is about 4 dB(A), which indicated that the closed-cell aluminum foam can play a certain noise reduction effect in the tunnel.

Keywords: highway tunnel; noise reduction; closed-cell aluminum foam; field test; sound absorption; insertion loss

Citation: Lin, Z.; Gong, W.; Wan, L.; Shen, J.; Zhang, H.; Huang, J.; Zhu, B. Field Measurements for Traffic Noise Reduction in Highway Tunnels Using Closed-Cell Aluminum Foam Board. *Appl. Sci.* **2022**, *12*, 538. https://doi.org/10.3390/app12020538

Academic Editor: Daniel Dias

Received: 22 November 2021
Accepted: 15 December 2021
Published: 6 January 2022

1. Introduction

Noise pollution has attracted more and more people's attention. Noise not only causes damage to people's hearing, but also affects people's physiology and endangers people's physical and mental health. In many situations, the noise pollution in the tunnel is particularly serious. In recent years, tunnel construction in China has developed rapidly. As a consequence, there were 21,316 graded highway tunnels with a total length about 21,999.3 km operated by the end of 2020. Compared with the previous two years, the number of graded highway tunnels increased by 40% and 49% respectively. Besides, more and more tunnels were constructed at a complex environment [1–3]. While the tunnel construction is highly developed, the noise pollution problem in the tunnel has also become more serious. Due to the closed environment of the tunnel, the noise generated during the driving process is difficult to dissipate, seriously affecting the driver's comfort [4], and causing harm to the staff on the shoulder.

Kang [5,6] believed that the sound field in the tunnel is similar to a sound field in long enclosures, as a consequent it is impossible to predict the propagation law of noise by classical room acoustic theory. Li et al. [7] established a coherent model that considers the interference caused by direct waves and reflected waves. Besides, the total sound field can be calculated by coherent summation of the contributions of image source. However, under the normal circumstances, the tunnel is not a simply long straight passage. Usually, there are many branches such as connecting channel and escape route in the tunnel. The existence of these branches will affect the propagation of acoustic waves, which is slightly different from the calculation of coherent model. In order to solve the above problems, Liu et al. [8] studied the propagation law of sound in long enclosures with vertical or inclined branches.

The noise in the tunnel mainly comes from the traffic flow [9], which is composed of the noise generated by the automobile engine and the noise generated by the friction between the wheel and the road surface. The former is difficult to improve from the sound source except improving the engine manufacturing technology level to enhance the quality of engine, while the latter can achieve the same effect by improving the pavement material. An et al. [10] found that asphalt pavements can effectively reduce noise between wheels. However, due to the fire protection requirements and construction maintenance considerations in highway tunnels, cement concrete pavement is more suitable than asphalt pavement in this condition. Fang et al. [11] found that the noise level of ordinary grooved cement concrete pavement is high, but the cement concrete pavement modified by polymer has a low noise level. Although the above method can achieve the effect of reducing vehicle noise in tunnels, it does not have strong feasibility in existing tunnels, and it has no effect on other noise sources in the tunnel. Laying sound-absorbing boards on the side walls of the tunnel is the most common and excellent method. Most of the initial sound-absorbing are flexible porous materials. Gupta et al. [12] studied the sound absorption properties of polyurethane, glass wool and melamine, three kinds of common flexible porous sound-absorbing materials. The above materials have excellent sound absorption properties, but due to the lack of fire resistance and high production cost, the prospect of extensive application in tunnels is not high. As a consequent, more and more scholars concentrated more on the metal foams. Xu et al. [13] studied the sound absorption properties of stainless steel foam. Liang et al. [14] compared the sound absorption properties of metal foams and flexible porous materials, and confirmed that aluminum foam materials have better sound absorption properties. However, it was found in the study that the internal holes of metal foams with good sound absorption performance were mostly through-holes, and such metal foams have low strength and high cost. Therefore, the research on sound absorption properties of closed-cell metal foams came into being. Opiela et al. [15] studied the drilled closed-cell aluminum foam and found that it has sufficient sound absorption capacity.

Although many scholars have confirmed that the closed-cell aluminum foam after treatment has enough sound absorption capacity, most of the conclusions obtained by laboratory experiments. These research results are lack of confirmation in practical application., not to mention the application of closed-cell aluminum foam boards in the tunnel. The noise reduction effect of closed-cell aluminum foam boards in tunnels was still unknown. This paper evaluates the effect of closed-cell aluminum foam boards in tunnel noise reduction under two different working conditions, and obtains the noise reduction effect of fixed pure sound sources and moving vehicle noise in the tunnel, which can provide theoretical basis for the promotion of closed-cell aluminum foam in tunnel application.

2. Possibility Analysis and Testing Method of the Structure Sound Absorption Performance

2.1. Analysis of the Possibility of Sound Absorption

Most studies have shown that viscous dissipation and resonant sound absorption [16,17] are the main sound absorption modes of porous metal foams. Viscous dissipation is due to the velocity gradient of air vibration in pores caused by the existence of pore walls, and acoustic energy is transformed into heat energy dissipation through internal friction [18,19]. Resonant sound absorption relies on the resonance of the resonator to acoustic energy dissipation. When the frequency of the incident acoustic wave is close to the natural frequency of the resonator, resonance occurs. At this time, the velocity of air vibration reaches the maximum, and the acoustic energy loss reaches the maximum, so the sound absorption effect is the best.

The existing research have shown that the closed-cell aluminum foam does not have good sound absorption performance [20], but according to mechanism of the above two sound absorption methods, it is feasible to improve its sound absorption performance by improving the structure of the closed-cell aluminum. Some scholars have found that the resonant sound absorption structure composed of holes can improve the sound absorption effect [21,22]. Other scholars have also found that by compressing to make small cracks appear through the pore wall, a better viscous dissipation effect can be achieved [23]. It can be seen that the installation structure of closed-cell aluminum foam board plays a sound absorption and noise reduction effect in the tunnel is reasonable.

2.2. Testing Method

The sound absorption coefficient is an important index to evaluate the sound absorption performance of materials [24–26]. Among the many methods of acoustic performance testing, the reverberation chamber method and tube measurement method are the most commonly used methods to measure the sound absorption coefficient of materials. Tube measurement method is divided into standing wave ratio method and transfer function method. Standing wave ratio method calculates the sound absorption coefficient of the material to the vertical incident wave by measuring the maximum and minimum values of the standing wave sound pressure. Transfer function method separates the incident wave and the reflected wave by processing the signal delay to obtain the sound absorption coefficient of the material. The above two methods measure the vertical sound absorption coefficient of the material. The reverberation chamber method is used to measure the random incident sound absorption coefficient of the material, which is obtained by the reverberation time before and after the material is placed.

There are a variety of noise sources in the tunnel, including fan noise and vehicle noise. The incident trajectory of acoustic waves is untraceable, so it cannot be measured by pipe measurement method. In addition, because of the unique structure of the tunnel, the distribution of the internal sound field is different, and it is impossible to simulate the sound field with a simple reverberation chamber. Therefore, it is necessary to seek a novel testing method, which requires accurate evaluation criteria and can be applied to tunnels. The testing method proposed in this paper refers to the loudspeaker test and typical tunnel measurements, which are the two typical methods in the existing tunnel noise research. The loudspeaker test is carried out in the tunnel by using a fixed sound source. And the fixed sound source is used to release several octave pure tone signals, then the sound pressure level changes along the longitudinal direction of the tunnel can be measured. Typical tunnel measurement is to compare the noise level before and after sound absorption treatment of the actual tunnel, and to effectively evaluate the actual effect of the sound absorption materials. This test is based on the actual measurement of a typical tunnel and combined with some ideas in the loudspeaker test, the new fashioned two test conditions Case I and Case II have been formulated. In Case I, a fixed sound source was selected to release several octave pure tones separately, and the sound pressure level changes were measured along the longitudinal direction of the tunnel before and after the installation of the aluminum foam boards. In Case II, the change of vehicle noise before and after the installation of the aluminum foam boards were measured, and the insertion loss of the closed-cell aluminum foam was analyzed and studied. The two test conditions will be described in detail below.

3. Field Test

3.1. Engineering Background

The Haoshanyu Tunnel of Qinglan Expressway is located at the junction of Zhangjiapo Town and Shiqiao Town in Yiyuan County. The Haoshanyu Tunnel is a separate one-way double-track tunnel. The starting and ending pile number of the left line of the tunnel are K221 + 597~K222 + 297, the length is 700 m. The starting and ending mileage of the right line is YK221 + 617~YK222 + 332, the length is 715 m. The Haoshanyu Tunnel belongs to the middle tunnel. As shown in Figure 1.

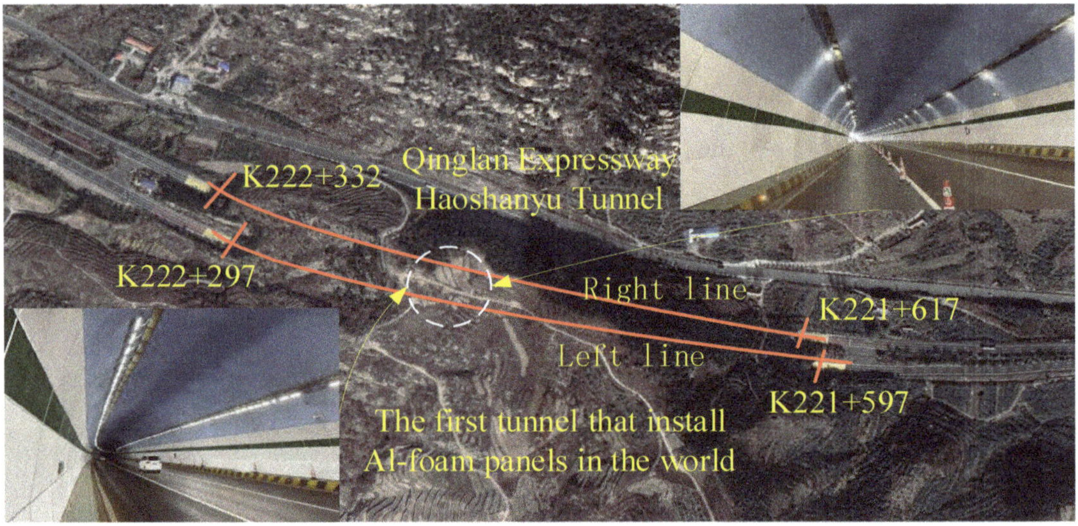

Figure 1. Spatial position of tunnel.

Fire prevention and noise reduction are two of the key points in the tunnel operation. The previous main fire prevention method used in the Haoshanyu Tunnel was spraying fire retardant coatings. However, due to the cool and humid environment in the tunnel and the lack of ventilation, the fire retardant coatings are prone to mildew and discoloration. During the review of the Haoshanyu tunnel, it was found that the average fire retardant coating of the Haoshanyu tunnel peeled off about 6 pcs/80 m^2, as shown in Figure 2. Therefore, measures need to be taken to enhance the tunnel fire performance.

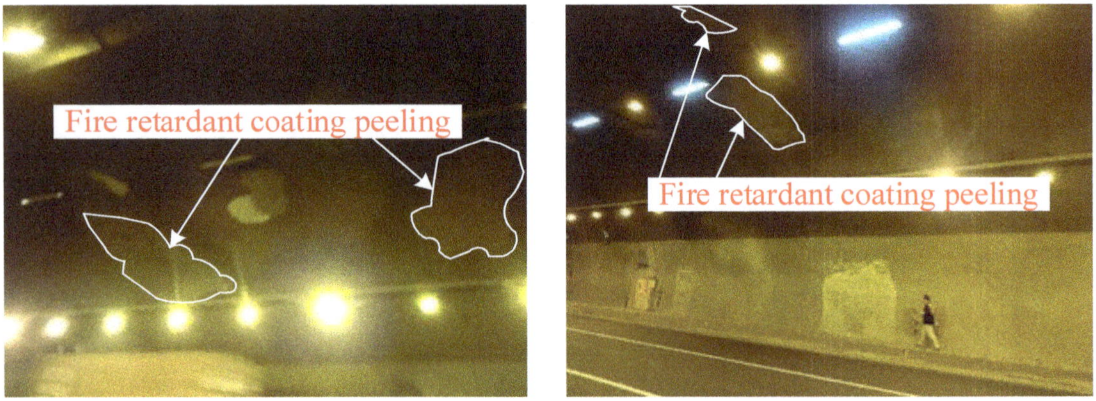

(**a**) Left line (**b**) Right line

Figure 2. The scene of the peeling off of the fire-retardant coating.

According to statistics from the transport sector, the traffic volume in the Haoshanyu tunnel has increased year by year, and the equivalent total traffic volume increasing from 23,550 vehicles per day in 2017 to 33,773 vehicles per day in 2020, as shown in Figure 3. With the increase of traffic flow, the noise pollution in the tunnel becomes more and more serious. Therefore, it is necessary to take noise reduction measures.

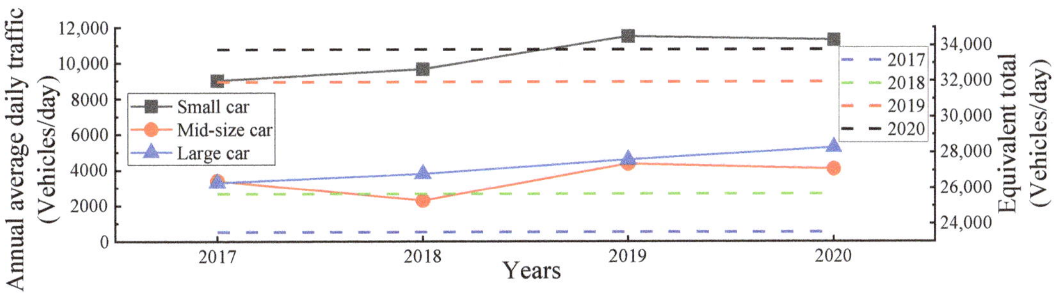

Figure 3. Traffic flow diagram of the Haoshanyu Tunnel.

The closed-cell aluminum foam board have great fire resistance and heat insulation performance [27–29]. The existing research shows that the thermal conductivity of aluminum foam is very low, and the aluminum foam will not produce toxic and harmful gases at high temperature. So, it can be used as a substitute for the traditional thermal insulation refractory in tunnels. As for the sound absorption properties of closed-cell aluminum foams, it has been preliminarily confirmed [17]. Since the closed-cell aluminum foam board has the properties of fire prevention and sound absorption, it meets the needs of the Haoshanyu Tunnel. The research on the sound absorption performance of closed-cell aluminum foam board in the tunnel was carried out based on the installing project of aluminum foam board in the Haoshanyu Tunnel. The Haoshanyu Tunnel has also become the first highway tunnel in China to install aluminum foam boards.

3.2. Installation Method

The aluminum foam board used in this project was prepared by melt foaming method, that is, adding tackifier and foaming agent successively in molten aluminum. The specific parameters of aluminum foam board are shown in Table 1. Due to the closed and disconnected pores in the closed aluminum foam board (Figure 4), the sound absorption performance is weaker than that of the open aluminum foam board. Therefore, it is necessary to improve its sound absorption performance through appropriate installation methods.

Table 1. Performance index of aluminum foam.

Parameter	Density g/cm^3	Compressive Strength MPa	Rigidity GPa	Thermal Expansion Coefficient/°C	Thermal Conductivity W/(m·K)	Noise Reduction Coefficient
Index	0.1~1.0	1~35	0.7~1.4	$(14\text{~}20) \times 10^{-6}$	0.2~0.8	0.4~0.8

The installation method of aluminum foam board will affect the punch rate, punch depth and cavity depth behind the board. It is introduced above that the resonant sound absorption structure can be formed by drilling, and the drilling rate, drilling depth and the cavity depth behind will affect the performance of resonant sound absorption [21,22]. Therefore, it is necessary to explain the installation method of closed-cell aluminum foam board in the tunnel.

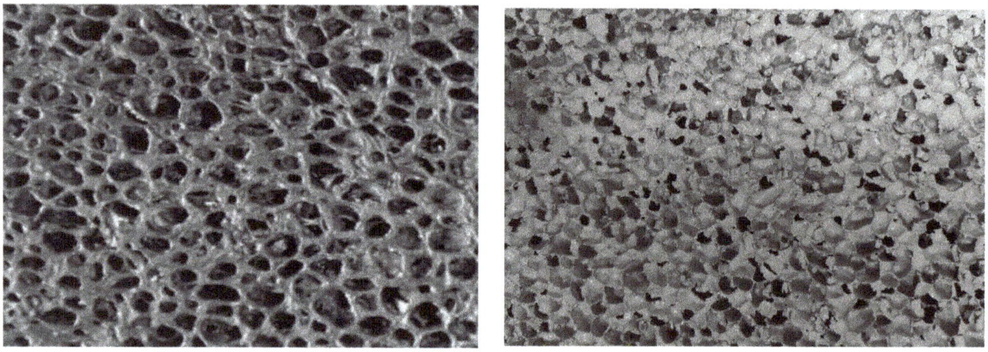

(**a**) open-cell Al-foam (**b**) closed-cell Al-foam

Figure 4. Aluminum foam sample diagram.

The thickness of aluminum foam board installed in the project is 10 mm, and the average pore diameter is 5 mm. The aluminum foam board was connected with light steel keel. During installation, the galvanized angle code was first fixed on the second lining with an expansion bolt, and then the light steel keel and the galvanized angle code were connected with bolts. Finally, the aluminum foam boards were inserted from the side into the groove between the light steel frame. The specific installation method is shown in Figure 5. Using the galvanized angle code as the connecting piece between the dragon frame and the second lining can effectively ensure the depth of the cavity behind the aluminum foam boards, so that effectively improve the resonance sound absorption effect of the Helmholtz resonator. However, due to the use of embedded groove fixation, the surface porosity of the aluminum foam board is not high, so it may suppress the sound absorption to a certain extent. In addition, the groove does not completely limit the aluminum foam boards, so when the car passes, the aluminum foam boards will produce vibration to play a certain active noise reduction effect, but this paper does not consider the effect of active noise reduction on the test.

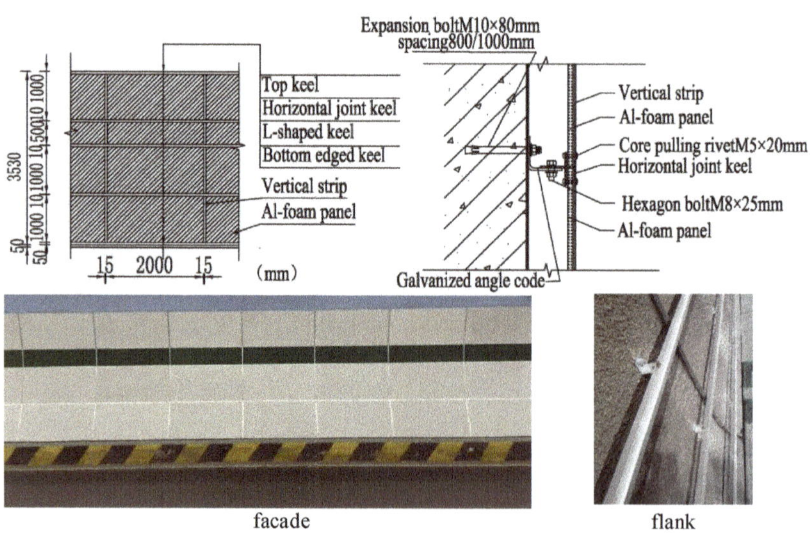

Figure 5. Installation of aluminum foam board in tunnel.

3.3. Case I

Case I mainly studies the sound absorption performance of closed-cell aluminum foam under a fixed pure sound source. By analyzing the sound pressure level changes along the longitudinal direction of the tunnel under the fixed-point pure sound source before and after the installation of the closed-cell aluminum foam boards, the effect of frequency, position and distance is explored. Because the human ear is more sensitive to mid-to-high frequency sounds (1–5 kHz), but not to low-frequency sounds (below 100 Hz), and the frequency response of the A-weighted network is close to the sensitivity of the human ear to wideband sounds, so A-weighted sound level has become the most widely used evaluation index. In addition, although the working condition is tested under stable pure tone, other noises such as testers and tunnel fans are inevitably generated inside the tunnel. Therefore, the equivalent continuous A sound level is selected as the evaluation index of Case I, which can effectively reduce the impact of unnecessary noise, and will not affect the results when there is no external noise.

This experiment was carried out following the installation progress of the aluminum foam boards in the tunnel. Repeated tests were carried out in three stages: before installation, one-sided installation was completed, and two-sided installation was completed. The details were shown in Figure 6. During the test, the power amplifier and loudspeaker were connected to the signal generator, and the loudspeaker was placed on the middle line of the tunnel at the central mileage of the tunnel to avoid the influence of the middle and short tunnel entrances as far as possible. Referring to the height of the car engine from the ground and set the height of the loudspeaker to 0.25 m from the ground. After the equipment is connected and placed, used a signal generator to release a 125 Hz octave pure tone signal, and adjusted the power amplifier so that the external sound can be heard clearly. At the same time, the setting value of the power amplifier was recorded, and the subsequent tests were based on this value. The pure tone signal was continuously released by the signal generator and placed outside the tunnel through the loudspeaker. After the sound was stable, the equivalent continuous A-weighted sound pressure level within 10S was measured by the sound level meter at the specified measuring point. Based on the requirements for the height of the sensor from the ground in the indoor measurement of noise-sensitive buildings and the requirements of the human ear for the height of noise acceptance, the height of the sound level meter from the ground was 1.2 m. Repeated the measurement for each measuring point 3 times, then changed the measuring point after the measurement was completed. After all measuring points were measured, stopped the signal generator from vibration. Finally, adjusted the frequency and restarted the vibration measurement.

The noise frequency in the tunnel is mainly distributed in 250–1000 Hz, and there is no need to analyze each frequency in the spectrum analysis of the sound signal. Therefore, in this experiment, the octave bandwidth was selected as the analysis bandwidth, which was 125 Hz, 250 Hz, 500 Hz, 1000 Hz, and 2000 Hz.

The groups affected by noise in the tunnel are mainly the staff at the shoulder of the road and the drivers in the car, so the measuring points were arranged at the center line of the tunnel and the shoulder of the road. In order to explore the distribution law of the sound absorption effect along the longitudinal direction of the tunnel, the measuring points were arranged at a longitudinal distance of 2 m, 4 m and 6 m from the sound source, as shown in Figure 7. This test was carried out following the installation progress of the aluminum foam boards in the tunnel. Repeated tests were carried out in three stages: before installation, one-sided installation is completed, and two-sided installation is completed. The detail can be seen in Figure 6.

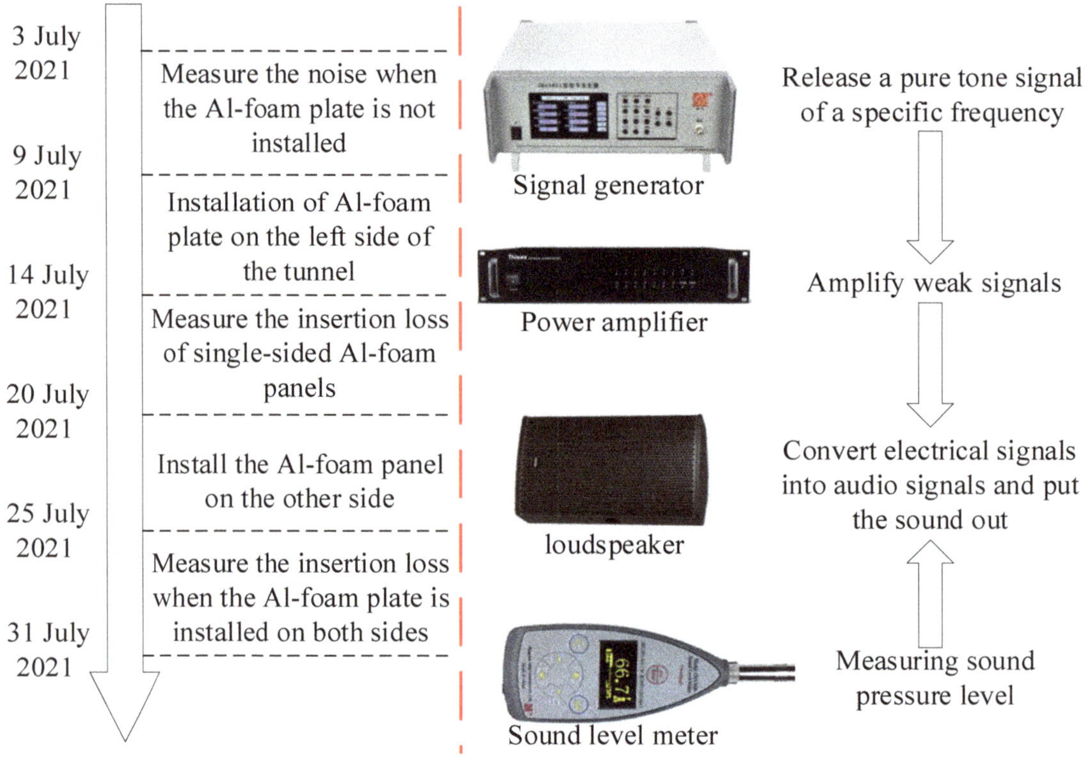

Figure 6. Test flow chart of Case I.

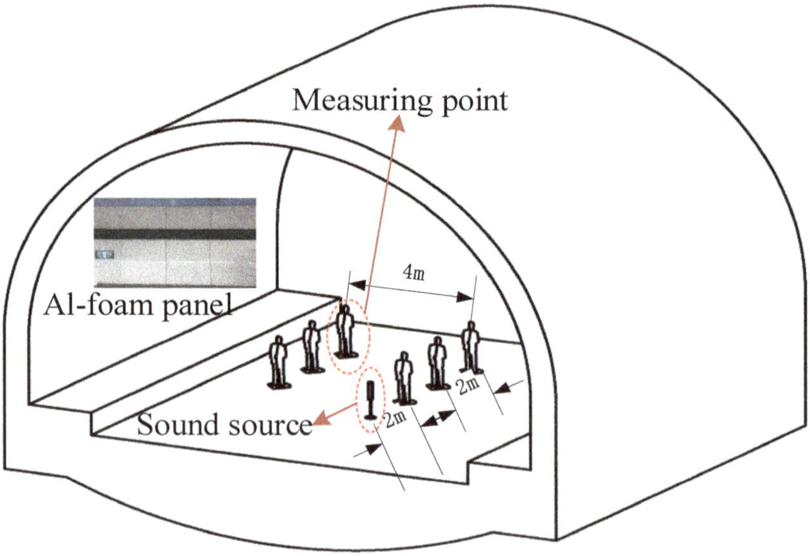

Figure 7. Sound source and measuring point layout.

Table 2 shows all 30 working conditions that need to be tested in Case I, and each working condition was tested repeatedly 3 times. In order to reduce the interference of external sound and ensure the accuracy of the data, when the deviation between the maximum value and the minimum value of the three data in a working condition exceeds 1 dB(A), the measurement shall be performed again.

Table 2. Test condition of Case I.

Condition.	Frequency/Position/ Distance	Condition	Frequency/Position/ Distance	Condition	Frequency/Position/ Distance
1	125 Hz/midline/2 m	11	250 Hz/shoulder/4 m	21	1000 Hz/midline/6 m
2	125 Hz/midline/4 m	12	250 Hz/shoulder/6 m	22	1000 Hz/shoulder/2 m
3	125 Hz/midline/6 m	13	500 Hz/midline/2 m	23	1000 Hz/shoulder/4 m
4	125 Hz/shoulder/2 m	14	500 Hz/midline/4 m	24	1000 Hz/shoulder/6 m
5	125 Hz/shoulder/4 m	15	500 Hz/midline/6 m	25	2000 Hz/midline/2 m
6	125 Hz/shoulder/6 m	16	500 Hz/shoulder/2 m	26	2000 Hz/midline/4 m
7	250 Hz/midline/2 m	17	500 Hz/shoulder/4 m	27	2000 Hz/midline/6 m
8	250 Hz/midline/4 m	18	500 Hz/shoulder/6 m	28	2000 Hz/shoulder/2 m
9	250 Hz/midline/6 m	19	1000 Hz/midline/2 m	29	2000 Hz/shoulder/4 m
10	250 Hz/shoulder/2 m	20	1000 Hz/midline/4 m	30	2000 Hz/shoulder/6 m

3.4. Case II

Case II mainly studies the sound absorption performance of closed-cell aluminum foam under moving sound sources, that is, the research on the insertion loss of closed-cell aluminum foam boards in tunnels. By comparing the maximum sound pressure level of the vehicle from entering the tunnel to leaving the tunnel before and after the installation of the aluminum foam board, then the effect of the aluminum foam board in the noise reduction application of the tunnel was obtained. The noise generated by the vehicle was affected by the type of vehicle, use time, brand and speed. In the field measurement of highway tunnels, it is impossible to adopt a unified standard for past vehicles. Therefore, the vehicle is divided into large, middle and small categories according to the size of the vehicle, limit the speed of the vehicle on the lane, collect large amounts of data and other methods to reduce the impact on the test results caused by the above reasons.

The test of Case II is similar to that of Case I. Following the installation progress of the aluminum foamed board, the test was also carried out in three stages before installation, one-sided installation is completed, and two-sided installation is completed, as shown in Figure 8. Before the test, the two lanes were blocked at 500 m from the entrance and only allowed to pass through the only lane, which can achieve the effect of limiting the speed of the vehicle and avoid some accidental noise interference caused by the lane change when the car is driving. During the test, turn on the continuous sampling function in the sound level meter when the vehicle enters the tunnel. The sampling interval of the sound level meter is 1S. When the vehicle exits the tunnel, stop sampling and read the maximum instantaneous A-weighted sound pressure level during the sampling period. For each vehicle type, at least 100 data samples were taken in each test, and the 100 data samples were divided into 10 groups evenly, and the average value of each group was calculated as the representative value of this group. Finally, the representative values were analyzed to explore the noise reduction law of closed-cell aluminum boards in tunnels.

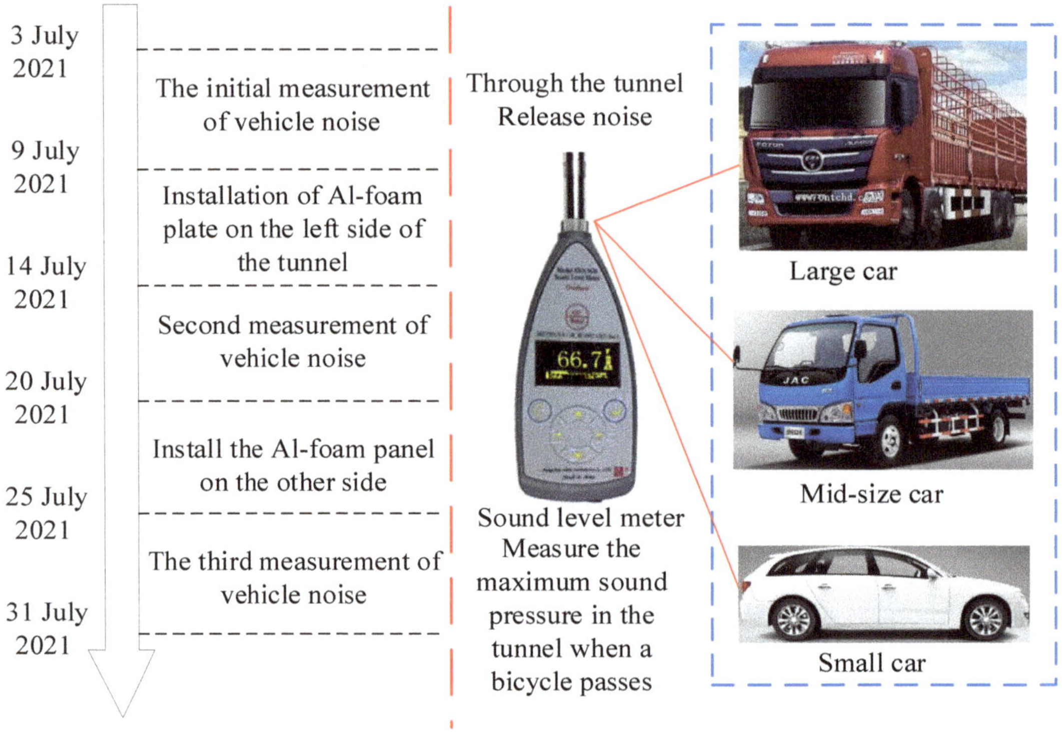

Figure 8. Test flow chart of Case II.

4. Results and Discussion

For different test purposes, this article gave two test methods under different working conditions. And in this section, we will analyze the results measured by the above method, and explain the sound absorption effects of the closed-cell aluminum foam board under the fixed pure sound source and the moving sound source. Furthermore, the sound absorption performance of the closed-cell aluminum foam board in the tunnel was comprehensively evaluated.

4.1. Test Results of Fixed-Point Pure Tone in the Tunnel

Case I studied the sound absorption performance of closed-cell aluminum foam board under fixed pure tone. The method adopted combined the loudspeaker test method with the typical tunnel measurement. Considering the process of field test and the key factors affecting the sound field distribution of the tunnel, the position, distance and frequency were taken as the key points to analyze which may have the effects on sound absorption under the fixed-point pure sound source.

(1) Effect of location and distance

Figure 9a–e not only show the distribution law of the sound field inside the tunnel, but also show the sound absorption effect of closed-cell aluminum foam boards under different sound field positions and distances. On the whole, the sound pressure at the center line of the tunnel is slightly greater than that at the tunnel shoulder. This is because the sound source of the tunnel is arranged on the center line of the tunnel. The linear distance between the measuring point on the center line is larger than the measuring point at the shoulder of the same cross section, so the sound energy has a longer dissipation distance. In addition, it can be found that the farther away from the sound source, the smaller the sound pressure level measured no matter the measurement is on the centerline of the tunnel or the shoulder of the road. This is because the longer the distance, the more the sound energy is dissipated. However, there are special cases that do not conform to the general rules in the data points measured this time. For example, the equivalent continuous A-weighted sound pressure level measured at the measurement which is on the center line of the tunnel and 4 m away from the sound source at 125 Hz is obviously greater than the distance is 2 m and 6 m; the sound level measured at the shoulder 6 m longitudinal distance from the sound source at 1000 Hz is higher than that measured at the tunnel midline. This is because the pure tone signal released is a sine wave. The positive superposition of the amplitude of the reflected sound wave and the direct sound wave at these measuring points enhances the sound pressure level, while the superposition of the reflected acoustic wave and the direct acoustic wave at other locations is very weak or the superposition is reversed.

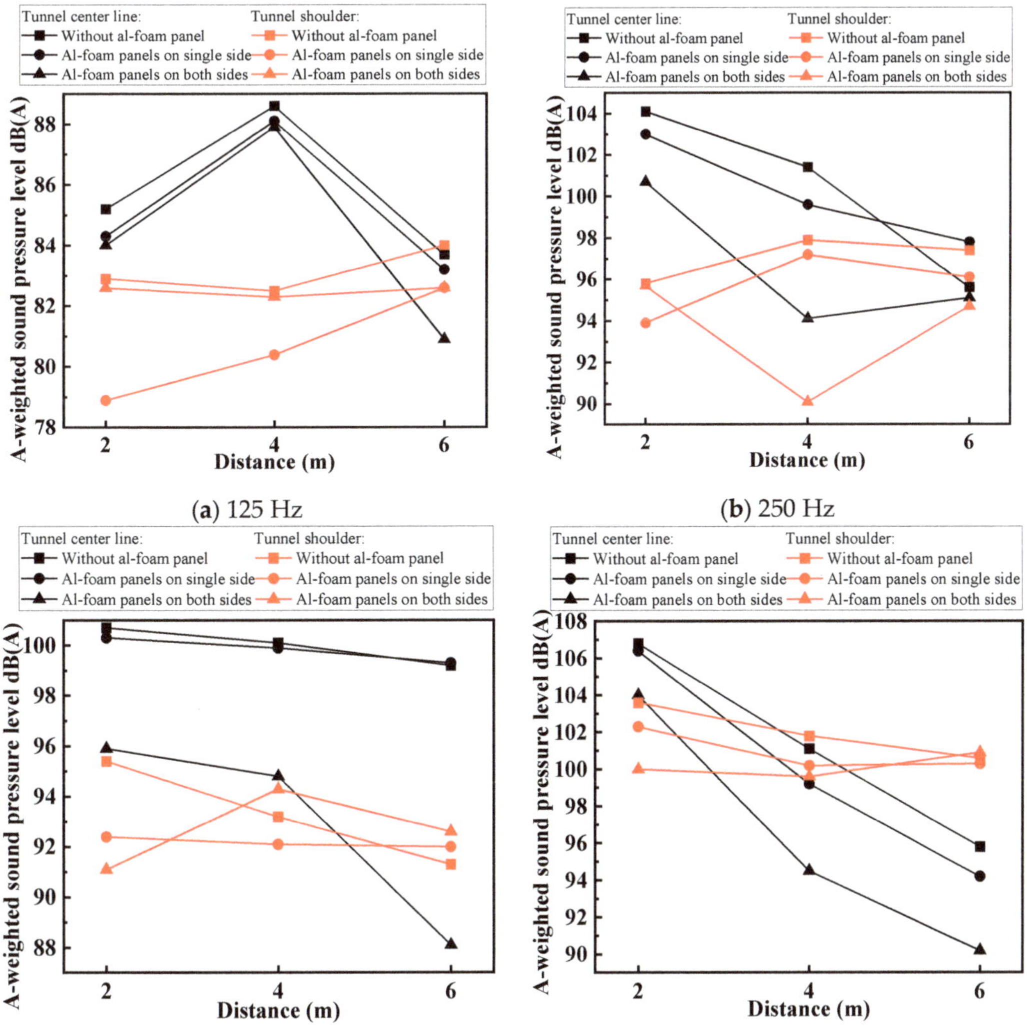

(**a**) 125 Hz (**b**) 250 Hz (**c**) 500 Hz (**d**) 1000 Hz

Figure 9. *Cont.*

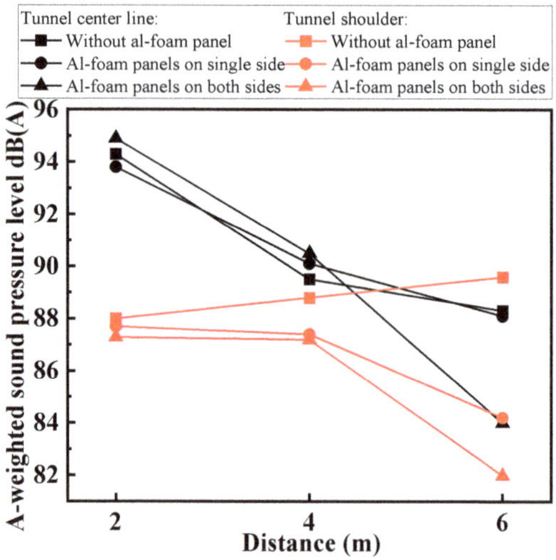

(**e**) 2000 Hz

Figure 9. Influence of location and distance.

From these five pictures, it can be seen that after the aluminum foamed boards were installed on one side, the results measured at each measuring point have a certain decrease compared with the unlaid aluminum foamed boards but the reduction is not large. After the double-sided aluminum foam boards were installed, the measured results at each measuring point were significantly reduced. This is because the sound waves reflected from the side where the aluminum foam boards were not installed, and interfere with the direct sound waves. When the double-sided aluminum foam boards were installed, the reflected sound waves on both sides were affected. Comparing the slopes of the curves of the three test results, it can be seen that the slope of the results measured after the aluminum foam boards were installed on both sides is significantly greater than that of the single-sided installation and non-installation. From this analysis, we can see that the equivalent continuous A-weighted sound pressure level in the tunnel that the aluminum foam boards were installed on both sides drops faster along the longitudinal direction of the tunnel, which means that the installation of aluminum foam boards on both sides has the best noise reduction effect on the sound in the tunnel.

Analyzed from the position, the measured results at the center line of the tunnel vary significantly more than the shoulder of the tunnel. From the results measured at the road shoulder, the closed-cell aluminum foam also has a certain sound absorption effect on the noise at the road shoulder, but the sound absorption capacity is limited, so the line graph shows the characteristics of small slope and flatness. However, the noise reduction is more obvious at the center line of the tunnel. The possible reason is that the interference between the reflected sound wave and the reflected sound wave at the shoulder of the road closer to the side wall is more obvious, and the interference between the reflected sound wave and the direct sound wave is also more obvious, but at the center line of the tunnel, the interference effect between the reflected sound wave and the reflected sound wave is weakened. From the distance point of view, the farther away from the sound source, the reflected sound wave can go through more reflections, and there is enough distance for the sound energy to be dissipated, whether it is in the center line of the tunnel or at the shoulder of the road.

In general, the noise reduction effect at the center line of the tunnel is better than that at the tunnel shoulder in the noise reduction of the tunnel, and the longer the distance, the more obvious the noise reduction effect. The noise reduction effect of the double-sided aluminum foam board is better than that of the single-sided aluminum foam board and the unmounted aluminum foam board. The single-sided aluminum foam board has a small noise reduction effect.

(2) Effect of frequency

Figure 10a–f reveal the sound absorption performance of closed-cell aluminum foam boards for pure sounds at different frequencies. In general, the equivalent continuous A-weighted sound pressure level measured in the tunnel presents a M-shaped distribution, that is, the equivalent continuous A-weighted sound pressure level is higher at the frequency of 250 Hz and 1000 Hz, and the results measured at the frequency of 125 Hz and 2000 Hz are very low.

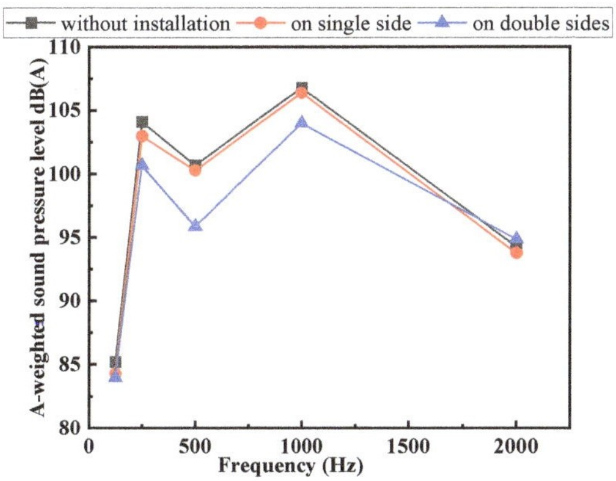

(**a**) Tunnel center line (2 m)

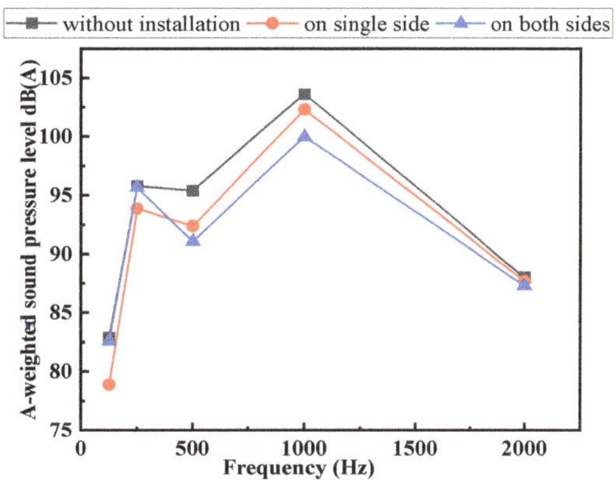

(**b**) Tunnel shoulder (2 m)

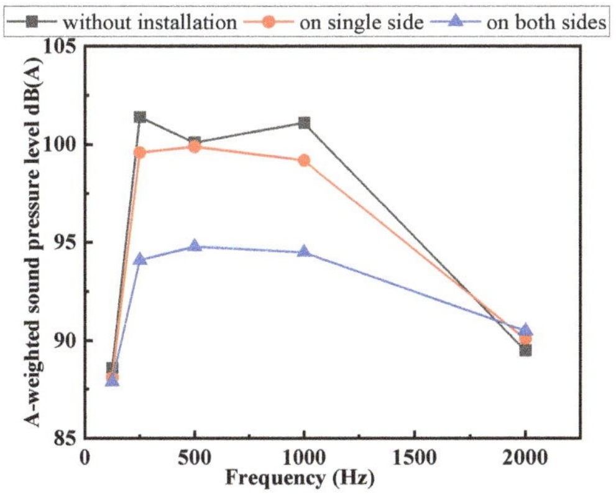

(**c**) Tunnel center line (4 m)

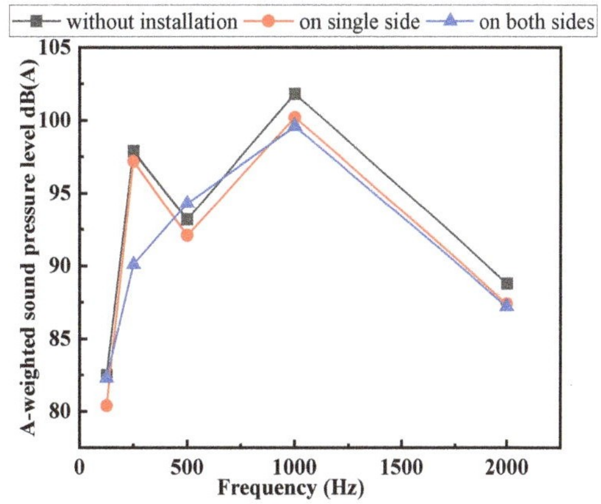

(**d**) Tunnel shoulder (4 m)

Figure 10. *Cont.*

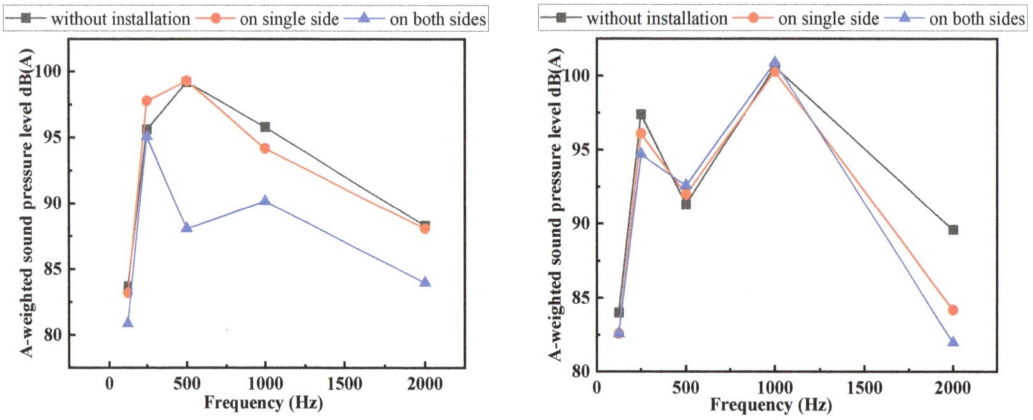

(**e**) Tunnel center line (6 m) (**f**) Tunnel shoulder (6 m)

Figure 10. Influence of frequency.

As for the insertion loss, the closed-cell aluminum foam board has the best sound absorption effect for pure sound with a frequency of 250–1000 Hz. According to the relationship between the speed of sound, frequency and wavelength, for pure sound with a frequency lower than 250 Hz, since the speed of sound is equal to the product of wavelength and frequency, the sound wave has a longer wavelength at low frequencies. However, the average pore diameter of closed-cell aluminum foam is only 5 mm, so the scattering effect of aluminum foam surface on low frequency acoustic wave is very weak. In addition, there are cavities behind the aluminum foam boards that were drilled, so the entire installation structure of the aluminum foam boards can be regarded as a structure in which multiple Helmholtz resonators are connected in parallel. According to the principle of resonance sound absorption, the natural frequency of the aluminum foam board is just close to the testing frequency, so when sound waves of the same frequency are transmitted to the aluminum foam boards, it can achieve the effect of sound absorption and noise reduction by resonate.

From the discount slope in Figure 10, it can be seen that the sound absorption performance of the closed-cell aluminum foam for pure tones with frequencies between 250–1000 Hz is relatively similar, while the sound absorption effect of closed-cell aluminum foam board increases rapidly from 125 Hz to 250 Hz. Besides, from 1000 Hz to 2000 Hz, the sound absorption effect of closed-cell aluminum foam board decreased obviously.

By comparing the figures of left column and right column, it can be seen that the influence of frequency on sound absorption effect of closed-cell aluminum foam board is more obvious at the center line of tunnel. At the shoulder, the influence of frequency on sound absorption effect is relatively small.

4.2. Test Results of Car Noise in the Tunnel

Case II mainly studies the sound absorption performance of closed-cell aluminum foam under moving sound sources. Analysis of the measurement results in Case II can more accurately reveal the noise reduction effect of closed-cell aluminum foam boards on vehicle noise in tunnels. Since the impact of the vehicle model on the noise generated was considered in the test, when analyzing the possible noise reduction effect of closed-cell aluminum foam in tunnel applications, the analysis was carried out according to the three common vehicle models in the tunnel, as shown in Figure 11a–c.

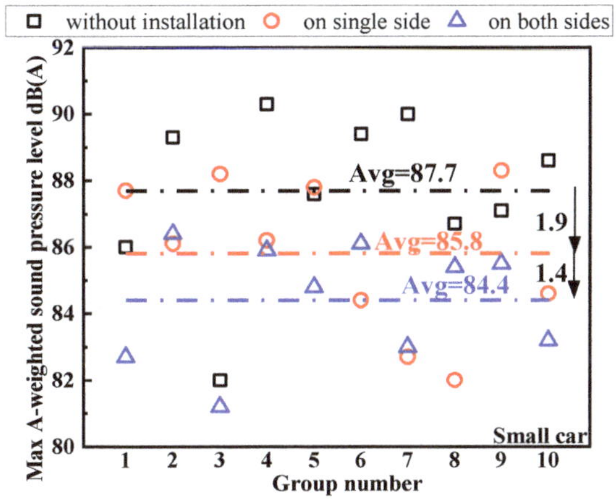

(a) Test results of small car

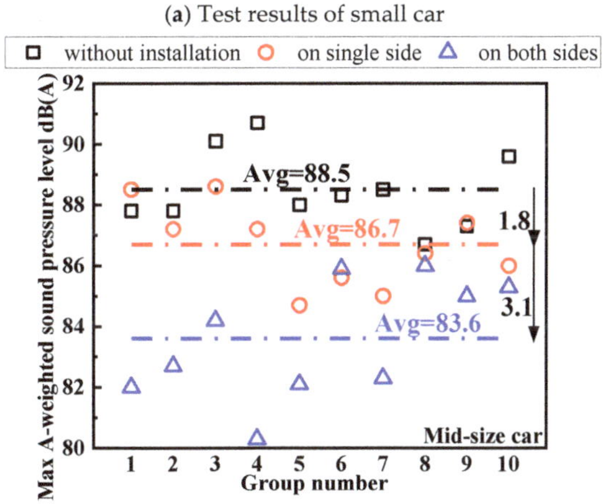

(b) Test results of mid-size car

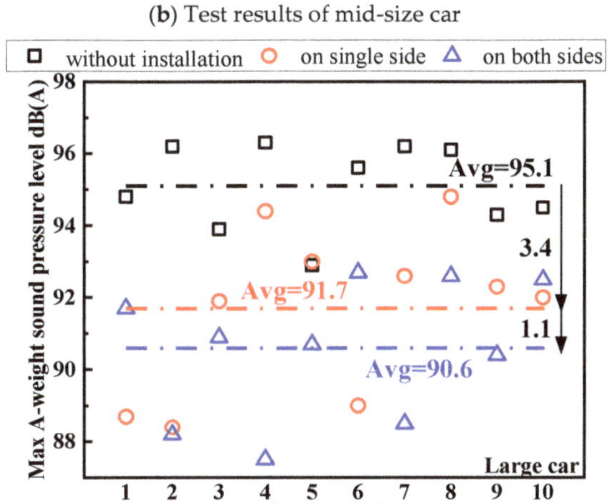

(c) Test results of large car

Figure 11. Test results of Case II.

It can be seen from the above that each point in the figure is the representative value of the average of 10 test data. It can be seen from the figure that when the aluminum foam board is not installed, the noise generated by the large car in the tunnel is the largest, and the maximum instantaneous A-weighted sound pressure level reaches 95.1 dB(A). The sound pressure level of noise produced by small cars is slightly lower than that of mid-sized cars, at 87.7 dB(A) and 88.5 dB(A) respectively. After the single-sided aluminum foam boards were installed, the noise generated by small and middle-sized cars is decreased to a limited extent. The maximum instantaneous A-weighted sound pressure level in the tunnel when a bicycle passes through the tunnel is reduced by less than 2 dB(A). However, the single-sided aluminum foam board has a more obvious reduction effect on the traffic noise generated by large vehicles, which is reduced by 3.4 dB(A). The attenuation of the vehicle noise by the double-sided aluminum foam boards is greater than the vehicle noise when installed on single side or not installed. After installation on both sides, the noise attenuation for medium-sized cars is the strongest, reaching 4.9 dB(A). The vehicle noise reduction for small cars and large cars are 3.3 dB(A) and 4.5 dB(A). This may be due to the fact that the frequency of the noise generated by the medium-sized car is closer to the natural frequency of the aluminum foam board structure.

5. Field Application Effect Evaluation

Closed-cell aluminum foam is a kind of material with both fire prevention and sound absorption and noise reduction functions. The Haoshanyu Tunnel on the Qinglan Expressway has become the first expressway tunnel in China to complete the whole line laying of aluminum foam board due to the needs of noise reduction and fire prevention improvement. The field performance of closed-cell aluminum foam board in the tunnel before installation, during installation and after installation were shown in Figure 12. Before the installation, the concrete on both side walls of the tunnel lacked fire protection treatment, which can be seen from the on-site photos before installation. According to the relevant regulations in China's "Code for Fire Protection of Building Design" GB 50016-2014, it is known that the interior decoration in the tunnel should be made of non-combustible materials, and the temperature of the concrete surface should not exceed 380 °C. The closed-cell aluminum foam has a good effective thermal conductivity at high temperatures [27], and its installation in the tunnel perfectly compensates for the lack of fire protection and meets the national fire protection standards.

(**a**) Before the installation of al-foam board

Figure 12. *Cont.*

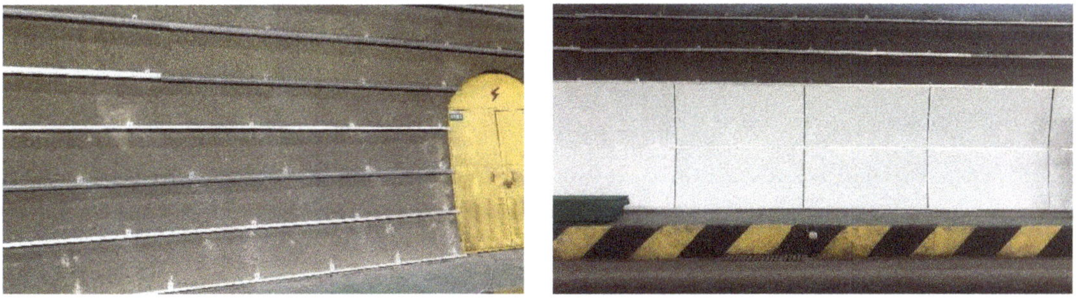

(**b**) During the installation of al-foam board

(**c**) After the installation of al-foam board

Figure 12. Comparison of the effect before and after installation of the al-foam board.

According to the relevant requirements in the "Regulations of the People's Republic of China on the Prevention and Control of Environmental Noise Pollution", "Urban Area Environmental Noise Standards" GB 3096-93, and "External Noise Limits and Measurement Methods for Accelerated Vehicles" GB 1495-2002, the maximum allowable noise level generated by trucks traveling in tunnels is 89 dB, which does not exceed 85 dB for small cars. According to the above field test results, before the aluminum foam boards were installed on both sides of the tunnel, the noise of vehicles in the tunnel is higher than the standard. After the installation of the aluminum foam board, the vehicle noise generated by small and medium-sized cars in the tunnel has been lower than the standard value. However, the noise generated by the large truck is still slightly higher than the standard value after noise reduction treatment, the possible reason is the influence of noise that is unavoidable in the test. Generally speaking, it has a good noise reduction effect.

In summary, the laying of aluminum foam boards on the entire line of the Haoshanyu Tunnel of the Qinglan Expressway not only improves the fire prevention and noise reduction capabilities in the tunnel, but also beautifies the tunnel environment, and play a very good effect in the field application. In addition, aluminum foam boards are more economical than fire-resistant panels and sound-absorbing panels. The aluminum foam boards have a service life of up to 50 years and can be recycled and reused with ultra-high value. Therefore, it has very good application prospects in tunnels.

6. Conclusions

Relying on the Qinglan Expressway Haoshanyu Tunnel full line installation of closed-cell aluminum board project. Aiming at the two different working conditions of fixed pure sound source and mobile sound source, combining the existing loudspeaker test and typical tunnel measurement, the two new methods are proposed to measure the sound absorption performance of closed-cell aluminum foam in the tunnel.

(1) Using the method that proposed in Case I in this paper, it is found that the closed-cell aluminum foam has a better sound absorption performance for sounds with a frequency of 250~1000 Hz, and the noise reduction effect at the center line of the tunnel is better than that at the shoulder of the tunnel. As the distance increases, the reflected sound waves are dissipated more, and the farther away from the sound source, the more obvious the sound absorption effect.

(2) The test results of the two test methods given in this paper are comprehensively analyzed. Overall, the closed-cell aluminum foam board can play a certain sound absorption and noise reduction effect in the tunnel. The application of closed-cell aluminum foam board as a new noise reduction material in the tunnel is feasible. However, due to the acoustic characteristics of the closed-cell aluminum foam board itself and the lack of installation methods, the sound absorption effect is not obvious. However, it can be changed by drilling to improve the sound absorption performance.

Author Contributions: Conceptualization, Z.L.; methodology, Z.L.; validation, W.G., L.W. and J.S.; investigation, H.Z.; resources, J.H.; data curation, B.Z.; writing—original draft preparation, Z.L.; writing—review and editing, B.Z.; supervision, J.H.; project administration, W.G.; funding acquisition, W.G. All authors have read and agreed to the published version of the manuscript.

Funding: This research was funded by Shandong Province Enterprise Technology Innovation Project 'Research and Application of Noise Reduction and Refractory Functional Materials for Highway Tunnels'.

Institutional Review Board Statement: Not applicable.

Informed Consent Statement: Not applicable.

Data Availability Statement: Some or all data, models, or code generated or used during the study are available from the corresponding author by request.

Acknowledgments: This research was funded by Shandong Province Enterprise Technology Innovation Project 'Research and Application of Noise Reduction and Refractory Functional Materials for Highway Tunnels'.

Conflicts of Interest: The authors declare no conflict of interest.

References

1. Gong, C.; Ding, W.; Mosalam, K.M.; Günayc, S.; Sogac, K. Comparison of the structural behavior of reinforced concrete and steel fiber reinforced concrete tunnel segmental joints. *Tunn. Undergr. Space Technol.* **2017**, *68*, 38–57. [CrossRef]
2. Ding, W.; Gong, C.; Mosalam, K.M.; Soga, K. Development and application of the integrated sealant test apparatus for sealing gaskets in tunnel segmental joints. *Tunn. Undergr. Space Technol.* **2017**, *63*, 54–68. [CrossRef]
3. Lei, M.; Zhu, B.; Gong, C.; Ding, W.; Liu, L. Sealing performance of a precast tunnel gasketed joint under high hydrostatic pressures: Site investigation and detailed numerical modeling. *Tunn. Undergr. Space Technol.* **2021**, *115*, 104082. [CrossRef]
4. Zeng, S. Analysis of the influence of the characteristics of mountain soil and the noise in the tunnel on people: Active noise control system. *Arab. J. Geosci.* **2021**, *14*, 912. [CrossRef]
5. Kang, J. A method for predicting acoustic indices in long enclosures. *Appl. Acoust.* **1997**, *51*, 169–180. [CrossRef]
6. Kang, J. The Unsuitability of the Classic Room Acoustical Theory in Long Enclosures. *Archit. Sci. Rev.* **1996**, *39*, 89–94. [CrossRef]
7. Li, K.M.; Iu, K.K. Propagation of sound in long enclosures. *J. Acoust. Soc. Am.* **2004**, *116*, 2759–2770. [CrossRef]
8. Liu, J.C. Sound propagation in long enclosures with a vertical or inclined branch. *Appl. Acoust.* **2006**, *67*, 1022–1030. [CrossRef]
9. Wang, G.; Smith, G.; Shores, R. Pavement noise investigation on North Carolina highways: An on-board sound intensity approach. *Can. J. Civ. Eng.* **2012**, *39*, 878–886. [CrossRef]
10. An, D.; Lee, J.; Ohm, B.; Son, H.; Kwon, S. A study of pavement noise for asphalt pavements with different service life in national highway. In Proceedings of the INTER-NOISE and NOISE-CON Congress and Conference Proceedings, Kaist, Korea, 16–19 November 2014.

11. Fang, J.; Tu, J.; Wu, K. Analysis of Skid Resistance and Noise Characteristics for Varieties of Concrete Pavement. *Adv. Mater. Sci. Eng.* **2020**, *2020*, 1–8. [CrossRef]
12. Gupta, P.; Parey, A. To investigate the influence of sound-absorbing materials on the transmission loss of double-wall panel. *Mater. Today Proc.* **2021**, *44*, 1500–1503. [CrossRef]
13. Xu, X.B.; Liu, P.S.; Chen, G.F.; Li, C.P. Sound Absorption Performance of Highly Porous Stainless Steel Foam with Reticular Structure. *Met. Mater. Int.* **2021**, *27*, 3316–3324. [CrossRef]
14. Liang, L.S.; Wu, X.L.; Ma, N.N.; Du, J.J.; Liu, M.B. *The Sound Absorption Properties Comparison of Metal Foams and Flexible Cellular Materials. Materials Science Forum*; Trans Tech Publications Ltd.: Zürich, Switzerland, 2018; pp. 357–366.
15. Opiela, K.C.; Zieliński, T.G.; Dvorák, T.; Kudela, S., Jr. Perforated closed-cell aluminium foam for acoustic absorption. *Appl. Acoust.* **2021**, *174*, 107706. [CrossRef]
16. Umnova, O. Propagation of Sound in Porous Media: Modelling of Sound Absorbing Materials. *J. Sound Vib.* **2010**, *329*, 4333–4334.
17. Allard, J.F.; Daigle, G. Propagation of Sound in Porous Media: Modeling Sound AbsorbingMaterials. *J. Acoust. Soc. Am.* **1994**, *95*, 2785. [CrossRef]
18. Li, Y.; Li, Z.; Han, F. Air Flow Resistance and Sound Absorption Behavior of Open-celled Aluminum Foams with Spherical Cells. *Procedia Mater. Sci.* **2014**, *4*, 187–190. [CrossRef]
19. Lu, T.J.; Chen, F.; He, D. Sound absorption of cellular metals with semiopen cells. *J. Acoust. Soc. Am.* **2000**, *108*, 1697–1709. [CrossRef] [PubMed]
20. Lu, T.J.; Hess, A.; Ashby, M.F. Sound absorption in metallic foams. *J. Appl. Phys.* **1999**, *85*, 7528–7539. [CrossRef]
21. Akiwate, D.C.; Date, M.D.; Venkatesham, B.; Suryakumar, S. Acoustic characterization of additive manufactured perforated panel backed by honeycomb structure with circular and non-circular perforations. *Appl. Acoust.* **2019**, *155*, 271–279. [CrossRef]
22. Wang, K.; Wang, W.; Yan, X. Effect of hole on sound absorption coefficient of micro-perforated panels investigated by Melling computing. *Concurr. Comput. Pract. Exp.* **2019**, *31*, e4725. [CrossRef]
23. Byakova, A.; Gnyloskurenko, S.; Bezimyanniy, Y.; Nakamura, T. Closed-Cell Aluminum Foam of Improved Sound Absorption Ability: Manufacture and Properties. *Metals* **2014**, *4*, 445–454. [CrossRef]
24. Xia, X.; Zhang, Z.; Zhao, W.; Li, C.; Ding, J.; Liu, C.; Liu, Y. Acoustic properties of closed-cell aluminum foams with different macrostructures. *J. Mater. Sci. Technol.* **2017**, *33*, 1227–1234. [CrossRef]
25. Cheng, W.; Duan, C.-Y.; Liu, P.-S.; Lu, M. Sound absorption performance of various nickel foam-base multi-layer structures in range of low frequency. *Trans. Nonferrous Met. Soc. China* **2017**, *27*, 1989–1995. [CrossRef]
26. Kim, S.Y.; Park, S.H.; Um, Y.S.; Hur, B.Y. *Sound Absorption Properties of Al Foam. Materials Science Forum*; Trans Tech Publications Ltd.: Zürich, Switzerland, 2005; pp. 468–471, 486–487.
27. Ye, H.; Ma, M.; Ni, Q. An experimental study on mid-high temperature effective thermal conductivity of the closed-cell aluminum foam. *Appl. Therm. Eng.* **2015**, *77*, 127–133. [CrossRef]
28. Ye, H.; Ma, M.; Yu, J. Anomalies in mid-high-temperature linear thermal expansion coefficient of the closed-cell aluminum foam. *Chin. Sci. Bull.* **2014**, *59*, 3669–3675. [CrossRef]
29. Navacerrada, M.A.; Fernández, P.; Díaz, C.; Pedrero, A. Thermal and acoustic properties of aluminium foams manufactured by the infiltration process. *Appl. Acoust.* **2013**, *74*, 496–501. [CrossRef]

applied sciences

MDPI

Article

A New Design Method of Shield Tunnel Based on the Concept of Minimum Bending Moment

Dawei Huang [1,2], Hao Jiang [1,2], Changjie Xu [1,2,*], Wenbo Tu [1,2], Xue Li [3] and Wei Wang [1,2]

[1] Engineering Research Center of Railway Environmental Vibration and Noise, Ministry of Education, East China Jiaotong University, Nanchang 330013, China; gddthdw@126.com (D.H.); jiangh1997@163.com (H.J.); wenbotu@126.com (W.T.); rebwg05@163.com (W.W.)
[2] State Key Laboratory of Performance Monitoring and Protecting of Rail Transit Infrastructure, East China Jiaotong University, Nanchang 330013, China
[3] School of Geoscience and Technology, Southwest Petroleum University, Chengdu 610500, China; jialixue521@163.com
* Correspondence: xucj@zju.edu.cn

Abstract: As the soil-resistance coefficient in a soft soil area is small, overlarge bending moment may cause exceeding transverse deformation for the shield tunnel and cause structural diseases and waterproof failure at the longitudinal segment joints. Hence, a new idea of cross-section design for a minimum bending moment shield tunnel was proposed. This article has first put forward the concept of a zero bending moment shield tunnel. Then, based on rational and feasible hypotheses, a structural mechanical model and an analytical expression of axis for the cross-section of the zero bending moment shield tunnel was obtained, and computational formulas of internal force and key geometry parameters were given. Based on the case of the metro shield tunnel constructed in the Shanghai soft soil area, the zero bending moment shield tunnel was designed, and its characteristics were analyzed. Considering only one cross-section of shield tunnel can be adopted in one metro line, the design method and procedure of the minimum bending moment shield tunnel were put forward. Finally, taking one of the Shanghai metro lines as an example, a cross-section of a minimum bending moment shield tunnel was designed, and its bending moment was compared with the bending moment of a circular section shield tunnel, which had the same horizontal diameter. The comparison revealed that the cross-section of the minimum bending moment can significantly reduce the bending moment of shield tunnel.

Keywords: minimum bending moment; cross-section; shield tunnel; soft soil area; design method

Citation: Huang, D.; Jiang, H.; Xu, C.; Tu, W.; Li, X.; Wang, W. A New Design Method of Shield Tunnel Based on the Concept of Minimum Bending Moment. *Appl. Sci.* **2022**, *12*, 1082. https://doi.org/10.3390/app12031082

Academic Editors: Mingfeng Lei, Chenjie Gong and Xianda Shen

Received: 31 December 2021
Accepted: 18 January 2022
Published: 20 January 2022

1. Introduction

The cross-sections for most of the existing shield tunnels all over the world are generally circular [1–4]. Additionally, there are other cross-sections for shield tunnels, such as transverse oval, rectangular, quasi-rectangular (applied in Ningbo Metro Line 3, Zhejiang province, China), semicircle, horseshoe-shaped, double-circle-shaped, and tri-circle-shaped, as shown in Figure 1. The main load for a shield tunnel as an underground structure is the earth pressure [5]. Normally, the tunnel is subjected to larger vertical earth pressure than horizontal earth pressure, and both vertical and horizontal earth pressure increase as the depth of the ground increases. So, the cross-section of a shield tunnel inevitably undertakes bending moment. Especially in the soft soil area, there is little horizontal soil resistance at the side of the tunnel [6–8]. That means that when even lager transverse oval deformation occurs, the difference between the vertical earth pressure and the horizontal earth pressure is large. According to the design specification of metro [9,10], the largest oval deformation for the new constructed shield tunnel upon the acceptance check was 5D‰, where D represents the outer diameter of the tunnel. However, for some soil conditions, the oval deformation will fail to meet this code, especially in the soft soil area.

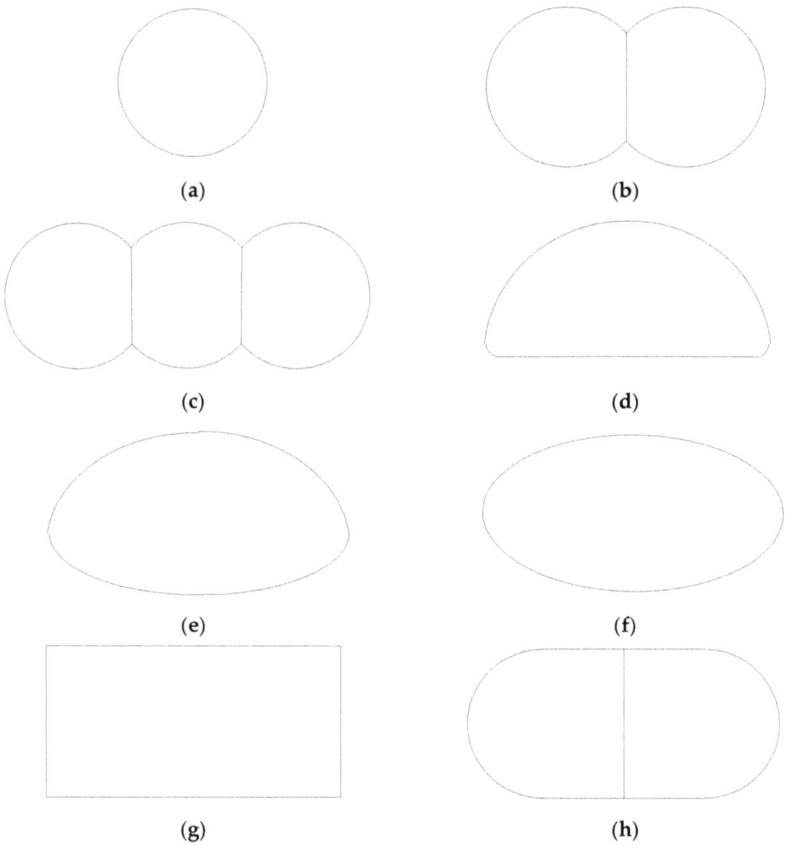

Figure 1. Typical cross-section of shield tunnels:(**a**) Circular; (**b**) Double-circle-shaped; (**c**) Tri-circle-shaped; (**d**) Semicircle; (**e**) Horseshoe-shaped; (**f**) Transverse oval; (**g**) Rectangular; (**h**) Quasi-rectangular.

The tunnel is a slender underground structure, so the shield tunnel can be considered as a curved beam structure from the perspective of plane strain, and the length of the curved beam is far larger than the height of the curved beam. The deformation of the beam structure is mainly caused by bending moment. For the cross-section of the shield tunnel, the rotation of the segment longitudinal joint and the bending moment are the main factors that cause transverse oval deformation [11,12]. Besides, when the shield tunnel is under the bending moment, the segment longitudinal joints are prone to segment corner damages and will stretch up [13,14], which will reduce the pressure stress on the waterproof sealing washer or, even worse, the waterproof sealing washer will completely open, causing joint waterproof failure [15–18]. Figure 2 shows the excessive deformation of the segmental joints and the related water-leakage issues. Besides this, under bending moment, the connecting bolts at the segment longitudinal joints will be stretched. Under the over-large bending moment, plastic deformation will occur for the thread of the connecting bolts, causing damages to the segment longitudinal joints [19–22]. One study [23] found that the segment joints were the weak parts of the test structure, through a full-scale test analysis of the bearing capacity of the shield tunnel structure assembled by joints, and the plastic development of the segment joints easily caused structural damage. It illustrates that most of the diseases are caused by the bending moment of the cross-section of the shield tunnel [19,24–26]. In all of the existing studies, the properties of a determined section of shield tunnel, as well as for other underground structures, such as culvert pipe and underground pipe (most of the pipe in the ground is circular), have been analyzed, but whether the cross-section is suitable for resisting transverse deformation has not been taken into consideration. From the above-mentioned diseases caused by bending moment, it is necessary to design a rational cross-section in order to decrease the bending moment as much as possible.

(a) (b) (c)

Figure 2. The excessive deformation and related water-leakage of longitudinal segment joints: (**a**) The excessive deformation of the segmental joint; (**b**) Longitudinal segment joint water-leakage; (**c**) Segment corner water-leakage.

Hence, if a kind of shield tunnel cross-section can be designed so that the theoretical bending moment for any section is zero (namely "zero bending moment shield tunnel"), the above-mentioned diseases of shield tunnels can be solved, which also contributes to the reduction in segment reinforcement. In this article, a new idea of cross-section design for a minimum bending moment shield tunnel is proposed. First, we put forward the concept of a zero bending moment shield. Then, based on rational and feasible hypotheses, a structural mechanical model and analytical expression of the axis for the cross-section of the zero bending moment shield tunnel were obtained, and computational formulas of internal force and key geometry parameters were given. Then, based on the case of the metro shield tunnel constructed in the Shanghai soft soil area, the zero bending moment shield tunnel was designed, and its characteristics were analyzed. Considering that only one cross-section of shield tunnel can be adopted in one metro line, the design method and procedure of the minimum bending moment shield tunnel have been put forward. Finally, taking one of the Shanghai metro lines as an example, the cross-section of the minimum bending moment shield tunnel has been designed. The pioneering work performed here shows that a rational cross-section can greatly decrease the bending moment for underground structures. For shield tunnels or buried pipe, the minimum bending moment cross-section is similar to a vertical ellipse. The research results can be used as a reference for the design of a shield tunnel or other underground structures.

2. Analysis of the Pressure Exerted on the Cross-Section of the Shield Tunnel

2.1. Hypothesis of the Earth Pressure Acting on the Shield Tunnel

According to structural mechanics, the rational axis (any section of the axis without bending moment) of the arch is related to the loads that it bears. Since the vertical earth pressure in the stratum is larger than the horizontal earth pressure, when the bending moment of the cross-section of the shield tunnel is zero, the horizontal diameter should not be equal to the vertical diameter. Therefore, when designing the reasonable axis of a shield tunnel section (the bending moment at any section is zero), only the normal condition is considered, which is convenient for analysis and calculation. The hypotheses on the loading of the shield tunnel have been made, as follows:

(1) There is no horizontal soil resistance.

There is zero bending moment in the designed shield tunnel cross-section. Hence, it can be approximately considered that no deformation takes place in the shield tunnel and there is no horizontal soil resistance at the two sides of the tunnel.

(2) Do not consider the weight of the shield tunnel structure.

As an underground structure bearing the soil pressure, the weight of the shield tunnel is much smaller than the surrounding earth pressure that it bears. According to the design specification of metro [10], the lining thickness should be 0.04~0.06 times larger than the outer contour diameter of the tunnel and the buried depth is much larger than the thickness of the segment. Hence, the bending moment of the tunnel structure that is caused by the weight of tunnel is much smaller than the bending moment of the tunnel structure that is caused by the surrounding earth pressure.

(3) Calculate the vertical earth pressure based on the earth-pillar theory and do not consider the water pressure.

The soil is soft in the soft soil area, and it is impossible to make an arch. Also, the permeability coefficient is small.

Based on the above hypotheses, the earth pressure surrounding the shield tunnel is demonstrated in Figure 3. The pressure can be calculated as Equations (1)–(3).

$$P_1 = \gamma H \tag{1}$$

$$P_2 = \gamma H k = P_1 k \tag{2}$$

$$P_3 = \gamma a k \tag{3}$$

where a is the vertical diameter of the shield tunnel, b is the horizontal diameter at the center of the vertical diameter of the shield tunnel, P_1 is the vertical earth pressure, P_2 is the horizontal earth pressure on top of the tunnel, P_3 is the part of earth pressure that occurs where the earth pressure at the bottom of the tunnel exceeds the earth pressure on top of the tunnel, γ is the volume weight of the soil in the stratum. For the layers of soil, take the average volume weight, H is the buried depth of the tunnel, k is the lateral earth pressure coefficient.

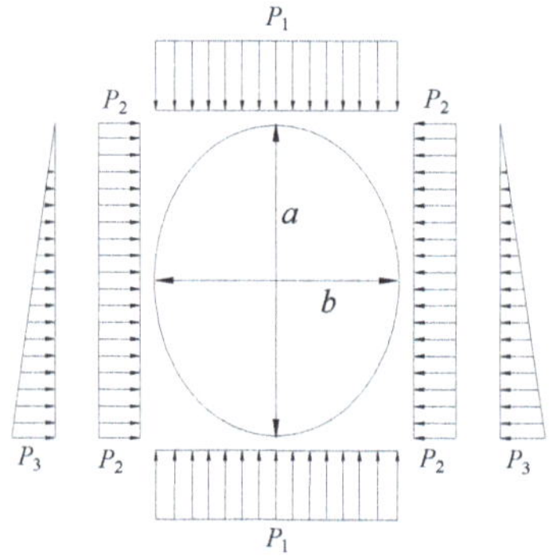

Figure 3. Load distribution of shield tunnel.

2.2. The Analysis of the Pressure Exerted on the Shield Tunnel

The simplified earth pressure unit is line loading (such as kN/m), the value of which equals to the earth pressure value in a 3D condition (the unit is the unit for plane loading, such as kN/m^2). It can be learnt from Figure 3 that the structural mechanics model of the way that the shield tunnel plane is subjected to pressure is the same as how symmetrical structure subjects to symmetrical loading, which means analysis of one part will be enough. In order to facilitate the structural mechanics analysis, the semi-structural analysis of Figure 3, after clockwise rotation of 90 degrees, was analyzed, and the corresponding structural mechanics model is shown in Figure 4. From the symmetrical relationship of the structure, loading, and $\Sigma X = 0$, it can be assumed that the shear force at cross-section A Q_1 and B Q_2 are both zero. Hence, it is required to design and calculate the axial equation for the zero bending moment of the shield tunnel. Assuming the bending moment of cross-section A and cross-section B are both zero, the calculation model of the structural mechanics of the rational axis of the shield tunnel cross-section is demonstrated in Figure 5.

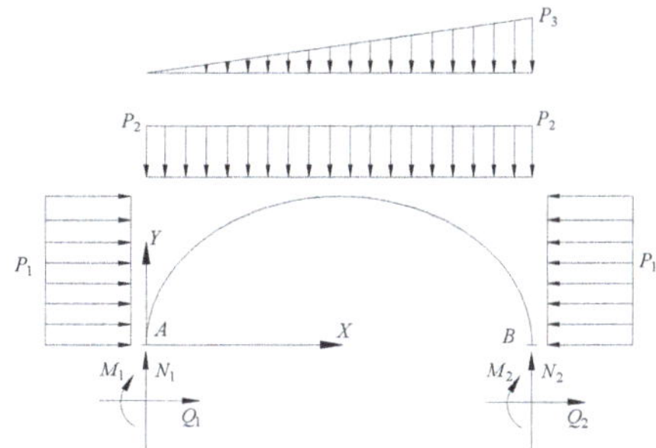

Figure 4. Structural mechanical model of semi-shield tunnel cross-section (the semi-structure of Figure 3, after rotating clockwise 90 degrees).

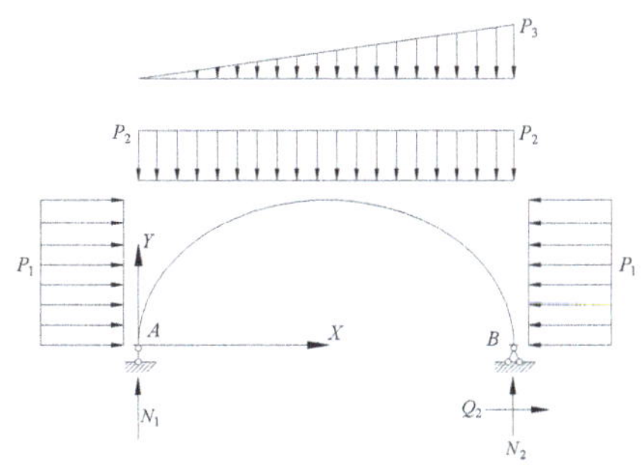

Figure 5. Structural mechanical model of rational axis for zero bending moment shield tunnel cross-section.

Based on $\sum M_A = 0$ and $\Sigma Y = 0$, the axial force can be calculated as Equations (4) and (5).

$$N_1 = \frac{1}{2}P_2a + \frac{1}{6}P_3a \quad (4)$$

$$N_2 = \frac{1}{2}P_2a + \frac{1}{3}P_3a \quad (5)$$

where N_1 is the axial force of cross-section A and N_2 is the axial force of cross-section B.

3. The Design and Calculation of the Zero Bending Moment Shield Tunnel Cross-Section

3.1. The Calculation of the Rational Axis for Shield Tunnel Cross-Section

The expression of the rational axis is required to be calculated when designing the rational axis for the shield tunnel cross-section. In Figure 5, the left cross-section of the shield tunnel is cross-section S, and its coordinate is (x, y). As in Figure 6, cross-section S is composed of four parts, as follows:

$$M_s = N_1x - \frac{1}{2}P_2x^2 - \frac{1}{6a}P_3x^3 - \frac{1}{2}P_1y^2 \quad (6)$$

where M_s is the bending moment of cross-section S; x is the horizontal coordinate of cross-section S; y is the vertical coordinate of cross-section S.

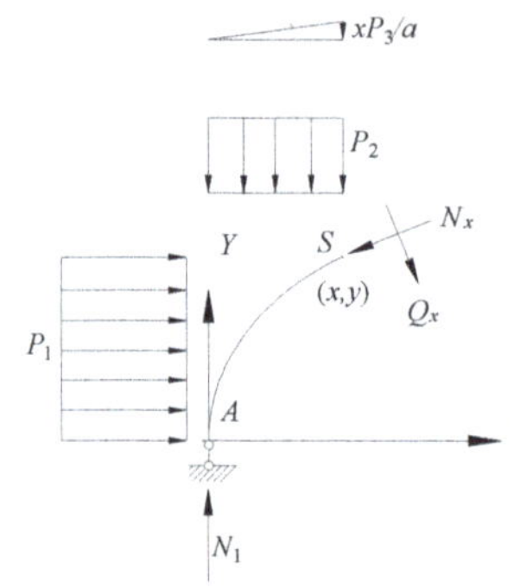

Figure 6. Internal force analysis for a section of zero bending moment shield tunnel cross-section 1.

Equation (6) shows that the first three bending moments M_s are all caused by the loading on the Y-axis and the values are only associated the value of x; the fourth M_s is caused by the loading along the X-axis, the value of which is only related with y. The first three bending moments and the corresponding simply supported beam (as shown in Figure 7) share the shame bending moment expression when subjected to the same loading along the Y-axis. Assuming the bending moment expression of the simply supported beam at horizontal coordinate x is as follows:

$$M_x = N_1 x - \frac{1}{2}P_2 x^2 - \frac{1}{6a}P_3 x^3 \tag{7}$$

where M_x is the bending moment at the position of horizontal coordinate x.

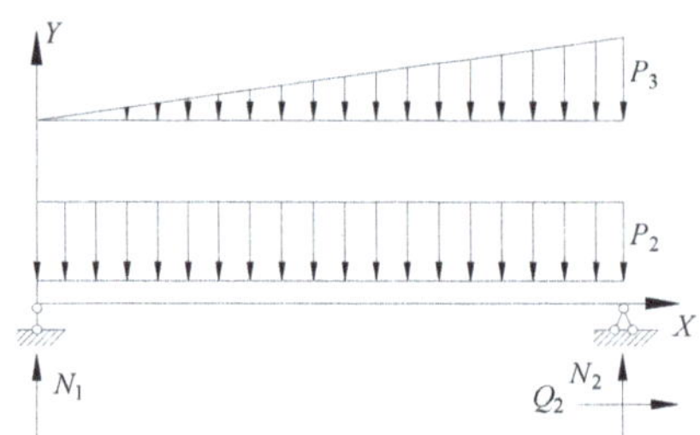

Figure 7. Corresponding simply supported beam model for zero bending moment shield tunnel cross-section.

In order to realize the zero bending moment of the cross-section S, the vertical coordinate for cross-section S can be calculated as Equation (8).

$$y = \sqrt{2M_x/P_1} = \sqrt{\left(2N_1 x - P_2 x^2 - \frac{1}{3a}P_3 x^3\right)/P_1} \tag{8}$$

However, the bending moment of the right cross-section S is composed of five parts, as shown in Figure 8. The bending moment of the cross-section S on the right of Figure 8 caused by the P_1 loading in the left direction exerted on curve CS shares the same value as the bending moment of the cross-section S caused by the P_1 loading in the right direction exerted on curve DC on the left of Figure 8. The two bending moments are in opposite directions so they can offset each other. Therefore, Equation (8) can be applied to any section in Figure 5.

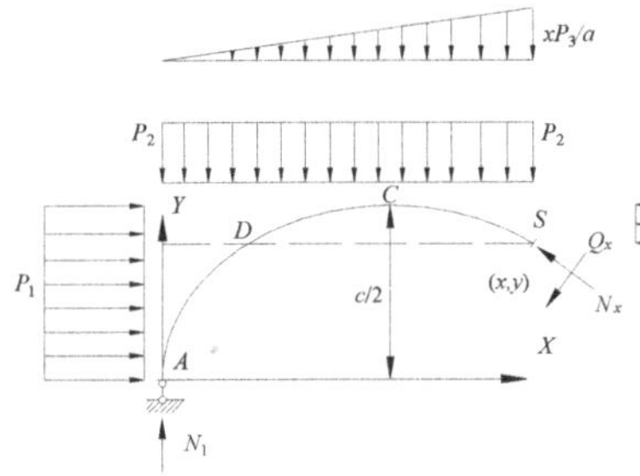

Figure 8. Internal force analysis for a section of zero bending moment shield tunnel cross-section 2.

3.2. The Calculation of the Horizontal Diameter of the Zero Bending Moment Shield Tunnel

The horizontal diameter at the center of the vertical diameter of the shield tunnel is known as the center horizontal diameter, represented by b. From the calculating equation of the rational axis of the shield tunnel cross-section and the calculating equation of the bending moment of the corresponding simply supported beam, it can be learnt that, since the lateral earth pressure of the shield tunnel is larger at the top than at the bottom, the largest horizontal diameter of the shield tunnel does exist at the vertical diameter center but somewhere lower than the center itself. The largest horizontal diameter of the shield tunnel is represented as c.

When $x = a/2$, the corresponding $y = b/2$, according to Equation (8), the calculating equation of b be calculated as Equation (9).

$$b = 2\sqrt{\left(N_1 a - \frac{1}{4}P_2 a^2 - \frac{1}{24}P_3 a^2\right)/P_1} \tag{9}$$

For the maximum of β, the corresponding $y = c/2$. The calculating equation of c can thus be obtained. Assume M_x is the derivative of x to x, as shown in Equation (10).

$$\beta = \frac{dM_x}{dx} = N_1 - P_2 x - \frac{1}{2a}P_3 x^2 \tag{10}$$

From Figure 7, it can be learnt that $x \in [0,\ a]$. When $\beta = 0$, the corresponding x will obtain the maximum M_x. Hence, the maximum of y will be obtained based on the corresponding x when $\beta = 0$, which is $c/2$. The value of x conditional on $\beta = 0$ can be calculated as Equation (11).

$$\overline{x} = \frac{a}{P_3}\left(\sqrt{P_2^2 + \frac{2}{a}N_1 P_3} - P_2\right) \tag{11}$$

where x is the value corresponding to the condition $\beta = 0$.

Put $\overline{x}$ obtained from Equation (11) into Equation (8), to obtain the calculating equation of c, as shown in Equation (12).

$$c = 2\sqrt{\left(2N_1\overline{x} - P_2\overline{x}^2 - \frac{1}{3a}P_3\overline{x}^3\right)/P_1} \tag{12}$$

where c is the largest horizontal diameter of the shield tunnel.

Name the distance difference between the position of the vertical diameter conditional on the maximum horizontal diameter and the position of the vertical diameter conditional on the center horizontal diameter as eccentricity Δ. It can be calculated as Equation (13).

$$\Delta = \overline{x} - \frac{a}{2} \tag{13}$$

where Δ is the difference of vertical diameter position under the different conditions.

3.3. The Calculation of the Internal Force of the Zero Bending Moment Shield Tunnel

For shield tunnels whose cross-section is rational axis, the internal force of the cross-section is only composed of the axial force and the shear force. Assume that the axial force of the cross-section S is Q_x and the shear force is Q_x. Given that the angle of the section will change as the coordinates change, the internal force of the cross-section S can be disintegrated as the horizontal loading F_{hx} and the vertical loading F_{vx} (in Figure 9 the leftward and the upward are set as positive), as shown in Figure 9. First calculate the position angle of the section S, then disintegrate the horizontal loading as F_{hx}, vertical loading F_{vx} as axial force N_x, and shear force Q_x (in Figure 8 the pressure exerted on the section S direction and the direction where the slider rotates clockwise are set as positive).

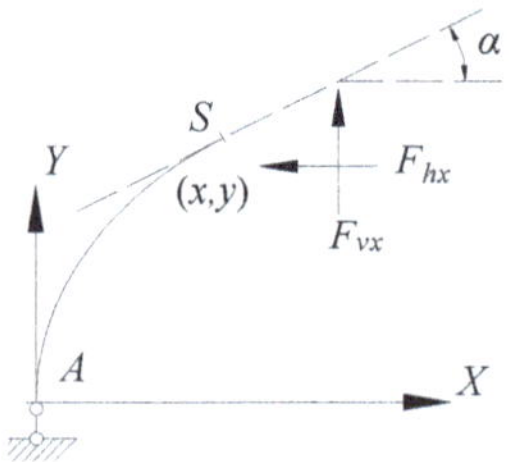

Figure 9. Internal force analysis for a section of zero bending moment shield tunnel cross-section 3.

$\sum X = 0$, F_{hx} is calculated as Equation (14).

$$F_{hx} = P_1 y \tag{14}$$

where F_{hx} is the loading of section S in a horizontal direction.

F_{vx}, F_{vx} is calculated as Equation (15).

$$F_{vx} = P_2 x + \frac{P_3}{2a} x^2 - N_1 \tag{15}$$

where F_{vx} is the loading of section S in a vertical direction.

Derivation of Equation (8) to x, ψ is calculated as Equation (16).

$$\psi = \frac{dy}{dx} = \frac{1}{2P_1} \cdot \sqrt{\frac{P_1}{2N_1 x - P_2 x^2 - \frac{1}{3a} P_3 x^3}} \cdot \left(2N_1 - 2P_2 x - \frac{1}{a} P_3 x^2\right) \tag{16}$$

The expression of the angle α can be calculated as Equation (17).

$$\alpha = \arctan\frac{dy}{dx} = \arctan\psi \tag{17}$$

where α is the angle between the tangent of cross-section S and the horizontal line.

According to the equivalent relation between F_{hx}, F_{vx}, and N_x, μ, N_x, Q_x can be calculated as Equations (18) and (19).

$$N_x = F_{hx}\cos\alpha - F_{vx}\sin\alpha \tag{18}$$

where N_x is the axial force of cross-section S.

$$Q_x = -F_{hx}\sin\alpha - F_{vx}\cos\alpha \tag{19}$$

where Q_x is the shear force of cross-section S.

4. Cross-Section Design and Case Analysis for Minimum Bending Moment Shield Tunnel in Soft Soil Area

4.1. Brief Introduction for Shanghai Metro Shield Tunnel

Taking the Shanghai soft soil area as an example, the minimum bending moment shield tunnel is designed according to its stratum characteristics, as shown in Figure 10. The circular cross-section tunnel lining of the shield tunnels with non-staggered installation, which is widely used in Shanghai's single-line metro, is assembled with non-staggered installation precast segments. The outer diameter of the segment ring is 6.2 m, and the thickness and width of the segment are 0.35 m and 1.2 m, respectively. The segment is precast using high-strength concrete (C55 grade), of which the Poisson's ratio and thr elastic modulus are 0.18 and 35.5 GPa, respectively. The segment ring is composed of six segments, which is shown in Figure 11. Straight bolts, with a mechanical property grade of 5.8, are adopted for the circumferential and longitudinal connection of the segment rings.

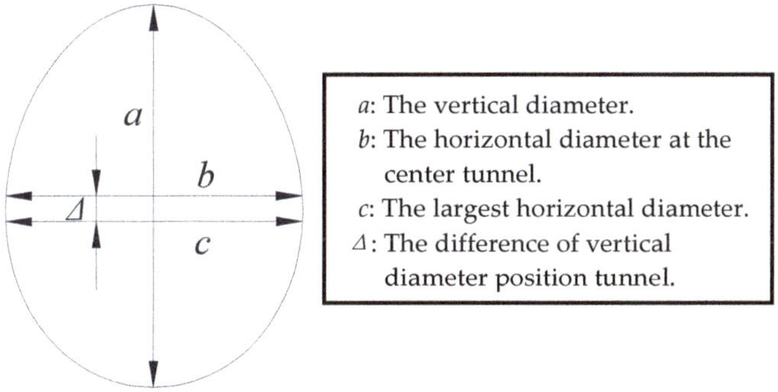

Figure 10. Critical parameters for vertical elliptical cross-section shield tunnel.

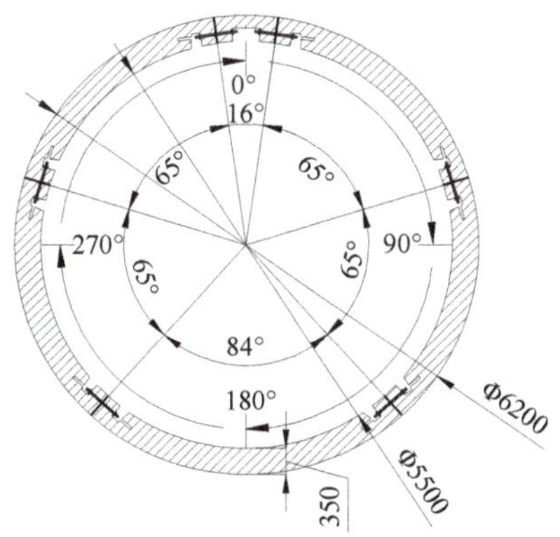

Figure 11. Diagram for structure parameters of shield tunnel in Shanghai (unit: mm).

Shanghai, as the representative saturated soft soil ground area in China, has shield tunnels that mainly go through soil ③, ④, ⑤, ⑥, especially soil ③, ④. Table 1 shows the physical parameters measured by some engineering test of corresponding stratum. These data indicate that the main soil that the shield tunnel goes through features large water content, large compressibility, and low strength. Such soil also features high sensibility. When a shield tunnel is constructed in such soil, the horizontal soil resistance is very small [27,28], which is one of the reasons why the beyond-limit oval deformation of the shield tunnel easily occurs. So, it is necessary to try to design a rational cross-section of shield tunnels constructed in soft soil with less bending moment.

Table 1. Shanghai geological soil parameters.

Soil Layer	Unit Weight/kN/m^3	Moisture Content/%	Void Ratio	Liquid Limit/%	Plastic Limit/%	Cohesion/kPa	Internal Friction Angle/°	Compression Modulus/MPa	Poisson Ratio
③ Mucky silty clay	19.6	39.7	1.123	35.4	20.5	9	16.5	3.36	0.3
④ Mucky clay	18.8	49.4	1.392	43.2	23	13	10.5	2.27	0.33
⑤ Silty clay	20.1	34.8	0.996	36.9	20.7	17	14	4.27	0.31
⑥ Clay	20.5	23.3	0.695	34	18.5	44	15.5	6.58	0.3

Suppose the coefficient of lateral earth pressure k is calculated as Equation (20).

$$k = \frac{\mu}{1-\mu} \tag{20}$$

where μ is the Poisson's ratio.

Then the coefficient of lateral earth pressure is 0.4286–0.4925.

4.2. Analysis for Cross-Section Key Parameters of Zero Bending Moment Shield Tunnel

For the properties of soft soil, the vertical earth pressure on the crown of the tunnel is considered as the earth-pillar theory earth pressure, calculated by Equation (1). Assume that the thickness of soil above the tunnel crown is 12 m, the vertical diameter of the cross-section axis of the tunnel is 6 m (the central axis position of the outer and inner diameter of the tunnel cross-section is taken as the design diameter). The average unit weight of soil is 19.5 kN/m^3. According to the proposed design method for the cross-section of the zero bending moment shield tunnel, cross-section key parameters of the zero bending moment are obtained based on different lateral earth pressure coefficients, as shown in Table 2. When the lateral earth pressure coefficient k increases, the eccentric distance Δ of the zero bending moment shield tunnel remains the same, and the center horizontal diameter b and the maximum horizontal diameter c both increase. Assume that the lateral earth pressure coefficient is 0.46, cross-section key parameters of the zero bending moment shield tunnel under a different thickness of soil above the tunnel crown is shown in Table 3. With the increase in the thickness of the soil above the tunnel crowns, the eccentric distance Δ, the center horizontal diameter b, and the maximum horizontal diameter c all decrease. The data in Tables 1 and 2 indicate that both the center horizontal diameter b and the maximum horizontal diameter c are smaller than the vertical diameter a. The calculation results indicate that there is zero shear force for any section of the cross-section of the zero bending moment, which means that there is no shear deformation at the segment longitudinal joints. The minimum axial force N_1 is apparent at the crown of the tunnel, and the maximum axial force N_3 is apparent at the level of maximum horizontal diameter. N_2 is the axial force at the tunnel invert.

Table 2. Key parameters for no bending moment shield tunnels with different coefficients of lateral earth pressure.

H/m	k	P_1/kPa	P_2/kPa	P_3/kPa	a/m	b/m	c/m	N_1/kN	N_2/kN	N_3/kN	Δ/m
12	0.42	222.00	93.24	46.62	6.00	4.347	4.350	326.34	372.96	482.83	0.0997
12	0.44	222.00	97.68	48.84	6.00	4.450	4.452	341.88	390.72	494.19	0.0997
12	0.46	222.00	102.12	51.06	6.00	4.550	4.552	357.42	408.48	505.30	0.0997
12	0.48	222.00	106.56	53.28	6.00	4.648	4.650	372.96	426.24	516.17	0.0997
12	0.50	222.00	111.00	55.50	6.00	4.743	4.746	388.50	444.00	526.81	0.0997

Table 3. Key parameters for no bending moment shield tunnels with different buried depths.

H/m	k	P_1/kPa	P_2/kPa	P_3/kPa	a/m	b/m	c/m	N_1/kN	N_2/kN	N_3/kN	Δ/m
6	0.46	111.00	51.06	51.06	6.00	4.984	4.992	204.24	255.30	277.03	0.1652
9	0.46	166.50	76.59	51.06	6.00	4.699	4.703	280.83	331.89	391.52	0.1244
12	0.46	222.00	102.12	51.06	6.00	4.550	4.552	357.42	408.48	505.30	0.0997
18	0.46	333.00	153.18	51.06	6.00	4.395	4.397	510.60	561.66	732.05	0.0713
28	0.46	518.00	238.28	51.06	6.00	4.282	4.282	765.90	816.96	1109.14	0.0483

The corresponding cross-sections for Tables 1 and 2 are in Figure 12, which are similar to a vertical ellipse. The axial forces for the zero bending moment tunnel are shown as Figure 13 (for the vertical symmetry of the axial forces, only axial forces at the range between 0° and 180° are given). The difference values of the horizontal diameters between any cross-sections of the zero bending moment tunnel are smaller than any difference value between the horizontal diameters of the zero bending moment tunnel and circular tunnel. So, any cross-section of the zero bending moment tunnel is adopted for the metro shield tunnel, and its maximum bending moment is smaller than the circular cross-section.

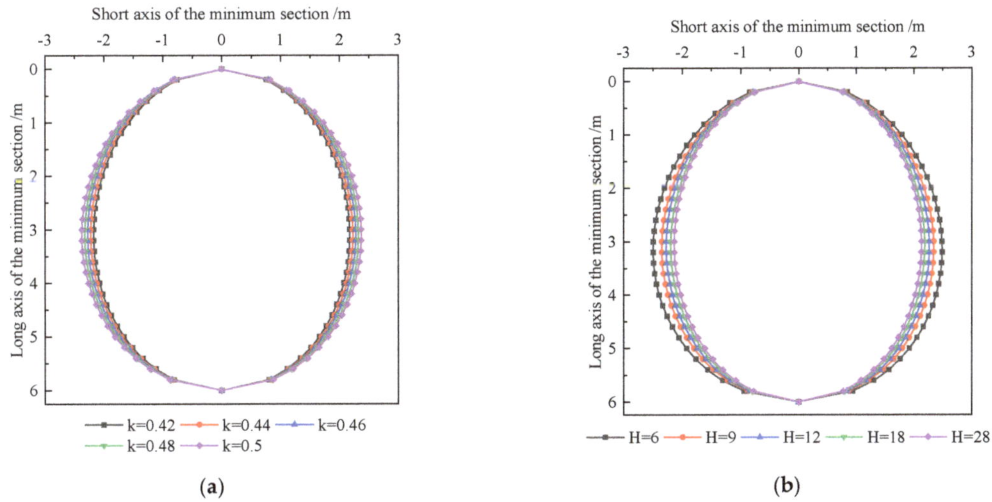

Figure 12. Zero bending moment shield tunnel cross-section with vertical diameter of 6 m: (**a**) Different coefficients of lateral earth pressure; (**b**) Different thickness of overlaying soil.

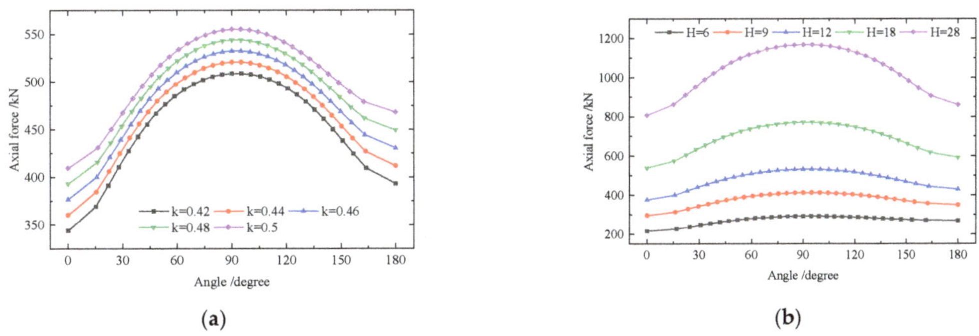

Figure 13. Axial force for zero bending moment shield tunnel with vertical diameter of 6 m: (**a**) Different coefficients of lateral earth pressure; (**b**) Different thickness of overlaying soil.

4.3. Cross-Section Design for the Minimum Bending Moment Shield Tunnel

The above analysis indicates that the cross-section of the zero bending moment shield tunnel is related to the assumed earth pressure. However, usually one metro line only adopts one tunnel cross-section in light of practical construction, and tunnel buried depth and soil properties of the ground are changed along the metro line. Therefore, it is necessary to evaluate the earth pressure mode and corresponding length for the shield tunnel. Suppose the shield tunnel has n kinds of engineering conditions, which are shown in Table 4. Taking every condition into account, the cross-section axis of the minimum bending moment shield tunnel can be calculated as Equation (21).

$$y = \frac{y_1 l_1 + y_2 l_2 + y_3 l_3 + \ldots\ldots + y_n l_n}{L} \tag{21}$$

in which L is the total length of the metro shield tunnel and is calculated as Equation (22).

$$L = l_1 + l_2 + l_3 + \ldots\ldots + l_n \tag{22}$$

where y_i refers to Equation (8) and can be calculated as Equation (23).

$$y_i = f(P_{1-i}, P_{2-i}, P_{3-i}, l_i) \tag{23}$$

Table 4. Design parameters for different engineering conditions.

Engineering Condition	P_1	P_2	P_3	l
1	$P_{1\text{-}1}$	$P_{2\text{-}1}$	$P_{3\text{-}1}$	l_1
2	$P_{1\text{-}2}$	$P_{2\text{-}2}$	$P_{3\text{-}2}$	l_2
3	$P_{1\text{-}3}$	$P_{2\text{-}3}$	$P_{3\text{-}3}$	l_3
……	……	……	……	……
n	$P_{1\text{-}n}$	$P_{2\text{-}n}$	$P_{3\text{-}n}$	l_4

In order to apply the cross-section design theory of the minimum bending moment shield tunnel, this proposed design flow chart is presented as Figure 14.

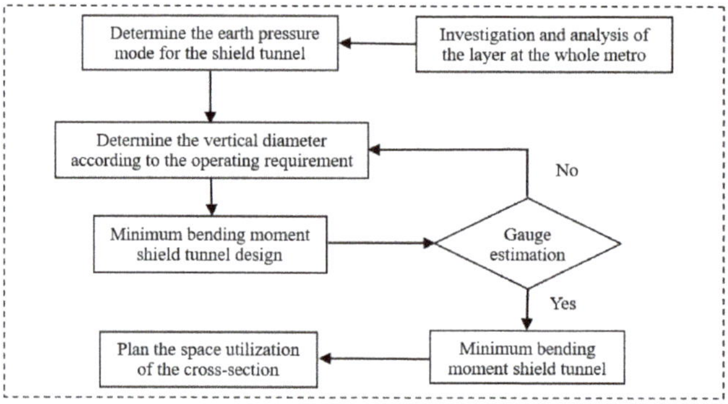

Figure 14. Design flow chart for shield tunnel using the concept of minimum bending moment.

4.4. Case Analysis for Minimum Bending Moment Shield Tunnel

The shield tunnel of one metro line in Shanghai passes through soft soil ③, ④, ⑤, and ⑥, as shown in Table 1, all of which feature a small soil-resistance coefficient, so it is important to consider reducing the bending moment as much as possible. According to the line design, the thickness of the overlying soil ranged between 6.5 and 28.8 m, combined with the geological investigation data of the tunnel, several designs have been made based on the different working conditions for the shield tunnel of this metro line. The cross-section parameters of the zero bending moment shield tunnel under different working conditions were obtained as presented in Table 5. Finally, the cross-section of the shield tunnel conditional on the minimum bending moment is shown in Figure 15. The shape of the cross-section of the shield tunnel conditional on the minimum bending moment was similar to a vertical ellipse.

Table 5. Design parameters for different conditions.

Engineering Condition	H/m	K	P_1/kPa	P_2/kPa	P_3/kPa	l/m	a/m	b/m	c/m	N_1/kN	N_2/kN	N_3/kN	Δ/m
1	7	0.429	136.50	58.56	50.19	2755	6.00	4.697	4.703	225.87	276.06	320.98	0.1489
2	10	0.429	195.00	83.66	50.19	5933	6.00	4.481	4.484	301.16	351.35	437.20	0.1149
3	10	0.493	195.00	96.14	57.68	4340	6.00	4.803	4.807	346.09	403.77	468.67	0.1149
4	13	0.429	253.50	108.75	50.19	2045	6.00	4.360	4.362	376.45	426.64	552.88	0.0935
5	13	0.493	253.50	124.98	57.68	6473	6.00	4.674	4.676	432.61	490.29	592.68	0.0935
6	16	0.493	312.00	153.82	57.68	5423	6.00	4.591	4.592	519.13	576.81	716.42	0.0788
7	16	0.449	312.00	140.09	52.53	2807	6.00	4.381	4.383	472.80	525.33	683.70	0.0788
8	19	0.493	370.50	182.66	57.68	680	6.00	4.533	4.534	605.65	663.33	840.00	0.0681
9	19	0.449	370.50	166.35	52.53	1470	6.00	4.326	4.327	551.60	604.13	801.64	0.0681
10	22	0.449	429.00	192.62	52.53	1620	6.00	4.286	4.287	630.40	682.93	919.49	0.0599
11	25	0.429	487.50	209.14	50.19	784	6.00	4.159	4.160	677.61	727.80	1013.92	0.0535
12	28	0.429	546.00	234.23	50.19	1980	6.00	4.135	4.136	752.90	803.09	1129.02	0.0483

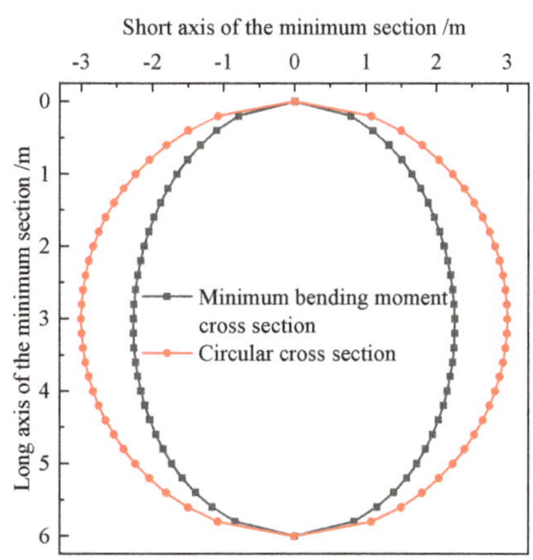

Figure 15. Comparison between minimum bending moment shield tunnel cross-section and circular shield tunnel cross-section in Shanghai.

The finite element software Midas GTS NX was adopted for the numerical simulation analysis in order to compare the bending moments of two differently shaped shield tunnel cross-sections due to the surrounding earth pressure shown in Figure 15 (take the bending moment for the tension on the inner side of the tunnel as positive). With reference to one metro line in Shanghai that travels through soft soil areas, simulation analysis was conducted for four working conditions, where the thickness of the overlying soil of the shield tunnel was 7 m, 13 m, 19 m, and 25 m. Due to the constant cross-section shape of the shield tunnel, the 2D model was utilized to simulate the cross-section bending moment [29,30]. The 2D plane strain model was utilized to simulate the different soils, and the soil constitutive model was a modified Mohr–Coulomb model. The soil parameters and material properties were the same as those of the soft area of the above-mentioned metro line in Shanghai, as shown in Table 6. The beam element model was utilized to simulate the tunnel lining with reference to the Shanghai metro shield tunnel in light of the dimension and mechanical parameters, as shown in Figure 11. The interaction between the tunnel and the soil was simulated using a compressed elastic spring, as shown in Figure 16, the vertical elliptical cross-section can be taken as an example.

Table 6. Soil mechanical parameters for different stratums.

Soil Layer	Thickness/m	Unit Weight /kN/m^3	Moisture Content/%	Void Ratio	Cohesion/kPa	Internal Friction Angle/°	Compression Modulus/MPa	Poisson Ratio
Clay	4.5	20.5	23.3	0.695	44	15.5	6.58	0.30
Mucky silty clay	6.5	19.6	39.7	1.123	9	16.5	3.36	0.30
Mucky clay	29	18.8	49.4	1.392	13	10.5	2.27	0.33

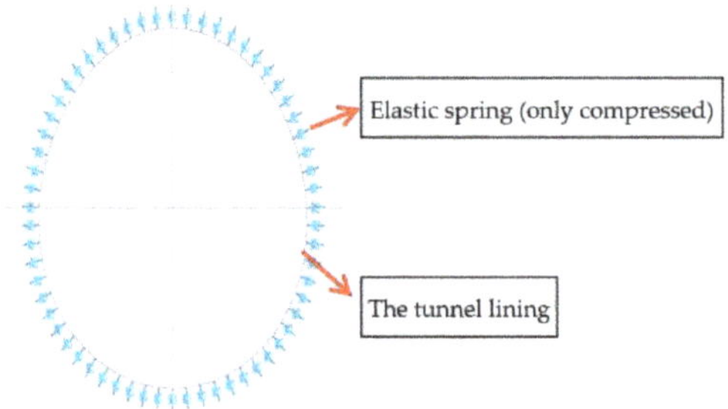

Figure 16. The interaction between the tunnel and the soil of the numerical simulation model.

The soil model of this article was designed to be 80 m wide and 40 m tall, constituted by three types of soil. From the top to the bottom, the soils were clay, mucky silty clay, and mucky clay and the thickness of the soils were 4.5 m, 11.5 m, and 29 m, respectively. The soil model was divided into 17,055 mixed grids. The radius of the circular cross-section tunnel model was 3 m and was divided into 38 grids through size control. The long-axis of the vertical elliptical cross-section tunnel model was 6 m long and the short-axis was 4.4 m long. It was divided into 62 grids through size control. The grid division of the numerical simulation model is shown in Figure 17.

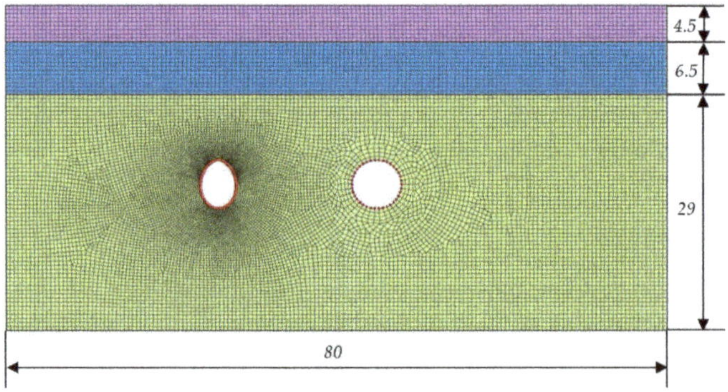

Figure 17. The grid division of the numerical simulation model (unit: m).

The maximum bending moment and the minimum bending moment of the two differently shaped cross-sections of the tunnel under the different working condition were obtained using the simulation analysis, as shown in Table 7. It can be learnt through Table 7 that the bending moments of the two differently shaped cross-sections increased as the thickness of the overlying soil increased, and the bending moment increased more obviously in the case of the circular cross-section. As for the conditional on the same working condition, compared to the circular cross-section, the vertical elliptical cross-section featured a much smaller bending moment. Also, as the thickness of the overlying soil increased, the decrease in the bending moment of the vertical elliptical cross-section was more obvious.

Table 7. Maximum and minimum bending moments of two different bending moment shield tunnel cross-sections for different conditions.

Engineering Conditions	H/m	Minimum Bending Moment Shield Tunnel Cross-Section		Circular Shield Tunnel Cross-Section	
		Max Bending Moment/kN·m	Min Bending Moment/kN·m	Max Bending Moment/kN·m	Min Bending Moment/kN·m
1	7	218.644	−164.328	362.403	−327.64
2	13	245.091	−208.737	567.907	−538.893
3	19	331.966	−275.233	779.613	−755.177
4	25	422.12	−350.606	995.103	−969.411

When the thickness of the overlying soil of the tunnel was 25 m, the bending moment diagram of two differently shaped cross-sections is shown in Figure 18. It can be seen from Figure 18 that the maximum values of the two differently shaped cross-sections both appear at the bottom of the tunnel arch, while the minimum bending moment values both appear at the waist of the tunnel. The maximum bending moment value of the circular cross-section tunnel was 995.103 kN·m and the minimum bending moment value was −969.411 kN·m. On the other hand, the maximum bending moment value of the vertical elliptical cross-section tunnel was 422.12 kN·m and the minimum value was −350.606 kN·m. When compared with the circular cross-section, the maximum bending moment of the vertical elliptical cross-section was 57.58% less while the minimum value was 67.83% less.

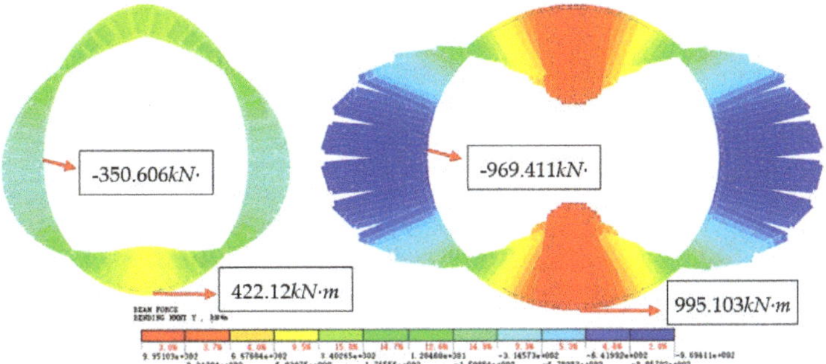

Figure 18. Comparison between minimum bending moment shield tunnel cross-section and circular shield tunnel cross-section in Shanghai.

The cross-section shape of the zero bending moment shield tunnel in this article was obtained by methods of rational assumptions. In addition, with the consideration that, normally, metro lines are constructed with only one shield tunnel of one cross-section, the weighted average method was used to obtain the minimum bending moment shield tunnel cross-section shape. As a result, the bending moment of the similar vertical elliptical cross-section shield tunnel obtained in the simulation was not zero. However, it can be learnt from the numerical simulation calculation, by means of the improvement of the cross-section shape of the tunnel, that the bending moment was significantly decreased compared to the circular tunnel, which makes it favorable both in the case of the transverse deformation of the tunnel structure and in the case of longitudinal segment joint failure. One of the Shanghai metro tunnels mainly passes through soft soil, so the soil resistance is small. It is necessary to design a rational cross-section to decrease the bending moment. According to the primary design drawing, the buried depth was from 6.5 to 28.8 m. According to the geological investigation data, the engineering conditions were divided into 12 types, and the critical parameters of the cross-section for the corresponding zero bending moment shield tunnels were calculated, which are shown in Table 5. According to Equation (21), the minimum bending moment shield tunnel was obtained, as shown in Figure 15. The cross-section of minimum bending moment shield tunnel was similar to a vertical elliptical cross-section, and it is obviously different from the circle. Table 7 shows that with this shield tunnel, with this peculiar cross-section, its bending moment can be greatly reduced in soft soil area.

5. Conclusions

(1) Given that the bending moment of the cross-section of the shield tunnel constructed in the soft soil area tends to easily lead to beyond-limit oval deformation of the cross-section and induce diseases and waterproof failure in terms of the segment joint structure, this article puts forward the concept of designing a zero bending moment shield tunnel for the first time.

(2) Based on the characteristics of the surrounding rocks of the tunnel and rational assumption conditions, this article obtains the structural and mechanical calculation model of the rational axis for the zero bending moment shield tunnel and the expression of the rational axis. In addition, the internal force and key parameters calculating the equations of the zero bending moment shield tunnel are advised here in this article.

(3) Taking the shield tunnel constructed in the Shanghai soft soil area as an example, we designed and analyzed the zero bending moment shield tunnel. The results indicate that if the vertical diameter a remains the same, as the lateral earth pressure coefficient k increases, the center horizontal diameter b and the maximum horizontal diameter c both increase, but still less than the vertical diameter a; however, if the center horizontal diameter b remains the same, as the buried height of the tunnel increases, the vertical diameter a increases and the shear force of the zero bending moment shield tunnel is zero.

(4) Normally, one metro line can only use one shield tunnel with one cross-section shape. Given this, the shield tunnel cross-section design methods and procedures based on the minimum bending moment are proposed here. We took the parameters of the soils that one metro line shield tunnel in Shanghai passes through as an example and used the weighted average to obtain the minimum bending moment tunnel cross-section of that metro line. The numerical simulation analysis indicates that the similar vertical elliptical cross-section shield tunnel features a significantly smaller bending moment compared to that of the circular shield tunnel.

Author Contributions: Investigation of the actual utilization for different shield tunnel cross-sections, D.H. and H.J.; the proposal of the concept of zero bending moment, D.H. and C.X.; the proposal of the concept of minimum bending moment, D.H. and H.J.; rational and feasible hypotheses about zero bending moment, W.T. and X.L.; theoretical analysis for structural mechanical model of zero bending moment shield tunnel, D.H. and C.X.; data analysis of zero bending moment shield tunnel, H.J. and W.W.; numerical simulation of Shanghai minimum bending moment shield tunnel, D.H. and H.J. All authors have read and agreed to the published version of the manuscript.

Funding: This study was supported by the National Natural Science Foundation of China, grant number 52078213(the funder: Dawei, Huang), U1934208(the funder: Changjie, Xu) and 52178333(the funder: Wenbo, Tu) and Natural Science Foundation of Jiangxi Province, grant number 20192ACBL21003 (the funder: Dawei, Huang) and 20212BAB204013(the funder: Wenbo, Tu).

Institutional Review Board Statement: Not applicable.

Informed Consent Statement: Not applicable.

Data Availability Statement: All data are available from the author.

Conflicts of Interest: The authors declare no conflict of interest.

References

1. Japan Society of Civil Engineers. *Japanese Standard for Shield Tunneling*; Japan Society of Civil Engineers: Tokyo, Japan, 1996.
2. Japan Society of Civil Engineers. *Standard Specifications for Tunneling: Shield Tunnels*; Japan Society of Civil Engineers: Tokyo, Japan, 2006.
3. Yukinori, K. Present status and technology of shield tunneling method in Japan. *Tunn. Undergr. Space Technol.* **2003**, *18*, 145–159. [CrossRef]
4. He, C.; Wang, B. Research progress and development trends of highway tunnels in China. *J. Mod. Transp.* **2013**, *21*, 209–223. [CrossRef]
5. Mashimo, H.; Ishimura, T. Evaluation of the load on shield tunnel lining in gravel. *Tunn. Undergr. Space Technol.* **2003**, *18*, 233–241. [CrossRef]
6. Blom, C.B.M. Design Philosophy of Concrete Linings for Tunnels in Soft Soils. Ph.D. Thesis, Technische Universiteit Delft, Delft, The Netherlands, 2002.
7. Bian, X.; Hong, Z.S.; Ding, J.W. Evaluating the effect of soil structure on the ground response during shield tunnelling in Shanghai soft clay. *Tunn. Undergr. Space Technol.* **2016**, *58*, 120–132. [CrossRef]
8. Vu, M.N.; Broere, W. Structural design model for tunnels in soft soils: From construction stages to the long-term. *Tunn. Undergr. Space Technol.* **2018**, *78*, 16–26. [CrossRef]
9. Working Group No. 2; International Tunnelling Association. Guidelines for the design of shield tunnel lining. *Tunn. Undergr. Space Technol.* **2000**, *15*, 303–331. [CrossRef]
10. General Administration of Quality Supervision, Inspection and Quarantine of the People's Republic of China. *Code for Design of Metro GB 50157-2013*; China Standards Press: Beijing, China, 2013.
11. Ding, W.Q.; Gong, C.J.; Mosalam, K.M.; Soga, K. Development and application of the integrated sealant test apparatus for sealing gaskets in tunnel segmental joints. *Tunn. Undergr. Space Technol.* **2017**, *63*, 54–68. [CrossRef]
12. Gong, C.J.; Ding, W.Q.; Mosalam, K.M.; Selim, G.; Soga, K. Comparison of the structural behavior of reinforced concrete and steel fiber reinforced concrete tunnel segmental joints. *Tunn. Undergr. Space Technol.* **2017**, *68*, 38–57. [CrossRef]
13. Do, N.A.; Dias, D.; Oreste, P.; Djeran-Maigre, I. The behaviour of the segmental tunnel lining studied by the hyperstatic reaction method. *Eur. J. Environ. Civ. Eng.* **2014**, *18*, 489–510. [CrossRef]
14. Huang, Z.; Fu, H.L.; Chen, W.; Zhang, J.B.; Huang, H.W. Damage detection and quantitative analysis of shield tunnel structure. *Autom. Constr.* **2018**, *94*, 303–316. [CrossRef]
15. Huang, D.W.; Zhou, S.H.; Feng, Q.S.; Liu, L.Y.; Zhang, P.F. Optimization for the longitudinal joint of metro shield tunnel under straight joint assembling in soft soli area. *China Railw. Sci.* **2017**, *38*, 62–69. (In Chinese) [CrossRef]
16. Gong, C.J.; Ding, W.Q.; Soga, K.; Mosalam, K.M. Failure mechanism of joint waterproofing in precast segmental tunnel linings. *Tunn. Undergr. Space Technol.* **2019**, *84*, 334–352. [CrossRef]
17. Gong, C.J.; Ding, W.Q.; Xie, D.W. Parametric investigation on the sealant behavior of tunnel segmental joints under water pressurization. *Tunn. Undergr. Space Technol.* **2020**, *97*, 103231. [CrossRef]
18. Lei, M.F.; Zhu, B.B.; Gong, C.J.; Ding, W.Q.; Liu, L.H. Sealing performance of a precast tunnel gasketed joint under high hydrostatic pressures: Site investigation and detailed numerical modeling. *Tunn. Undergr. Space Technol.* **2021**, *115*, 104082. [CrossRef]
19. Teachavorasinskun, S.; Chub-Uppakarn, T. Influence of segmental joints on tunnel lining. *Tunn. Undergr. Space Technol.* **2010**, *25*, 490–494. [CrossRef]
20. Guan, Z.C.; Deng, T.; Wang, G.; Jiang, Y.J. Studies on the key parameters in segmental lining design. *J. Rock Mech. Geotech. Eng.* **2015**, *7*, 674–683. [CrossRef]
21. Caratelli, A.; Meda, A.; Rinaldi, Z.; Giuliani-Leonardi, S.; Renault, F. On the behavior of radial joints in segmental tunnel linings. *Tunn. Undergr. Space Technol.* **2018**, *71*, 180–192. [CrossRef]
22. Wang, F.; Shi, J.K.; Huang, H.W.; Zhang, D.M. Modified analytical solution of shield tunnel lining considering nonlinear bending stiffness of longitudinal joint. *Tunn. Undergr. Space Technol.* **2020**, *106*, 103625. [CrossRef]
23. Bi, X.L.; Liu, X.; Wang, X.Z.; Lu, L.; Yang, Z.H. Experimental investigation on the ultimate bearing capacity of continuous-jointed segmental tunnel linings. *China Civ. Eng. J.* **2014**, *47*, 117–127. (In Chinese) [CrossRef]
24. Möller, S. Tunnel Induced Settlements and Structural Forces in Linings. Ph.D. Thesis, University of Stuttgart, Stuttgart, Germany, 2006.
25. Li, X.J.; Yan, Z.G.; Wang, Z.; Zhu, H.H. Experimental and analytical study on longitudinal joint opening of concrete segmental lining. *Tunn. Undergr. Space Technol.* **2015**, *46*, 52–63. [CrossRef]
26. Arnau, O.; Molins, C. Three dimensional structural response of segmental tunnel linings. *Eng. Struct.* **2012**, *44*, 210–221. [CrossRef]
27. Huang, D.W.; Zhou, S.H.; Lai, G.Q.; Feng, Q.S.; Liu, L.Y. Mechanism and character for deterioration of shield tunnel under surface surcharge. *Chin. J. Geotech. Eng.* **2017**, *39*, 1173–1181. (In Chinese) [CrossRef]

28. Di, H.G.; Zhou, S.H.; Yao, X.P.; Tian, Z.Y. In situ grouting tests for differential settlement treatment of a cut-and-cover metro tunnel in soft soils. *Bull. Eng. Geol. Environ.* **2021**, *80*, 6415–6427. [CrossRef]
29. Hao, X.J.; Zhang, Q.; Sun, Z.W.; Wang, S.H.; Yang, K.; Ren, B.; Yu, G.F.; Zhou, W.; Chen, B.L.; Zhang, X.Y. Effects of the major principal stress direction respect to the long axis of a tunnel on the tunnel stability: Physical model tests and numerical simulation. *Tunn. Undergr. Space Technol.* **2021**, *114*, 103993. [CrossRef]
30. Kumar, J.; Jain, H. Elasto-plastic ground settlement response and stability of single and twin circular unsupported and supported tunnels. *Transp. Geotech.* **2021**, *30*, 100620. [CrossRef]

applied sciences

MDPI

Article

A Suitability Evaluation Method of Urban Underground Space Based on Rough Set Theory and Conditional Entropy: A Case Study in Wuhan Changjiang New Town

Defu Tong [1], Fei Tan [1,*], Bangchuang Ma [2], Yu-Yong Jiao [1] and Jun Wang [3]

1 Faculty of Engineering, China University of Geosciences, Wuhan 430074, China; defu@cug.edu.cn (D.T.); yyjiao@cug.edu.cn (Y.-Y.J.)
2 China Railway First Survey & Design Institute Group Co., Ltd., Xi'an 710043, China; mabangchuang1219@163.com
3 Shanghai Investigation, Design & Research Institute Co., Ltd., Shanghai 200434, China; wang_jun29@ctg.com.cn
* Correspondence: tanfei@cug.edu.cn

Abstract: In order to make full use of underground space resources and reasonably determine the development function and scale, it is important to evaluate the suitability evaluation of urban underground space (UUS). The development and utilization of underground space is affected by many factors, such as topography, geomorphology, geotechnical characteristics, hydrogeology and so on. Taking the starting area of Changjiang New Town in Wuhan as a case study, this paper introduces the rough set theory and conditional entropy and establishes a suitability evaluation model of UUS. Rough set theory is used to construct a decision information table, preprocess sample data and classify the knowledge base, while conditional entropy is employed to calculate the attributes' own and relative importance. Then, the fuzzy comprehensive evaluation method, the most unfavorable grade discrimination method and the exclusive method are used to evaluate the model, and the partition map of regional underground space development suitability is obtained. In this method, any prior knowledge and additional information will not be needed except the sample data, and the obtained weights are more objective and reasonable. The results show that the overall suitability of underground space development and utilization in the starting area of Wuhan Changjiang New Town is good.

Keywords: urban underground space (UUS); suitability assessment; rough set theory; conditional entropy

Citation: Tong, D.; Tan, F.; Ma, B.; Jiao, Y.-Y.; Wang, J. A Suitability Evaluation Method of Urban Underground Space Based on Rough Set Theory and Conditional Entropy: A Case Study in Wuhan Changjiang New Town. *Appl. Sci.* **2022**, *12*, 1347. https://doi.org/10.3390/app12031347

Academic Editor: Edyta Plebankiewicz

Received: 27 December 2021
Accepted: 21 January 2022
Published: 27 January 2022

1. Introduction

With the development of the economy, the continuous advance of urbanization and the rapid agglomeration of urban population, current urban development is facing a severe test. A series of problems such as congestion of ground space, traffic jams, environmental pollution and ecological deterioration appear one after another. At present, the scale of cities cannot meet the demand of people. Under the constraint of total land supply, the development and utilization of urban underground space has become an indispensable part of urban construction and development in the 21st century. The development and utilization of underground space can effectively relieve the pressure of population and traffic on surface space, thus saving land resources, protecting urban green space, improving urban environment and improving the quality of urban life. In addition, due to its own structure, underground space has unique advantages in urban disaster prevention and reduction. However, the development and utilization of underground space is a complex system of engineering that is affected by many factors, and its reversibility is poor. Once destroyed, it is difficult to restore the original condition. Therefore, scientific and reasonable evaluation of the suitability of underground space development is an important basic condition for making full use of underground space resources, reasonably determining the development function and scale and compiling underground space planning [1,2].

The research on the suitability evaluation of underground space mainly focuses on the evaluation of underground space resources, engineering geology suitability evaluation, spatial planning evaluation and underground space vision evaluation. As for the evaluation of underground space resources, based on the geological conditions of sandstone as the main body in Minneapolis, Sterling and Nelson [3] adopted the method of comprehensive superposition to give the distribution range and the appropriate space form of development and utilization of underground space resources in Minneapolis. Boivin [4] employed the transparency superposition method and determined the spatial capacity, distribution range and development difficulty level of Quebec in Canada. Peng and Peng [5,6] proposed an improved UUS resource evaluation model, synthetically considering major factors and applying GIS, AHP, the most unfavorable grading method, and the exclusive method is applied to the UUS master planning of Chinese cities. Rienzo et al. [7] introduced a 3D geological and underground structure model based on GIS, which can provide great help for the planning and management of underground space development. Zhu et al. [8] used digital technology to handle problems existing in traditional evaluation tools, elaborated relevant concepts and main technologies of digitization and established an engineering geological intelligence system based on GIS. Regarding the visual evaluation of underground space, Zhou et al. [9] proposed a measurable visual comfort evaluation method for underground space based on human perception and machine learning methods and applied it to the underground space of Pentagon Square in Shanghai, China.

As for the suitability evaluation of engineering geology, many scholars have carried out relevant studies and evaluated the suitability of engineering geology of underground space in many cities from the perspective of the analytic hierarchy process (AHP). Lu et al. [10] presented a multilayer UUS exploitation engineering geological suitability evaluation framework, employed a fuzzy set analytic hierarchy process and TOPSIS for the basic layer evaluation and evaluated the geological suitability of multilayer underground space combining with the transferring coefficient matrix. Hou et al. [11] presented a process for quality assessment of the underground space resources in Foshan City by introducing a coupled 3D geological model with borehole data. Chen et al. [12] collected and normalized the data of 56 representative cities at different levels in China on the premise of summarizing the indicators affecting the suitability of underground space development and utilization. Youssef et al. [13] established the suitability evaluation model of engineering geological conditions by using the analytic hierarchy process. Wang et al. [14] used the structural equation model (SEM) to evaluate the suitability of urban underground space by taking Gulou District in Nanjing as an example. In addition, scholars at home and abroad have used the expert questionnaire survey method, entropy weight method and other methods to study the influencing factors of engineering geology suitability and achieved a lot of results. Based on the time sequential weighted average (TOWA) operator, Liu et al. [15] combined the classical entropy weight method with the time dimension weight method and proposed an entropy time weighted mixed weight allocation model for UUSR evaluation. Duan et al. [16] used the index scale analytic hierarchy process (AHP) to analyze the geological conditions and proposed an evaluation system suitable for Kunming, China. To sum up, the key points of UUS suitability evaluation mainly focus on three aspects [17]: 1. the purpose of evaluation; 2. selection, quantification and weight determination of indicators; 3. evaluation model and method. This paper argues that the attributes of the suitability of UUS depend on three types of factors: basic geological conditions, adverse geological phenomena and restricted development zones. The better the basic geological conditions are, the less adverse geological phenomena are developed, the smaller the restrictive development zones are and the better the comprehensive engineering geological suitability is. However, at the present stage, the index weighting method of UUS engineering geological suitability evaluation is relatively simple, mostly using some fuzzy-based approaches, such as analytic hierarchy process and entropy method, and less involving other subjective and objective weighting methods, such as the correlation matrix method (Gulin method) [14], sequential scoring method [18], principal component analysis [19], CRITIC method [20], rough set theory [21,22], etc. So, there are many subjectivities and randomness, and the task is time-consuming.

In this paper, based on the status quo of underground space development and utilization in the starting area of Wuhan Changjiang New Town and relevant data, a suitability evaluation model is developed. Rough set theory is used to construct a decision information table, preprocess sample data and classify the knowledge base. The concept of conditional entropy is introduced to calculate the attributes' own and relative importance. Finally, the attribute weight of each classification index is obtained through standardization. On this basis, the fuzzy comprehensive evaluation method, the most unfavorable grade discrimination method and the exclusive method are used to evaluate the model, and the partition map of regional underground space development suitability is obtained.

2. Description of the Study Area

Wuhan is located in the middle of China, which is the transition zone from the southeast of Hubei and the eastern margin of Jianghan plain to the south of Dabie mountain. Wuhan is surrounded by the hills and ridges in the north and south portions, but in the middle portion the terrain is relatively low, which is the only mega-city in the six central provinces. With the rapid development of urban economy, urban population has further increased, and the accompanying traffic congestion, scarcity of land resources and other urban diseases have become further highlighted. It is an inevitable choice to fully develop and utilize underground space.

Changjiang New Town is situated in the northeast of Wuhan, with a planned construction area of about 560 km^2 (see Figure 1a). The construction will be carried out in three phases, i.e., the starting area, the medium-term development area and the long-term construction area. The starting area of Changjiang New Town (see Figure 1b) is mainly located in the Chenjiaji-Wuhu area, and it is about 70 km^2.

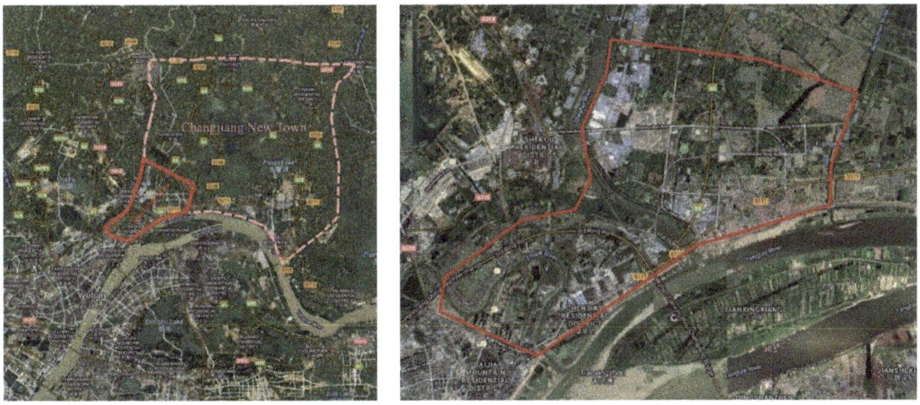

(a) (b)

Figure 1. (**a**) Planning study area of Wuhan Changjiang New Town. (**b**) The starting area of Wuhan Changjiang New Town.

Due to its adjacence to the Yangtze River, Fuhe River, Sheshui River, Daoshui River and many other rivers and lakes, the water resources in the starting area are rich. The area is dominated by a plain landform, and the residual hills are separated from each other. The NW fault in the area is characterized by the Xiangguang fault, while the NE fault is marked by the Yangtze River fault. The limestone and dolomite in the area are developed with karst; thus, there is the possibility of a karst collapse geological disaster.

3. Study Methods and Evaluation Process

According to the data collected on the suitability of underground space development and utilization in five regions, combined with the engineering geological conditions of Wuhan [23,24], statistical analysis was carried out. This paper evaluates the suitability of underground space engineering geology in the starting area of Wuhan Changjiang New Town from six aspects: geotechnical characteristics, geomorphology, geological structure, hydrogeology, adverse geological phenomena and restricted development zone. The evaluation path is shown in Figure 2.

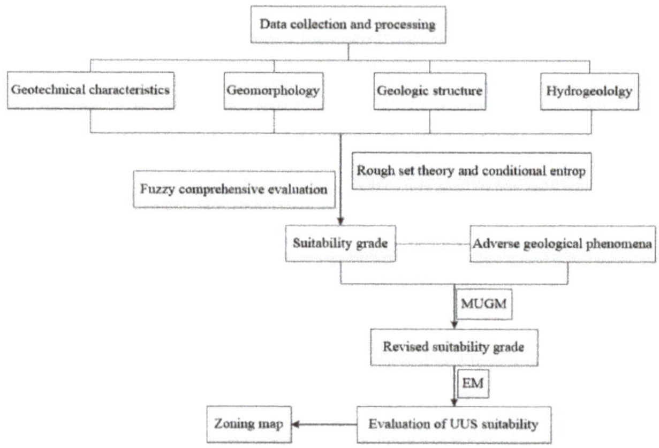

Figure 2. Flow chart of UUS suitability evaluation.

3.1. Selection and Quantification of Indicators

There are many factors that affect the suitability of UUS. Firstly, the underground space takes the rock and soil mass as a carrier. The bearing capacity of rock and soil mass and the compressibility of soil mass determine the strength and stability of the layer of soil mass, which are related to the construction difficulty of underground space and the sensitivity to the influence of surface disturbance and deformation. Secondly, the study area is rich in water resources, which not only increases the difficulty of construction but also the long-term leakage problem that underground engineering still faces after construction completion. In addition, soft soil with different thickness and middle developed karst are distributed in the area, which will cause great harm to underground space development and ground construction. Finally, in ecological control areas, historical relics and monuments and underground water conservation areas, underground construction should meet the policy requirements and control the scale of development. According to the specific situation of the study area, the factors with strong correlation are selected, and the qualitative and quantitative indicators are evaluated by using the numerical method to divide them into five grades of excellent, good, medium, poor and very poor. The indicators are quantified by the following methods:

1. According to the membership function adopted in this paper, its quantitative score is selected from the interval of (0,100) by linear interpolation method;
2. For measurable indicators, the indicators are divided into five levels and quantified by referring to relevant literature and similar cases;
3. For the qualitative evaluation indexes that cannot be measured, expert evaluation is adopted to determine the membership degree of each grade.

Through the above methods, the quantitative and grading results of indicators are shown in Table 1, and the membership degree of qualitative indicators is shown in Table 2.

Table 1. Index selection and quantification classification.

	Criterion Layer B_i	Index Layer C_i	Underground Space Development Geological Suitability Grade I ≥90	II $90 > x \geq 80$	III $80 > x \geq 70$	IV $70 > x \geq 60$	V <60
Target layer D	Geotechnical characteristics B_1	Compression coefficient of soil mass/MPa^{-1} C_1	<0.1	0.1~0.3	0.3~0.5	0.5~0.7	≥0.7
		Rock/soil bearing capacity/Mpa C_2	≥0.4	0.3~0.4	0.2~0.3	0.1~0.2	<0.1
		Soil uniform C_3	Uniform	More uniform	Less uniform	Nonuniform	-
	Geomorphology B_2	Landform unit C_4	Hill	Erosion and accumulation of low ridge	Alluvial plain	Alluvial lacustrine plain	Lacustrine plain
		Topographic slope/% C_5	<10	10~20	20~30	30~50	≥50
	Geologic structure B_3	Fault/m C_6	≥1000	1000~800	800~500	500~200	<200
		Seismic basic intensity C_7	≤VI	VII	VIII	IX	≥X
	Hydrogeology B_4	The depth of confined water roof/m C_8	≥18	16~18	14~16	10~14	<10
		Water inflow of single wellt·d^{-1} C_9	<100	100~400	400~700	700~1000	≥1000
		Groundwater corrosion C_{10}	No	Slight	Weak	Medium	Strong
	Adverse geological phenomenon B_5	Karst ground collapse C_{11}	No	Weak	Medium	-	Strong
		Soft soil settlement C_{12}	No	Weak	Medium	-	Strong

Table 2. Membership degree of qualitative indicators.

Rank	Membership
I	(0.75, 0.25, 0, 0, 0)
II	(0.15, 0.7, 0.15, 0, 0)
III	(0, 0.15, 0.7, 0.15, 0)
IV	(0, 0, 0.15, 0.7, 0.15)
V	(0, 0, 0, 0.25, 0.75)

3.2. Determination of Index Weight

Different indexes have different influences on the evaluation results, that is, they are endowed with different weights in the calculation process. Accurate determination of the importance of each factor is an important premise for the evaluation results to conform to the reality. In previous studies, the analytic hierarchy process (AHP) and expert scoring method were mostly adopted. This method based on the decision makers' subjective information to determine the weight of each factor relies too much on the level and ability of the decision makers, and the weight is not necessarily reasonable. So, the evaluation results may appear to have large differences. In this paper, the method of combining rough set and conditional entropy is adopted, which can fully reflect the objectivity of the data without providing any prior information outside the sample data set. Thus, the weight determination is more reasonable, and the evaluation results are more in line with the reality.

3.2.1. Rough Set Theory

Rough set (*Rs*) theory is a theoretical method proposed by Pawlak et al. [25] to study the expression, learning and generalization of incomplete and uncertain knowledge and data. It can mine potential and useful knowledge from a large amount of data and reduce the unnecessary workload of calculation and classification caused by redundant knowledge. In Rs theory, membership is no longer an initial concept but an objectively calculated one, which is only related to known knowledge, thus avoiding the influence of subjective factors [26–28].

In rough set theory, an information table is a basic tool to express and process knowledge [29]. The basic component of the information table knowledge expression system is the collection of research objects. The knowledge about these objects is described by specifying the attributes (characteristics) of objects and their attribute values. Generally, an information table knowledge expression system S can be expressed as:

$$S =< U, R, V, f > \quad (1)$$

where U is the set of objects, also known as the domain, $R = C \cup D$, $C \cap D = \Phi$, C is the set of conditional attributes and D is the set of decision attributes. $V = V_r$ is the set of attribute values, V_r is the range of values of the attribute r and $r \in R$. $f : U \times R \to V$ is an information function, which specifies the properties of each object X in U. The decision table is a knowledge expression system with conditional attributes and decision attributes.

Knowledge in the knowledge base is not equally important, and some knowledge can be derived from other knowledge [30]. For knowledge base $K =< U, R >$, P and Q belong to R. If indiscernible relation $ind(P)$ belongs to $ind(Q)$, then knowledge Q depends on knowledge P. The dependence of knowledge Q on P is defined as:

$$\gamma_P(Q) = card(POS_p(Q))/card(U) \quad (2)$$

where *card* represents the cardinality of the set and $POS_p(Q)$ represents the positive region of the set P in $U/ind(Q)$ and

$$POS_p(Q) = \cup\{Y_n \subseteq E_t\} \quad (3)$$

where Y_n and E_t represent the basic sets of U/P and U/Q, respectively.

In the decision table, when a certain attribute C_i in the attribute set C is removed, the classification of the decision table changes greatly, indicating that the removed attribute is of high importance and vice versa. The concept of attribute importance is defined by the difference of attribute dependence:

$$\sigma_{CD}(C_i) = \gamma_C(D) - \gamma_{C-C_i}(D) \quad (4)$$

The above equation shows how the positive domain of classification U/C is affected when the attribute subset C_i is removed from set C. When $0 \le \sigma_{CD}(C_i) \le 1$, the larger $\sigma_{CD}(C_i)$ is, the more important C_i is in the entire conditional attribute set. If $\sigma_{CD}(C_i) = 0$, C_i is a redundant attribute and can be reduced.

The importance of each attribute is normalized to obtain the weight W_i of each attribute.

$$W_i = \frac{\sigma_{CD}(C_i)}{\Sigma\sigma_{CD}(C_i)} \quad (5)$$

3.2.2. Method for Determining Attribute Weight Based on Rough Set and Conditional Entropy

In the process of using rough set theory to calculate the attribute weight, sometimes the attribute importance degree is zero, that is, the attribute weight is zero. The reason for this phenomenon is that the rough set theory only considers the importance of a single attribute to the entire attribute set, without considering the importance of the attribute itself, and ignores the practical significance of the attribute. To solve this problem, Bao and Liu [31] introduced the concept of conditional entropy to improve the method. The calculation process is shown in Figure 3.

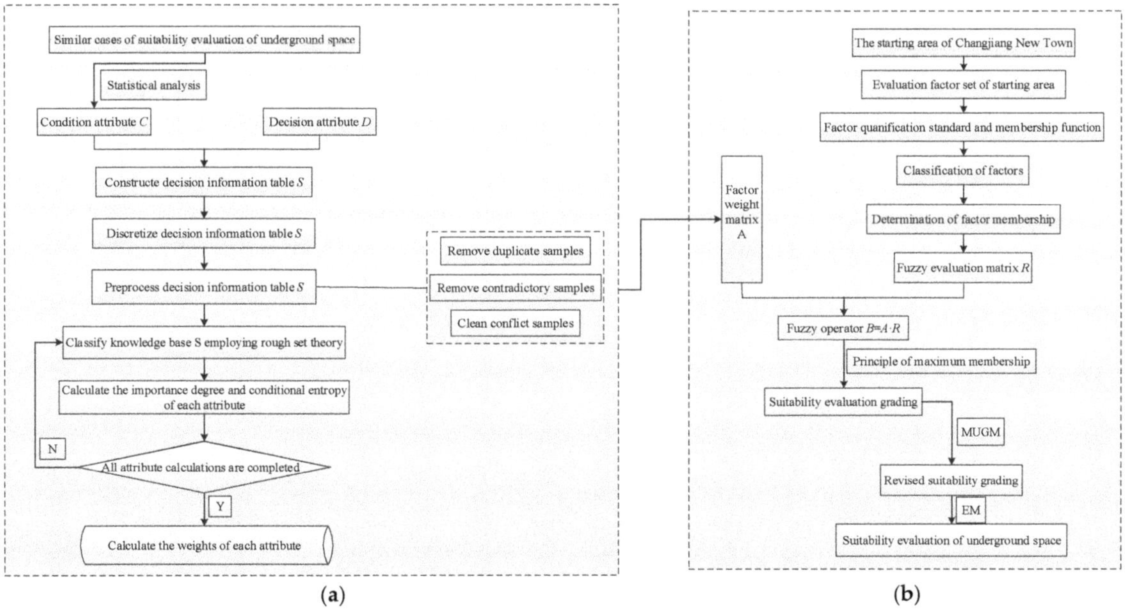

Figure 3. Technical flow chart of underground space suitability evaluation based on rough set and conditional entropy: (**a**) flow chart of weight calculation based on rough set and conditional entropy; (**b**) flow chart of fuzzy comprehensive evaluation.

In the decision information table $S = (U, C, D, V, f)$, the conditional entropy of the decision attribute set $D(U/D = \{D_1, D_2, \ldots, D_k\})$ relative to the conditional attribute set $C(U/C = \{C_1, C_2, \ldots, C_m\})$ is expressed as:

$$I(D|C) = \sum_{i=1}^{m} \frac{|C_i|^2}{|U|^2} \sum_{j=1}^{k} \frac{|D_j \cap C_i|}{|C_i|} \left[1 - \frac{|D_j \cap C_i|}{|C_i|}\right] \quad (6)$$

In the decision information table, $\forall C_i \in C$, and the importance degree of conditional attribute (index) C_i is expressed as:

$$New\ Sig(C_i) = I(D|C - \{C_i\}) - I(D|C) \quad (7)$$

where $New\ Sig(C_i)$ indicates the importance of the conditional attribute C_i in the entire conditional attribute set, and $I(D|\{C_i\})$ indicates the importance of the conditional attribute C_i itself in the system.

Considering these two aspects comprehensively and carrying out standardization processing, the weight of each conditional attribute can be obtained:

$$W(C_i) = \frac{New\ Sig(C_i) + I(D|\{C_i\})}{\sum_{i=1}^{m} \{New\ Sig(C_i) + I(D|\{C_i\})\}} \quad (8)$$

However, when the value of conditional attribute C_i is less than three, the weight is always 1/2 according to the above method, which is too average. In this paper, it will be corrected and the importance degree of conditional attribute (index) C_i is expressed as:

$$New\ Sig(C_i) = I(D|C - \{C_i\}) - I(D|C) + \frac{\sum\limits_{a \in C} |a(x)| - \sum\limits_{a \in C - \{C_i\}} |a(x)|}{\sum\limits_{a \in C} |a(x)|} \tag{9}$$

where the conditional entropy and the normalized weight of C_i are still calculated according to Equations (6) and (8), respectively.

3.2.3. Examples of Index Weight Calculation

Fourteen partitions are selected as a sample set from the evaluation results of five regions. In terms of the criteria layer factors in this paper, the decision table is constructed where the geotechnical characteristics (B_1), geomorphology (B_2), geological structure (B_3) and hydrogeology (B_4) are considered as the conditional attributes, and the actual classification of underground space suitability is considered as the decision attribute (D). Table 3 shows the discretization results of the sample data.

Table 3. Sample data discretization of criterion layer.

No. of Sample (X)	B_1	B_2	B_3	B_4	D
x_1	1	2	1	2	2
x_2	2	3	2	4	3
x_3	2	1	2	2	3
x_4	2	3	3	3	3
x_5	3	3	2	3	3
x_6	1	2	1	2	1
x_7	3	2	2	3	4
x_8	2	5	3	5	3
x_9	3	3	2	4	3
x_{10}	3	3	2	3	3
x_{11}	3	4	3	5	4
x_{12}	4	4	4	3	4
x_{13}	3	5	3	5	4
x_{14}	4	5	4	4	4

The calculation process of index weight is as follows:

1. According to the process shown in Figure 3a, the sample data are processed from three aspects.
   (1) Removing duplicate samples: the fifth group and the tenth group are duplicate samples, thus the tenth group is removed;
   (2) Removing contradictory samples: the level of each conditional attribute of the third group is two, one, two and two, respectively, which are all higher than the level three of decision attribute. It is not consistent with the reality, so let us dispose of the third group. Similarly, the seventh group is removed;
   (3) Cleaning conflict samples: the sixth group and the first group have the same level of conditional attributes, but different levels of decision attributes are contradictory, so the sixth group is removed.

Through data pre-processing, sample data x_1, x_2, x_4, x_5, x_8, x_9, x_{11}, x_{12}, x_{13} and x_{14} are retained to form the decision table.

2. Calculation according to rough set theory and conditional entropy,

$$U/(B) = \{\{x_1\},\{x_2\},\{x_4\},\{x_5\},\{x_8\},\{x_9\},\{x_{11}\},\{x_{12}\},\{x_{13}\},\{x_{14}\}\}$$
$$= \{X_1, X_2, X_3, X_4, X_5, X_6, X_7, X_8, X_9, X_{10}\}$$
$$U/(D) = \{\{x_1\},\{x_2, x_4, x_5, x_8, x_9\},\{x_{11}, x_{12}, x_{13}, x_{14}\}\} = \{D_1, D_2, D_3\}$$
$$U/(B_1) = \{\{x_1\},\{x_2, x_4, x_8\},\{x_5, x_9, x_{11}, x_{13}\},\{x_{12}, x_{14}\}\}$$
$$U/(B - B_1) = \{\{x_1\},\{x_2, x_9\},\{x_4\},\{x_5\},\{x_8, x_{13}\},\{x_{11}\},\{x_{12}\},\{x_{14}\}\}$$
$$X_1 \cap D_1 = \{x_1\}, X_1 \cap D_2 = \Phi, X_1 \cap D_3 = \Phi$$
$$\cdots$$
$$X_{10} \cap D_1 = \Phi, X_{10} \cap D_2 = \Phi, X_{10} \cap D_3 = \{x_{14}\}$$
$$\therefore I(D|B) = \left(\frac{1}{10}\right)^2\left[\frac{1}{1} \times \left(1 - \frac{1}{1}\right) + \frac{0}{1} \times \left(1 - \frac{0}{1}\right) + \frac{0}{1} \times \left(1 - \frac{0}{1}\right)\right] \times 10 = 0$$

Similarly, $I(D|B_1) = 0.08$ and $I(D|B - B_1) = 0.02$ can be calculated.

According to Equation (7), the importance of condition attribute B_1 is calculated as follows:

$$New\ Sig(B_1) = I(D|B - B_1) - I(D|B) = 0.02$$
$$New\ Sig(B_1) + I(D|B_1) = 0.1$$

In the same way, the conditional entropy and importance of geomorphology (B_2), geological structure (B_3) and hydrogeology (B_4) are calculated, and the weight of each attribute is calculated according to Equation (8). The calculation results are shown in Table 4.

Table 4. Conditional entropy, importance degree and attribute weight of criterion layer.

	$I(D \mid B_i)$	$I(D \mid B-B_i)$	$New\ Sig(B_i)$	$W(B_i)$
B_1	0.08	0.02	0.02	0.294
B_2	0.04	0	0	0.118
B_3	0.08	0	0	0.235
B_4	0.12	0	0	0.353

3. When the number of conditional attributes is less than three, fault (C_6) and seismic basic intensity (C_7) are taken as conditional attributes and geological structure (B_3) as the decision attribute according to index layer factors. After the above data preprocessing, the decision table is constructed (see in Table 5).

Table 5. Sample data discretization of index layer.

	C_6	C_7	B_3		C_6	C_7	B_3
x_1	1	3	2	x_6	5	2	4
x_2	2	2	2	x_7	3	3	3
x_3	3	5	4	x_8	3	4	3
x_4	2	3	2	x_9	2	4	3
x_5	4	2	3	x_{10}	1	5	3

According to the above calculation process, the conditional entropies of C_6 and C_7 are 0.12 and 0.18, respectively, and the importance degree is obtained according to Equation (9).

$$\sum_{a \in C} |a(x)| = 9 \quad \sum_{a \in C-\{C_6\}} |a(x)| = 4 \quad \sum_{a \in C-\{C_7\}} |a(x)| = 5$$
$$New\ Sig(C_6) = I(B_3|C - \{C_6\}) - I(B_3|C) + \frac{\sum_{a \in C} |a(x)| - \sum_{a \in C-\{C_6\}} |a(x)|}{\sum_{a \in C} |a(x)|} = 0.736$$
$$New\ Sig(C_7) = I(B_3|C - \{C_7\}) - I(B_3|C) + \frac{\sum_{a \in C} |a(x)| - \sum_{a \in C-\{C_7\}} |a(x)|}{\sum_{a \in C} |a(x)|} = 0.564$$

According to Equation (8), $W(C_6)$ and $W(C_7)$ are 0.535 and 0.465, respectively.

Based on the above decision table construction, sample data processing and the calculation process of attribute weight, the conditional entropy, importance degree and attribute weight of other factors in the index layer are calculated, respectively. The calculation results are shown in Table 6.

Table 6. Condition entropy, importance degree and attribute weight of index layer.

	$I(B\,|\,C_i)$	$I(B\,|\,C-C_i)$	*New Sig*(C_i)	$W(C_i)$
C_1	0.10	0.02	0.02	0.316
C_2	0.12	0.02	0.02	0.368
C_3	0.12	0	0	0.316
C_4	0.10	0.20	0.825	0.578
C_5	0.20	0.10	0.475	0.422
C_8	0.10	0.02	0.02	0.333
C_9	0.10	0.06	0.06	0.444
C_{10}	0.08	0	0	0.223

3.3. Fuzzy Comprehensive Evaluation

According to fuzzy comprehensive theory, the membership degree of each measurable single factor to each grade is determined by membership function. For ease of calculation, the membership function shown in Figure 4 is adopted for all factors. At the same time, in order to give consideration to the evaluation of the overall factors, the weighted average type of synthesis operator is used to perform the fuzzy calculation. Finally, the grade is determined by the principle of maximum membership degree. The evaluation calculation process is shown in Figure 3b.

$$A_V(x) = \begin{cases} 1 & x \le 60 \\ \frac{x-60}{10} & 60 < x \le 70 \\ 0 & x > 70 \end{cases} \qquad A_{IV}(x) = \begin{cases} 0 & x \le 50 \\ \frac{x-50}{10} & 50 < x \le 60 \\ 1 & 60 < x \le 70 \\ \frac{80-x}{10} & 70 < x \le 80 \\ 0 & x > 80 \end{cases}$$

$$A_{III}(x) = \begin{cases} 0 & x \le 60 \\ \frac{x-60}{10} & 60 < x \le 70 \\ 1 & 70 < x \le 80 \\ \frac{90-x}{10} & 80 < x \le 90 \\ 0 & x > 90 \end{cases} \qquad A_{II}(x) = \begin{cases} 0 & x \le 70 \\ \frac{x-70}{10} & 70 < x \le 80 \\ 1 & 80 < x \le 90 \\ \frac{100-x}{10} & 90 < x \le 100 \\ 0 & x > 100 \end{cases}$$

$$A_I(x) = \begin{cases} 0 & x \le 80 \\ \dfrac{x-80}{10} & 80 < x \le 90 \\ 1 & x > 90 \end{cases}$$

Figure 4. Fuzzy comprehensive evaluation membership function.

3.4. The Most Unfavorable Grade Discrimination Method

According to the evaluation level of index factors, the lowest level is selected as the evaluation result. In the evaluation of adverse geological phenomena, if the superposition calculation is adopted for karst ground collapse and soft soil settlement, the scores of areas with adverse geological phenomena will be too average, which is different from the actual impact results. Therefore, the indexes of adverse geological phenomena are degraded by using the most unfavorable grade discrimination method.

4. Case Study

The Yangtze River is the highest river in the Changjiang New Town section. During the flood season, the water level of the Yangtze River is often higher than that of Wuhu Lake. When the Binjiang area in the Changjiang New Town is developed, attention should be paid to the protection of shoreline tidal wetlands. The ecological buffer zone should be set up within a certain range of river edge to avoid digging river sand, building artificial islands and large underground buildings, etc., because these will strongly change the hydraulic conditions and have a negative effect on the internal security of Changjiang New Town. Therefore, a restricted development zone should be set up within a certain range along the river.

On the basis of collecting various data, this paper imports the screened and pre-processed data into the geographic database. According to the established index system and evaluation methods, the properties of each zoning unit are evaluated, classified, superimposed and calculated, and finally the evaluation results of the suitability zoning of underground space in the starting area of Wuhan Changjiang New Town are generated, as shown in Figure 5. The summary table of the evaluation results is shown in Table 7.

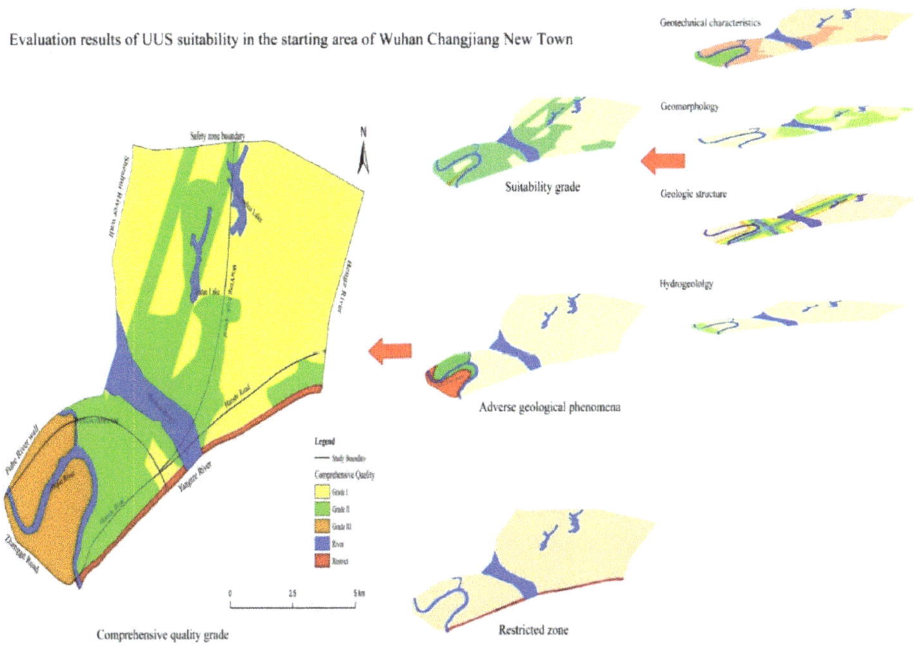

Figure 5. Evaluation results of UUS suitability in the starting area of Wuhan Changjiang New Town.

Table 7. Summary of suitability evaluation results of underground space.

Class	I	II	III	Water Area	Restricted Area
Area/km^2	34.42	21.80	7.34	6.69	1.4
Percentage/%	48.04	30.43	10.24	9.34	1.95

Combined with the suitability zoning map and the evaluation summary table, it can be known that:

(1) On the whole, the suitability of underground space development and utilization in the starting area of Wuhan Changjiang New Town is good. The evaluation results mainly include excellent suitability (class I) and good suitability (class II), and the proportion of medium suitability (class III) is small. According to the evaluation criteria and methods in this paper, there are no areas with poor suitability or very poor suitability in the starting area of Changjiang New Town.

(2) In the starting area, most of the areas north of the Sheshui River have better geotechnical conditions and geomorphological conditions, but the water content in this area

is abundant, and single well yield can be up to 467.467–666.399 m^3/d. Therefore, during the development and utilization of this area, the main restriction condition is groundwater.

(3) The areas with relatively poor suitability are mainly concentrated in the southwest side of the starting area. The hydrogeological conditions in this area are complex, and there is the phenomenon of soft soil surface subsidence and karst ground collapse medium development. The development and utilization of underground space is relatively unfavorable.

5. Conclusions

(1) According to the collected regional data and combined with the engineering geological conditions of the starting area of Wuhan Changjiang New Town, this paper chooses the main factors that affect the suitability of underground space development and utilization. The weight is more objective and reasonable by using rough set and conditional entropy for determination. At the same time, the fuzzy comprehensive evaluation method, the most unfavorable grade discrimination method and the exclusion method are used to evaluate the model. Finally, the evaluation results are more accurate and realistic.

(2) In terms of the analysis of relatively poor suitability areas, the main factors affecting the development and utilization of underground space in the starting area are the settlement of soft soil layer, karst ground collapse and groundwater. Therefore, in the process of development and utilization, we should focus on identifying and preventing these factors.

(3) This paper evaluates the development and utilization of underground space based on the suitability of engineering geology. Factors such as potential value of underground space development and utilization, number of resources and developable volume are not considered in the model. Therefore, future research will be based on geological factors evaluation, fully consider the role of economic value, resource amount, etc., to make the evaluation more comprehensive.

Author Contributions: D.T.: methodology, writing—original draft preparation. F.T.: conceptualization, methodology, writing—reviewing and editing. B.M.: investigation. Y.-Y.J.: conceptualization, supervision. J.W.: data curation. All authors have read and agreed to the published version of the manuscript.

Funding: This research was funded by the National Natural Science Foundation of China (Grant nos. 41920104007 and 51879245) and the Fundamental Research Funds for the Central Universities, China University of Geosciences (Wuhan) (No. CUGCJ1821 and CUG170645).

Institutional Review Board Statement: Not applicable.

Informed Consent Statement: Not applicable.

Data Availability Statement: Some or all data, models or code that support the findings of this study are available from the corresponding author upon reasonable request.

Conflicts of Interest: The authors declare no conflict of interest.

References

1. Rönkä, K.; Ritola, J.; Rauhala, K. Underground space in land-use planning. *Tunn. Undergr. Space Technol.* **1998**, *13*, 39–49. [CrossRef]
2. Bobylev, N.; Sterling, R. Urban underground space: A growing imperative. Perspectives and current research in planning and design for underground space use. *Tunn. Undergr. Space Technol.* **2016**, *55*, 1–342. [CrossRef]
3. Sterling, R.L.; Nelson, S.R. Planning for underground space: A case study for Minneapolis. *Undergr. Space* **1982**, *7*, 86–103.
4. Boivin, D.J. Underground space use and planning in the Québec City area. *Tunn. Undergr. Space Technol.* **1990**, *5*, 69–83. [CrossRef]
5. Peng, J.; Peng, F.L. A GIS-based evaluation method of underground space resources for urban spatial planning: Part 1 methodology. *Tunn. Undergr. Space Technol.* **2018**, *74*, 82–95. [CrossRef]
6. Peng, J.; Peng, F.L. A GIS-based evaluation method of underground space resources for urban spatial planning: Part 2 application. *Tunn. Undergr. Space Technol.* **2018**, *77*, 142–165. [CrossRef]

7. Rienzo, F.D.; Oreste, P.; Pelizza, S. 3D GIS supporting underground urbanisation in the city of Turin (Italy). *Geotech. Geol. Eng.* **2009**, *27*, 539–547. [CrossRef]
8. Zhu, H.; Huang, X.; Li, X.; Zhang, L.; Liu, X. Evaluation of urban underground space resources using digitalization technologies. *Undergr. Space* **2016**, *1*, 124–136. [CrossRef]
9. Zhou, B.; Gui, Y.; Xie, X.; Li, W.; Li, Q. A measurable evaluation method of visual comfort in underground space by intelligent sorting and classification algorithms. *Undergr. Space*, 2021, *in press*.
10. Lu, Z.; Wu, L.; Zhuang, X.; Rabczuk, T. Quantitative assessment of engineering geological suitability for multilayer urban underground space. *Tunn. Undergr. Space Technol.* **2016**, *59*, 65–76. [CrossRef]
11. Hou, W.; Yang, L.; Deng, D.; Ye, J.; Clarke, K.; Yang, Z.; Zhuang, W.; Liu, J.; Huang, J. Assessing quality of urban underground spaces by coupling 3D geological models: The case study of Foshan city, South China. *Comput. Geosci.* **2016**, *89*, 1–11. [CrossRef]
12. Chen, Z.L.; Chen, J.Y.; Liu, H.; Zhang, Z.F. Present status and development trends of underground space in Chinese cities: Evaluation and analysis. *Tunn. Undergr. Space Technol.* **2018**, *71*, 253–270. [CrossRef]
13. Youssef, A.M.; Pradhan, B.; Tarabees, E. Integrated evaluation of urban development suitability based on remote sensing and GIS techniques: Contribution from the analytic hierarchy process. *Arab. J. Geosci.* **2011**, *4*, 463–473. [CrossRef]
14. Wang, X.; Zhen, F.; Huang, X.; Zhang, M.; Liu, Z. Factors influencing the development potential of urban underground space: Structural equation model approach. *Tunn. Undergr. Space Technol.* **2013**, *38*, 235–243. [CrossRef]
15. Liu, D.; Wu, L.; Yang, Y. A hybrid weight assignment model for urban underground space resources evaluation integrated with the weight of time dimension. *Appl. Sci.* **2020**, *10*, 5152. [CrossRef]
16. Duan, Y.; Xie, Z.; Zhao, F.; Zeng, H.; Lin, M.; Chen, H.; Zuo, X.; He, J.; Hou, Z. Suitability of underground space development in plateau cities based on geological environment analysis: Case study in Kunming, China. *J. Urban. Plan. Dev.* **2021**, *147*, 05021014. [CrossRef]
17. Zhou, D.; Li, X.; Wang, Q.; Wang, R.; Wang, T.; Gu, Q.; Xin, Y. GIS-based urban underground space resources evaluation toward three-dimensional land planning: A case study in Nantong, China. *Tunn. Undergr. Space Technol.* **2019**, *84*, 1–10. [CrossRef]
18. Sinha, S.K.; McKim, R.A. Probabilistic based integrated pipeline management system. *Tunn. Undergr. Space Technol.* **2007**, *22*, 543–552. [CrossRef]
19. Peng, J.; Peng, F.L.; Yabuki, N.; Fukuda, T. Factors in the development of urban underground space surrounding metro stations: A case study of Osaka, Japan. *Tunn. Undergr. Space Technol.* **2019**, *91*, 103009. [CrossRef]
20. van der Hoeven, F.; Juchnevic, K. The significance of the underground experience: Selection of reference design cases from the underground public transport stations and interchanges of the European Union. *Tunn. Undergr. Space Technol.* **2016**, *55*, 176–193. [CrossRef]
21. Ulucay, V. A new similarity function of trapezoidal fuzzy multi-numbers based on multi-criteria decision making. *J. Inst. Sci. Technol.* **2020**, *10*, 1233–1246. [CrossRef]
22. Uluçay, V.; Deli, I.; Şahin, M. Trapezoidal fuzzy multi-number and its application to multi-criteria decision-making problems. *Neural Comput. Appl.* **2018**, *30*, 1469–1478. [CrossRef]
23. Li, X.; Xu, H.; Li, C.; Sun, L.; Wang, R. Study on the demand and driving factors of urban underground space use. *Tunn. Undergr. Space Technol.* **2016**, *55*, 52–58. [CrossRef]
24. Tan, F.; Wang, J.; Jiao, Y.Y.; Ma, B.; He, L. Suitability evaluation of underground space based on finite interval cloud model and genetic algorithm combination weighting. *Tunn. Undergr. Space Technol.* **2021**, *108*, 103743. [CrossRef]
25. Pawlak, Z.; Wong, S.K.M.; Ziarko, W. Rough sets: Probabilistic versus deterministic approach. *Int. J. Man-Mach. Stud.* **1990**, *29*, 81–95. [CrossRef]
26. Munakata, T. *Fundamentals of the New Artificial Intelligence: Neural, Evolutionary, Fuzzy and More*; Springer: London, UK, 2008.
27. Zhenfei, Z.; Guangdao, H.; Mingguo, Y.; Xing, L.; Zhenhai, W.; Wenhui, L.; Xiaobing, Z.; Li, H. An approach to spectral discrimination of rocks using ASAI and rough sets. *J. Earth Sci.-China* **2003**, *5*, 257–260.
28. Bakbak, D.; Uluçay, V.; Şahin, M. Intuitionistic trapezoidal fuzzy multi-numbers and some arithmetic averaging operators with their application in architecture. In Proceedings of the 6th International Multidisciplinary Studies Congress (Multicongress'19), Gaziantep, Turkey, 26–27 April 2019.
29. Pawlak, Z. Rough sets and intelligent data analysis. *Inf. Sci.* **2002**, *147*, 1–12. [CrossRef]
30. Hu, X.; Cercone, N. Learning in relational databases: A rough set approach. *Comput. Intell.* **1995**, *11*, 323–338. [CrossRef]
31. Bao, X.; Liu, Y. A new method of ascertaining attribute weight based on rough sets theory. In Proceedings of the International Conference on Electronic Commerce & Business Intelligence, Beijing, China, 6–7 June 2009; pp. 507–510.

applied sciences

MDPI

Article

Verification and Comparison of Direct Calculation Method for the Analysis of Support–Ground Interaction of a Circular Tunnel Excavation

Yu-Lin Lee [1,*], Wen-Kuei Hsu [2], Po-Yu Chou [1], Pei-Wen Hsieh [1], Chi-Huang Ma [1] and Wei-Cheng Kao [1]

[1] Department of Civil Engineering, Chung Hua University, No. 707, Sec. 2, Wufu Rd., Hsinchu 30012, Taiwan; yvelinelee@msn.com (P.-Y.C.); joseph0928@gmail.com (P.-W.H.); joseph0928lee@gmail.com (C.-H.M.); leejoseph0928@gmail.com (W.-C.K.)

[2] Department of Civil and Disaster Prevention Engineering, National United University, No. 2 Lienda Rd., Miaoli 36003, Taiwan; hsuwen918@gmail.com

* Correspondence: rosalee@g.chu.edu.tw; Tel.: +886-03-5186701

Abstract: A direct algorithmic process can deal with the solution of the support–ground interaction in a circular tunnel excavation through the convergence-confinement method (CCM) with the concept of increment. This process is the so-called direct calculation method (DCM) that can find solutions, the mobilized support pressure and the convergence, in the analysis of CCM. To achieve the solution, using two linear equations in the elastic region and Newton's recursive method to find the roots in the plastic region are proposed and realized by a calculated spreadsheet. The validity of the algorithmic process for the analytical solutions was investigated and verified by the finite element computation, and compared with the published results, Rocksupport (2004), Oreste (2009), and Gschwandtner-Galler (2012). The results obtained between DCM and related studies show no significant differences.

Keywords: tunnel analysis; direct algorithmic process; support-ground interaction; confinement loss; Newton's recursive method; convergence-confinement method

Citation: Lee, Y.-L.; Hsu, W.-K.; Chou, P.-Y.; Hsieh, P.-W.; Ma, C.-H.; Kao, W.-C. Verification and Comparison of Direct Calculation Method for the Analysis of Support–Ground Interaction of a Circular Tunnel Excavation. *Appl. Sci.* **2022**, *12*, 1929. https://doi.org/10.3390/app12041929

Academic Editor: Mingfeng Lei

Received: 15 January 2022
Accepted: 11 February 2022
Published: 12 February 2022

1. Introduction

The convergence-confinement method (CCM) adopts the assumption of the plane strain and is used to simulate the interaction between support and ground of a circular tunnel, and can analyze the displacements/stresses generated around the tunnel. This method is an effective calculation method for designing underground excavation support and consists of a combination of three different curves [1–3]. As shown in Figure 1, there is the longitudinal displacement profile (LDP) or a so-called confinement loss curve (CLC), the support confining curve (SCC), and the ground reaction curve (GRC). Numerous publications discuss the CCM and its use in predicting and applying underground excavation, and behavioral analysis of supporting structures [4–9]. For design considerations and applications, the method becomes an initiation tool for the supports design of tunnel excavation [10–13]. As shown in Figure 1, the curve ABCEG represents the GRC, as the tunnel continues to excavate, the surrounding rock stress decreases, and the radial displacement increases gradually. Therefore, this curve determines when the support is installed and the stiffness of the support plays an important role [14–17]. Numerous studies have been investigating this approach, using empirical and mathematical expressions to develop GRCs under different behavioral assumptions and generally applicable to tunnels excavated in isotropic stress state [18–21]. The curve DEI denotes the SCC that characterizes the deformation imposed on the support and its ability to withstand the radially converging pressure (point *D*) that has occurred since the support was installed. There are many studies on SCC that focus on the transient conditions of supports, progressive hardening, time-dependent properties, and non-linearities [5,6,15,19,22–24]. Numerous studies have explored the analytical solutions of SCC and GRC at equilibrium (point *E*) under different hypotheses [25–28].

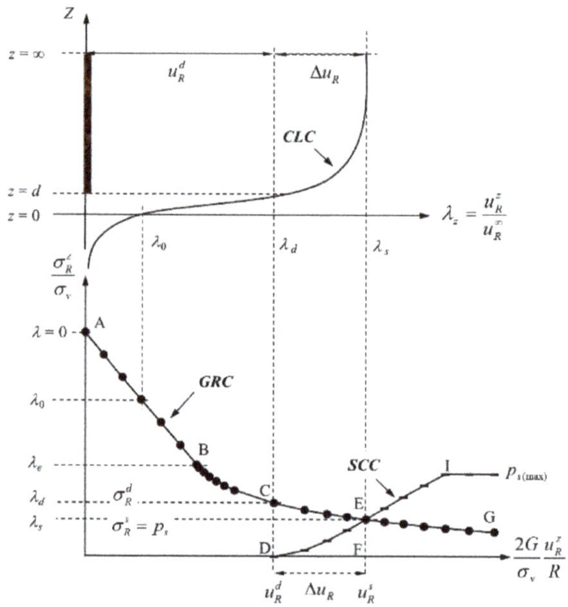

Figure 1. Graphical concept of the support-ground interaction in the analysis of convergence-confinement method (CCM).

As shown in the upper part of Figure 1, the third curve of this method is called the longitudinal displacement profile (LDP) [11] or the confinement loss curve (CLC) [2–4]. Recent studies have shown some divergent views on the assumptions and definitions of confinement loss in the CCM analysis [29–33]. This principle suggests that the confinement loss increases while the radial stress decreases during the advancing excavation of the tunnel face and uses the concept of increment to simulate this effect, and its value is between 0 and 1 [3,34,35].

The direct algorithmic procedure proposed in this paper, called the direct calculation method (DCM), is used to solve the displacement/stress at the equilibrium (point *E*). From a theoretical point of view, this method is proposed to investigate the interaction behavior of the support and ground due to the excavation of a circular tunnel in the isotropic stress field was rigorously derived. The feasibility of a direct algorithmic process for the analytical solution is examined by numerical analysis steps, in particular the finite element method (FEM). Various study cases, including Oreste [6], Gschwandtner-Galler [23], and Rocksupport [36], are selected to compare with DCM, respectively. Therefore, the purpose of this paper is to provide computational concepts for tunnel analysis and to comprehensively discuss solutions for the support–ground interaction behavior in tunneling, and to provide the complete equations for support design and details of their derivation.

2. Equations Derivation of the Direct Calculation Method (DCM)

To distinguish whether the stress at the points *C* and *E* are in the elastic or the plastic regions, this study presents two analysis situations that include that stress state of points *C* and *E* are both (1) in the elastic region (Case I, $\lambda_d < \lambda_s < \lambda_e$), and (2) in the plastic region (Case II, $\lambda_e < \lambda_d < \lambda_s$). In addition, λ_e, λ_d, and λ_s are the confinement loss in the elastic limit state, at the moment of installing support (points *C*), and at the equilibrium state (points *E*), respectively. Therefore, λ_e and λ_d can be given as the following [37–39]:

$$\lambda_e = \frac{(K_p - 1) + 2N}{K_p + 1} \tag{1}$$

$$\lambda_d = 1 - \left(\frac{m}{m + (d/R)}\right)^4 \tag{2}$$

where R is the radius of tunnel excavation, d is the unsupported span, K_p is the coefficient of passive lateral pressure, N is the stability number, and the parameter of the function m can be found by the regression analysis with the tunnel convergence data. Note that this confinement loss is not only a function of the property geo-material but also of the unsupported span.

2.1. Solution for the Support-Ground Interaction under the Elastic Condition

When the stresses of the surrounding rock at the intrados of the tunnel are in the elastic region (Case I), the analytical solution of the displacement/stress at the equilibrium point (point *E*) can be obtained according to the proposed process of DCM in the elastic medium [40]. Then, the radial displacement (u_R^s) and the mobilized support pressure (p_s) can be obtained as:

$$p_s = \frac{k_s}{2G + k_s}(1 - \lambda_d)\sigma_v \tag{3}$$

$$\frac{2G}{\sigma_v}\frac{u_R^s}{R} = \frac{2G + \lambda_d k_s}{2G + k_s} = \lambda_s \tag{4}$$

where G, k_s, σ_v and are the shear modulus of the ground, the stiffness of the support, and the vertical stress of the ground, respectively.

2.2. Solution for the Support–Ground Interaction under the Plastic Condition

When the stresses of the surrounding rock at the intrados of the tunnel are in the plastic region (Case II) [40]. Therefore, at the equilibrium state (point *E*), the mobilized support pressure (p_s) and the radial displacement (u_R^s) can be obtained as:

$$P_s = k_s\left(\frac{u_R^s}{R} - \frac{u_R^d}{R}\right) = \sigma_R^s = \frac{2}{(K_p - 1)}\left[\lambda_e\left(\frac{R}{R_R^s}\right)^{K_p - 1} - N\right]\sigma_v \tag{5}$$

$$\frac{2G}{\sigma_o}\frac{u_R^s}{R} = \lambda_e\left[C_1 + C_2\left(\frac{R}{R_p^s}\right)^{K_p - 1} + C_3\left(\frac{R_p^s}{R}\right)^{K_\psi + 1}\right] \tag{6}$$

where K_ψ is the coefficient of the plastic flow, and C_1, C_2, and C_3 are the coefficients of ground. In addition, the plastic zone radius at the moment of installing support (R_p^d) can be represented as:

$$\frac{R_p^d}{R} = \left[\frac{2\lambda_e}{(K_p + 1)\lambda_e - (K_p - 1)\lambda}\right]^{\frac{1}{K_p - 1}} \tag{7}$$

The mobilized pressure of the support system blocks the continuous expansion of the plastic zone radius and maintains a fixed value (R_p^s) at the equilibrium state (point *E*). For the expansion of the plastic zone, the calculation of the plastic zone radius due to the advancing excavation of the tunnel at different stress state (between point *C* and *E*) plays an important role in the analysis of DCM.

As the stresses are in the plastic state, through the function of the DCM algorithmic process, a numerical method called Newton's recursive method is proposed, which can find the solution of the intersection between the two curves (SCC and GGC) in the equilibrium state (point E). Therefore, this function can be represented as:

$$f\left(\frac{R_p^s}{R}\right) = \left[C_2 - \frac{4G}{k_s(K_p-1)}\right]\left(\frac{R_p^s}{R}\right)^{1-K_p} + C_3\left(\frac{R_p^s}{R}\right)^{K_\psi+1} - \left[C_2\left(\frac{R}{R_p^d}\right)^{K_p-1} + C_3\left(\frac{R_p^d}{R}\right)^{K_\psi+1}\right] - \frac{4GN}{\lambda_e k_s(K_p-1)} \quad (8)$$

$$f'\left(\frac{R_p^s}{R}\right) = \left[C_2 - \frac{4G}{k_s(K_p-1)}\right](1-K_p)\left(\frac{R_p^s}{R}\right)^{-K_p} + C_3(K_\psi+1)\left(\frac{R_p^s}{R}\right)^{K_\psi} \quad (9)$$

where the derivative of the function $f(R_p^s/R)$ is represented by $f'(R_p^s/R)$. Therefore, the plastic zone radius (R_p^s), can be given as:

$$\left(\frac{R_p^s}{R}\right)_{n+1} = \left(\frac{R_p^s}{R}\right)_n - \frac{f\left(\left(\frac{R_p^s}{R}\right)_n\right)}{f'\left(\left(\frac{R_p^s}{R}\right)_n\right)} \quad (10)$$

where the incremental step in the recursion is represented by n. For substituting Equation (10) into Equations (5) and (6), the radial displacement and the mobilized support pressure at the equilibrium state can finally be obtained.

3. Procedure of Numerical Analysis for the Direct Calculation Method (DCM) and Finite Element Method (FEM)

3.1. Procedure of Numerical Analysis for the DCM

The algorithmic process used in the analysis of DCM is to solve the displacement/stress at the equilibrium state (point E). The computation flowchart of DCM is represented in Figure 2. The process and steps of the calculation performed by DCM include (1) input of calculation data (including tunnel geometry, in situ stress, surrounding rock material, support member material, unsupported span, and other related parameters), (2) according to the unsupported span calculate the value of λ_d, λ_e, and λ_s, (3) determine whether the stress change of surrounding rock (i.e., the corresponding λ value) caused by the continuous excavation of the tunnel is in the elastic zone or in the plastic zone, (4) according to Equations (3)–(6), calculate the solution at the equilibrium point, and (5) record the displacement/stress corresponding to all incremental values and plot the graphs or tables.

In the analysis of DCM, the unknown parameters obtained by the calculation with the known parameters shown in Table 1 are the radial stress (σ_R), the radial displacement (u_R), and the tangential stress (σ_θ). Corresponding to a specified value of the confinement loss, these unknown parameters are the behavior at the intrados of the tunnel in different steps of excavation. In this study, the Mohr–Coulomb failure criterion is used to determine whether the stress reaches the plastic region, and the plastic strain adopts the plastic associated flow rule.

3.2. Procedure of Numerical Analysis for the FEM

To examine the results obtained by DCM, a numerical analysis that provides rigorous results for comparison, such as the finite element method (FEM), is investigated in this study. The finite element computing processor developed in the laboratory includes (1) the pre-processor (matrix optimization, boundary condition needed, mesh modeling, applied force, input data of material used, and geometry), (2) the main program (mechanic models, tolerance used, and the core of calculation including the increment steps), and (3) the post-processor (iso-value illustrations, drawing the contour line of displacement/stress, and output data).

The mesh made by finite element analysis includes 1971 total nodes and 658 elements (118 T6 elements and 540 Q8 elements), using three components of calculation (ground, excavation, and lining), and the analysis boundary of the roller support is 20 times the tunnel excavation radius. The simulation results obtained by the FEM comprise the displacements/stresses of the ground around the periphery of the tunnel. In addition, Figure 3 shows a plot of simulation results for tunnel excavation under initial isotropic stress condition and includes the plastic zone, the stress field, the total displacement, the minor principal stress, the major principal stress, and the iso-value illustrations of the initial stress. The input data used for the calculation are shown in Table 1.

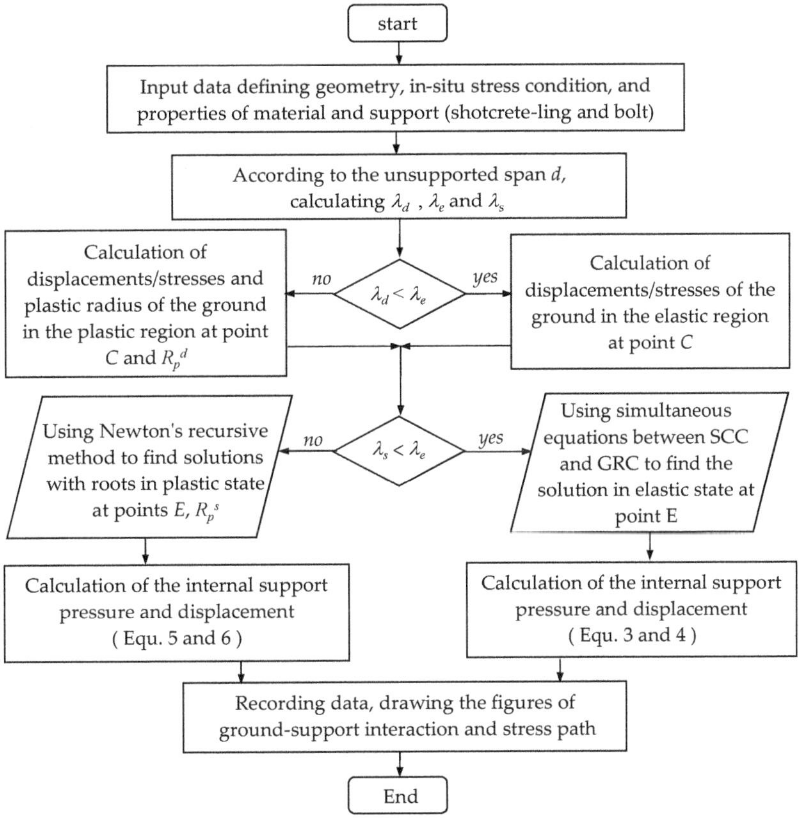

Figure 2. Computation flowchart of the algorithmic process of direct calculation method (DCM).

Table 1. Input data of the numerical computation of DCM and finite element method (FEM).

Ground		Support (Shotcrete-Lining)	
Parameter	Value	Parameter	Value
Vertical stress, σ_v (MPa)	1.0	Poisson's ratio, v_{shot}	0.2
In-situ stress ratio, K_o	1.0	Elastic modulus, E_{shot} (GPa)	25.0
Cohesion, c (MPa)	0.1	Unit weight, γ_{shot} (MPa/m)	0.025
Friction angle, φ (°)	30.0	Uniaxial compression strength, $\sigma_{c(shot)}$ (MPa)	20.0
Dilation angle, ψ (°)	30.0	Thickness, t_{shot} (m)	0.2
Poisson's ratio, ν	0.25	Unsupported span, d (m)	0.53, 1.37
Elastic modulus, E (MPa)	300.0	Coefficients, m	0.75
Confinement loss, λ	0.0–1.0	Tunnel radius, R (m)	5.2

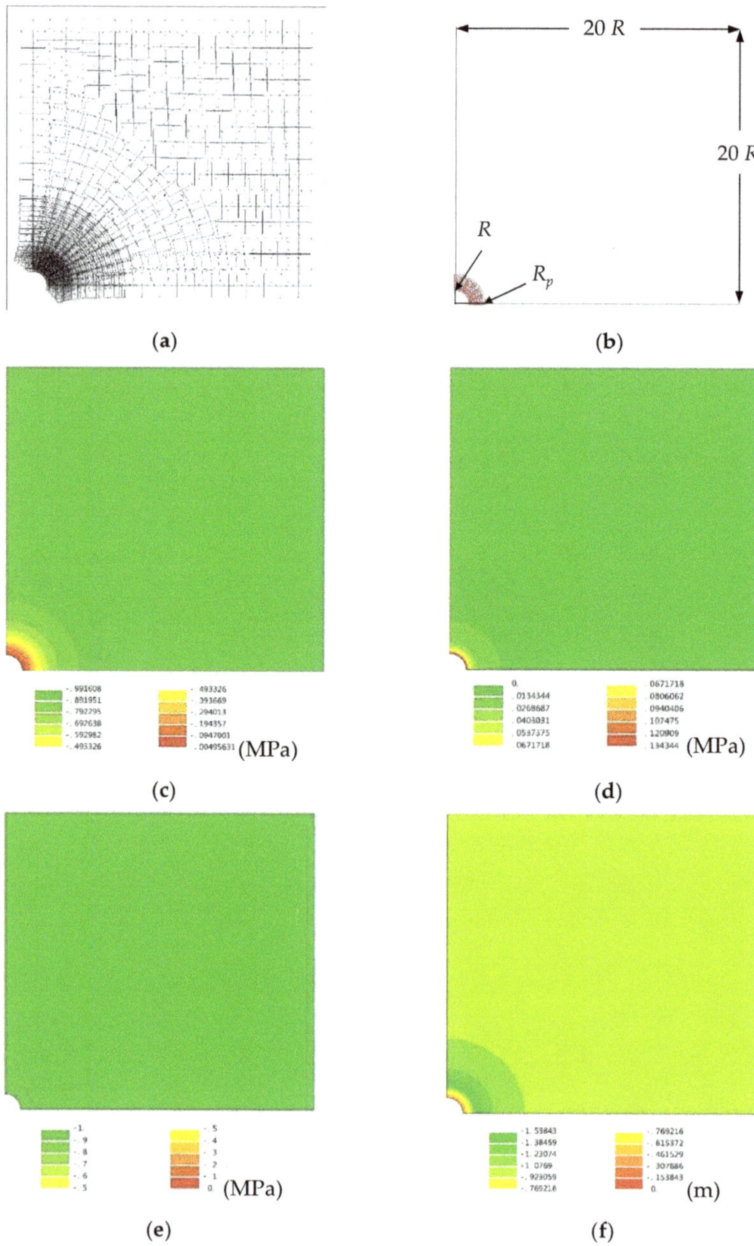

Figure 3. Simulation results of a circular tunnel excavation calculated by finite element analysis. (**a**) the stress field, (**b**) the plastic zone, (**c**) the major principal stress, (**d**) the minor principal stress, (**e**) the initial stress, and (**f**) the total displacement.

4. Verification and Comparison of Results between This Study and Other Research

4.1. Verification and Comparison of Results between FEM and DCM

The feasibility of the developed direct algorithmic process (DCM) is verified by numerical analysis with finite element analysis (FEM). The comparison of analytical and numerical results, including the displacement/stress change of GRC and SCC in each excavation stage, and especially the influence of radial displacement and mobilized support pressure at the equilibrium point is investigated.

Considering the influence of tunnel-advancing excavation, the unsupported span d (or unsupported distance) is usually used to consider the influence of the stress state. If the tunnel is excavated without considering the support system, it is clear that the in situ stress will drop rapidly to a plastic state. Therefore, under the isotropic stress condition, two unsupported spans were selected including (1) short unsupported span (λ_d = 0.4, d = 0.53 m) so that the surrounding rock stress falls in the elastic zone, and (2) long unsupported span (λ_d = 0.7, d = 1.37 m) so that the stress falls in the plastic zone.

According to the results obtained by FEM and DCM for the support–ground interaction, the displacement/stress at the equilibrium state is different in the elastic region (Figure 4a) and the plastic region (Figure 4b). From the comparison results, it can be found that the mobilized support pressure in the elastic zone shows a relatively high value, while the radial displacement indicates a lower value, but the situation in the plastic zone shows the opposite. Furthermore, it must be noted that the trajectory of the support–ground interaction follows the ground reaction curve so, in this case, the interaction curve (IC) and the ground reaction curve (GRC) coincide with each other, i.e., the isotropic stress field.

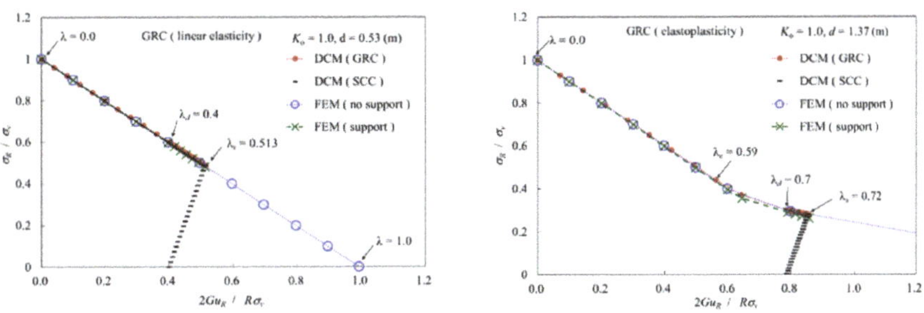

(a) **(b)**

Figure 4. Comparison of the interaction behavior of support-ground between FEM and DCM. (**a**) short unsupported span (d = 0.53 m), and (**b**) long unsupported span (d = 1.37 m).

The mechanical behavior is demonstrated in this study for the displacement/stress continuity at the elastoplastic interface as shown in Figure 5, that is, the continuous connection between the interface (between the elastic and plastic regions). For the comparison of the distribution of displacement/stress around circular tunnel proximity, the radial displacement and the plastic zone radius of the ground under no support condition are larger than the support condition. In other words, the support installation in tunneling blocks the development of the plastic zone (Figure 5a). On the comparison between FEM and DCM in numerical analysis, the results obtained at the intrados of the tunnel are approximatively the same as shown in Table 2, and that around the tunnel proximity illustrates a consistent trend as shown in Figure 5.

4.2. Comparison of Results between DCM and Other Studies

The case study compared with DCM includes research such as Rocksupport (2004) [36], Oreste (2009) [6], and Gschwandtner-Galler (2012) [23]. First of all, for analyzing the ground reaction caused by the tunnel excavation under the no support condition, the input data of the numerical calculation can be found in the aforementioned articles. According to the analysis results between DCM and the listed articles, the percentage error of the plastic zone radius is from 0.79% to 1.31%, and the percentage error of the radial displacement is from 0.24% to 1.70%. In particular, compared with the results of Rocksupport (2004), this study presents values with little error, as shown in Figure 6 and Table 3.

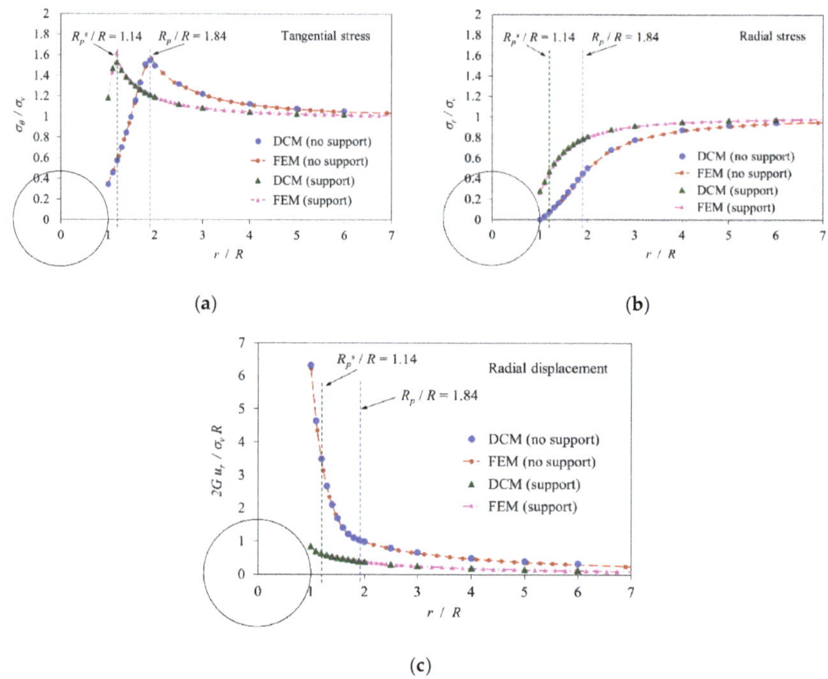

Figure 5. Distribution of the displacement/stress around a circular tunnel proximity under the support and no support condition by the comparison results obtained between FEM and DCM. (**a**) the tangential stress (σ_θ), (**b**) the radial stress (σ_r), and (**c**) the radial displacement (u_r).

Table 2. Comparison of results between FEM and DCM.

	FEM			DCM		
Support condition	Mobilized support pressure, p_s (MPa)	Radial displacement, u_R^s (mm)	Plastic zone radius, R_p^s (m)	Mobilized support pressure, p_s (MPa) (error %) *	Radial displacement, u_R^s (mm) (error %) *	Plastic zone radius, R_p^s (m) (error %) *
No support (Elasticity)	n/a	21.665	n/a	n/a	21.667 (0.01%)	n/a
No support (Elastoplasticity)	n/a	136.73	9.62	n/a	136.83 (0.07%)	9.57 (0.53%)
Support $\lambda_d = 0.53$ [a]	0.481	11.074	n/a	0.487 (1.1%)	11.124 (0.5%)	n/a
Support $\lambda_d = 1.37$ [b]	0.262	18.534	5.972	0.279 (6.6%)	18.482 (0.3%)	5.923 (0.8%)

[a] short unsupported span, [b] long unsupported span; * error (%) = 100%(| FEM-DCM | /FEM).

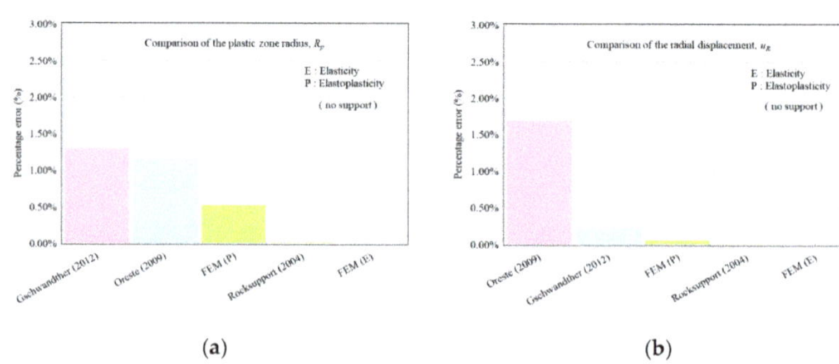

Figure 6. Percentage error of results between DCM and other studies under no support condition. (**a**) the plastic zone radius (R_p), and (**b**) the radial displacement (u_R).

Table 3. Comparison of results between DCM and other studies (no support condition).

Other Research Results (No Support Condition)	Radial Displacement, u_R (mm)	Plastic Zone Radius, R_p (m)	DCM Radial Displacement, u_R (mm) (Error %)	DCM Plastic Zone Radius, R_p^s (m) (Error %)
Rocksupport (2004)	121.56	13.77	119.75 (1.49%)	13.66 (0.79%)
Oreste (2009)	6.2	4.65	6.306 (1.70%)	4.704 (1.16%)
Gschwandtner-Galler (2012)	160.0	13.0	160.39 (0.24%)	12.83 (1.31%)

Regarding the comparison of DCM calculation results with published research articles, firstly, the input data of the calculation proposed by the tutorial manual of Rocksupport (2004) [36] includes two support systems that are only rock bolts and rock bolt plus shotcrete lining. The results calculated by DCM are as shown in Table 4, and one can observe that the range of percentage error of the plastic zone radius is 2.6–2.7%, and the range of the percentage error is 2.2–5.2% for the radial displacement, and 8.7~10.0% for the mobilized support pressure.

Table 4. Comparison of results between DCM and Rocksupport (2004) [36] (support condition).

	Rocksupport (2004)			DCM		
Support condition	Mobilized support pressure, p_s (MPa)	Radial displacement, u_R^s (mm)	Plastic zone radius, R_p^s (m)	Mobilized support pressure, p_s (MPa) (error %)	Radial displacement, u_R^s (mm) (error %)	Plastic zone radius, R_p^s (m) (error %)
Rock bolts	0.19	60.0	10.0	0.171 (10.0%)	61.29 (2.15%)	10.27 (2.70%)
Rock bolt and shotcrete-lining	0.21	55.56	9.74	0.191 (8.7%)	58.42 (5.15%)	9.99 (2.57%)

Secondly, the comparison with the study of Oreste (2009), the results as shown in Table 5 that displays a short unsupported span, deep tunnel location, good quality of rock mass, and general support system. The percentage error of the mobilized support pressure, the radial displacement, and the plastic zone radius are 7%, 0.7%, and 1.7%, respectively.

Table 5. Comparison of results between DCM and Oreste (2009) [6] (support condition).

Oreste (2009)			DCM		
Mobilized support pressure, p_s (MPa)	Radial displacement, u_R^s (mm)	Plastic zone radius, R_p^s (m)	Mobilized support pressure, p_s (MPa) (error %)	Radial displacement, u_R^s (mm) (error %)	Plastic zone radius, R_p^s (m) (error %)
0.16	4.68	4.65	0.144 (7.0%)	4.65 (0.7%)	4.57 (1.7%)

Thirdly, for a comparison with the study of Gschwandtner-Galler (2012), the results are shown in Table 6. The percentage error of the mobilized support pressure, the radial displacement, and the plastic zone radius are 7.6%, 1.9%, and 4.0%, respectively. Finally, for the comparison of results shown in Figure 7, the more obvious difference is the mobilized support pressure calculated by DCM, and the range of percentage error is between 7% and 10%.

Table 6. Comparison of results between DCM and Gschwandtner-Galler (2012) [23] (support condition).

Gschwandtner-Galler (2012)			DCM		
Mobilized support pressure, p_s (MPa)	Radial displacement, u_R^s (mm)	Plastic zone radius, R_p^s (m)	Mobilized support pressure, p_s (MPa) (error %)	Radial displacement, u_R^s (mm) (error %)	Plastic zone radius, R_p^s (m) (error %)
0.84	75.2	7.69	0.904 (7.56%)	73.72 (1.97%)	8.00 (4.03%)

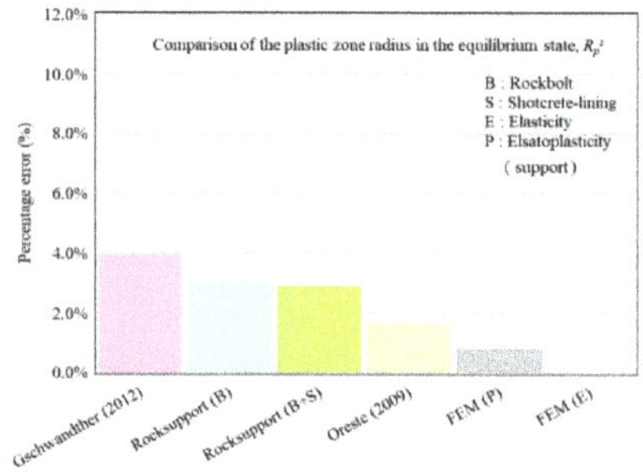

(a)

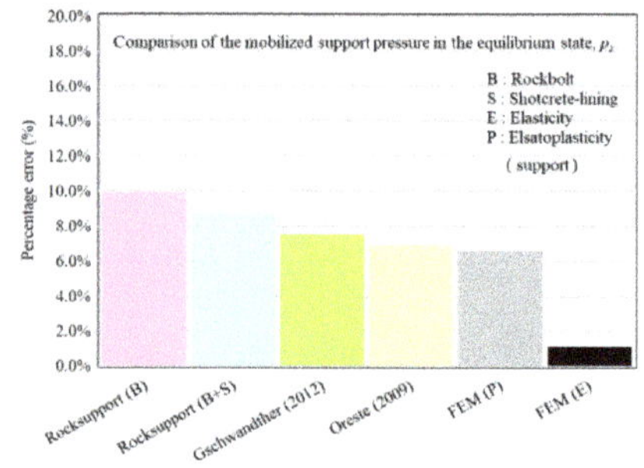

(b)

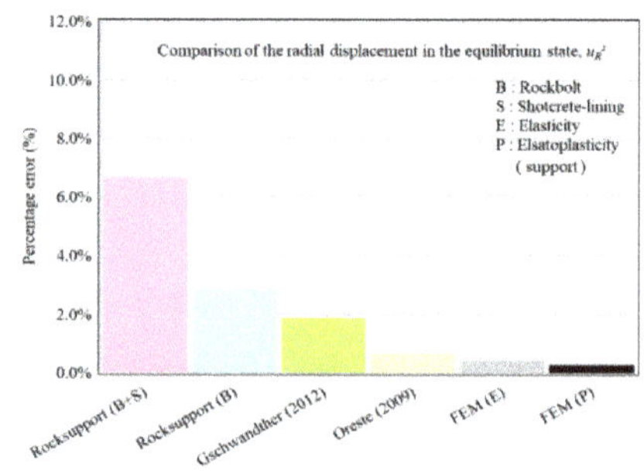

(c)

Figure 7. Percentage error of results at the equilibrium state between DCM and other studies under the support condition. (**a**) the plastic zone radius (R_p^s), (**b**) the mobilized support pressure (p_s), and (**c**) the radial displacement (u_R^s).

5. Conclusions

Through a series of formula derivation, algorithm implementation, numerical verification, and comparison with three published research results, this study can draw the following conclusions:

(1) The direct calculation method (DCM) is used to study the theoretical explanation and numerical analysis of the support–ground interaction caused by a circular tunnel excavation in the isotropic stress field.

(2) The DCM is proposed to provide a special algorithmic process to solve the support-ground interaction solution at the equilibrium state. The roots are obtained by applying the method of simultaneous equations in the elastic region and Newton's recurrence method in the plastic region. DCM investigated the solution for the interaction between GRC and SCC, namely the mobilized support pressure and the radial displacement.

(3) To resolve the theoretical analysis into an executable numerical program, a simple spreadsheet of calculations is used to realize the application of DCM.

(4) The confinement loss as the incremental factor defining the situation of tunnel advancing excavation is classified by two cases (Case I and II) and proposed to distinguish whether the stress state is in the plastic or the elastic regions.

(5) Good validation results are obtained between FEM and DCM for tunnel excavation simulations under support or no support conditions.

(6) Three research results, Rocksupport (2004) [36], Oreste (2009) [6], and Gschwandtner-Galler (2012) [23], are used to compare with that obtained by DCM. In the no support condition, the consistency of the results is also very good. In the support condition,

the maximum percentage error is 10%, 6%, and 4% for the mobilized support pressure, the radial displacement, and the plastic zone radius, respectively.

Author Contributions: Y.-L.L.: analysis, validation, and main editing and reviewing; W.-K.H. and P.-Y.C.: formula derivation and verification; P.-W.H., C.-H.M. and W.-C.K.: software programming and computation. All authors have read and agreed to the published version of the manuscript.

Funding: This research received no external funding.

Institutional Review Board Statement: Not applicable.

Informed Consent Statement: Not applicable.

Data Availability Statement: Not applicable.

Conflicts of Interest: The authors declare no conflict of interest. The sponsors had no role in the design, execution, interpretation, or writing of the study.

References

1. Hoek, E.; Brown, E.T. *Underground Excavations in Rock*; The Institution of Mining and Metallurgy: London, UK, 1980.
2. AFTES (Association Française des Tunnels et de l'Espace Souterrain). Recommandations pour l'emploi de la méthode convergence-confinement. *Tunn. Ouvrages Souterr.* **1983**, *59*, 119–138.
3. Panet, M. *Le Calcul des Tunnels par la Méthode de Convergence-Confinement*; Presses de l'Ecole Nationale des Ponts et Chaussées: Paris, France, 1995.
4. Panet, M. *Recommendations on the Convergence-Confinement Method*; Association Française des Tunnels et de l'Espace Souterrain (AFTES): Paris, France, 2001; pp. 1–11.
5. Oreste, P. Analysis of structural interaction in tunnels using the convergence-confinement approach. *Tunnell. Undergr. Space Technol.* **2003**, *18*, 347–363. [CrossRef]
6. Oreste, P. The convergence–confinement method: Roles and limits in modern geomechanical tunnel design. *Am. J. Appl. Sci.* **2009**, *6*, 757–771. [CrossRef]
7. Sandrone, F.; Labiouse, V. Analysis of the evolution of road tunnels equilibrium conditions with a convergence-confinement approach. *Rock Mech. Rock Eng.* **2010**, *43*, 201–218. [CrossRef]
8. Brown, E.T.; Bray, J.W.; Ladanyi, B.; Hoek, E. Ground response curves for rock tunnels. *J. Geotech. Eng. ASCE* **1983**, *109*, 15–39. [CrossRef]
9. Brady, B.H.G.; Brown, E.T. *Rock Mechanics for Underground Mining*; Chapman & Hall: London, UK, 1993.
10. Wang, Y. Ground response of a circular tunnel in poorly consolidated rock. *J. Geotech. Eng. ASCE* **1996**, *122*, 703–708. [CrossRef]
11. Carranza-Torres, C.; Fairhurst, C. Application of the convergence-confinement method of tunnel design to rock masses that satisfy the Hoek–Brown failure criterion. *Tunnell. Undergr. Space Technol.* **2000**, *15*, 187–213. [CrossRef]
12. Carranza-Torres, C. Elasto-plastic solution of tunnel problems using the generalized form of the Hoek-Brown failure criterion. *Int. J. Rock Mech. Min. Sci.* **2004**, *41*, 629–639. [CrossRef]
13. Sharan, S.K. Exact and approximate solutions for displacements around circular openings in elastic-brittle plastic Hoek-Brown rock. *Int. J. Rock Mech. Min. Sci.* **2005**, *42*, 542–549. [CrossRef]
14. Park, K.H.; Kim, Y.J. Analytical solution for a circular opening in an elastic–brittle–plastic rock. *Int. J. Rock Mech. Min. Sci.* **2006**, *43*, 616–622. [CrossRef]
15. Guan, Z.; Jiang, Y.; Tanabasi, Y. Ground reaction analyses in conventional tunnelling excavation. *Tunnell. Undergr. Space Technol.* **2007**, *22*, 230–237. [CrossRef]
16. Alejano, L.R.; Rodriguez-Dono, A.; Alonso, E.; Fdez-Manín, G. Ground reaction curves for tunnels excavated in different quality rock masses showing several types of post-failure behaviour. *Tunnell. Undergr. Space Technol.* **2009**, *24*, 689–705. [CrossRef]
17. Serrano, A.; Olalla, C.; Reig, I. Convergence of circular tunnels in elastoplastic rock masses with non-linear failure criteria and non-associated flow laws. *Int. J. Rock Mech. Min. Sci.* **2011**, *48*, 878–887. [CrossRef]
18. Zhang, Q.; Jiang, B.S.; Wang, S.L.; Ge, X.R.; Zhang, H.Q. Elasto-plastic analysis of a circular opening in strain-softening rock mass. *Int. J. Rock Mech. Min. Sci.* **2012**, *50*, 38–46. [CrossRef]
19. Oreste, P.; Spagnoli, G. A combined analytical and numerical approach for the evaluation of radial loads on the lining of vertical shaft. *Geotech. Geol. Eng.* **2016**, *34*, 1057–1065. [CrossRef]
20. Lee, Y.L. Explicit procedure and analytical solution for the ground reaction due to advance excavation of a circular tunnel in an anisotropic stress field. *Geotech. Geol. Eng.* **2018**, *36*, 3281–3309. [CrossRef]
21. Lee, Y.L. Incremental procedure method for the analysis of ground reaction due to excavation of a circular tunnel by considering the effect of overburden depth. *Tunnell. Undergr. Space Technol.* **2019**, *93*, 1–16. [CrossRef]
22. Oreste, P.; Peila, D. Modelling progressive hardening of shotcrete in convergence-confinement approach to tunnel design. *Tunnell. Undergr. Space Technol.* **1997**, *12*, 425–431. [CrossRef]

23. Gschwandtner, G.G.; Galler, R. Input to the application of the convergence confinement method with time-dependent material behavior of the support. *Tunnell. Undergr. Space Technol.* **2012**, *27*, 13–22. [CrossRef]
24. Rodríguez, R.; Díaz-Aguado, M.B. Deduction and use of an analytical expression for the characteristic curve of a support based on yielding steel ribs. *Tunnell. Undergr. Space Technol.* **2013**, *33*, 159–170. [CrossRef]
25. Cui, L.; Zheng, J.; Zhang, R.; Lai, H. A numerical procedure for the fictitious support pressure in the application of the convergence-confinement method for circular tunnel design. *Int. J. Rock Mech. Min. Sci.* **2015**, *78*, 336–349. [CrossRef]
26. Carranza-Torres, C.; Engen, M. The support characteristic curve for blocked steel sets in the convergence-confinement method of tunnel support design. *Tunnell. Undergr. Space Technol.* **2017**, *69*, 233–244. [CrossRef]
27. Oke, J.; Vlachopoulos, N.; Diederichs, M. Improvement to the convergence-confinement method: Inclusion of support installation proximity and stiffness. *Rock Mech. Rock Eng.* **2018**, *51*, 1495–1519. [CrossRef]
28. De La Fuente, M.; Taherzadeh, R.; Sulem, J.; Nguyen, X.S.; Subrin, D. Applicability of the convergence-confinement method to full-face excavation of circular tunnels with stiff support system. *Rock Mech. Rock Eng.* **2019**, *52*, 2361–2376. [CrossRef]
29. Bernaud, D.; Rousset, G. The new implicit method for tunnel analysis. *Int. J. Numer. Anal. Methods Geomech.* **1996**, *20*, 673–690. [CrossRef]
30. González-Nicieza, C.; Álvarez-Vigil, A.E.; Menéndez-Díaz, A.; González-Palacio, C. Influence of the depth and shape of a tunnel in the application of the convergence-confinement method. *Tunnell. Undergr. Space Technol.* **2008**, *23*, 25–37. [CrossRef]
31. Mousivand, M.; Maleki, M.; Nekooei, M.; Msnsoori, M.R. Application of Convergence-Confinement method in analysis of shallow non-circular tunnels. *Geotech. Geol. Eng.* **2017**, *35*, 1185–1198. [CrossRef]
32. Mousivand, M.; Maleki, M. Constitutive models and determining methods effects on application of Convergence-Confinement method in underground excavation. *Geotech. Geol. Eng.* **2018**, *36*, 1707–1722. [CrossRef]
33. Vlachopoulos, N.; Diederichs, M. Improved longitudinal displacement profiles for convergence confinement analysis of deep tunnels. *Rock Mech. Rock Eng.* **2009**, *42*, 131–146. [CrossRef]
34. Panet, M.; Guenot, A. Analysis of convergence behind the face of a tunnel. In Proceedings of the 3rd International Symposium, Brighton, UK, 7–11 June 1982; pp. 197–204.
35. Humbert, P.; Dubouchet, A.; Fezans, G.; Remaud, D. CESAR-LCPC, un progiciel de calcul dédié au génie civil. *Bull. Des Lab. Ponts Chaussées* **2005**, *256*, 7–37.
36. Rocksupport. *Rock Support Interaction and Deformation Analysis for Tunnels in Weak Rock*; Tutorial Manual of Rocscience Inc.; Rocscience Inc.: Toronto, ON, Canada, 2004; pp. 1–76.
37. Lee, Y.L.; Lin, M.Y.; Hsu, W.K. Study on relationship between the confinement loss and the longitudinal deformation curve by using three-dimensional finite element analysis. *Chin. J. Rock Mech. Eng.* **2008**, *27*, 258–265.
38. Lee, Y.L.; Hsu, W.K.; Lin, M.Y. Analysis of the advancing effect and the confinement loss by using deformation measurement in tunneling. *Chin. J. Rock Mech. Eng.* **2009**, *28*, 39–46. [CrossRef]
39. Lee, Y.L. Prise en Compte des Non-Linéarité de Comportement des Sols et des Roches dans la Modélisation du Creusement d'un Tunnel. Ph.D. Thesis, École Nationale des Ponts et Chaussées, Paris, France, 1994. Available online: https://pastel.archives-ouvertes.fr/pastel-00569423 (accessed on 28 September 2021).
40. Lee, Y.L. Explicit analysis for the ground-support interaction of circular tunnel excavation in anisotropic stress fields. *J. Chin. Inst. Eng.* **2020**, *43*, 13–26. [CrossRef]

applied sciences

MDPI

Article

Behaviour of Square and Rectangular Tunnels Using an Improved Finite Element Method

Chi Thanh Nguyen [1,2], Ngoc Anh Do [1,2], Daniel Dias [3,4,*], Van Vi Pham [2] and Gospodarikov Alexandr [5]

[1] Department of Underground and Mining Construction, Faculty of Civil Engineering, Hanoi University of Mining and Geology, Hanoi 100000, Vietnam; nguyenchithanh@humg.edu.vn (C.T.N.); dongocanh@humg.edu.vn (N.A.D.)
[2] Sustainable Development in Underground Engineering Research Team, Hanoi University of Mining and Geology, Hanoi 100000, Vietnam; phamvanvi0410@gmail.com
[3] Laboratory 3SR, Grenoble Alpes University, 38000 Grenoble, France
[4] Geotechnical Expert, Antea Group, 92160 Antony, France
[5] Department of High Mathematics, Saint Petersburg Mining University, 199106 Saint Petersburg, Russia; kafmatem@spmi.ru
* Correspondence: daniel.dias@3sr-grenoble.fr

Abstract: Due to its advantages (fast and accurate calculations), the Hyperstatic Reaction Method (HRM) was used to calculate the internal forces of circular tunnel linings in former works. This paper presents an improved HRM method that is developed to estimate the internal forces induced in square and rectangular tunnel linings. Based on the comparison of the internal forces induced in these linings obtained from the HRM method and the finite element method (FEM), the improved HRM method was validated. An extensive parametric analysis of the tunnel lining and ground parameters was then carried out using both the HRM and FEM. The results indicated a great influence of the lateral earth pressure coefficient K_0, and the tunnel lining flexibility ratio F on the internal forces induced. Accordingly, the bending moments M, normal forces N, and shear forces T, induced in the tunnel lining decrease when the flexibility ratio of tunnel lining F increases. The maximum bending moment is observed at the tunnel sides that are perpendicular with the larger principal stress direction.

Keywords: tunnel; flexibility; rectangular; square; numerical method

Citation: Nguyen, C.T.; Do, N.A.; Dias, D.; Pham, V.V.; Alexandr, G. Behaviour of Square and Rectangular Tunnels Using an Improved Finite Element Method. *Appl. Sci.* **2022**, *12*, 2050. https://doi.org/10.3390/app12042050

Academic Editors: Mingfeng Lei, Chenjie Gong and Xianda Shen

Received: 8 January 2022
Accepted: 14 February 2022
Published: 16 February 2022

1. Introduction

Nowadays, circular tunnels are commonly used when mechanized tunnelling is used. The design of circular tunnels was thoroughly studied by many authors [1–5]. However, some other tunnel shapes such as rectangular and square tunnels are also used in practice. The main advantage of the square and rectangular tunnels compared to the circular ones is that they have a greater space utilization ratio. Design methods of these tunnels were developed and could be categorized in analytical methods [1,6,7] and numerical methods [7–17]. Analytical methods are very effective because they can give results quickly. However, the drawback of analytical methods lies in their simplified assumptions. They consider that the behaviour of the soil and tunnel lining is isotropic and homogeneous [6,7]. In addition, the interaction of the tunnel lining and the surrounding soil was not thoroughly considered. It is mainly modelled on two critical cases of no-slip and/or full slip conditions [1]. Meanwhile, numerical methods allow the modelling of all the factors that influence on the tunnel behaviour, such as discontinuous linings, soil anisotropy and inhomogeneity, and interaction of the tunnel lining-ground [4,5,8,10]. However, it should be noted that most of the numerical research was conducted using commercial software and they required time to build the numerical model and also to obtain the results. It is therefore interesting to develop free and simple tools that can help users to calculate the internal forces induced in rectangular and/or squared tunnels in a short time.

The Hyperstatic Reaction Method (HRM) is a numerical method that was developed based on the finite element method and originally used to design circular tunnels [4,18], horseshoe-shaped tunnels [12,19], and sub-rectangular tunnels [17]. In the HRM method, the interaction of the soil and tunnel lining is simulated through tangential and normal springs assigned at the nodes of the tunnel lining. Based on the advantages of the method such as its fast calculation, high accuracy of displacements, and internal forces induced in the tunnel lining, the HRM method is appropriate for preliminary tunnel designs.

The purpose of this study is to develop the HRM method for the lining design of square and rectangular tunnels. In previous HRM models, the tunnel lining was divided into 360 elements. Each element corresponding to one degree, and therefore the lengths of elements were constant for curved tunnel lining parts having the same radius. However, when applying to square and rectangular tunnels, the length of elements is fixed while the angle made by two consecutive nodes is changed depending on the node's position of the tunnel boundary. The developed HRM method is validated by comparing with results obtained by FEM analysis. The HRM method is then used to investigate the effects of different parameters of the tunnel lining and the surrounding soil, including the flexibility ratio of the tunnel lining F, and the coefficient of lateral earth pressure K_0 on the internal forces induced in the tunnel lining.

2. Methods Calculation

2.1. Improved (Hyperstatic Reaction Method) HRM Method

In the HRM method, the tunnel lining is segmented into 1D beam elements. These beam elements are linked to each other by nodes assigned at two ends of the beam element. The tunnel lining element is linked to the surrounding ground by normal and tangent springs at the defined element nodes. Loads of the surrounding ground are applied to the tunnel lining through the springs mentioned above. In the HRM method, when the displacements of the nodes on the tunnel lining elements are determined, internal forces induced in the tunnel lining can be calculated through the global stiffness matrix of the tunnel lining elements. The detail of the HRM method applied to circular tunnels was introduced by Oreste [18] and Do et al. [4]. In this paper, some important improvements of the HRM method applied in square and rectangular tunnels are proposed.

In the case of square and rectangular tunnels, the length of tunnel lining elements was recalculated according to the corresponding coordinates of these elements. The tunnel lining is divided into 360 elements. The lengths of elements are fixed while the angle made by two consecutive nodes and the centre of the tunnel, i.e., point O in Figure 1, is changed depending on the node's position on the tunnel boundary. On the cross-section of the square and rectangular tunnel lining, tunnel lining was divided into 6 regions specified by points A_j (j counted from 0 to 6) (see Figure 1).

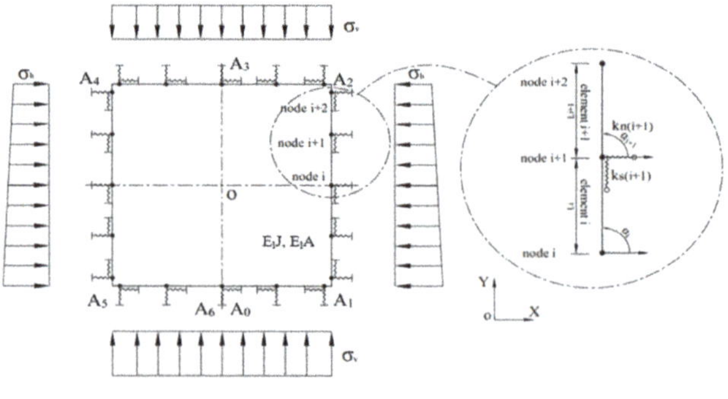

Figure 1. Diagram of the square tunnel in the HRM method.

The tunnel lining beam element "i" has two nodes, with L_i is the length of the beam element. Bending (E_lJ) and normal stiffness of tunnel lining (E_lA) (see Figure 2).

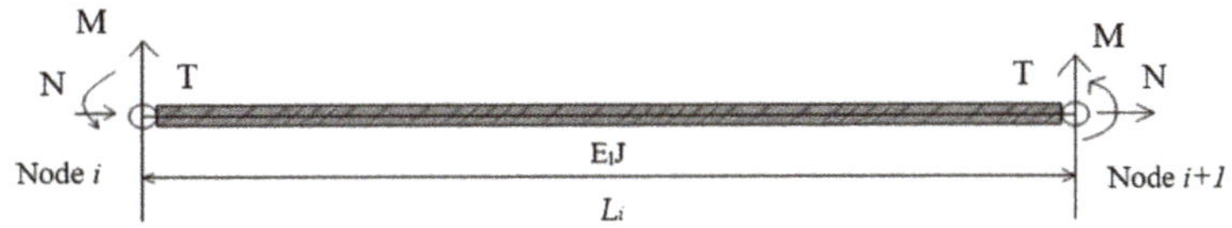

Figure 2. Beam-type finite element in the tunnel lining with reference to the local Cartesian coordinates (i is initial node; i + 1 is final node; L_i is element length).

The Ground-Support Interaction Impact

In the HRM method, the interaction between the tunnel lining and the surrounding soil is considered through the active vertical load and the horizontal load applied at all sides of the tunnel. Springs at the element nodes are specified by their normal stiffnesses (k_n) and shear stiffnesses (k_s), (Figure 1) [4,18]. The global stiffness matrix of the tunnel lining element is defined as follows:

$$\begin{aligned}
K^*_{3i-2,3i-2} &= K_{3i-2,3i-2} + k_{n,i} \times \cos^2\left(\tfrac{\alpha_{i+1}}{2} + \tfrac{\alpha_i}{2} - \tfrac{\pi}{2}\right) + k_{s,i} \times \sin^2\left(\tfrac{\alpha_{i+1}}{2} + \tfrac{\alpha_i}{2} - \tfrac{\pi}{2}\right) \\
K^*_{3i-1,3i-1} &= K_{3i-1,3i-1} + k_{n,i} \times \sin^2\left(\tfrac{\alpha_{i+1}}{2} + \tfrac{\alpha_i}{2} - \tfrac{\pi}{2}\right) + k_{s,i} \times \cos^2\left(\tfrac{\alpha_{i+1}}{2} + \tfrac{\alpha_i}{2} - \tfrac{\pi}{2}\right) \\
K^*_{3i-1,3i-2} &= K_{3i-1,3i-2} + (k_{n,i} - k_{s,i}) \times \sin\left(\tfrac{\alpha_{i+1}}{2} + \tfrac{\alpha_i}{2} - \tfrac{\pi}{2}\right) \times \cos\left(\tfrac{\alpha_{i+1}}{2} + \tfrac{\alpha_i}{2} - \tfrac{\pi}{2}\right) \\
K^*_{3i-2,3i-1} &= K_{3i-2,3i-2} + (k_{n,i} - k_{s,i}) \times \sin\left(\tfrac{\alpha_{i+1}}{2} + \tfrac{\alpha_i}{2} - \tfrac{\pi}{2}\right) \times \cos\left(\tfrac{\alpha_{i+1}}{2} + \tfrac{\alpha_i}{2} - \tfrac{\pi}{2}\right)
\end{aligned} \tag{1}$$

where i is the number of the generic node; $k_{n,i}$ is the stiffness of the normal spring linked to node i; $k_{s,i}$ is the stiffness of the tangential spring linked to node i; α_i and α_{i+1} is the angle between the local and global reference systems, of element i and element (i + 1).

The reaction pressure p of the surrounding ground is represented by a nonlinear relationship (hyperbolic) with the tunnel lining deformation δ (Figure 3).

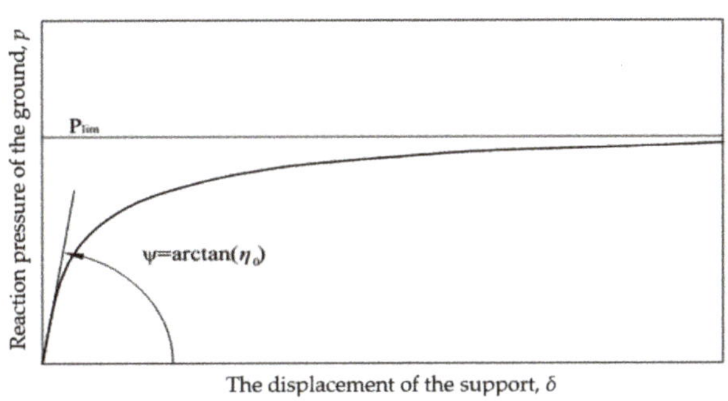

Figure 3. Relationship between the reaction pressure p caused by the surrounding soil and the tunnel lining deformation δ.

The parameters representing the soil properties surrounding the tunnel are presented through the following equations [4,18]:

$$\eta^* = \frac{p_{\lim}}{\delta} \times \left(1 - \frac{p_{\lim}}{p_{\lim} + \eta_0 \delta}\right) \tag{2}$$

$$\eta_{n,0} = \beta \times \left(\frac{1}{1+\nu}\right) \times \left(\frac{E}{R_i}\right) \tag{3}$$

$$\eta_s = \frac{1}{3} \times \eta_n \tag{4}$$

$$p_{n,\lim} = \left(\frac{2 \times c \times \cos\varphi}{1 - \sin\varphi}\right) + \left(\frac{1 + \sin\varphi}{1 - \sin\varphi}\right) \times \Delta\sigma_{conf} \quad (5)$$

$$\Delta\delta_p = \left(\frac{\sigma_h + \sigma_v}{2}\right) \times \left(\frac{\nu_s}{1 - \nu_s}\right) \quad (6)$$

$$p_{s,\lim} = \left(\frac{\sigma_h + \sigma_v}{2}\right) \times tg\varphi \quad (7)$$

where η^* is the apparent stiffness of the surrounding ground (N/m^3); $p_{\lim}$ is the maximum reaction pressure (MPa); η_0 is the initial stiffness of the surrounding ground (N/m^3); $\eta_{n,0}$ is the initial normal ground stiffness (N/m^3); ν_s is the Poisson's ratio of the surrounding ground; E is Young's modulus of ground (MPa); R_i is the distance from the centre of the square that is the cross-section tunnel to the element "i" of the tunnel lining (m); $p_{n,\lim}$ is the maximum normal reaction pressure (MPa); $p_{s,\lim}$ is the maximum shear reaction pressure (MPa); c is the ground cohesion (MPa); φ is the surrounding ground friction angle (degrees); $\Delta\delta_p$ is the confining pressure (MPa); and β is a dimensionless factor.

The β value was assessed by researchers to vary depending on the parameters of the tunnel lining and the surrounding soil. β could be equal to 1 [8–10,19–21] or 2 [4]. In this study, β = 2 is adopted based on the fitting of the results obtained by using the finite element method Plaxis^{2D}.

The stiffnesses $k_{n,i}$ and $k_{s,i}$ of each spring are given by the formula:

$$k_{n,i} = \frac{p_{n,\lim}}{\delta_{n,i}} \times \left(1 - \frac{p_{n,\lim}}{p_{n,\lim} + \eta_{n,0} \times \delta_{n,i}}\right) \times \frac{(L_{i-1} + L_i)}{2} \quad (8)$$

$$k_{s,i} = \frac{p_{s,\lim}}{\delta_{s,i}} \times \left(1 - \frac{p_{s,\lim}}{p_{s,\lim} + \eta_{s,0} \times \delta_{s,i}}\right) \times \frac{(L_{i-1} + L_i)}{2} \quad (9)$$

Active Loads in the HRM Method

The vertical load σ_v is estimated depending on the tunnel depth. When the tunnel is shallow:

$$\sigma_{v(i)} = \gamma_i \times h_i \quad (10)$$

where h_i is the depth measured from the soil surface to the lining point (m); γ_i is the soil density (MN/m^3).

In the deep tunnel case, the vertical loads could be calculated using Terzaghi's formula [19,22]:

$$\sigma_v = \gamma \times h_0 \quad (11)$$

$$h_0 = \frac{D_1 \times [1 - (c/D_1 \times \gamma)]}{K_0 \times \tan\varphi} \times (1 - e^{-K_0 \times \tan\varphi(H/D_1)}) + \frac{P_0}{\gamma} \times (e^{-K_0 \times \tan\varphi(H/D_1)}) \quad (12)$$

$$D_1 = D + H_t \times \cot\left(\frac{(\pi/4) + (\varphi/2)}{2}\right) \quad (13)$$

where γ is the ground density surrounding the tunnel (MN/m^3); φ is the soil internal friction angle (degrees); D and H_t are the width and height of the tunnel, respectively, (m); H is the distance from the surface ground to the tunnel crown (m); P_0 is the overload on the ground surface (MPa); and K_0 is the coefficient of lateral earth pressure.

The horizontal load acting on the sidewall of the square or rectangular tunnel is determined through the equation:

$$\sigma_h = K_0 \times \sigma_v \quad (14)$$

It should be mentioned that the necessary time for each calculation using the HRM method is very short and more less equal to 5 s. This allows the HRM method to be useful for parametric analyses.

2.2. FEM Calculation

In this study, ground parameters of the 3rd line Nhon—Kim Ma belonging to the Hanoi metro system are used as a reference case [23–26]. The size of the square tunnel cross-section is 5.5 m. The rectangular tunnel is 6.0 m in width and 5.0 m in height. The tunnel is located 20 m below the ground surface. The tunnel is assumed to be supported by a continuous lining. The other parameters of the tunnel lining and the surrounding ground are shown in Table 1.

Table 1. Parameters of the soil and tunnel lining.

Parameters	Unit	Value
Parameters of tunnel lining		
Young's modulus, E_l	MPa	35,000
Poisson's ratio, ν_l	-	0.15
Tunnel lining thickness, t	m	0.35
Overburden, H	m	20
Parameters of soil		
Young's modulus, E	MPa	10
Poisson's ratio, ν	-	0.34
The unit weight, γ	kN/m^3	18.1
Lateral earth pressure factor, K_0	-	0.5
Cohesion, c	kPa	22.5
Internal friction angle, φ	degrees	33

Square and rectangular tunnels are simulated using the plain strain FEM model. While Mohr Coulomb's constitutive model is adopted for the ground, the behaviour of the tunnel lining is assumed to be linear elastic. The ground surrounding the tunnel is simulated using a 15 nodes element. Plate elements are used to model the tunnel lining. An interface is also modelled to simulate the interaction between the tunnel lining and the surrounding ground [27–30].

The dimensions of the model built in FEM are 100 m in wide and 40 m in height. The model includes 33,185 nodes and 4092 zones (see Figure 4). While the top of the model is free, the bottom of the model was fixed in the horizontal direction and the vertical sides of the model were fixed in the vertical direction. The influence of groundwater is not taken into consideration.

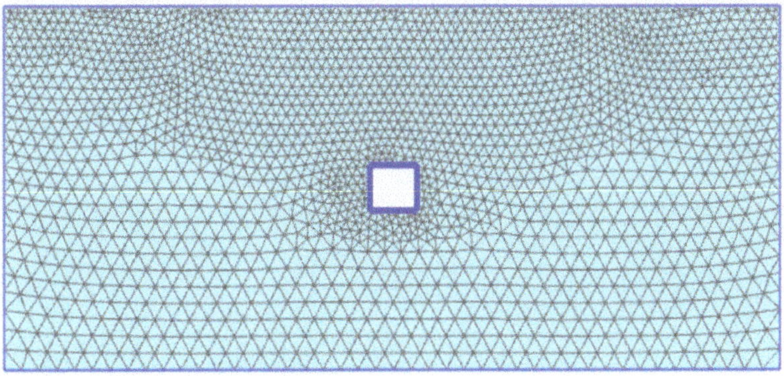

Figure 4. Geometry of the FEM model.

The construction process is divided into the following phases:

Phase 1—Setup the model: build up the model's mesh and set up the boundary conditions and initial stress state;

Phase 2—Construction phase: the soil in the tunnel area is removed;

Phase 3—Setting of the tunnel lining: the tunnel lining was installed on the tunnel periphery. It should be mentioned that the relaxation process induced in the ground surrounding the tunnel after the excavation and before the tunnel lining installation is not considered in this study. This case means the worst situation of active loads applied on the tunnel lining.

2.3. Validation of the HRM Method

The application of the HRM method in the case of square and rectangular tunnels is verified by comparing the results of internal forces (bending moment M, normal force N, and shear force T) induced in the tunnel lining obtained in this method with those determined by a FEM using Plaxis^{2D}. In this paper, the ground and tunnel lining parameters of the Hanoi metro system (3rd line Nhon—Kim Ma) are used as input data of the HRM method and FEM. The parameters of the tunnel lining and the soil surrounding the tunnel, i.e., the tunnel lining thickness t = 0.35 m, Young's modulus of soil E = 10 MPa, and the coefficient of lateral earth pressure K_0 = 0.5 were adopted.

From the results in Figure 5 and Table 2, in the case of the square tunnel, the following comments can be made: the internal forces in the square tunnel lining calculated by the improved HRM method are in good agreement with results obtained in the FEM. The differences of the maximum bending moment M_{max} and the minimum bending moment M_{min} between the two methods are 2.03% and 0.24%, respectively. Meanwhile, the differences of the maximum normal force N_{max} and minimum normal force N_{min} by the HRM method and FEM are correspondingly 0.55% and 7.89%. Those of the maximum shear force T_{max} and minimum normal shear force T_{min} are 1.91%.

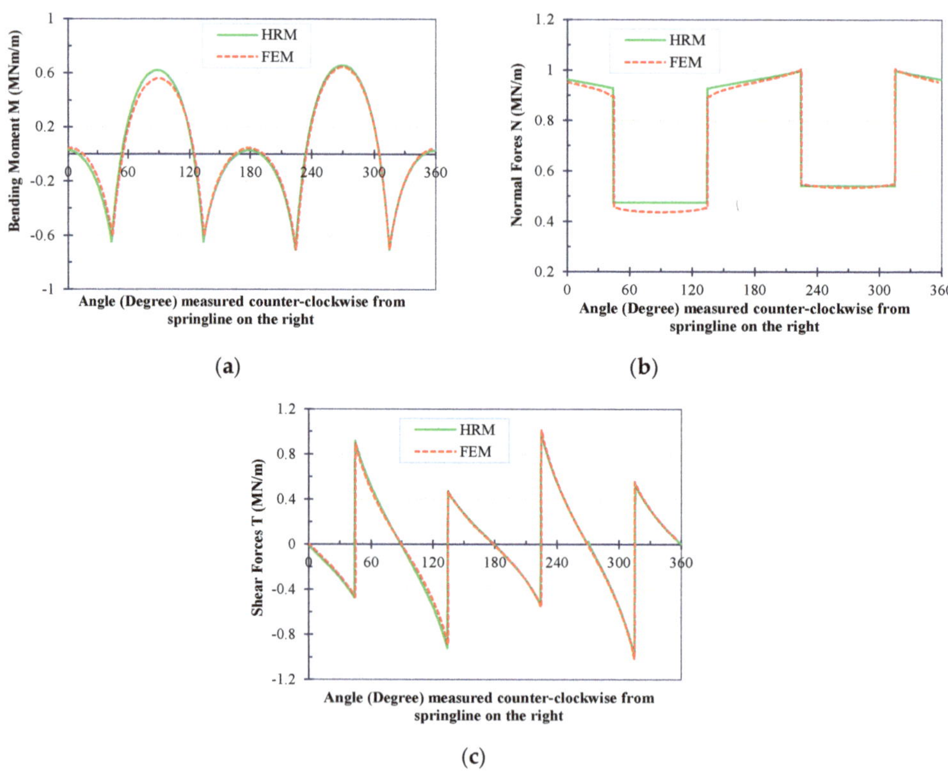

Figure 5. Internal forces in the square tunnel lining: (**a**) Bending moments M; (**b**) Normal forces N; (**c**) Shear forces T.

Table 2. Comparison of the internal forces by the HRM method and FEM.

Tunnel Case/Calculation Method	Value	M (MNm/m)	N (MN/m)	T (MN/m)
Case 1—square tunnel				
FEM	Max	0.644	1.003	1.004
	Min	−0.702	0.437	−1.004
HRM	Max	0.657	0.998	0.985
	Min	−0.701	0.474	−0.985
Difference (%)	Max	2.03	0.55	1.91
	Min	0.24	7.89	1.91
Case 2—rectangular tunnel				
FEM	Max	0.781	1.073	1.071
	Min	−0.764	0.391	−1.071
HRM	Max	0.788	1.052	1.040
	Min	−0.754	0.441	−1.040
Difference (%)	Max	0.85	1.95	2.93
	Min	1.33	12.74	2.93

Figure 6 and Table 2 show the comparative results in the case of the rectangular tunnel. The maximum bending moment M_{max} and the minimum bending moment M_{min} obtained in the HRM method are 0.85% and 1.33% different from the corresponding values of FEM. In the HRM method, the maximum normal force N_{max} and minimum normal force N_{min} are 1.95% and 12.74%, respectively, smaller than those of the FEM model. The differences of maximum shear force T_{max} and minimum normal shear force T_{min} between the HRM method and FEM are both 2.93%.

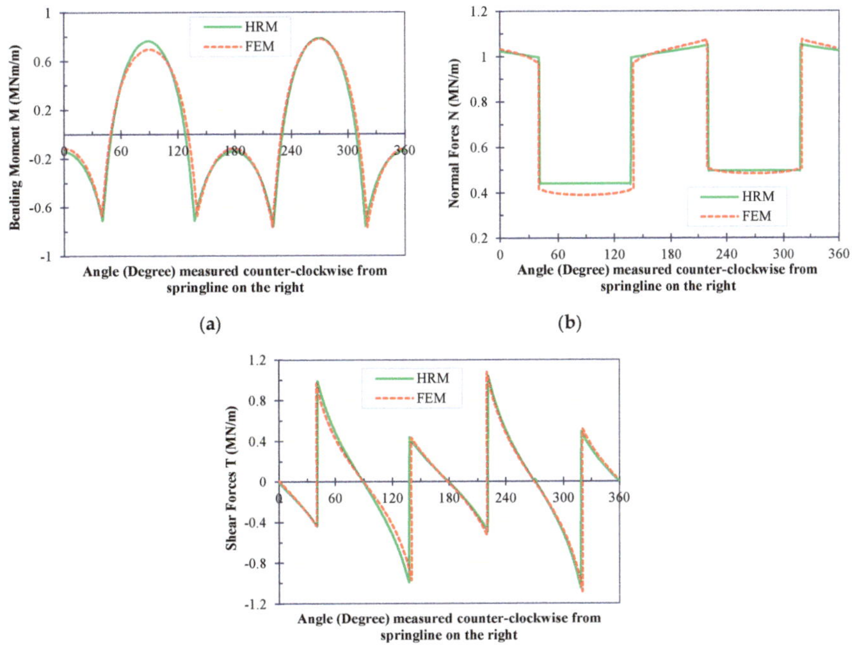

Figure 6. Internal forces in the rectangular tunnel lining: (**a**) Bending moments M; (**b**) Normal forces N; (**c**) Shear forces T.

The internal forces differences induced in square and rectangular tunnel linings in the HRM method and the FEM could be explained by the fact that the internal forces and deformations of the tunnel lining in the HRM method are mainly affected by the active loads and the soil-lining interaction. These factors are considered through the springs stiffnesses between the ground and the tunnel lining (normal and tangential springs) assigned at the lining beam nodes. In FEM, the ground–structure interaction is taken into account through the interface embedded between the lining and the ground elements. In this study, when using the FEM to simulate square and rectangular tunnels, without an interface between the tunnel lining and the ground elements, the structure and the soil are tied together. It means that no relative displacements (slipping/gapping) are possible between the structure and soil [27].

Based on the above comparison and analysis, it is reasonable to conclude that the HRM method can be effectively used to estimate the internal forces induced in linings of square and rectangular tunnels.

3. Parametric Study

In this part, a parametric investigation using both the HRM method and FEM is conducted to highlight the effects of the flexibility ratio (F) of the lining and the coefficient of lateral earth pressure (K_0) on the behaviour of the tunnel. The square tunnel with dimensions of 5.5 × 5.5 m was adopted. The other parameters of the tunnel lining and the soil are shown in Table 1.

Influence of the flexibility ratio of the tunnel lining (F)

For investigation purposes, the coefficient of lateral earth pressure K_0 is kept as constant, K_0 = 0.5 and the thickness of the square tunnel lining t changes from 0.35 m to 0.5 m. It is assumed that Young's modulus of the ground E is changed in the range from 10 MPa to 100 MPa. The flexibility ratio of the tunnel lining F is determined by the following equation [6,7]:

$$F = \frac{G_m}{24} \times \left(\frac{H^2 W}{E_l I_w} + \frac{H W^2}{E_l I_R} \right) \tag{15}$$

where E_l is Young's modulus of the tunnel lining (MPa); G_m is the shear modulus of ground (MPa); I_w is inertia moment of lining at the sidewall (m^4); I_R is inertia moment of lining at the lining roof (m^4); W is the tunnel lining width (m); and H is tunnel height (m).

The lining internal forces calculated by the HRM method and FEM were presented in Figure 7 and Table 3. It is reasonable to conclude that:

- The internal forces, including, bending moments M, normal forces N, and shear forces T, induced in the tunnel lining decrease when the flexibility ratio of tunnel lining F increases. This could be explained by the fact that a larger F value means a more flexible or softer lining and/or a stronger ground. As a consequence, a smaller yielded ground zone that causes active loads acting on the tunnel lining could be predicted. In addition, due to the higher flexibility of the lining, which implies a larger movement/deformation possibility of the lining, a greater relaxation process is induced in the ground surrounding the tunnel and therefore a decrease in active pressure applied in the tunnel. Internal forces induced in the tunnel lining are decreased as mentioned above.
- In the range of the small flexibility ratio of tunnel lining ($F \leq 1$) (i.e., the thicker tunnel lining and the smaller Young's modulus of soil E), the deviation of the internal forces obtained using the improved HRM method and FEM is small. For $F > 1$, an increase in the F value is followed by a larger difference in the results of the two methods. While the absolute bending moments obtained by the HRM method are always greater than the ones of the FEM model, the absolute normal forces and shear forces observed in the HRM results are smaller than those determined in the FEM model. It means that the internal forces estimated in the HRM lead to a more unstable situation of the tunnel lining, which can be acceptable in terms of the tunnel design. The observed larger deviation at the greater flexibility ratio of tunnel lining F could be concerned

with the difference in the simulation of soil–lining interaction of the two methods as mentioned above.

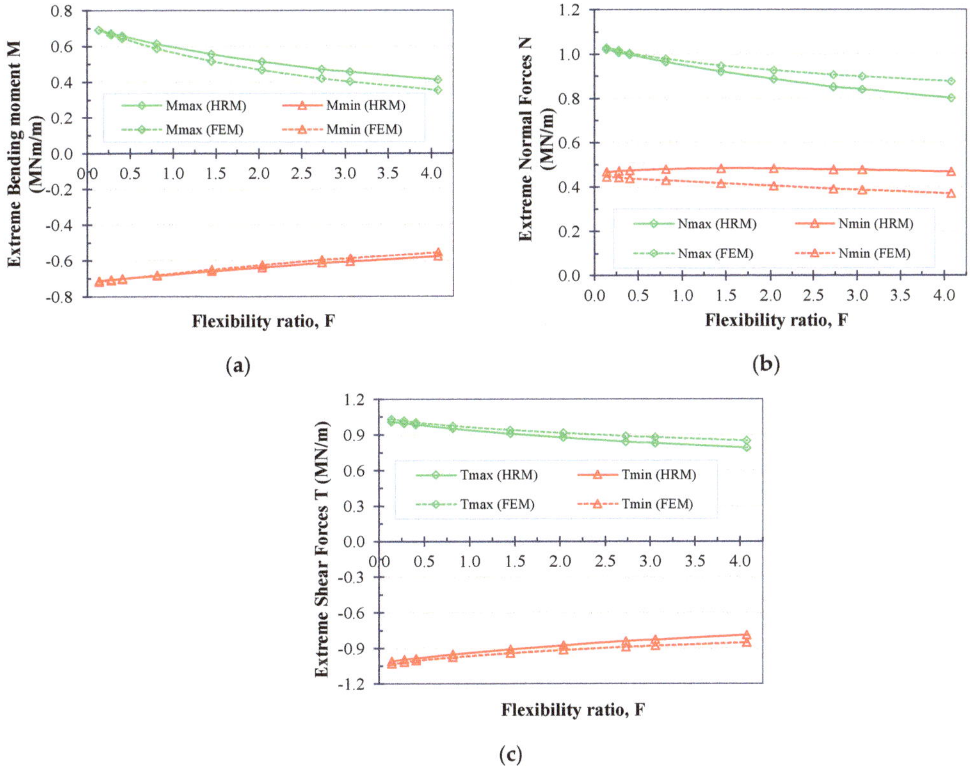

Figure 7. Effect of flexibility ratios F on internal forces in the square tunnel: (**a**) Extreme bending moments M; (**b**) Extreme normal forces N; (**c**) Extreme shear forces T.

Table 3. Analysis of the effect of the flexibility ratio of tunnel lining F on the internal forces.

The Flexibility Ratio F	Values	M (MNm/m)			N (MN/m)			T (MN/m)		
		HRM	FEM	Difference (%)	HRM	FEM	Difference (%)	HRM	FEM	Difference (%)
$F = 0.14$	Max Min	0.691 −0.712	0.690 −0.719	−0.116 0.902	1.022 0.468	1.029 0.446	0.624 −5.136	1.010 −1.010	1.031 −1.031	2.056 2.063
$F = 0.28$	Max Min	0.672 −0.706	0.664 −0.710	−1.148 0.626	1.008 0.472	1.017 0.443	0.890 −6.606	0.996 −0.996	1.018 −1.018	2.182 2.190
$F = 0.41$	Max Min	0.657 −0.701	0.703 −0.0586	−2.077 0.239	0.998 0.045	1.004 0.437	−6.167 0.553	0.986 −0.986	1.005 −1.005	1.880 1.887
$F = 0.82$	Max Min	0.612 −0.684	0.681 −0.626	−4.516 −0.417	0.965 0.481	0.979 0.429	−8.573 1.392	0.953 −0.953	0.976 −0.976	2.394 2.399
$F = 1.45$	Max Min	0.556 −0.660	0.517 −0.651	−7.595 −1.357	0.922 0.484	0.948 0.417	−11.91 19.101	0.910 −0.910	0.941 −0.941	3.301 3.305
$F = 2.04$	Max Min	0.514 −0.639	0.467 −0.626	−10.004 −1.129	0.889 0.483	0.928 0.405	2.775 −16.139	0.876 −0.876	0.915 −0.915	4.190 4.192
$F = 2.73$	Max Min	0.471 −0.614	0.419 −0.597	−12.346 −2.961	0.853 0.478	0.907 0.391	4.216 −19.313	0.840 −0.840	0.888 −0.888	5.354 5.354
$F = 3.06$	Max Min	0.457 −0.606	0.402 −0.587	−13.541 −3.234	0.841 0.477	0.899 0.386	5.946 −22.412	0.829 −0.829	0.879 −0.879	5.704 5.703
$F = 4.08$	Max Min	0.413 −0.577	0.354 −0.555	−16.600 −4.098	0.802 0.468	0.877 0.369	6.483 −23.574	0.790 −0.790	0.851 −0.851	7.142 7.142

Influence of the coefficient of lateral earth pressure K_0

The coefficient of lateral earth pressure K_0 is varied from 0.5 to 2.0. The ground deformation modulus E = 10 MPa, the tunnel lining thickness t = 0.35 m, and the flexibility ratio of the tunnel lining F = 0.41 are adopted. Figure 8 illustrates the distribution of internal forces along the square tunnel periphery. Figure 9 and Table 4 indicate the dependency of the extreme internal forces in the tunnel lining obtained by the HRM method and FEM on the K_0 value. Some following observations can be given:

- In most investigated cases of K_0 values, the HRM gives internal forces results that are in good agreement with those of the FEM model. The deviation is always smaller than 5%;
- An increase in the K_0 value is followed by a decrease in the bending moments at the top and tunnel bottom but an increase in the bending moment at the sidewalls (Figure 8a). In other words, the maximum bending moment is observed at the tunnel sides that are perpendicular with the larger principal stress direction, i.e., at angles of 90 and 270 degrees when K_0 is smaller than unity and at angles of 0 and 180 degrees when K_0 is larger than unity;
- The smallest maximum bending moments are seen for the K_0 value of unity. The smaller and larger K_0 values cause an increase in the maximum bending moments (Figure 9a). It could be explained by the fact that for a K_0 value of unity, the deviation between the vertical and lateral active loads that are the origin of the bending moments induced in the tunnel lining is the smallest;
- The change in the K_0 value mainly causes the variation of normal forces at the top and bottom parts of the lining. Its influence on the normal forces at the tunnel sidewalls is negligible. An increase in the K_0 value causes a corresponding increase in the normal forces at the top and bottom of the lining (see Figure 8b). It is related to the larger lateral active load applied from the surrounding ground on the sidewall and then transferring to the top and bottom lining parts when the K_0 value increases. Generally, the greater the K_0 value, the higher the maximum normal forces induced in the tunnel lining (see Figure 9b);
- The value of maximum shear force T_{max} and the minimum shear force T_{min} do not change when K_0 is smaller than one. When the K_0 is greater than one, the absolute extreme shear force is increased rapidly (Figure 9c).

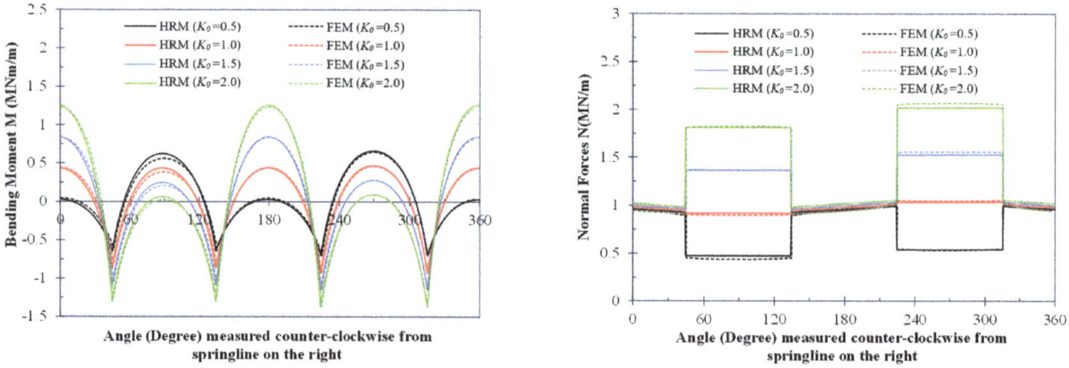

(a) **(b)**

Figure 8. *Cont.*

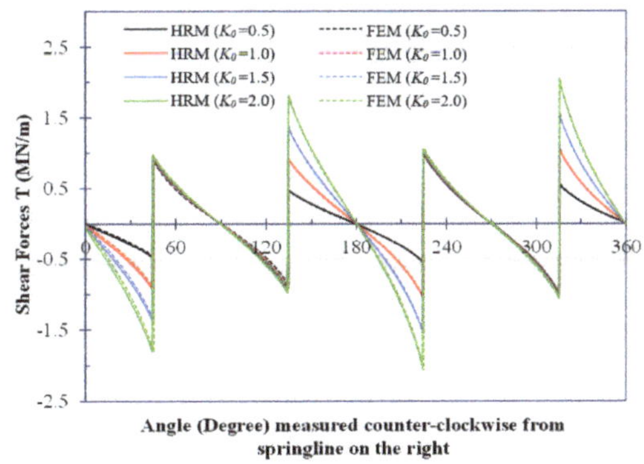

(c)

Figure 8. Internal forces in the squared tunnel lining with different coefficients of lateral earth pressure (**a**) Bending moments M; (**b**) Normal forces N; (**c**) Shear forces T.

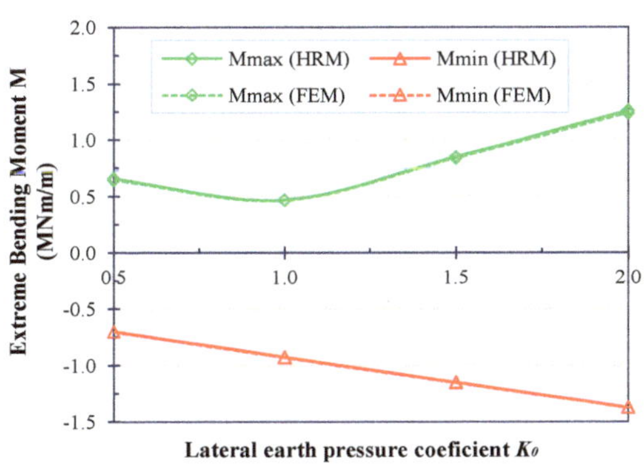

(a)

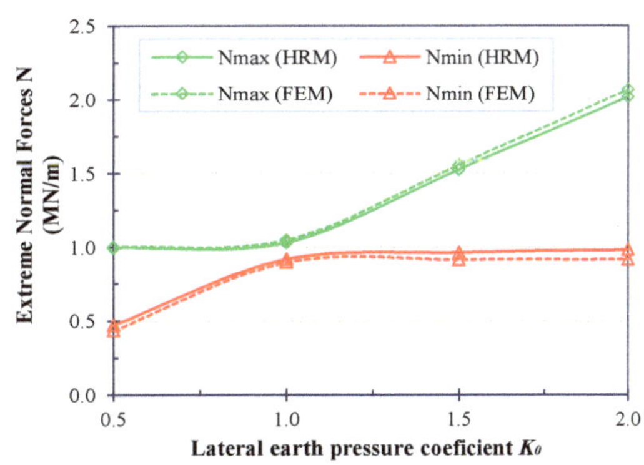

(b)

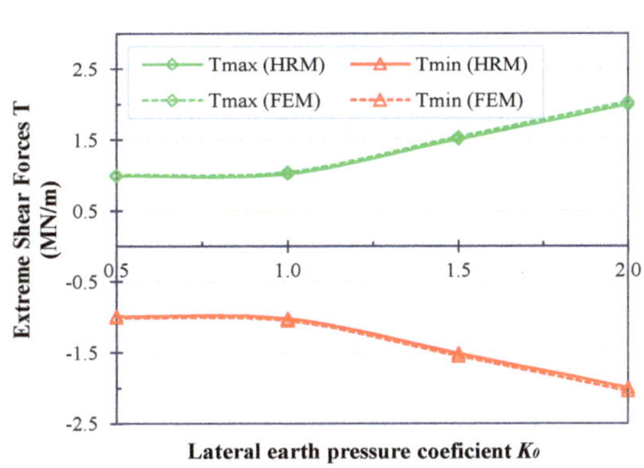

(c)

Figure 9. Effect of K_0 on the internal forces in the squared tunnel lining: (**a**) Extreme bending moments M; (**b**) Extreme normal forces N; (**c**) Extreme shear forces T.

Table 4. Influence of the coefficient of lateral earth pressure K_0 on the internal forces.

The Lateral Earth Pressure Coefficient K_0	Values	M (MNm/m)			N (MN/m)			T (MN/m)		
		HRM	FEM	Difference (%)	HRM	FEM	Difference (%)	HRM	FEM	Difference (%)
$K_0 = 0.5$	Max	0.657	0.644	−2.018	0.998	1.004	0.597	0.986	1.005	1.890
	Min	−0.701	−0.703	0.284	0.457	0.437	−4.576	−0.986	−1.005	1.890
$K_0 = 1$	Max	0.468	0.465	−0.645	1.034	1.051	1.617	1.022	1.047	2.483
	Min	−0.926	−0.930	0.430	0.920	0.900	−2.222	−1.021	−1.047	2.483
$K_0 = 1.5$	Max	0.850	0.840	−1.190	1.527	1.557	1.926	1.509	1.542	2.204
	Min	−1.151	−1.154	0.256	0.965	0.918	−5.119	−1.508	−1.542	2.269
$K_0 = 2.0$	Max	1.261	1.238	−1.857	2.021	2.068	2.272	1.996	2.037	2.061
	Min	−1.376	−1.375	−0.072	0.984	0.919	−7.072	−1.994	−2.037	2.110

4. Conclusions

In this research, the HRM method was developed and improved to be applied to square and rectangular tunnels. The comparison of the internal forces induced in the tunnel lining when using the HRM method and FEM model indicated that the HRM method can be effectively used to calculate internal forces for square and rectangular tunnels. The influences of the flexibility ratio of the tunnel lining F and the coefficient of lateral earth pressure K_0 on the square tunnel behaviour were then investigated. Based on the obtained results, the following conclusions can be drawn:

- The internal forces, including bending moments M, normal forces N, and shear forces T, induced in the tunnel lining decrease when the flexibility ratio of tunnel lining F increases;
- The maximum bending moment is observed at the tunnel sides that are perpendicular with the larger principal stress direction. The smallest maximum bending moment is seen for the K_0 value equal to unity. The smaller and larger K_0 values cause an increase in the maximum bending moment;
- The change in the K_0 value mainly causes a normal forces variation at the top and bottom parts of the lining. Its influence on the normal forces at the tunnel sidewalls is negligible.

Author Contributions: Conceptualization, N.A.D., D.D. and C.T.N.; methodology, N.A.D. and V.V.P.; software, C.T.N. and V.V.P.; formal analysis, D.D. and N.A.D.; investigation, N.A.D. and C.T.N.; writing—original draft preparation, C.T.N. and N.A.D.; writing—review and editing, D.D. and N.A.D.; supervision, D.D. and G.A. All authors have read and agreed to the published version of the manuscript.

Funding: This research was funded by Vietnam National Foundation for Science and Technology Development (NAFOSTED) under grant number 17/2020/STS02, Vietnamese Ministry of Education and Training under grant number B2021-MDA-09.

Institutional Review Board Statement: Not applicable.

Informed Consent Statement: Not applicable.

Data Availability Statement: The data presented in this study are available on request from the corresponding author. The data are not publicly available due to privacy restrictions.

Conflicts of Interest: The authors declare no conflict of interest.

References

1. Wang, J.N. *Seismic Design of Tunnels: A State-of-the-Art Approach*; Brinkerhoff Quade and Douglas Inc.: New York, NY, USA, 1993.
2. Penzien, J.; Wu, C. Stresses in Linings of Bored Tunnels. *J. Earthq. Eng. Struct. Dyn.* **1998**, *27*, 283–300. [CrossRef]
3. Penzien, J. Seismically Induced Racking of Tunnel Linings. *Int. J. Earthq. Eng. Struct. Dyn.* **2000**, *29*, 683–691. [CrossRef]
4. Do, N.A.; Dias, D.; Oreste, P.P.; Maigre, I.D. A New Numerical Approach to the Hyperstatic Reaction Method for Segmental Tunnel Linings. *Int. J. Numer. Anal. Methods Geomech.* **2014**, *38*, 1617–1632. [CrossRef]
5. Do, N.A.; Dias, D.; Oreste, P.P.; Maigre, I.D. The behaviour of the segmental tunnel lining studied by the hyperstatic reaction method. *Eur. J. Environ. Civ. Eng.* **2014**, *18*, 489–510.

6. Wood, J.H. *Earthquake Design Procedures for Rectangular Underground Structures*; Project Report to Earthquake Commission, EQC Project No 01/470: Lower Hutt City, New Zealand, 2004.
7. Wood, J.H. Earthquake Design of Rectangular Underground Structures. *Bull. N. Z. Soc. Earthq. Eng.* **2007**, *40*, 1–6. [CrossRef]
8. Mashimo, H.; Ishimura, T. Numerical modelling of the behavior of shield tunnel lining during assembly of a tunnel ring. In Proceedings of the Fifth International Symposium on Geotechnical Aspects of Underground Construction in Soft Ground, Amsterdam, The Netherlands, 15–17 June 2005.
9. Moller, S.C. Tunnel Induced Settlements and Structural Forces in Linings. Ph.D. Dissertation, Stuttgart University, Stuttgart, Germany, 2006.
10. Moller, S.C.; Vermeer, P.A. On numerical simulation of tunnel installation. *Tunn. Undergr. Space Technol.* **2008**, *23*, 461–475. [CrossRef]
11. Zhang, D.; Huang, H.; Hu, Q.; Jiang, F. Influence of multi-layered soil formation on shield tunnel lining behavior. *Tunn. Undergr. Space Technol.* **2015**, *47*, 123–135. [CrossRef]
12. Du, D.C.; Dias, D.; Do, N.A.; Oreste, P.P. Hyperstatic reaction method for the design of U-shaped tunnel supports. *Int. J. Geomech.* **2018**, *18*, 04018030. [CrossRef]
13. Du, D.C.; Dias, D.; Do, N.A.; Vo, T.H. U-shaped tunnel lining design using the Hyperstatic Reaction Method–Influence of the invert. *Soils Found.* **2020**, *60*, 592–607. [CrossRef]
14. Du, D.; Dias, D.; Do, N.A. Lining performance optimization of sub-rectangular tunnels using the Hyperstatic Reaction Method. *Comput. Geotech.* **2020**, *117*, 103279.
15. Sahoo, J.P.; Kumar, B. Support pressure for stability of circular tunnels driven in granular soil under water table. *Comput. Geotech.* **2019**, *109*, 58–68. [CrossRef]
16. Weng, X.; Sun, Y.; Zhang, Y.; Niu, H.; Liu, X.; Dong, Y. Physical modeling of wetting-induced collapse of shield tunneling in loess strata. *Tunn. Undergr. Space Technol.* **2019**, *90*, 208–219. [CrossRef]
17. Do, N.A.; Dias, D.; Zhang, Z.; Huang, X.; Nguyen, T.T.; Pham, V.V.; Ouahcène, N.R. Study on the behavior of squared and sub-rectangular tunnels using the Hyperstatic Reaction Method. *Transp. Geotech.* **2020**, *22*, 100321. [CrossRef]
18. Oreste, P.P. A Numerical Approach to the Hyperstatic Reaction Method for the Dimensioning of Tunnel Supports. *Tunn. Undergr. Space Technol.* **2007**, *22*, 185–205. [CrossRef]
19. Nguyen, T.C.; Gospodarikov, A.P. Hyperstatic reaction method for calculations of tunnels with horseshoe shaped cross-section under the impact of earthquakes. *Earthq. Eng. Eng. Vib.* **2020**, *19*, 179–188.
20. ITA. ITA guidelines for the design of tunnels. *Tunn. Undergr. Space Technol.* **1998**, *3*, 237–249.
21. Oreste, P.P.; Spagnoli, G.; Ramos, C.A.L.; Sebille, L. The hyperstatic reaction method for the analysis of the spraryed concrete linings behavior in tunneling. *Geotech. Geol. Eng.* **2018**, *36*, 2143–2169. [CrossRef]
22. Takano, Y.H. Guidelines for the Design of Shield Tunnel Lining. *Tunn. Undergr. Space Technol.* **2000**, *15*, 303–331.
23. Systra. *Hanoi Pilot LRT Line Feasibility Study*; Executive Summary: Hanoi, Vietnam, 2005.
24. Gospodarikov, A.P.; Nguyen, T.C. The Impact of Earthquakes of Tunnel Linings: A Case Study from the Hanoi Metro System. *Int. J. GEOMATE* **2018**, *14*, 151–158.
25. Gospodarikov, A.P.; Nguyen, T.C. Behaviour of Segmental Tunnel Linings under the Impact of Earthquakes: A Case Study from the Tunnel of Hanoi Metro System. *Int. J. GEOMATE* **2018**, *15*, 91–98.
26. Protosenya, A.G.; Verbilo, P.E. Research of Compression Strength of Fissured Rock Mass. *J. Min. Inst.* **2017**, *223*, 51–57.
27. Kees, V. *PLAXIS. CONNECT Edition V20.04*; Delft University of Technology, Civil Engineering (NL): Delft, The Netherlands, 2020.
28. Pang, R.; Bin, X.; Yang, Z.; Song, L. Seismic time-history response and system reliability analysis of slopes considering uncertainty of multi-parameters and earthquake excitations. *Comput. Geotech.* **2021**, *136*, 104245. [CrossRef]
29. Bin, X.; Pang, R.; Yang, Z. Verification of stochastic seismic analysis method and seismic performance evaluation based on multi-indices for high CFRDs. *Eng. Geol.* **2020**, *264*, 105412.
30. Pang, R.; Xu, B.; Zhou, Y.; Zhang, X.; Wang, X. Fragility analysis of high CFRDs subjected to mainshock-aftershock sequences based on plastic failure. *Eng. Struct.* **2020**, *206*, 110152. [CrossRef]

applied sciences

MDPI

Article

Intelligent Classification of Surrounding Rock of Tunnel Based on 10 Machine Learning Algorithms

Siguang Zhao [1,2], Mingnian Wang [1,2], Wenhao Yi [1,2], Di Yang [1,2] and Jianjun Tong [1,2,*]

1 Key Laboratory of Transportation Tunnel Engineering of Ministry of Education, Southwest Jiaotong University, Chengdu 610031, China; 2019310044@my.swjtu.edu.cn (S.Z.); wangmingnian@swjtu.edu.cn (M.W.); 1256635900@my.swjtu.edu.cn (W.Y.); 2020210212@my.swjtu.edu.cn (D.Y.)

2 School of Civil Engineering, Southwest Jiaotong University, Chengdu 610031, China

* Correspondence: jjtong@swjtu.edu.cn; Tel.: +86-135-5031-1331

Abstract: The quality evaluation of the surrounding rock is the cornerstone of tunnel design and construction. Previous studies have confirmed the existence of a relationship between drilling parameters and the quality of surrounding rock. The application of drilling parameters to the intelligent classification of surrounding rock has the natural advantages of automatic information collection, real-time analysis, and no extra work. In this work, we attempt to establish the intelligent surrounding rock classification model and software system driven by drilling parameters. We collected 912 samples containing four drilling parameters (penetration velocity, hammer pressure, rotation pressure, and feed pressure) and three surrounding rock (grade-III, grade-IV, and grade-V). Based on the python machine learning toolkit (Scikit-learn), 10 types of supervised machine learning algorithms were used to train the intelligent surrounding rock classification model with the model parameter selection technology of grid search cross validation. The results show that the average accuracy is 0.82, which proves the feasibility of this method. Finally, the tunnel surrounding rock intelligent classification system was established based on three models with better comprehensive performance among them. The classification accuracy of the system was 0.87 in the tunnel test section, which indicates that the system has good generalization performance and practical value.

Keywords: drill and blast tunnel; machine learning; measure-while-drilling; drilling parameters; intelligent surrounding rock classification model

Citation: Zhao, S.; Wang, M.; Yi, W.; Yang, D.; Tong, J. Intelligent Classification of Surrounding Rock of Tunnel Based on 10 Machine Learning Algorithms. *Appl. Sci.* **2022**, *12*, 2656. https://doi.org/10.3390/app12052656

Academic Editors: Mingfeng Lei, Chenjie Gong and Xianda Shen

Received: 7 February 2022
Accepted: 27 February 2022
Published: 4 March 2022

1. Introduction

Unlike aboveground engineering, in tunnel engineering, tunnels are built underground. Tunnel design and construction are closely related to the stratum where it is buried. The quality evaluation of the surrounding rock is the cornerstone of tunnel design [1–3] and construction. Surrounding rock classification is a main evaluation method for the surrounding rock quality. It usually collects the information of surrounding rock by one or more means, and finally gives a comprehensive evaluation index based on specific rule. It can reflect the strength characteristics and deformation characteristics of the surrounding rock and stability characteristics of the tunnel face, and can be directly used to guide the tunnel design and construction. The classification of surrounding rock is a common method for the surrounding rock quality evaluation of tunnels in various countries. This method was started in Europe in 1774. After more than two hundred years of development, there have been hundreds of methods employed for this purpose, such as the Q-value method [4], the rock mass rating (RMR) method [5], and the surrounding rock basic quality index (BQ) method [6].

At present, the methods used to classify the rock surrounding tunnels are mainly qualitative, with quantitative verification using laboratory tests requiring extensive professional knowledge and engineering experience. Therefore, there is some randomness in the results. With the development of information technology, artificial intelligence, the Internet of Things, big data, and other technologies, the trend of tunnel construction automation and unmanned is becoming more and more obvious, and the intelligent classification of surrounding rock is the most important part. Thus, an automatic, rapid and accurate intelligent classification technology of surrounding rock is required for the construction of tunnel with high quality and high efficiency.

Machine learning is an important method of artificial intelligence research [7–9] that has been applied in many types of tunnel analyses, such as deformation prediction [10–14], prediction of energy consumption of cutter head drives [15], rock burst prediction [16], reliability analysis [17,18], stability analysis [19], optimization of blasting parameters [20], support pattern selection [21], the prediction of blast-induced ground vibrations [22], tunneling risk prediction and assessment [23,24], diagnosing tunnel collapse sections [25], and TBM tunneling construction and management [26–30]. Machine learning is also an important method in the intelligent classification of surrounding rock. The physical and mechanical parameters of rock mass have been applied to the RMR value prediction using a neural network [31,32]. These parameters include the bulk density, compressive strength, ingress of water, rock quality designation (RQD), average distance between leak, and seismic velocity. The prediction of RMR has also been realized by using a neuro-fuzzy inference system based on the uniaxial compressive strength, RQD, joint or discontinuity spacing, joint condition, and groundwater condition [33]. In addition, the geophysical parameters, such as the seismic velocity and resistivity [34,35], have been used to classify the surrounding rocks [36–38].

The machine learning algorithms used in these studies include a variety of single basic algorithms, optimization algorithms, and integrated algorithms. These previous studies all show that the appropriate machine learning algorithms have excellent performance in different fields when there is an internal connection between input index and output index and the number of samples is sufficient. Especially for the highly nonlinear problems, the machine learning method often has better performance and higher computational efficiency than traditional statistical analysis methods. More importantly, the machine learning methods have the intelligent characteristics of automatic analysis and continuous learning, which provides effective help for this study.

However, such methods generally require manual field testing of classification indices, such as rock strength and rock mass integrity, followed by manual input into the system. None of these parameters applied to the intelligent classification of surrounding rocks can achieve real-time automatic collection in the tunneling process.

The emergence of measure-while-drilling (MWD) technology provides a good solution to this problem. The correlation between the drilling parameters and the surrounding rock quality parameters was studied and explored by scholars as early as the 1960s and 1970s [39–43]. Using field experiments with statistical analysis, the correlations between drilling parameters and surrounding rock quality parameters, such as the uniaxial compressive strength [44–46], shear strength [47], Schmidt rebound hardness [48], cutting performance (Kerf angle d and specific energy) [49,50], RQD [51], and zones of volcanic weathering and decomposition grades [52] have been studied. In recent years, drilling parameters have been used for surrounding rock classification based on the Q method [53] or RMR method [54].

Although these previous studies have confirmed the existence of a relationship between drilling parameters and the quality of surrounding rock, the correlation based on current research of the drilling parameters and the surrounding rock quality parameters mostly refers to a certain lithology. Furthermore, the samples do not cover common rock lithology and are not universal. This is mainly because the use of an intelligent drill jumbo (which refers to a drill jumbo that can automatically collect and transmit drilling parameters) for tunnel construction is low, which makes it more difficult to collect sufficient drilling parameter samples for the surrounding rock classification. Thus, the classification method with more objective, intelligent, and efficient evaluation requires further study.

The purpose of this research is to introduce 10 machine learning algorithms to predict the quality of surrounding rock using MWD data (drilling parameters) obtained from five tunnels of the Zhengzhou–Wanzhou line of the high-speed railway project in China. Through comparative analysis, three machine learning models with better comprehensive performance among them were selected to establish the tunnel surrounding rock intelligent classification system by the drill and blast method. The results of this study lay a solid foundation for the dynamic design and intelligent construction of tunnels.

2. Materials and Methods

2.1. Proposed Methods and Procedures

In this study, we firstly collected sample data, and then conducted a range of data processing, including sample data cleaning, sample imbalance treatment, sample feature analysis, and sample data splitting. Finally, we trained some intelligent classification models of the surrounding rock, and selected the better of them.

The research flow chart about main procedures and proposed methods of this study is shown in Figure 1.

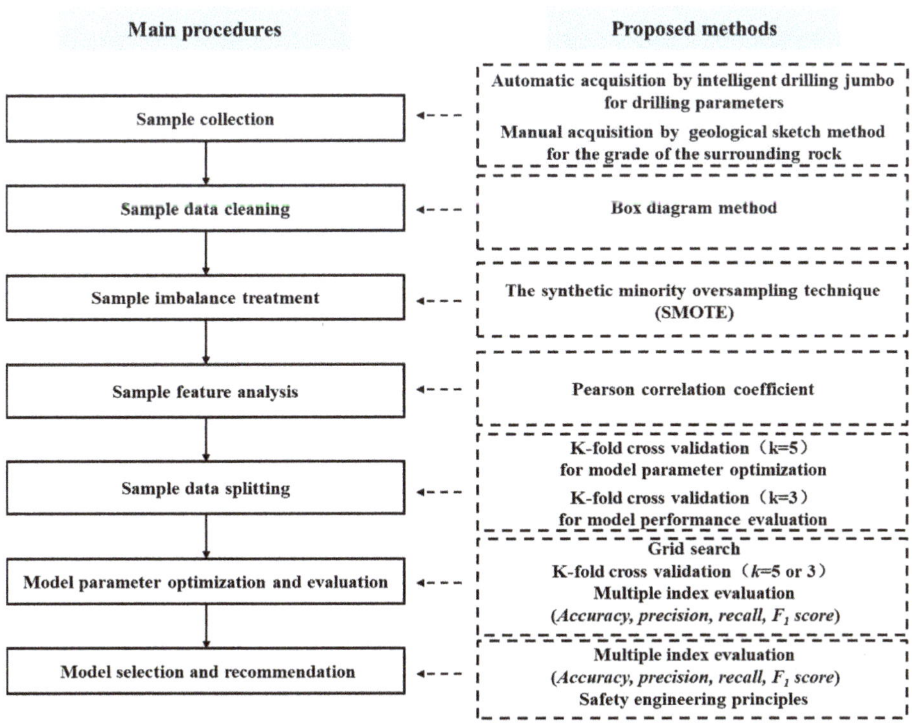

Figure 1. The research flow chart about main procedures and proposed methods.

2.2. Sample Collection

The sample of this study was obtained from some tunnels of the Zhengzhou–Wanzhou line of the high-speed railway project in China. This railway line runs from Zhengzhou East Railway Station to Wanzhou North Railway Station, with a total length of 818 km. It has 18 stations and a designed speed of 350 km/h, which connects Henan province, Hubei province, and Chongqing province. There are 32.5 tunnels in the Hubei province section, with a total length of 167.6 km, and the lithologies of stratum exposed by these tunnels are mainly dolomite, sandstone, limestone, shale, and mudstone. As shown in Figure 2, the sample collection was carried out in five tunnels in the Hubei province section. The New Austrian Tunneling Method was employed in these tunnels.

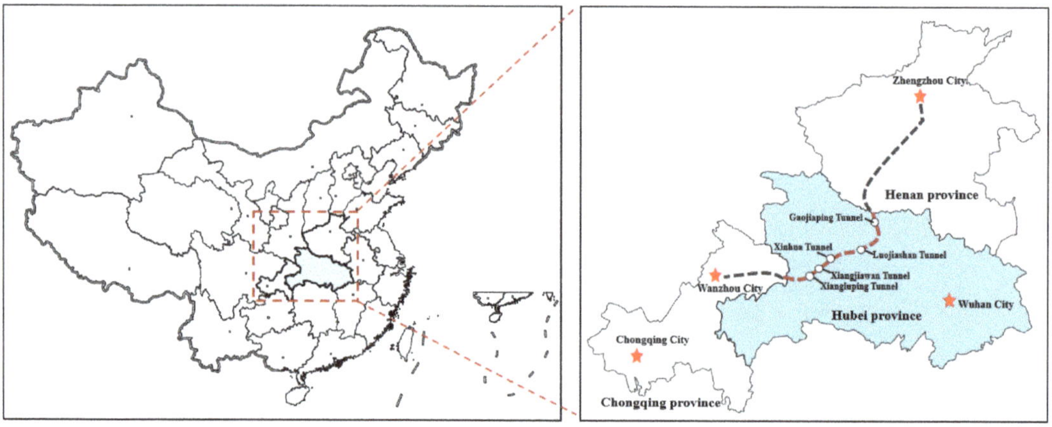

Figure 2. Locations of Zhengzhou–Wanzhou high-speed railway line and the tunnels where the samples were collected.

The sample of intelligent surrounding rock classification is composed of the drilling parameters and the surrounding rock grade of the tunnel face.

In this study, the drilling parameters were collected by using the intelligent drill jumbo (Figure 3), which was made by China Railway Construction Heavy Industry Corporation Limited.

Figure 3. Intelligent drill jumbo (made by China Railway Construction Heavy Industry Corporation Limited).

The intelligent drill jumbo has the functions of automatic positioning, automatic drilling and automatic recording of log. It can complete the drilling operation of pipe roof, anchor bolt and blast hole. In the process, the integrated sensors of the intelligent drill jumbo are used to automatically collect the drilling parameters (penetration velocity, hammer pressure, rotation pressure, and feed pressure) during the drilling of the borehole in the tunnel face.

The drilling parameters are described below:

1. Penetration velocity (V_p, m/min): the rate of penetration of the drill bit through the rock mass.

2. Hammer pressure (P_h, bar): the measurement of the impact pressure of the bit against the rock mass.
3. Rotation pressure (P_r, bar): the pressure of the bit against the rock to maintain the required rotation.
4. Feed pressure (P_f, bar): the hydraulic pressure inside the cylinders required to keep the bit in contact with the bottom of the hole.

All of these parameters are recorded at equal depth intervals of 20 mm by the help of displacement transducer.

The drilling parameters data in this study were collected through the process of blasting hole drilling in the tunnel face (Figure 4a), and the bit used was a cemented carbide bit with a diamond content of 7%, spherical shape, and nine teeth (Figure 4b).

(**a**) (**b**)

Figure 4. Field collection of the drilling parameters. (**a**) Blasting hole drilling in the tunnel face; (**b**) the cemented carbide bit.

The area of the tunnel face is about 150 m^2, containing about 200~300 blasting holes. The value of each drilling parameter for each sample is the average of all the drilling holes in the tunnel face.

The typical layout of the blasting hole is shown in Figure 5.

The grade of the surrounding rock in this study is specified in the current Code for Design of Railway Tunnel (TB10003-2016). According to the degree of hardness, integrity, groundwater state, crustal stress state, and major weak structural surface, the surrounding rocks are divided into Grades I~VI according to the quality.

The approximate correspondence between the surrounding rock grade in this study and the indices of other classification methods, such as the Q method, RMR method, and BQ method, is shown in Table 1 [55].

Table 1. Approximate correspondence between the surrounding rock grade in this study and other indices (Q, RMR, and BQ).

Method	Surrounding Rock Grade				
	I	II	III	IV	V
Q value	(10,000,40)	(40,10)	(10,1)	(1,0.1)	(0.1,0.001)
RMR value	(100,80)	(80,60)	(60,40)	(40,20)	(20,0)
BQ value	(700,550)	(550,450)	(450,350)	(350,250)	(250,0)

Note: Grade-VI generally refers to the special geology, such as the powdery fault fracture zone, aeolian sand, and seriously collapsible loess.

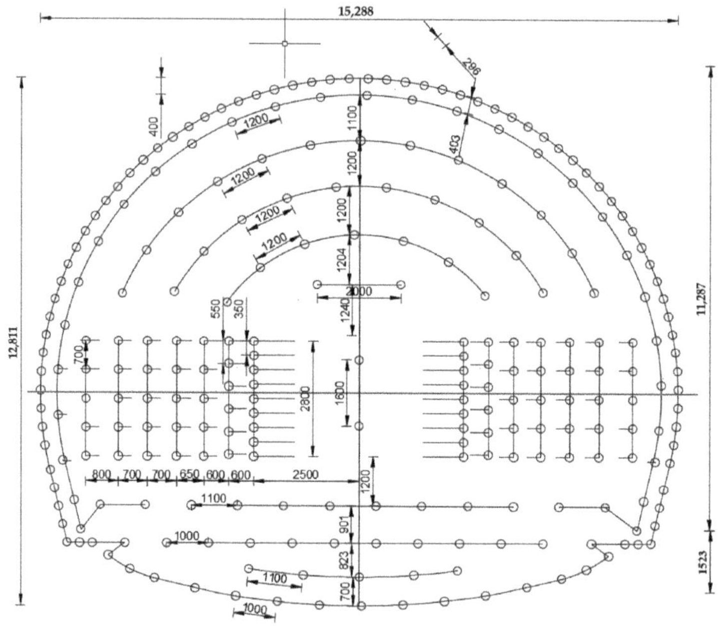

Figure 5. The typical layout of the blasting hole (mm).

The geological sketch method is usually adopted to determine the grade of the surrounding rock in the tunnel, which is analyzed and determined by professional geological engineers. The geological sketch record card of the tunnel face is shown in Table 2.

Table 2. Geological sketch record card of the tunnel face.

The Geological Sketch Record Card of Tunnel Face								
Project name: XXX Date: XXX		Mileage: XXX Construction unit: XXX						
No.	Item	State description						
1	Tunnel face and support type	Width (m)	Area (m^2)	Excavation method	Depth (m)	Advanced support	Primary support	Secondary lining
2	Lithology							
3	Weathering degree	Slightly		Weakly		Strongly		Totally
4	Degree of hardness (Mpa)	Extremely hard (>60)		Hard (30~60)	Relatively soft (15~30)		Soft (5~15)	Extremely soft (<5)
5	Number of structural surfaces	1		2		3		>3
6	Average spacing of the structural surface (m)	>1		0.4~1		0.2~0.4		0~0.2
7	Degree of structural surface development	Not developed		Relatively developed		Developed		Extremely developed
8	Crack width (mm)	0~1		1~3			>3	
9	Crack filling	None	Siliceous	Calcium		Argillaceous	Rock debris	Mud clamps rock debris

Table 2. *Cont.*

The Geological Sketch Record Card of Tunnel Face								
10	Degree of structural surface bonding	Good		Fair		Poor		Extremely poor
11	Integrity	Integrated		Relatively integrated	Relatively broken	Broken		Extremely broken
12	Groundwater state	None		Drip outflow		Linear outflow		Inrush outflow
13	The major weak structural surface	Occurrence				Angle with tunnel axis		
14	Crustal stress state	Low			High		Extremely high	
15	Surrounding rock grade	I	II	III		IV	V	VI
	Tabulator: XXX				Reviewer: XXX			

By the method discussed above, 912 intelligent surrounding rock classification samples were collected in five tunnels of the Zhengzhou–Wanzhou high-speed railway. These samples cover three surrounding rock grades (grade-III, grade-IV, and grade-V) and five main lithologies (dolomite, sandstone, limestone, shale, and mudstone).

More details are shown in Table 3.

Table 3. Intelligent surrounding rock classification sample statistics.

Surrounding Rock Grade	Tunnel	Lithology	Sample Size	Total
III	Luojiashan Tunnel	Dolomite	110	325
	Chufeng Tunnel	Dolomite	28	
	Xinhua Tunnel	Sandstone	31	
	Xiangjiawang Tunnel	Limestone	156	
IV	Gaojiaping Tunnel	Shale/Sandstone	105	420
	Luojiashan Tunnel	Dolomite	75	
	Chufeng Tunnel	Dolomite	44	
	Xinhua Tunnel	Sandstone	125	
	Xiangjiawang Tunnel	Limestone	30	
	Xiangluping Tunnel	Mudstone/Sandstone	41	
V	Gaojiaping Tunnel	Shale	84	167
	Luojiashan Tunnel	Dolomite	62	
	Chufeng Tunnel	Dolomite	4	
	Xinhua Tunnel	Dolomite	15	
	Xiangjiawan Tunnel	Limestone	2	
Total				912

2.3. Sample Data Cleaning

To analyze the sample data outliers, four box diagrams are shown in Figure 6 according to the drilling parameter class and surrounding rock grade.

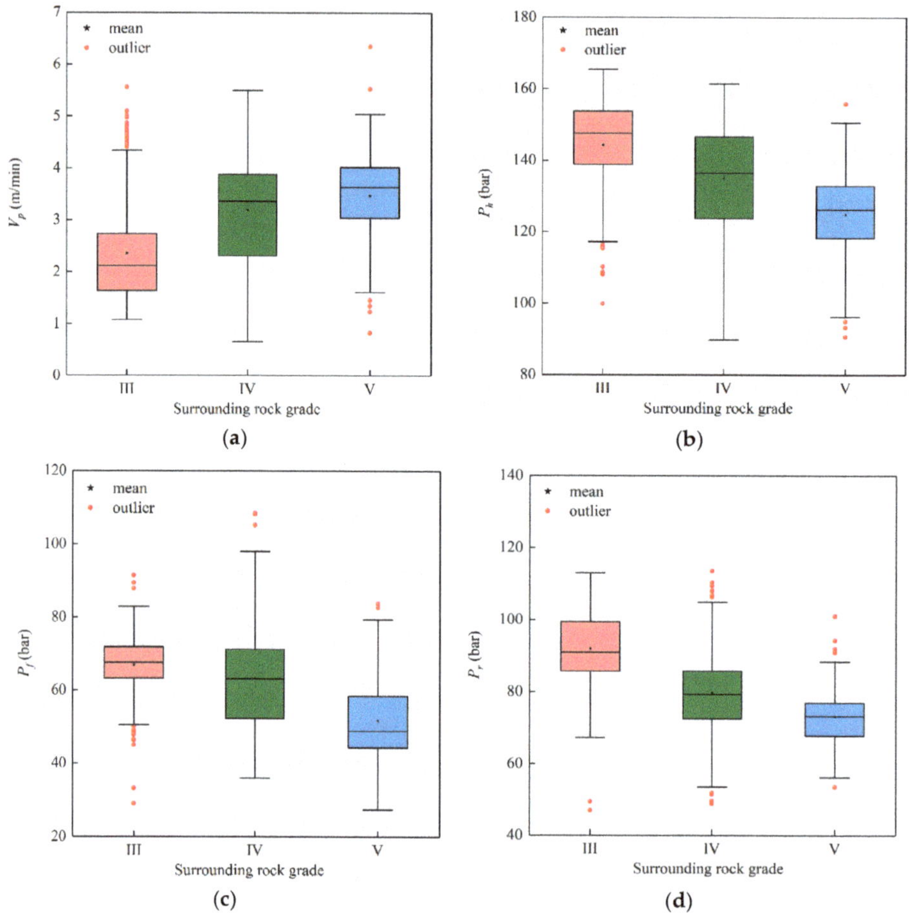

Figure 6. Drilling parameter box diagrams for each surrounding rock grade. (**a**) Penetration velocity (V_p, m/min); (**b**) hammer pressure (P_h, bar); (**c**) feed pressure (P_f, bar); (**d**) rotation pressure (P_r, bar).

According to the box diagram method, when the value is not within the interval represented by the following formula Equation (1), it is regarded as an outlier

$$[\mathrm{Q1} - 1.5 \times \mathrm{IQR}, \mathrm{Q3} + 1.5 \times \mathrm{IQR}] \quad (1)$$

where Q1 is the first quartile, namely the equal to the 25th percentile of all values in the sample from small to large, Q3 is the third quartile, namely the equal to the 75th percentile of all values in the sample from small to large, IQR is the interquartile range, namely IQR is equal to Q3 minus Q1.

In Figure 6, under different surrounding rock grades, all four drilling parameters contained outliers. To eliminate the effects of the outliers, the average value of each parameter was used to replace the outlier of each surrounding rock in this study.

The changes in data characteristics before and after cleaning are compared in Table 4.

Table 4. Comparison of the data characteristics before and after cleaning.

Surrounding Rock Grade	Index	V_p		P_h		P_f		P_r	
		Before	After	Before	After	Before	After	Before	After
III	mean	2.35	2.04	144.32	148.85	66.95	67.69	91.91	92.18
	std	0.97	0.51	12.59	5.98	8.02	5.91	10.29	9.69
	min	1.08	1.08	99.95	132.88	29.05	53.70	46.94	67.27
	max	5.56	3.39	165.41	165.41	91.45	81.44	112.97	112.97
IV	mean	3.19	3.19	135.06	135.06	63.17	62.86	79.68	79.25
	std	1.03	1.03	14.00	14.00	13.21	12.67	11.24	9.59
	min	0.66	0.66	89.79	89.79	35.97	35.97	48.84	55.38
	max	5.49	5.49	161.38	161.38	107.41	97.95	113.53	102.29
V	mean	3.47	3.54	124.96	125.87	51.75	50.72	73.18	72.62
	std	0.85	0.67	11.68	9.85	11.06	9.59	7.59	5.65
	min	0.83	1.97	90.82	101.39	27.43	27.43	53.47	57.70
	max	6.35	5.05	155.88	150.65	83.75	73.60	101.17	86.72

Note: The full name of each index and corresponding abbreviated in the table is the mean value (mean), the standard deviations (std), the minimum value (min) and the maximum value (max).

In Table 4, after cleaning, under different surrounding rock grades, the distributions of these drilling parameters are more centralized. Specifically, the maximum value is smaller, the minimum value is larger, and the standard deviation is smaller.

2.4. Sample Imbalance Treatment

Referring to machine learning classification, when the difference in the number of samples of all the classes is too large, the training model will pay too much attention to the sample characteristics with a greater proportion, so the classification effect of the samples with a smaller proportion is not ideal. This is called the problem of sample imbalance.

These data sets cover three classes of surrounding rock (namely, grade-III, -IV, and -V), and each class contains 325, 420, and 167 samples belonging to the imbalance sample set.

There are three common approaches to deal with unbalanced samples in machine learning:

1. Over-sampling

The over-sampling method achieves sample balance by increasing the number of minority samples in the classification. The most direct method is to simply copy the minority samples to form multiple records. The improved over-sampling methods produce new composite samples by adding random noise, interference data to a few classes, or certain rules such as the synthetic minority over-sampling technique (SMOTE) and adaptive synthetic sampling (ADASYN).

2. Under-sampling

The under-sampling method achieves sample balance by reducing the number of majority samples in the classification. The most direct method is to randomly remove some majority class samples. The disadvantage of this method is that some important information from the majority class samples may be lost.

3. Sample weight adjustment

This method is used to guide models to learn more features of the minority samples in the classification by assigning higher weights to them during the machine learning model training. Generally, the weights of various samples are set to be inversely proportional to the sample size.

Considering the small size of the sample data set, to sufficiently learn the characteristics of all types of samples during the machine learning model training and to improve the universality of the model, the SMOTE over-samples to replenish samples in grade-III and grade-V of the surrounding rock.

The scatter diagram of the data set after over-sampling is shown in Figure 7.

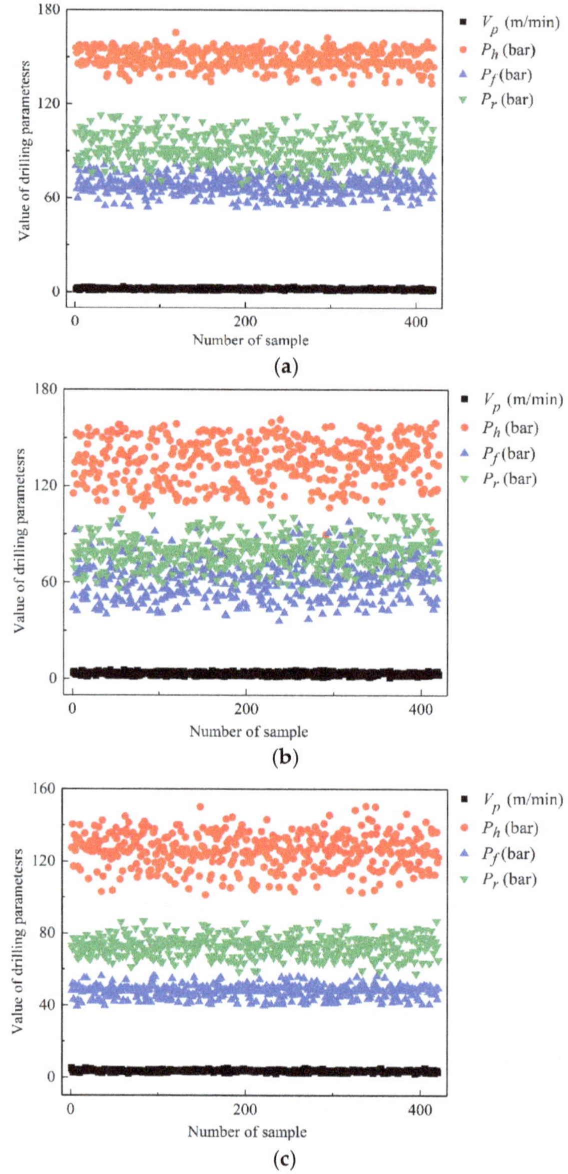

Figure 7. Scatter diagram of the data set after over-sampling (1260). (**a**) Drilling parameters of surround rock grade-III (420); (**b**) drilling parameters of surround rock grade-IV (420); (**c**) drilling parameters of surround rock grade-V (420).

2.5. Sample Feature Analysis

Figure 6 shows that all median, mean, and upper and lower quartile values of each drilling parameter in the box diagram monotonically change with the surrounding rock grade, which preliminarily indicates a correlation between the drilling parameters and the surrounding rock grade.

Pearson correlation coefficient [56] was used to analyze the correlations between the surrounding rock grade and the drilling parameters in 1260 samples, and the correlation coefficients are shown in Table 5.

Table 5. Correlation coefficients between the surrounding rock grade and each drilling parameter.

Index	V_p	P_h	P_f	P_r
R	0.59	−0.64	−0.52	−0.65

In Table 5, all the absolute value of the correlation coefficients between the four drilling parameters and the surrounding rock grades are above 0.5, which is a weak correlation, and the correlation coefficients are close to each other. Therefore, these four indices were selected for the surrounding rock classification.

2.6. Sample Data Splitting

According to the above 1260 data samples, the intelligent surrounding rock classification sample database was established.

In order that the selected parameters can represent the majority of samples, the majority samples are selected for training and the minority samples for prediction in the parameter optimization stage (validation process). As to the model evaluation stage (testing process), we selected more samples for prediction than the previous stage so that the performance evaluation of the model is more convincingly.

Thus, the sample database after random sequencing was divided into the training set and prediction set using the five-fold cross verification method in the validation process, and as to the testing process, the ratio is three-fold.

The specific distribution is shown in Table 6.

Table 6. Intelligent surrounding rock classification sample database distribution.

Item		Surrounding Rock			
		III	IV	V	Total
Validation process	Training set	336	336	336	1008
	Prediction set (Validation set)	84	84	84	252
Testing process	Training set	280	280	280	840
	Prediction set (Testing set)	140	140	140	420

In the field of machine learning, different evaluation indexes (that is, different features in feature vectors are described as different evaluation indexes) often have different dimensional and dimensional units, which will affect the results of data analysis. In order to eliminate the dimensional influence between indexes, data normalization is required.

Normalization means that the input data are limited to a certain range, and this time the data are limited 0 to 1. The data are normalized using the min–max normalization method, as defined in Equation (2).

$$x' = \frac{x - x_{min}}{x_{max} - x_{min}} \tag{2}$$

where x is the value of the original data, x_{max} is the max value of the original data, x_{min} is the min value of the original data, x' is the normalized value.

The normalization needs to be carried out in the training set in Table 6 first, and then the normalization of the prediction set is completed by taking the normalization parameters (x_{max} and x_{min}) of the training set.

2.7. Model Parameter Optimization and Evaluation

Based on the python machine learning toolkit (Scikit-learn), 10 types of supervised machine learning algorithms were used to train the intelligent surrounding rock classification model: the support vector machine (SVM), back propagation neural network (BP) [55,56], radial basis function neural network (RBF), K-nearest neighbor (KNN), Gaussian naive Bayes (GNB), decision tree (DT), random forest (RF), extra trees (ET), bootstrap aggregating (Bagging) and gradient boosting (GB) algorithms.

The model parameters were optimized and evaluated by the grid search K-flod cross validation method ($k = 5$), and the model performance was externally unbiased when using the method of K-flod cross validation ($k = 3$).

In the model performance optimization process, the average accuracy was adopted to determine the optimum hyperparameter combinations. The average accuracy was calculated by first averaging the accuracy of each validation set over the three surrounding rock grades, and then averaging the averages over the five validation sets.

In the final evaluation of the model performance, the precision, recall, F_1 score, and accuracy were adopted for a comprehensive evaluation. And these indexes are the average of the results of the three test sets.

The process is shown in Figure 8.

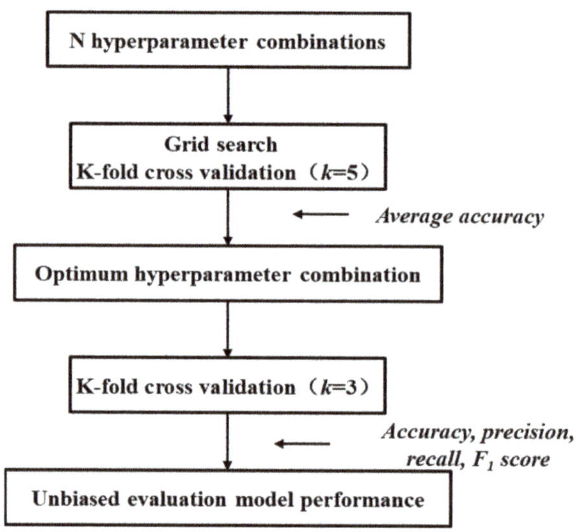

Figure 8. Model parameter optimization and evaluation process.

The performance of the models after training is shown in Figure 9 and Table 7.

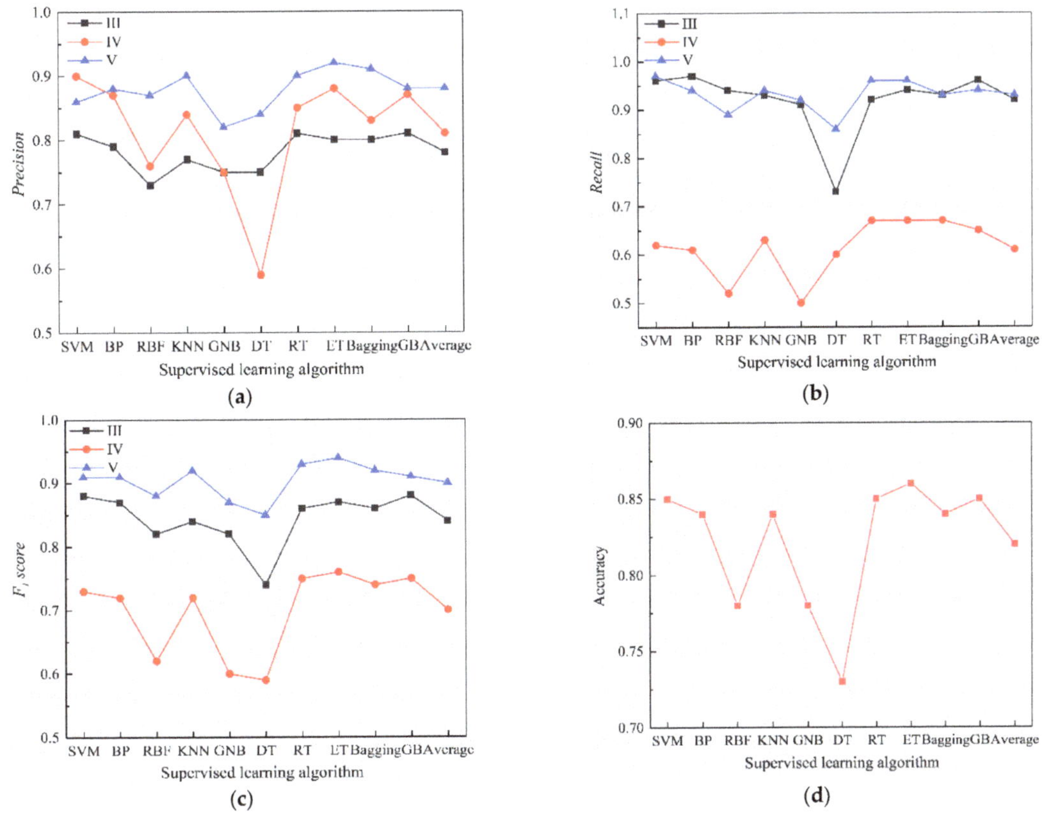

Figure 9. Model parameter optimization and evaluation process: (**a**) precision; (**b**) recall; (**c**) F_1; (**d**) accuracy.

Table 7. Performance of the intelligent surrounding rock models.

Performance Index	Rock Grade	Supervised Learning Algorithm										Average
		SVM	**BP**	**RBF**	**KNN**	**GNB**	**DT**	**RT**	**ET**	**Bag-ging**	**GB**	
Precision	III	0.81	0.79	0.73	0.77	0.75	0.75	0.81	0.80	0.80	0.81	0.78
	IV	0.90	0.87	0.76	0.84	0.75	0.59	0.85	0.88	0.83	0.87	0.81
	V	0.86	0.88	0.87	0.90	0.82	0.84	0.90	0.92	0.91	0.88	0.88
Recall	III	0.96	0.97	0.94	0.93	0.91	0.73	0.92	0.94	0.93	0.96	0.92
	IV	0.62	0.61	0.52	0.63	0.50	0.60	0.67	0.67	0.67	0.65	0.61
	V	0.97	0.94	0.89	0.94	0.92	0.86	0.96	0.96	0.93	0.94	0.93
F_1 score	III	0.88	0.87	0.82	0.84	0.82	0.74	0.86	0.87	0.86	0.88	0.84
	IV	0.73	0.72	0.62	0.72	0.60	0.59	0.75	0.76	0.74	0.75	0.70
	V	0.91	0.91	0.88	0.92	0.87	0.85	0.93	0.94	0.92	0.91	0.90
Accuracy	-	0.85	0.84	0.78	0.84	0.78	0.73	0.85	0.86	0.84	0.85	0.82

The following conclusions can be drawn from Figure 9 and Table 7.

Among 10 machine learning algorithm models, the average precision, recall, F_1 score and accuracy were above 0.7, except for the recall of grade-IV surrounding rock. In particular, the average recalls of grade-III and -V were greater than 0.90.

Thus, these machine learning models established by the drilling parameters are feasible and reliable in the intelligent classification of surrounding rocks.

2.8. Model Selection and Recommendation

When comprehensively considering the precision, recall, F_1 score, and accuracy, three types of models (SVM, RT, and ET) had better performance among the 10 machine learning algorithm models. Their average precision, recall, F_1 score, and accuracy were above 0.8, except for the recall of grade-IV surrounding rock. In particular, the recall of the grade-V surrounding rock, which has a great influence on the safety of the tunnel, was above 0.95.

Thus, these machine learning models based on SVM, RT, and ET have some degree of safety and high practical value.

3. Case Study

3.1. Intelligent Surrounding Rock Classification System Software

The tunnel surrounding rock intelligent classification system by the drill and blast method was established based on the intelligent surrounding rock classification models based on SVM, RT, and ET. It can automatically record and transmit the drilling parameters and intelligently classify the surrounding rock with carriers of the intelligent drill jumbo.

Specifically, the surrounding rock grade of the tunnel face is determined by the votes of three models (SVM, RT, and ET). When all types of the surrounding rock levels are not superior, the higher grade is taken as the surrounding rock grade of the tunnel face in consideration of safety.

In addition, these samples for the models do not require consideration of the modification conditions (i.e., low crustal stress, without groundwater or drip outflow of groundwater, and without a main weak structure surface). Therefore, in consideration of safety and popularization, the grade determined from these models is the basic surrounding rock grade; the grade amended by the groundwater state, major weak structural surface, and crustal stress state is the final surrounding rock grade.

The specific process is shown in Figure 10.

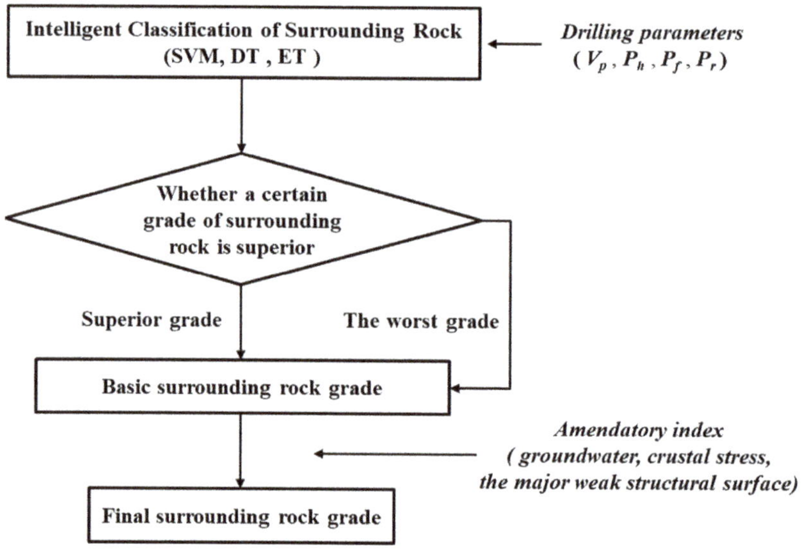

Figure 10. Classification process of the intelligent surrounding rock classification system.

The system interface is shown in Figure 11.

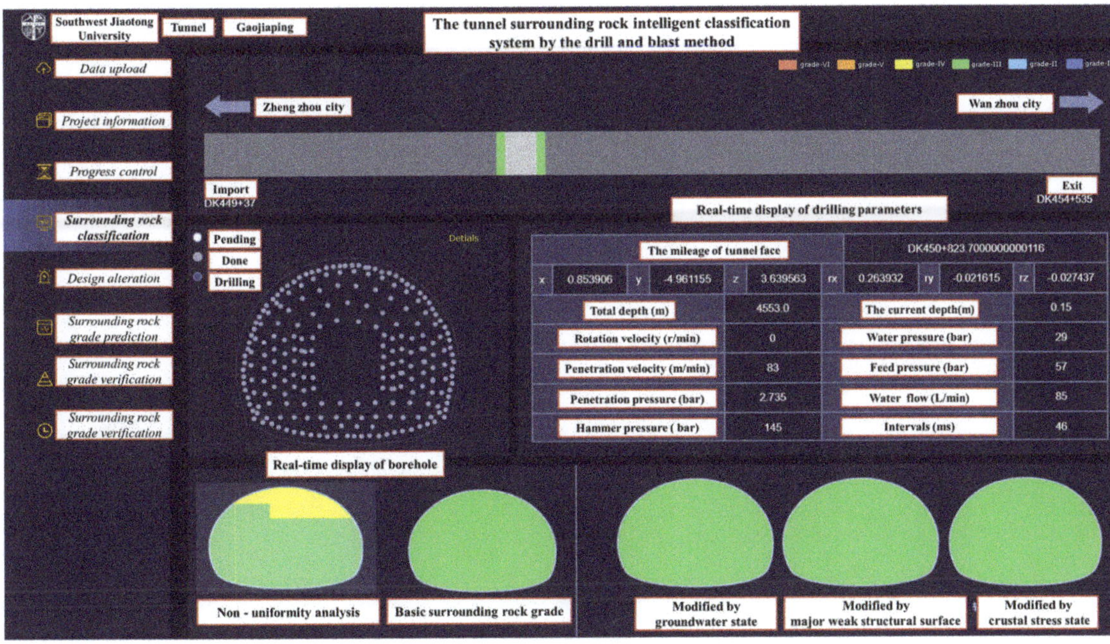

Figure 11. Interface of the 'tunnel surrounding rock intelligent classification system by the drill and blast method'.

3.2. Project Overview of Test Tunnel

The Gaojiaping Tunnel of the Zhengzhou–Wanzhou high-speed railway is located in Nanzhang county, Xiangyang city, Hubei province, China. The tunnel is 5498 m in length with a maximum buried depth of 320 m. The main lithologies of the surrounding rock are limestone and shale, and the grades of the surrounding rocks are III (1493 m), IV (1050 m), and V (720 m).

The method of construction of this tunnel is the full-section method with large machinery, i.e., the tunnel face (150 m^2) is excavated and formed by blasting once, and the primary support is closed once. With this method, the circular footage is 2–4.8 m, and the monthly progress is 60–150 m in this tunnel.

The tunnel position is shown in Figure 12.

This test section is DK450 + 834~DK451 + 126, which is 292 m in total length. The excavation revealed that the surrounding rock was grade-III; the lithology was bluish grey, a massive structure, and hard limestone rock. The stability of the tunnel face was good without developed joints, fractures, or groundwater.

The test section of this tunnel is shown in Figure 13, and the tunnel face is shown in Figure 14.

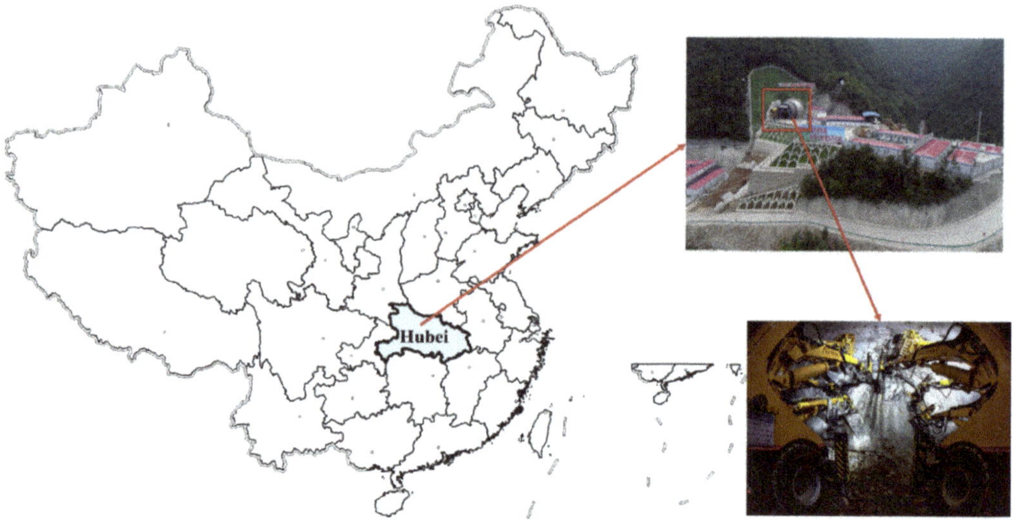

Figure 12. Interface of the position of the Gaojiaping Tunnel of the Zhengzhou–Wanzhou high-speed railway.

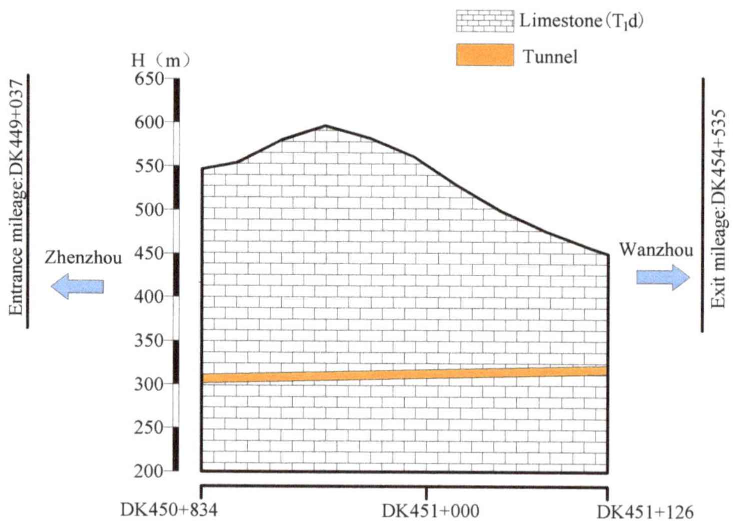

Figure 13. Test section of Gaojiaping Tunnel.

Figure 14. Tunnel face of this test section of the Gaojiaping Tunnel.

3.3. Intelligent Surrounding Rock Classification in Test Tunnel

Among the test section, drilling parameter information from 30 tunnel faces was collected by the intelligent surrounding rock classification system. The intelligent classification results of the surrounding rock on each tunnel face are shown in Table 8.

Table 8. Intelligent classification of surrounding rock on each tunnel face.

Number	Mileage	Predicted Results	Actual Results	Right/Error
1	DK450+834	III	III	Right
2	DK450+837	III	III	Right
3	DK450+840	III	III	Right
4	DK450+916	III	III	Right
5	DK450+919	III	III	Right
6	DK450+922	III	III	Right
7	DK450+926	III	III	Right
8	DK451+014	III	III	Right
9	DK451+017	III	III	Right
10	DK451+021	III	III	Right
11	DK451+024	III	III	Right
12	DK451+027	III	III	Right
13	DK451+030	III	III	Right
14	DK451+034	III	III	Right
15	DK451+037	III	III	Right
16	DK451+042	III	III	Right
17	DK451+046	III	III	Right
18	DK451+051	III	III	Right
19	DK451+055	III	III	Right
20	DK451+060	III	III	Right
21	DK451+064	III	III	Right
22	DK451+100	III	III	Right
23	DK451+103	III	III	Right
24	DK451+106	III	III	Right
25	DK451+109	III	III	Right
26	DK451+113	IV	III	Error
27	DK451+116	IV	III	Error
28	DK451+119	IV	III	Error
29	DK451+123	IV	III	Error
30	DK451+126	III	III	Right
		Accuracy		86.7%

In Table 8, the classification accuracy of the surrounding rock intelligent classification system on-site is 86.7%, which indicates that the system has good generalization.

However, due to the single lithology and surrounding rocks (limestone, grade-III) in the field test section, its generalization performance must be further verified.

4. Conclusions

Based on 912 drilling parameters of the Zhengzhou–Wanzhou high-speed railway tunnel project, 10 intelligent surrounding rock classification models were established using multiple machine learning algorithms. With three models (SVM, RT, and ET), an intelligent surrounding rock classification system was established and verified by the field test section.

The main conclusions of this study are:

1. All the absolute values of the correlation coefficients between the four drilling parameters (V_p, P_h, P_r, and P_f) and the surrounding rock grades (III, IV, and V) are above 0.5.

Thus, there is a correlation between the drilling parameters and the surrounding rock grade, so the method of surrounding rock classification by drilling parameters is feasible.

2. The 10 models based on supervised machine learning algorithms all have good performance. The average accuracy of them is 0.82. In particular, the average recalls of grades-III and -V were greater than 0.90.

Thus, these machine learning models established by drilling parameters are feasible and reliable in the intelligent classification of surrounding rocks.

3. By comprehensively considering precision, recall, F_1 score, and accuracy, we observe that three types of models containing SVM, RT, and ET have better performance among 10 machine learning algorithm models. In particular, the recall of grade-V surrounding rock, which has great influence on the safety of the tunnel, was above 0.95.

Thus, these three machine learning models have a degree of safety and high practical value.

4. The classification accuracy of the surrounding rock intelligent classification system on-site is 86.7%, which indicates that these models have good generalization.

This study found that the drilling parameters V_p, P_h, P_f, and P_r can be used to grade the surrounding rock directly by training the machine learning model. Compared with the traditional manual method, the proposed method is faster and has fewer indices, higher classification accuracy, and better stability. When carried on an intelligent jumbo, it can realize automatic recording and transmission of drilling parameters and intelligent classification of surrounding rock grade by using the intelligent surrounding rock classification system established in this study. This is what most previous studies have failed to do.

However, there is still room to improve the classification accuracy of this system. In this study, three models have been used to determine the most unfavorable results to ensure the safety of site construction, and its essence is a management means. Therefore, the research on data preprocessing and algorithm optimization should be strengthened to further improve the reliability of the technology. In addition, the samples collected in this study only covered five lithologies and three surrounding rock grades. Thus, the sample types and numbers should be further expanded in the future to improve the application range of the technique.

Author Contributions: Conceptualization, S.Z. and J.T.; methodology, S.Z. and J.T.; software, W.Y.; data curation, D.Y.; writing—original draft preparation, S.Z. and J.T.; writing—review and editing, M.W. and J.T.; visualization, D.Y.; funding acquisition, M.W. All authors have read and agreed to the published version of the manuscript.

Funding: The authors are grateful for the support from the National Natural Science Foundation of China (51878567, 51878568) and China Railway Corporation Science and Technology Research and Development Program (K2018G014, K2020G035).

Conflicts of Interest: The authors declare no conflict of interest.

References

1. Zhang, X.; Wang, M.; Li, J.; Wang, Z.; Tong, J.; Liu, D. Safety factor analysis of a tunnel face with an unsupported span in cohesive-frictional soils. *Comput. Geosci.* **2020**, *117*, 103–221. [CrossRef]
2. Dla, D.; Yl, B.; Jc, C.; Hm, A. An analytical model for axial performance of rock bolts under constant confining pressure based on continuously yielding criterion. *Tunn. Undergr. Space Technol.* **2021**, *113*, 103955.
3. Li, D.; Cai, M.; Masoumi, H. A constitutive model for modified cable bolts exhibiting cone shaped failure mode. *Int. J. Rock Mech. Min.* **2021**, *145*, 104855. [CrossRef]
4. Barton, N.; Lien, R.; Lunde, J. Engineering classification of rock masses for the design of tunnel support. *Rock Mech.* **1974**, *6*, 189–236. [CrossRef]
5. Bieniawski, Z.T. The geomechanics classification in rock engineering applications. In Proceedings of the 4th ISRM Congress, Montreux, Switzerland, 2–8 September 1979.
6. Yunmei, L. Study of BQ formula in national standard of qunatitative classfication for basic quality of rock mass. *Chin. J. Geotech. Eng.* **1999**, *4*, 481–485.
7. Reich, Y. Machine learning techniques for civil engineering problems. *Comput.-Aided Civ. Infrastruct. Eng.* **1997**, *12*, 295–310. [CrossRef]
8. Elshaboury, N.; Abdelkader, E.M.; Al-Sakkaf, A.; Alfalah, G. Teaching-Learning-Based Optimization of Neural Networks for Water Supply Pipe Condition Prediction. *Water* **2021**, *13*, 3546. [CrossRef]
9. Elshaboury, N.; Mohammed Abdelkader, E.; Al-Sakkaf, A.; Alfalah, G. Predictive Analysis of Municipal Solid Waste Generation Using an Optimized Neural Network Model. *Processes* **2021**, *9*, 2045. [CrossRef]
10. Xue, X.; Xiao, M. Deformation evaluation on surrounding rocks of underground caverns based on PSO-LSSVM. *Tunn. Undergr. Space Technol.* **2017**, *69*, 171–181. [CrossRef]
11. Mahdevari, S.; Torabi, S.R. Prediction of tunnel convergence using artificial neural networks. *Tunn. Undergr. Space Technol.* **2012**, *28*, 218–228. [CrossRef]
12. Torabi-Kaveh, M.; Sarshari, B. Predicting Convergence Rate of Namaklan Twin Tunnels Using Machine Learning Methods. *Arab. J. Sci. Eng.* **2019**, *45*, 3761–3780. [CrossRef]
13. Qi, C.; Fourie, A.; Zhao, X. Back-Analysis Method for Stope Displacements Using Gradient-Boosted Regression Tree and Firefly Algorithm. *J. Comput. Civil. Eng.* **2018**, *32*, 04018031. [CrossRef]
14. Zhang, L.; Wu, X.; Ji, W.; Abourizk, S.M. Intelligent Approach to Estimation of Tunnel-Induced Ground Settlement Using Wavelet Packet and Support Vector Machines. *J. Comput. Civil. Eng.* **2016**, *31*, 04016053. [CrossRef]
15. Zhou, C.; Ding, L.; Zhou, Y.; Zhang, H.; Skibniewski, M.J. Hybrid Support Vector Machine Optimization Model for Prediction of Energy Consumption of Cutter Head Drives in Shield Tunneling. *J. Comput. Civil. Eng.* **2019**, *33*, 04019019. [CrossRef]
16. Xue, Y.; Bai, C.; Qiu, D.; Kong, F.; Li, Z. Predicting rockburst with database using particle swarm optimization and extreme learning machine. *Tunn. Undergr. Space Technol.* **2020**, *98*, 103287. [CrossRef]
17. Zhao, H.; Ru, Z.; Chang, X.; Yin, S.; Li, S. Reliability analysis of tunnel using least square support vector machine. *Tunn. Undergr. Space Technol.* **2014**, *41*, 14–23. [CrossRef]
18. Liu, W.; Cai, L.; Chen, J.; Wang, Y.; Wu, H. Reliability Analysis of Operational Metro Tunnel Based on a Dynamic Bayesian Copula Model. *J. Comput. Civil. Eng.* **2020**, *34*, 05020002. [CrossRef]
19. Leu, S.-S.; Chen, C.-N.; Chang, S.-L. Data mining for tunnel support stability: Neural network approach. *Automat. Constr.* **2001**, *10*, 429–441. [CrossRef]
20. Liu, K.; Liu, B. Optimization of smooth blasting parameters for mountain tunnel construction with specified control indices based on a GA and ISVR coupling algorithm. *Tunn. Undergr. Space Technol.* **2017**, *70*, 363–374. [CrossRef]
21. Liu, J.; Jiang, Y.; Ishizu, S.; Sakaguchi, O. Estimation of tunnel support pattern selection using artificial neural network. *Arab. J. Geosci.* **2020**, *13*, 1–12. [CrossRef]
22. Rana, A.; Bhagat, N.; Jadaun, G.; Rukhaiyar, S.; Pain, A.; Singh, P. Predicting Blast-Induced Ground Vibrations in Some Indian Tunnels: A Comparison of Decision Tree, Artificial Neural Network and Multivariate Regression Methods. *Mining. Metall. Explor.* **2020**, *37*, 1039–1053. [CrossRef]
23. Yazdani-Chamzini, A. Proposing a new methodology based on fuzzy logic for tunnelling risk assessment. *J. Civ. Eng. Manag.* **2014**, *20*, 82–94. [CrossRef]
24. Liu, Y.; Chen, H.; Zhang, L.; Wang, X. Risk prediction and diagnosis of water seepage in operational shield tunnels based on random forest. *J. Civ. Eng. Manag.* **2021**, *27*, 539–552. [CrossRef]
25. Chen, Z.; Zhang, Y.; Li, J.; Li, X.; Jing, L. Diagnosing tunnel collapse sections based on TBM tunneling big data and deep learning: A case study on the Yinsong Project, China. *Tunn. Undergr. Space Technol.* **2020**, *108*, 103700. [CrossRef]
26. Wen, Z.; Rong, X.; Wang, Z.; Han, S.; Shi, Y. A Fast Estimation Method of Soil Discharged by an Earth Pressure Balanced Shield Machine. *KSCE J. Civ. Eng.* **2021**, *25*, 2239–2249. [CrossRef]
27. Hu, M.M.; Li, B.; Zhang, B.; Wang, R.R.; Chen, L. Improved SVR Method for Predicting the Cutting Force of a TBM Cutter Using Linear Cutting Machine Test Data. *KSCE J. Civ. Eng.* **2021**, *25*, 4425–4442. [CrossRef]

28. Jung, J.H.; Chung, H.; Kwon, Y.S.; Lee, I.M. An ANN to Predict Ground Condition ahead of Tunnel Face using TBM Operational Data. *KSCE J. Civ. Eng.* **2019**, *23*, 3200–3206. [CrossRef]
29. Lee, S.S. Prediction of Subsidence during TBM Operation in Mixed-Face Ground Conditions from Realtime Monitoring Data. *Appl. Sci.* **2021**, *11*, 12130. [CrossRef]
30. Xu, H.; Zhou, J.; Asteris, P.G.; Armaghani, D.J.; Tahir, M.M. Supervised Machine Learning Techniques to the Prediction of Tunnel Boring Machine Penetration Rate. *Appl. Sci.* **2019**, *9*, 3715. [CrossRef]
31. Zafirovski, Z.; Jovanovski, M.; Moslavac, D.; Krakutovski, Z. *Application of Neural Networks in Analyzing of Rock Mass Parameters in Tunneling*; ResearchGate: Split, Croatia, 2014.
32. You, K. A case study on the utilization of tunnel face mapping data for a back analysis based on artificial neural network. *KSCE J. Civ. Eng.* **2014**, *18*, 751–759. [CrossRef]
33. Jalalifar, H.; Mojedifar, S.; Sahebi, A.; Nezamabadi-Pour, H. Application of the adaptive neuro-fuzzy inference system for prediction of a rock engineering classification system. *Comput. Geosci.* **2011**, *38*, 783–790. [CrossRef]
34. Hasegawa, N.; Hasegawa, S.; Kitaoka, T.; Ohtsu, H. Applicability of Neural Network in Rock Classification of Mountain Tunnel. *Mater. Trans.* **2019**, *60*, 758–764. [CrossRef]
35. Lee, H.L.; Kim, J.S.; Hong, C.H.; Cho, D.K. Ensemble Learning Approach for the Prediction of Quantitative Rock Damage Using Various Acoustic Emission Parameters. *Appl. Sci.* **2021**, *11*, 4008. [CrossRef]
36. Han, M.-S.; Yang, I.-J.; Kim, K.-M. A prediction of the rock mass rating of tunnelling area using artificial neural networks. *J. Korean Tunn. Undergr. Space Assoc.* **2002**, *4*, 277–286.
37. Bressan, T.S.; de Souza, M.K.; Girelli, T.J.; Junior, F. Evaluation of machine learning methods for lithology classification using geophysical data. *Comput. Geosci.* **2020**, *139*, 104475. [CrossRef]
38. Sun, D.; Lonbani, M.; Askarian, B.; Armaghani, D.J.; Huynh, V.V. Investigating the Applications of Machine Learning Techniques to Predict the Rock Brittleness Index. *Appl. Sci.* **2020**, *10*, 1691. [CrossRef]
39. Schimazek, J.; Knatz, H. Der Einfluß des Gesteinsaufbaus auf die Schnittgeschwindigkeit und den Meißelverschleiß von Streckenvortriebsmaschinen. *Glückauf* **1970**, *106*, 274–278.
40. Schimazek, J.; Knatz, H. Die beurteilung der bearbeitbarkeit von gesteinen durch schneid-und rollenbohrwerkzeuge. *Erzmetall* **1976**, *29*, 113–119.
41. Gehring, K. Möglichkeiten zur Beurteilung des Arbeitsverhaltens von Werkzeuge zur schneidenden Gesteinsbearbeitung. *Berg-Und Hüttenmännische Mon.* **1973**, *118*, 319–327.
42. Becker, H.; Lemmes, F. Gesteinsphysikalische Untersuchungen im Streckenvortrieb. *Tunnel* **1984**, *2*, 71–76.
43. Thuro, K. *Bohrbarkeit Beim Konventionellen Sprengvortrieb: Geologisch-Felsmechanische Untersuchungen Anhand Sieben Ausgewählter Tunnelprojekte;* Inst. für Allg. u. Angewandte Geologie d. Ludwig-Maximilians-Univ.: Berlin, Germany, 1996.
44. Torno, S.; Torao, J.; Menéndez, M.; Gent, M.; Velasco, J. Mathematical and fuzzy logic models in prediction of geological and geomechanical properties of rock mass by excavation data on underground works. *J. Civ. Eng. Manag.* **2011**, *17*, 197–206. [CrossRef]
45. Mostofi, M.; Rasouli, V.; Mawuli, E. An estimation of rock strength using a drilling performance model: A case study in blacktip field, Australia. *Rock Mech. Rock Eng.* **2011**, *44*, 305. [CrossRef]
46. Naville, C.; Serbutoviez, S.; Throo, A.; Vincké, O.; Cecconi, F.J.O. Seismic while drilling (SWD) techniques with downhole measurements, introduced by IFP and its partners in 1990–2000. *Oil Gas Sci. Technol.* **2004**, *59*, 371–403. [CrossRef]
47. Kalantari, S.; Hashemolhosseini, H.; Baghbanan, A. Estimating rock strength parameters using drilling data. *Int. J. Rock Mech. Min.* **2018**, *104*, 45–52. [CrossRef]
48. Mehmet, C.; Osman, Y.A.; Serdar, Y. Relationships between the drilling rate index and physicomechanical rock properties. *Bull. Eng. Geol. Environ.* **2017**, *76*, 253–261.
49. Aydin, G.; Karakurt, I.; Hamzacebi, C. Performance Prediction of Diamond Sawblades Using Artificial Neural Network and Regression Analysis. *Arab. J. Sci. Eng.* **2015**, *40*, 2003–2012. [CrossRef]
50. Aydin, G.; Karakurt, I.; Hamzacebi, C. Artificial neural network and regression models for performance prediction of abrasive waterjet in rock cutting. *Int. J. Adv. Manuf. Technol.* **2014**, *75*, 1321–1330. [CrossRef]
51. He, M.; Li, N.; Yao, X.; Chen, Y. A New Method for Prediction of Rock Quality Designation in Borehole Using Energy of Rotary Drilling. *Rock Mech. Rock Eng.* **2020**, *53*, 3383–3394. [CrossRef]
52. Yue, Z.Q.; Lee, C.F.; Law, K.T.; Tham, L.G. Automatic monitoring of rotary-percussive drilling for ground characterization—Illustrated by a case example in Hong Kong. *Int. J. Rock Mech. Min.* **2004**, *41*, 573–612. [CrossRef]
53. Eldert, J.V.; Schunnesson, H.; Johansson, D.; Saiang, D. Application of measurement while drilling technology to predict rock mass quality and rock support for tunnelling. *Rock Mech. Rock Eng.* **2019**, *53*, 1349–1358. [CrossRef]
54. Galende-Hernández, M.; Menéndez, M.; Fuente, M.; Sainz-Palmero, G. Monitor-While-Drilling-based estimation of rock mass rating with computational intelligence: The case of tunnel excavation front. *Automat. Constr.* **2018**, *93*, 325–338. [CrossRef]
55. Chen, L.; Chen, S.; Tu, P.; He, T. Study on Mutual Relationships between Surrounding Rock Classifications by Q Value, RMR and BQ Method for Underground Cavern. *Subgrade Eng.* **2017**, *6*, 107–112. (In Chinese)
56. Aydin, G.; Karakurt, I.; Aydiner, K. Investigation of the surface roughness of rocks sawn by diamond sawblades. *Int. J. Rock Mech. Min.* **2013**, *61*, 171–182. [CrossRef]

applied sciences

MDPI

Article

Asymmetric Force Effect and Damage Analysis of Unlooped Segment of Large-Diameter Shield under Synchronous Propulsion and Assembly Mode

Zeyu Dai [1], Peinan Li [1,*], Xi Wang [2], Jun Liu [3], Jie Fan [4] and Xiaoyong Kou [4]

1 College of Environmental Science and Engineering, Donghua University, Shanghai 201620, China; 2201977@mail.dhu.edu.cn
2 College of Civil Engineering, Tongji University, Shanghai 200092, China; pnhylw2021@163.com
3 College of Urban Railway Transportation, Shanghai University of Engineering Science, Shanghai 201620, China; junliurs@sues.edu.cn
4 Shanghai Tunnel Engineering Co., Ltd., Shanghai 200032, China; zyhylw2021@163.com (J.F.); kxyyxk@126.com (X.K.)
* Correspondence: lipeinan@dhu.edu.cn

Abstract: This article outlines the merits of the proposed synchronous propulsion and assembly method in the construction of a long-distance tunnel. In traditional shield construction, the stress of the unlooped segment during the assembly process is systematically overlooked. However, in the novel tunnel construction method, the advancing force of the shield directly acts on the unlooped segment, so the safety of the unlooped segment is unknown. In this paper, focusing on the safety of the segment assembly process, the segment interactions and stress concentrations under asymmetric force effects are analysed in detail via a suite of finite element models. The results show that in the synchronous propulsion and assembly mode the segments will rotate inward. A clamping effect will gradually appear during the assembly process, which makes segment deflection decrease and the stress distribution more uniform. Under asymmetrical stress, the damage to longitudinal segments is highly correlated with the types of assembly errors. The damage position that is deflected radially inward will change with the deflection angle, and the outer joint of the segment is the largest. Based on the numerical outputs, guidelines for the application of synchronous propulsion and assembly technology in practical engineering are provided.

Keywords: synchronous propulsion and assembly mode; segment interaction; assembling error; concrete segment; segment damage

Citation: Dai, Z.; Li, P.; Wang, X.; Liu, J.; Fan, J.; Kou, X. Asymmetric Force Effect and Damage Analysis of Unlooped Segment of Large-Diameter Shield under Synchronous Propulsion and Assembly Mode. *Appl. Sci.* **2022**, *12*, 2850. https://doi.org/10.3390/app12062850

Academic Editors: Mingfeng Lei, Chenjie Gong and Xianda Shen

Received: 9 February 2022
Accepted: 7 March 2022
Published: 10 March 2022

1. Introduction

With the development of underground spaces in cities and surrounding areas, the construction of shield tunnels is becoming more and more frequent. However, the development of a regional economy means that the length of the shield tunnel continues to extend. There is an urgent need to change the traditional construction technology to adapt to the rapid growth of tunnel demand. In terms of improving tunnel construction speed, this mainly depends on: (i) improving the shield tunneling speed; and (ii) simultaneous shield propulsion and segment assembly, i.e., synchronous propulsion and assembly technology. In the case of normal shield construction, the shield stops advancing. After the segments are assembled into rings, the hydraulic jack propels the newly assembled assembly ring to provide the support of the tunnel surface and the thrust of the shield. In the synchronous propulsion and assembly mode, the shield does not need to stop, and the jack acts directly on the unlooped segments to propel while assembling. The synchronous propulsion and assembly technology systematically reduces the suspension time during tunnel segment assembly [1]. As we all know, the shield propulsion force is borne by the lining ring during the construction of the traditional shield method, and the impact of shield propulsion on the lining ring does not need to be considered during the assembly process of the segment. The new construction method exposes the segments to propulsion loads before they are looped, which puts the segments at greater risk of damage [2–5]. It is necessary to study the safety of this construction method to ensure the smooth progress of construction.

In fact, the load and working state of shield tunnel segments in the construction stage are complex and uncertain [6]. Focusing on the engineering situation, many scholars have conducted in-depth research on segment damage caused by assembly error in the construction stage. The main research method involves using the integral design models, such as homogeneous ring, elastic hinge ring, and multi-hinge angle ring models. Chen Junsheng studied the concrete cracks detected at different locations [7]. Combined with finite element model analysis, it was shown that the shield machine and segment had an influence in terms of an inclined and uneven jacking load on the assembly quality of the segment. The Japanese Society of Civil Engineers statistically studied the damage frequency of segments in 50 tunnel sites during construction [8]. Tsumoto found that the damage types of segments focused on longitudinal cracks and corner spalling before or during shield propulsion. SHP Cavalaro studied the causes of the above damage. The results show that when a jack is loaded, the longitudinal dislocation of the upper ring can easily cause a longitudinal crack in the next ring segment, and the collision and extrusion of the adjacent segments of the current ring are important causes of corner damage [9].

In addition to the direct relationship between assembly errors in segments and the damage of segments, the excessive vertical loadings caused by jacks during shield propulsion are also an important reason for segment cracking (prefabricated structures), spalling, and water leakage [10–13]. In the construction process, due to the deformation and contraction of the segment or the low surface smoothness of the segment, the uneven support or load eccentricity in the construction process will further deepen the cracking of the segment [14]. Xu Guowen's research shows that segment torsion produces additional shear stress when the segment is supported unevenly, which causes oblique cracks. Under the action of axial thrust, the pipe section appears in the compression and tensile stress area, and longitudinal cracks may appear near the vault, inverted arch, and waist [15]. Saleta Gil Lorenzo carried out a detailed study on the situation of segments subjected to radial force. Under symmetric radial force, a newly assembled tube deflects radially around the previously assembled tube [16]. When subjected to asymmetric radial force, the hoop tension at the inversion increases locally [17]. The circumferential torque is caused by the radial component of the reaction force of the pad in the fixed section or by the gradual adoption of an ellipse facing the back of the R1 ring. The difference of ring deformation between the front and back of the segment will lead to serious torsion.

When the lining is subjected to continuous radial load, the lining ring will appear 'tulip'-shaped, that is, the diameter of the front end of the lining will be larger than the uneven shape of the compressed area at the back end [18–20]. This 'tulip' shape will cause contact defects on the bearing surface of the longitudinal joint, and plastic deformation will occur in the compression zone, which may even cause permanent damage to some tunnel linings. With the lining out of the shield tail, the tunnel segment will gradually show horizontal diameter increase and vertical diameter decrease in the self-weight stage of the segment, and the overall shape will resemble a duck egg; this has a serious impact on the force of the segment. At the same time, the lining of the tulip shape is a huge challenge [21,22].

Nowadays, a large number of studies focus on the performance of lining structures under different loads. The purpose is to understand the stress response and cracking mechanism of linings under different external concentrated loads. Usually, loads are applied on segments with different positions and shapes through large-scale bench tests, full-scale tests, field measurements, and finite element numerical modelling [23–28]. Some scholars have also introduced assembly tolerance and penalty functions, and the principle of minimum potential energy is applied to study the analytical solution of segment internal force considering the pre-deformation and inaccuracy of lining segments [29]. The segment structure based on the traditional shield construction method is studied in detail. Most of the research objects are complete lining rings and segment structures with constraints on both sides.

However, with the development of shield construction technology, the stress of traditional shield linings is completely different from that of segments under synchronous propulsion and assembly technology. First, the absence of the hoop effect on both sides of a newly assembled segment allows the segment to have a higher degree of freedom when advancing. Second, compared with the traditional shield method, the newly assembled segments will bear the jack thrust directly during propulsion under the synchronous propulsion and assembly mode. Third, when there is an assembly error or the jack is inclined, this will greatly affect the ultimate load of the segment. There is also the technical difficulty relating to synchronous propulsion and assembly technology which needs to be overcome urgently.

Based on the background of synchronous propulsion and assembly technology, this paper studies the structural response of segments in different assembling positions and under the condition of asymmetrical propulsion exerted by jacks. By establishing a finite element model of segment assembly, we explore the interaction and stress concentration of unlooped segments under asymmetric stress in the synchronous propulsion and assembly mode. This study determines the assembly error control range of the synchronous propulsion and assembly mode applied in practical projects. The theoretical basis and technical reference are established to warrant future synchronous propulsion and assembly projects.

2. Project Overview and the Synchronous Propulsion and Assembly Method

2.1. Project Overview

The airport connecting line (west section) project of the Shanghai rail transit city line is the trunk line connecting Shanghai's international airports. The tunnel runs east–west as a whole, with a total length of about 68.6 km, including an underground section of 56.7 km. The maximum overburden of the tunnel top is 33 m, the minimum is 14 m, the minimum radius is R520, the maximum longitudinal slope is 2%, and the length below the radius of R600 is about 660 m. The construction sequence of the tunnel is shown by the red arrow in Figure 1. The main crossing strata are ④ Silt clay, ⑤$_1$ silty clay, ⑥$_1$ silty clay, and ⑦$_1$ silty sand mixed with silty clay. The specific soil layer structure is shown in Figure 1.

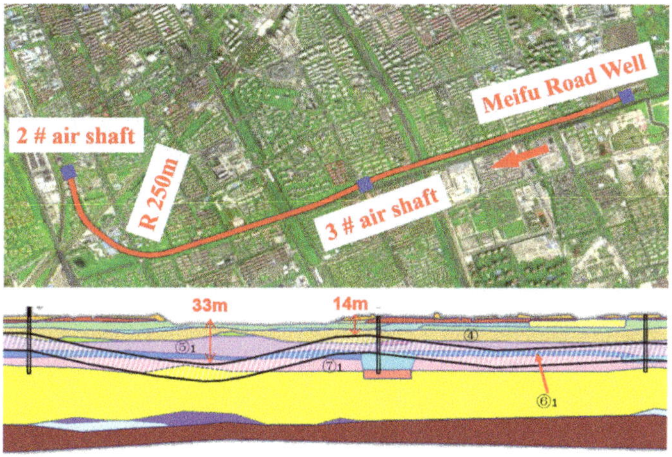

Figure 1. Longitudinal profile of excavation soil layer in synchronous propulsion and assembly mode.

The outer diameter of the tunnel lining ring is Φ13.6 m, the inner diameter is Φ12.5 m, and the thickness of the segment is 550 mm. The lining adopts a universal wedge-shaped ring, in the form of a 6 + 2 + 1 (1/2) block, which consists of six standard segments, two adjacent segments, and one capping block. The average length of the lining ring is about 2 m. The segment is made of high-performance concrete with a strength of C60 and an impermeability grade of P12. Taking the standard block as an example, the outer side of the segment contact surface is wrapped with an elastic gasket covering 42.352°. The inner side of the contact surface of the segment is covered with a packer, and the packer coverage angle is 40°. Each standard block is directly loaded by four jacks, and the contact area between the jack and the segment is protected by a shoe plate.

2.2. Synchronous Propulsion and Assembly Method

Based on the above engineering requirements, the synchronous propulsion and assembly technology is proposed in the hope that it will reduce the impact of long tunnel distances on construction periods. The synchronous propulsion and assembly technology is mainly used in the straight-line section of Meifu Road Well to 3# air shaft construction. The synchronous propulsion and assembly technology is adopted under the condition of suitable geology, as shown in Figure 2.

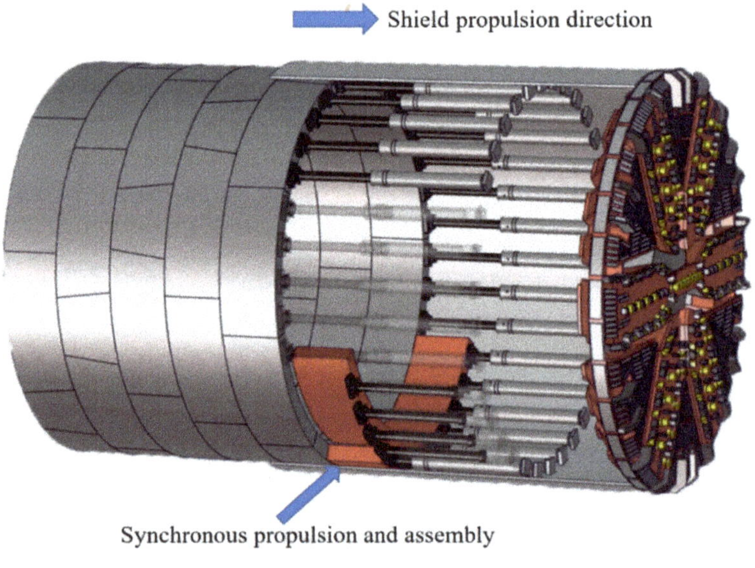

Figure 2. Schematic diagram of the synchronous propulsion and assembly method.

Synchronous propulsion and assembly method is realized by intelligently controlling the pressure of the shield propelling cylinder. The difference with respect to the traditional shield is that the shield machine of the synchronous propulsion and assembly method lengthens the length of the jacking cylinder and realizes the synchronous assembly of some segments by using the surplus of the stroke of the push cylinder. The construction efficiency of the synchronous propulsion and assembly method is directly proportional to the remaining distance of the jack, which can shorten the construction time of each ring segment by 60% at most. The shield machine used in this project has a margin time for the assembly of three segments. Therefore, the focus of this paper is also on the stress of the three segments in the synchronous propulsion and assembly mode. The specific operation methods of the synchronous propulsion and assembly method are as follows:

(a) The first segment is assembled, the jack acting on the first segment shrinks, and the thrust of the jack smoothly transitions to the thrust of the corresponding working condition;

(b) The assembly of the first segment is completed; the jack acting on the first segment is advanced. The second segment is retracted corresponding to the jack, and the thrust of the jack smoothly transitions to the thrust of the corresponding working condition.

The above working conditions are to be continuously cycled until the assembly of three segments is completed, as shown in Figure 3. During construction, the shield machine will switch to the synchronous propulsion and assembly mode. The contraction of the jack and the setting of the thrust will be automatically carried out by the shield machine. Different from traditional shield propulsion, in the synchronous propulsion and assembly mode, the jack propulsion force will be reasonably distributed to the non-assembled segments.

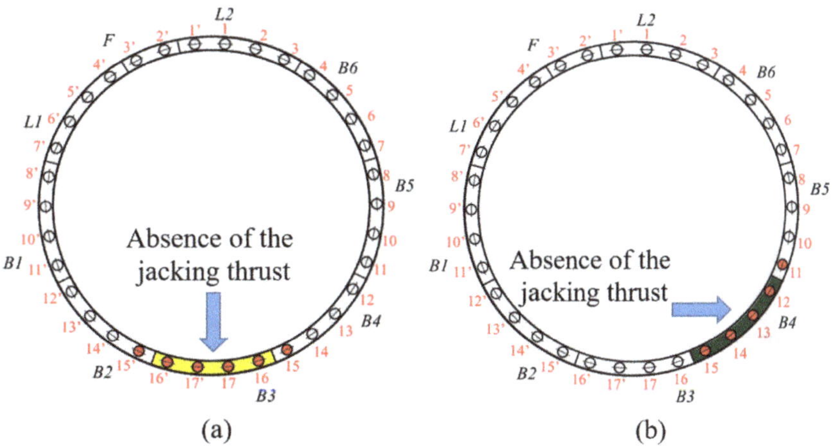

Figure 3. Steps of jack jacking in synchronous propulsion and assembly mode: (**a**) The working condition of the jack when the first segment is assembled; (**b**) The working condition of the jack when the second segment is assembled.

3. Research Methods and Parameters

The segment in the assembly has a certain asymmetry, which is mainly reflected in two aspects: (i) the asymmetry caused by the segment structure at different positions; (ii) stress asymmetry caused by the inclination of the external load direction. In the synchronous propulsion and assembly mode, the assembly state of the assembled segment in the asymmetric state has certain unknowns. Therefore, it is urgent to understand the stress effect of the segment during the assembly process under the condition of asymmetric position, and given that force may be asymmetric, the contact quality between segments is also one of the key subjects of this paper.

By establishing a numerical analysis of the actual engineering segment, the stress response and damage to the tunnel segment in the synchronous propulsion and assembly mode are determined. Based on the assembly sequence of the three segments, the segment assembly model is established, and different assembly errors are introduced when assembling segments. The solid model of segment, packer, and elastic gasket is established by finite element software, and the damage to the segment is described by a concrete elastoplastic damage constitutive model.

3.1. General Description of the Models

The tunnel segment model is established based on the airport connecting line (west section) project of the Shanghai metro line, as shown in Figure 4. The study focuses on the stress between the assembled segment and the completed lining, so the model can be simplified as the interactive simulation of the assembled segment and the adjacent ring contact segment. Some minor details are ignored in modellings, such as bolt holes, wedges of segments, grooves, and bolts. Setting springs between the newly assembled segments simulates the role of bolts between segments. In the assembly, the default assembled segments experience no assembly errors, and the assembly of the segments will follow the longitudinal seam bonding principle, that is, default perfect bonding between adjacent segments of the same ring segment. The geometric parameters and mesh properties of the segment are shown in Tables 1 and 2.

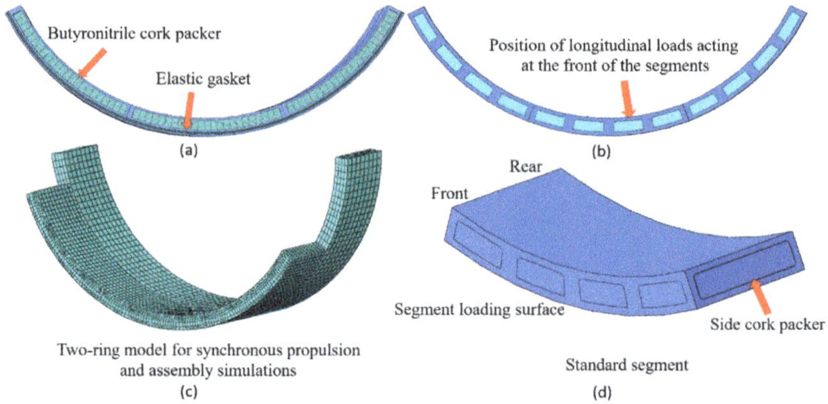

Figure 4. Geometry and mesh of finite element models: (**a**) The location of the butyronitrile cork packer and elastic gasket; (**b**) Position of longitudinal loads acting; (**c**) Mesh of standard segments; (**d**) Description of segment orientation.

Table 1. Input geometrical parameters for the finite element models.

Input Parameters	Value
Tunnel radius (m)	13.6
Standard segment thickness (mm)	550
Length of standard segment (m)	2.03
Angle of standard segment (°)	42.35
No. of gasket pairs for each segment (-)	1
No. of packer pairs for each segment (-)	1
Gasket depth and thickness (mm)	70 and 10
Packer depth and thickness (mm)	300 and 5

Table 2. Mesh properties for the FE modelling.

Feature	Number of Nodes	Number of Elements	Element Type
Standard segment	12,225	9144	C3D8R
Packer	729	416	C3D8R
Gasket	390	200	C3D8R
Bolts between the segments	-	162	SPRING

The segments, packers, and gaskets are modeled with first-order hexahedral elements, which can reduce the calculation time of the 3D numerical model and ensure the accuracy of the model calculation. To reflect the inelastic behavior of concrete, the finite element software adopts the isotropic elastic damage and tensile-compression plasticity theory. In order to show the elastic-plastic results of concrete under compression, the isotropic elastic damage and tensile-compression plasticity theory are used in the model. The Mooney–Rivlin strain potential energy model [30,31] is used for the gasket and packer. The physical parameters of the materials are shown in Table 3.

Table 3. Constitutive parameters for the finite element models.

Input Parameters	Value
Concrete density (kg/m^3)	2450
Concrete Young's modulus (GPa)	36
Concrete Poisson's ratio (-)	0.20
Butyronitrile cork packer Young's modulus (MPa)	7.29
Butyronitrile cork packer C10 and C01 (MPa)	0.74 and 0.18
Elastic gasket (EPDM) Young's modulus (MPa)	5.32
Elastic gasket (EPDM) C10 and C01 (MPa)	0.56 and 0.14
Bolting system stiffness (kN/mm)	9.18

When the jack acts on the segment, the segment has been fixed at the assembly position by the robotic arm, and the gasket contacts with the adjacent segment. In the simulation, the default gasket will not have relative displacement, and the gasket is fixed with two adjacent segments by binding. The hard contact relationship is adopted between the assembled segments to minimize the relative positional movement between them. One side of the packer is bound to the new assembly ring. The penalty function is used to simulate the finite sliding of circumferential gasket–segment and segment–segment surface contact. Based on the shear stiffness test of the circular tunnel joint [32], the friction coefficient between fine concrete and liner can be taken as 0.5.

3.2. Loading and Boundary Conditions

During the assembly process of the segment, the position of the first segment will form a periodic rotation due to the wedge shape of the segment itself, which is the main reason for the position asymmetry. The propulsion force of the synchronous propulsion and assembly mode will increase in gradient with different depths, so different boundary conditions and constraints need to be set for segments at different positions. According to the actual engineering situation, three models were established to simulate the assembly process of the three segments. Gravity is introduced in the assembly, and the different positions of the segments are simulated by setting different directions of gravity.

In the synchronous propulsion and assembly mode, the external loads borne by the segment during the assembly process are the thrust of the jack and its gravity. The main reaction of the asymmetric force of the segment is the angle error formed between the segment and the jack, as shown in Figure 5. The main reasons for this are the assembly error of the segment and the angle formed by the shield and the segment during the construction process. In this paper, it is assumed that the initial position of the segment is suspended in parallel with the segment that has been looped, and the assembly error is transformed into the radial and tangential components of the jack. The deflection error range of the segment in the study mainly refers to the maximum assembly error between the shield and the segment and the maximum stroke difference of the shielding jack.

The assembly process of the segment consists of a series of complex segment–liner–segment interaction processes. Compared with the unassembled segment, the assembled segment will form a hoop effect due to the interaction of the annular structure, and the relative displacement is smaller. In the study, the back push surface of the ringed segment and the two sides of the ringed segment are used as boundaries. The longitudinal displacement of the looped segment is eliminated by applying hinge constraints to the back push surface of the looped segment and both sides of the segment.

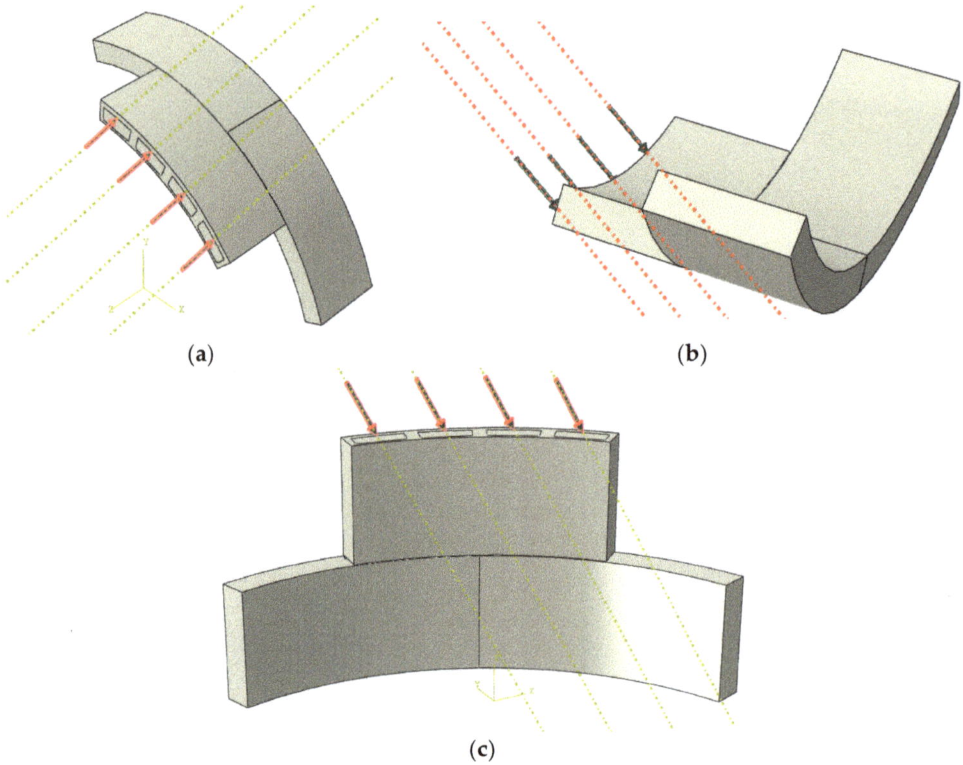

Figure 5. Assembly error of segment: (**a**) inward deflection; (**b**) outward deflection; (**c**) lateral deflection. (Take the assembly of one segment as an example.)

3.3. Modelling Sequence

The modeling order of the model changes as the segments are assembled. The assembly sequence of segments is as shown in Figure 6. Taking the third segment assembly model as an example, displacement constraints are first imposed on both sides of the looped segment. The back push surface and the global gravity is activated according to the working conditions to initialize the segment's positional information. Next, the thrust of the jack on the face of the two segments that have been assembled is activated. Before the radial or tangential component force of the jack is activated, a certain longitudinal thrust is applied to form a certain contact between the gasket and the surface of the ring-shaped segment to reduce the relative slip. To maintain the boundary conditions of the ring segment, the radial or tangential component force of the jack can be added as required. Finally, the jack thrust and error component of the third segment are activated, and the component force is consistent with that of the other two segments. By adjusting the component force of the jack accordingly, the response of the segment under different assembly errors is simulated.

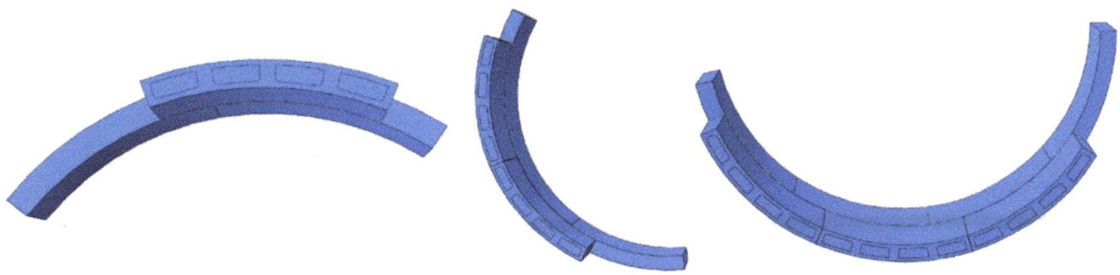

(a) (b) (c)

Figure 6. The assembly sequence of segments: (**a**) assembly of the first segment (take the segment at the top as an example); (**b**) assembly of the second segment (take the lateral segment as an example); (**c**) assembly of the third segment (take the segment at the bottom as an example).

4. Evaluation of Concrete Segment Damage

The damage–plastic constitutive relation of concrete is based on the elastic–plastic constitutive relation of concrete, and the damage factor is introduced. Non-associative hardening was introduced into the elastoplastic constitutive relation of concrete to better simulate the compressive elastoplastic behavior of concrete.

The strength of the super-large diameter shield segment concrete is C60. According to the Code for Design of Concrete Structures GB50010-2002 [33], the compressive stress–strain relationship of the C60 concrete segment is expressed as:

$$\sigma_c = \begin{cases} f_c\left[1-\left(1-\frac{\varepsilon_c}{\varepsilon_0}\right)^n\right] & (\varepsilon_c \le \varepsilon_0) \\ f_c & (\varepsilon_0 \le \varepsilon_c \le \varepsilon_{cu}) \end{cases} \tag{1}$$

$$n = 2 - \frac{1}{60}\left(f_{cu,k} - 50\right) \le 2.0 \tag{2}$$

$$\varepsilon_0 = 0.002 + 0.5 \times \left(f_{cu,k} - 50\right) \times 10^{-5} \ge 0.002 \tag{3}$$

$$\varepsilon_{cu} = 0.0033 - \left(f_{cu,k} - 50\right) \times 10^{-5} \le 0.0033 \tag{4}$$

where σ_c is the stress in concrete; f_c is the compressive strength of concrete; ε_c is the strain in concrete; $f_{cu,k}$. is the characteristic value of cubic concrete compressive strength; ε_0 is the strain when σ_c reaches f_c; and ε_{cu} is the maximum compressive strain in concrete.

In addition to compressive stress causing surface damage, tensile stress concentration under combined stress is also one of the main reasons for segment failure. Under tensile stress, the crack opening of the segment starts at a critical cohesive traction and ends when a critical opening displacement is attained [34]. The tensile stress–strain relationship of the C60 concrete segment is expressed as:

$$\sigma_t = \begin{cases} E_c\varepsilon\ (\sigma_t \le f_t) \\ f_t \times [1 + (m \times c_1)]e^{-m \times c_2} - f_t \times m \times (1 + c_1^3)e^{-c_2}\ (\sigma_t > f_t) \end{cases} \tag{5}$$

$$m = \frac{w_t}{w_{cr}} \tag{6}$$

where σ_t is the concrete tensile stress; E_c is the modulus of elasticity of concrete; w_t is the concrete crack width; w_{cr} is the crack width of fully cracked concrete; f_t is the peak tensile stress of concrete; and c_1 and c_2 are the constants 3 and 6.93, respectively.

Considering the differences in the tensile and compression properties of concrete materials, a damage factor is introduced to describe the reduction of the elastic stiffness matrix of concrete materials under different stress conditions [35–38]. Based on the compression damage theory of concrete, the damage to a segment is evaluated by the relationship between inelastic strain and the compression damage factor. For tension, the theoretically determined crack width of concrete is the integral of tensile strain to the crack width. Assuming that the crack width is constant, the relationship curve between the crack width and the damage factor can be obtained (Figure 7). The crack width here is based on the average sense.

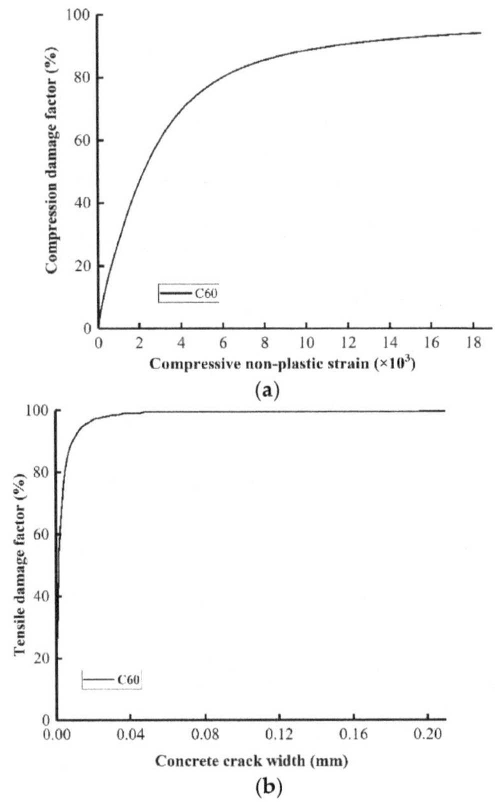

Figure 7. Damage factor definition curve: (**a**) compression damage factor; (**b**) tensile damage factor.

Finally, when the inelastic strain of concrete under compressive failure exceeds 0.002, the compressive strain damage factor is 46% to define the compressive damage of the segment. When the concrete tensile crack width exceeds 0.2 mm, the tensile strain damage factor is 99.61% to define the tensile damage of the segment.

5. Interactions between Segments along the Circumference

During assembly, the attitude of the segment will be deflected under the action of the jack, and the newly assembled segment will have a certain inclination with the previous segment. In addition to the thrust of the jack, the contact quality between the segments will directly have an important influence on the final posture of the segments. When the position of the segment is different, the rotation tendency of the arc segment under different gravity directions is also different. In addition to the posture of the segment, the segment will also have a certain bending deformation for the above-mentioned reasons. This section discusses the interaction results of the segment assembly process in the synchronous propulsion and assembly mode when there is no assembly error. At this time, the segment only bears the gravity and the thrust of the jack. In order to observe the extrusion between segments more intuitively, the deformation cloud map of the model is enlarged 20 times.

5.1. The Segment at the Top

When the initial segment is assembled on top of the tunnel, the segment will rotate inward. The main reason for the rotation of the segments concerns the packer and gaskets between the segments. The jack thrust in the synchronous propulsion and assembly mode will make the segment fit as close to the ring segment as possible, but the unevenness between segments makes the segment and segment contact form a deep beam. It is worth noting that, at this time, the deformation of the inner and outer sides of the packer and the gasket is different, and the inner side is deformed more seriously than the outer side.

When the first segment is assembled, the maximum rotation of the front of the segment is 9 mm, and the front surface rotates greatly as a whole under the action of the jack. At this time, the segment rotates around the two sides of the packer. When the segment is at the top, gravity further increases the inward rotation. When the second segment is assembled, since the support of the segment has been assembled, the rotation of the connecting part of the segment is significantly reduced. There is a certain extrusion between adjacent segments, mainly at the contact of the front of the segment, so the rotation of the free side of the push end of the first segment increases slightly, as shown in Figure 8. The longitudinal seam of the segment presents a small upper seam and a large lower seam. When the third segment is assembled, the rotation of the segment is significantly reduced due to the clamping of the segments on both sides, and the main rotation area is the unconstrained area. It is worth noting that, due to the continuous assembly of the segments, the direction of gravity is not the same. During the assembly process of the second and third segments, the direction of their gravity no longer contributes to the inward deflection of the segments. Compared with the assembly of the first segment, the degree of freedom of the rotation axis of the segment as a whole is constantly changing, so the degree of deflection of the segment is gradually reduced.

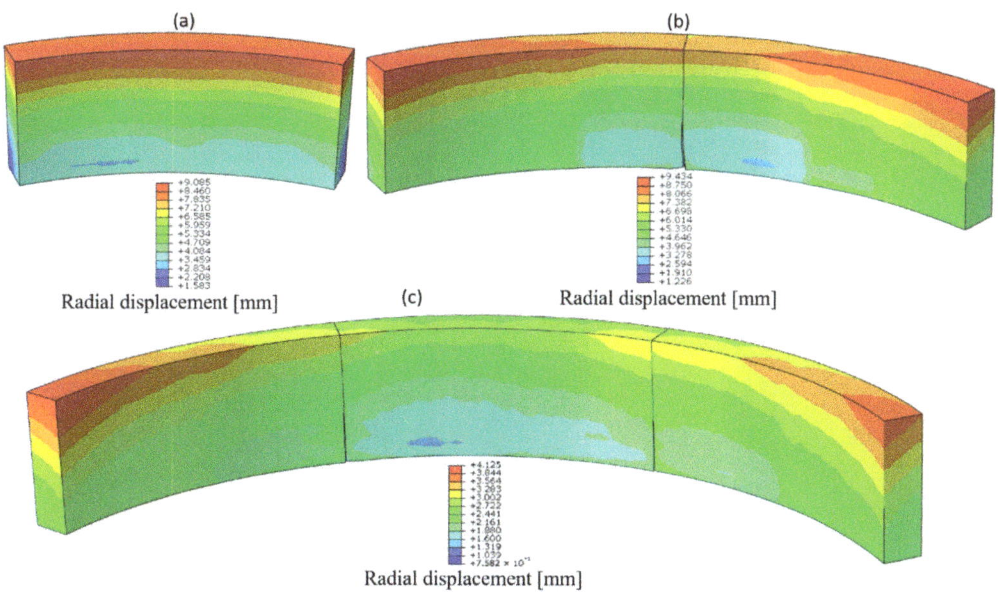

Figure 8. The interaction between segments when the segments are at the top: (**a**) first segment assembly; (**b**) second segment assembly; (**c**) third segment assembly.

Taking the free side edge of the assembled segment as the starting point, according to the angle, it can be seen that the stress of the front, the rear, and the centreline of the segment at different assembly stages presents different stress distributions (Figure 9a). When the first segment is assembled, the stress of the segment is mainly concentrated at the front and the rear of the segment and the maximum stress is 13 N/mm^2. When the segment is at the top, the segment has an obvious tendency to rotate inward, so the stress on the rear of the segment is relatively large. In particular, the stress on both sides of the rear increases rapidly. The main reason is that when the segment rotates, the rotation axis of the segment will autonomously choose to rotate along the edges on both sides of the segment. When the second segment is assembled, one side of the segment is supported by the assembled segment, so the stress at the rear of the segment is significantly reduced. At this time, the transmission of the stress of the segment is affected, and the stress at the front increases slightly. When the third segment is assembled, at least one side of all segments is supported, the stress distribution of segments is relatively uniform, and the overall stress is distributed within 3.5–10.5 N/mm^2. The inward rotation of segments is improved. When the third segment is assembled, the middle segment will be squeezed a little, so the stress will surge to a certain extent at the rear and centreline of the segment.

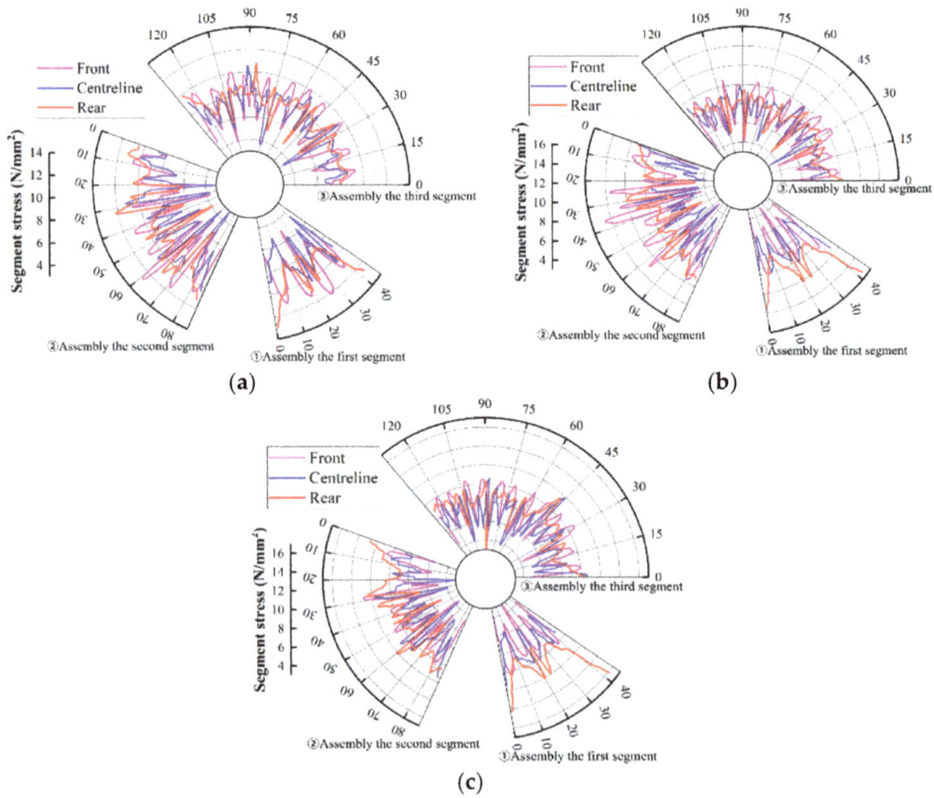

Figure 9. Internal forces in the model: (**a**) the segment at the top; (**b**) the lateral segment; (**c**) the segment at the bottom.

It can be seen from Figure 10a that the direction and size of the bending moment of the segment are constantly changing due to the assembly of segments. When the first segment is assembled, the bending moment of the segment reaches the maximum at the place where the jack directly acts, and the maximum is 480 kN·m. When the second segment is assembled, the bending direction of the segment changes, and its peak value reaches −590 kN·m. Since the two support each other when they are in contact, the bending moment is lowest at the contact position. When the third segment is assembled, the segments on both sides squeeze the first segment so that the bending direction of the segment is changed. At this time, the bending moment of the third assembled segment is significantly lower than the other two assembled segments, and it is difficult for the two assembled segments to move.

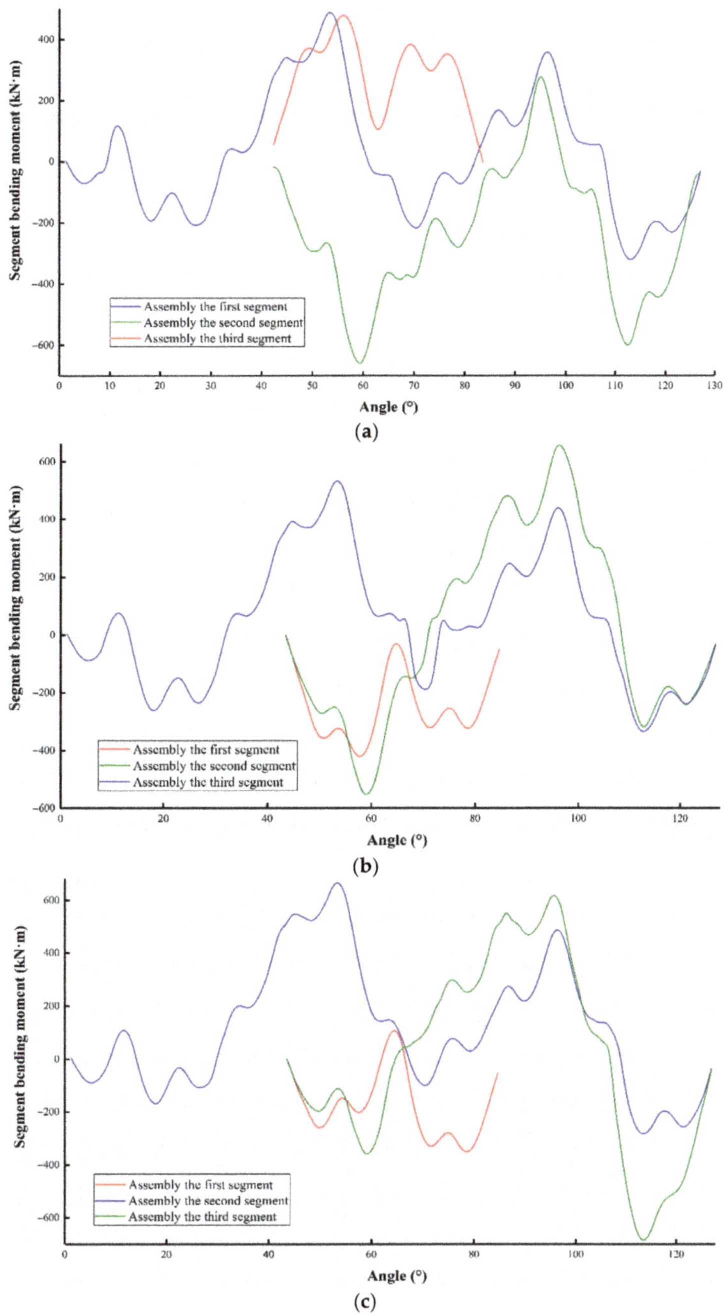

Figure 10. Bending moment in the model: (**a**) the segment at the top; (**b**) the lateral segment; (**c**) the segment at the bottom.

In short, with the assembly of segments, the maximum bending moment, maximum stress, and displacement of segments are significantly reduced, but there is a certain extrusion in the assembly process. This may mean a risk of damage in the subsequent assembly.

5.2. The Lateral Segment

When the initial assembled segment is lateral, the segment also deflects inward. The difference from the initial segment at the top is that the deflection of the segment is no longer symmetrical, and the symmetry of the segment is destroyed, as shown in Figure 11. When the first segment is assembled, under the action of gravity, the lower part of the segment deflects more inwardly, and the upper part deflects less inwardly. Since the right segment is the first to be assembled in the assembly sequence, when the second segment is assembled, the segment is below the first segment. At this time, a large inward deflection occurs on the free side of the lower segment. Unlike the top of the segment, the extrusion position of the segment is the front of the segment, and the rear of the segment will form a certain seam due to the extrusion of the segment. Due to the different deflection of the segment, it is easy to form a certain staggered seam in the segment at this time, which has a great impact on the assembly quality of the segment. After the assembly of the first two segments is completed, the third segment exhibits the same rotation trend as the first two segments, and a larger seam will be produced at the rear of the segment.

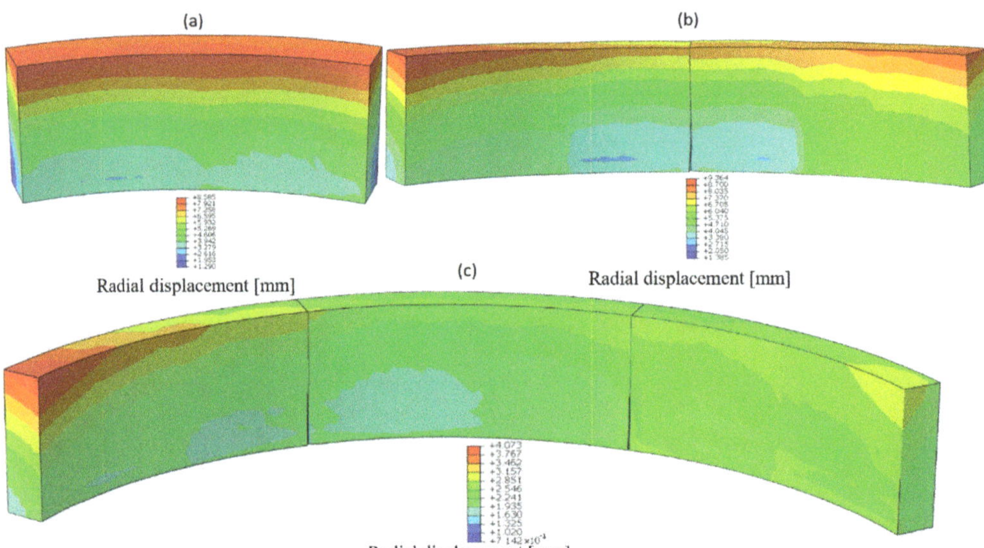

Figure 11. The interaction between the segments when the segments are lateral: (**a**) first segment assembly; (**b**) second segment assembly; (**c**) third segment assembly.

Unlike the initial assembled segment at the top, when the first segment is assembled, the stress distribution at the rear of the segment is asymmetric. The stress situation in the centreline and at the front of the segment is not much different. Obviously, the stress of the segment at the rear is uneven, and stress concentration is prone to occur (Figure 9b). When the second segment is assembled, the newly assembled segment will deflect inward first, which causes the contact surface at the front of the segment to contact first. When the jack thrust is gradually applied, the contact of the segment will generate a greater squeeze. Compared with a single assembled segment, since the rotation of the segment is supported by one side segment, the distribution of the thrust of the jack in other positions of the segment is more uniform, and the stress transmission is improved. When the third segment is assembled, the stress of the final segment is closer to that at the top of the segment. The stress of the first segment will increase slightly. The main reason is that the inclination of the first segment is completely different from the inclination of the latter two segments. Therefore, the segment will show the maximum inclination, making the stress transmission mode of the first segment different.

The bending moment of the first segment at the bottom is completely opposite to that when the segment is at the top (Figure 10b). The change in the direction of gravity makes the twisted posture of the segment change to a certain extent, and the segment bends outward in the longitudinal direction. After the second segment is assembled, the extrusion of the pushed-up end causes the bending moment of the segment to increase significantly. Since the position of the newly assembled segment is different from the first segment, the bending directions of the two segments are asymmetrical. In this case, the bending moment of the first segment to the outside is much larger than that in the case where the segment is at the top. The bending moment of the third segment has been greatly improved with the further assembly of the segment. Under the clamping of the third segment, the tendency of the first segment to bend outward changes and it bends inward.

5.3. The Segment at the Bottom

When the initially assembled segment is at the bottom, compared with the first segment assembled at the top and the lateral segment, the inward deflection tendency of the front of the segment is lowest. Under the action of gravity, the segment will fall to the outside of the segment, so the displacement of the segment is relieved to a certain extent. When the second segment is assembled, the first segment is less affected by the assembly and will not undergo a large displacement change, but a certain extrusion occurs at the inner corner of the bottom of the segment (Figure 12). When the third segment is assembled, the displacement is basically small, except that the first segment and the inner corners of the segments on both sides are squeezed to a certain extent. It can be seen that under the external shape characteristics of the segment, the clamping on both sides can well maintain the stability of the segment at the bottom.

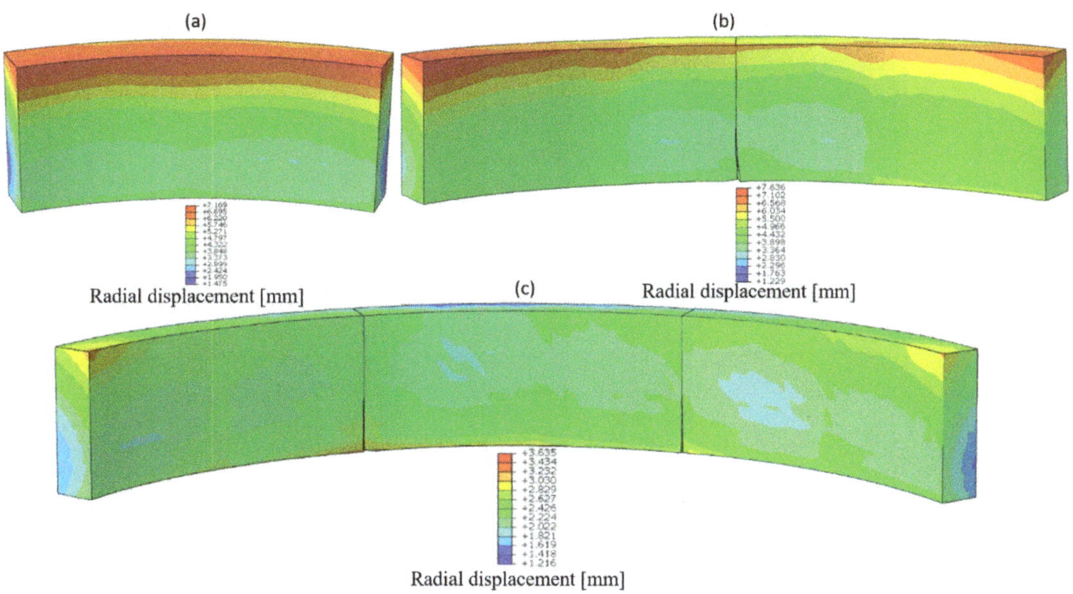

Figure 12. The interaction between the segments when the segments at the bottom: (**a**) first segment assembly; (**b**) second segment assembly; (**c**) third segment assembly.

When the initial segment is assembled, although the segment is at the bottom, the segment still rotates due to excessive freedom (Figure 9c). At this time, the maximum stress of the segment is 16 N/mm^2. When the second and third segments are assembled, the stress distribution of the segment is relatively uniform, and the maximum is only 0–11 N/mm^2. Compared with other positions of the segment, the stress distribution of the segment is smaller. Due to the close contact between the segments, the stress transfer between the segments is relatively uniform.

At this time, the bending moment of the segment is shown in Figure 10c. Compared with assembly of the lateral segment, when the first segment is assembled, the bending moment of the segment is slightly reduced. Under the smaller deflection, the angle formed between the segments is smaller, and the segment is less biased. After the second segment is assembled, since the newly assembled segment has a higher degree of freedom, the segment will deflect to a certain extent. At this time, under the action of gravity, the segment tends to bend outward, and the bending moment is slightly increased. When the synchronous propulsion and assembly mode is applied, the final bending moments of the segments in different positions are basically the same after the three segments are assembled. During the segment assembly process, as the segment's stability increases gradually, the influence of gravity on the segment will gradually decrease. It is necessary to focus on the possible damage to the segment when one or two segments are assembled and the segment has a large degree of freedom.

Obviously, during the construction of the synchronous propulsion and assembly mode, different positions of segments will impact the deflection of segments in the assembly process, which will have a great impact on the displacement, stress transfer, and bending of segments. When asymmetric stress acts on the segment, the assembly quality of the segment will be more compromised. It is necessary to explore possible segment damage in actual synchronous propulsion and assembly construction more finely.

6. Interactions between Segments along the Longitudinal Direction

The asymmetric stress between segments is affected by the rotation degree caused by the assembly error between segments, which can be divided into the radial rotation and lateral deflection of segments. The radial deflection of segments can be divided into radial inward deflection and radial outward deflection, which are mainly affected by the initial position of the shield machine and segments (i.e., assembly error). The tangential deflection of the segment is mainly caused by the stroke difference of the jack acting on the segment, resulting in the asymmetry of the force application device.

Different from most utilization studies, there is no contact between concrete segments in this study. This is close to the assembly of the actual project, and a certain gap will be formed between the segments due to the support of the gasket and packer.

6.1. Inward Radial Deflection Error

In the inward radial deflection model, the maximum compression damage factor of the newly assembled segment is almost linearly related to the assembly error. When the first segment is assembled, the maximum damage to the segment will be transferred from both sides of the gasket of the segment to the inner edge of the liner as the deflection angle increases. The inward radial deflection of the segment will aggravate the inward deflection during the natural assembly of the segment, which makes the segment packer more likely to generate contact stress concentration. It is also the main reason for the sudden increase in the maximum compression damage factor. This situation is gradually disintegrated due to the support on both sides during the assembly process of the segment. It can be seen from the maximum damage position of the second and third segments (Table 4) that with the assembly of the segments, the maximum damage position of the segment is transferred to the contact between the adjacent segment and the gasket place. Although there will be extrusion between the segments, the longitudinal contact quality of the segments is better than that of the rear surface, and the rotation that occurs is also smaller.

In the synchronous propulsion and assembly mode, the damage possibility and severity of the segment under the inward radial deflection error are evaluated by the compression damage factor. All simulations have reached the damage limit, and the maximum compression damage factor can reach 52%, as shown in Figures 13 and 14. The damage factor of 46% is marked by the dot line, so that the damage angle of the segment can be obtained intuitively. No matter where the initial segment is, the compression damage factor will form a certain mutation after a gentle rise to a certain deflection angle, which is especially obvious in the early stage of segment assembly. On the contrary, when the third segment is assembled, the compression damage factor of the segment will rise slowly and reach a certain limit. Compared with the initial position of the segment at the top and bottom, the effect of inward radial deflection on the lateral segment is smaller, and the damage growth of the segment is slower. When the segment reaches 1° deflection, the maximum damage to the segment is also smaller. The tensile damage factor of the segment undergoes little change in different positions and during the assembly process, and the peak value is 91.81%. The main tensile damage to the segment occurs at the symmetrical center of the jack, which is mainly caused by the transverse tensile stress caused by the thrust of the jack [39,40].

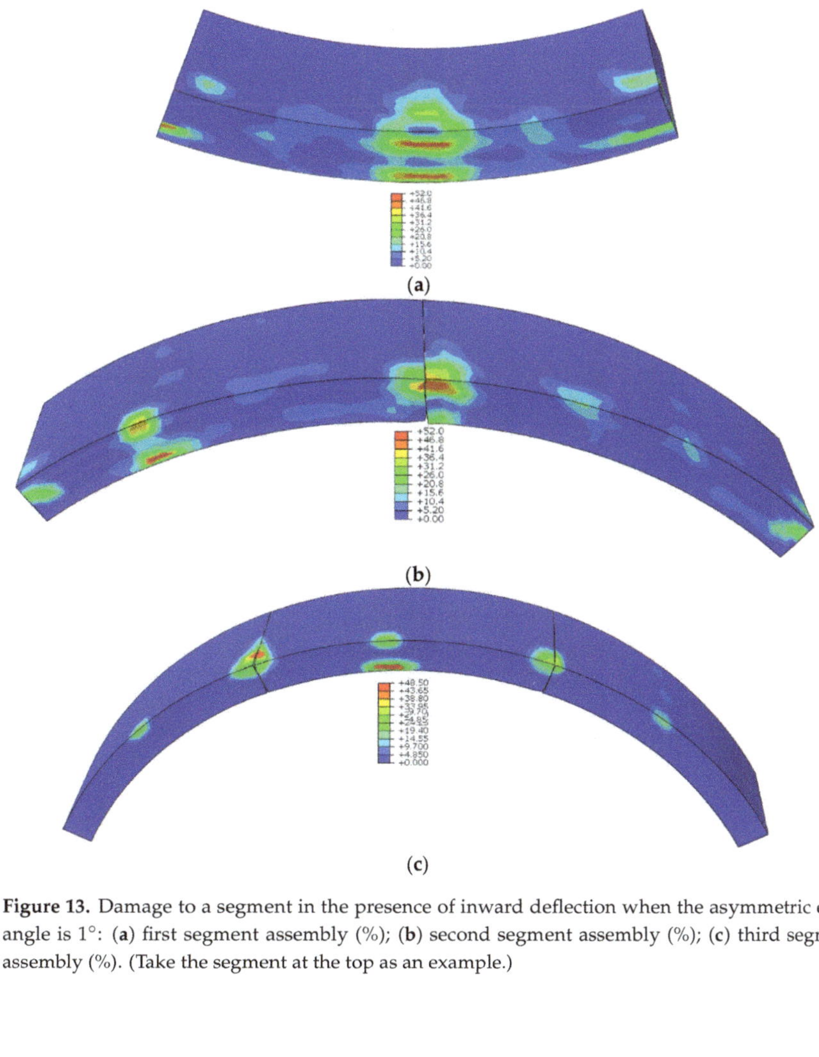

Figure 13. Damage to a segment in the presence of inward deflection when the asymmetric error angle is 1°: (**a**) first segment assembly (%); (**b**) second segment assembly (%); (**c**) third segment assembly (%). (Take the segment at the top as an example.)

Table 4. Damage to the segments.

Location	Assembly Process	Damage Angle (°)	Destruction Location
Top	Assembly of the first segment	0.60	Inside of packer
	Assembly of the second segment	0.71	Inside of packer
	Assembly of the third segment	0.60	Inside of packer and extrusion corner
Lateral	Assembly of the first segment	0.57	Inside of packer
	Assembly of the second segment	0.78	Inside of packer
	Assembly of the third segment	0.79	Inside of packer and extrusion corner
Bottom	Assembly of the first segment	0.64	Inside of packer
	Assembly of the second segment	0.74	Inside of packer
	Assembly of the third segment	0.69	Inside of packer

During assembly, the outer joint width of the segment will gradually increase with the maximum inward deflection of the segment. When the initial segment is at the top, the maximum joint of the segment is 10.44 mm, as shown in Figure 15. Similar to the changing trend in the compression damage factor, when the segment damage error changes abruptly, the maximum joint width of the segment increases sharply. The increase of the joint width means that the segment has a higher possibility of damage, which can be referred to as one of the most important sources of evidence when judging the quality of segment assembly in practical engineering. At the same time, the larger joint width will reduce the sealing performance of the gasket [41,42]. While the joint width of the top segment will undergo a larger increase with greater assembly error, the segments at the other positions will gradually tend to slow down. Obviously, under the action of the larger jack, the joints of the segments will not grow all the time and will gradually stabilize when the error reaches a certain level.

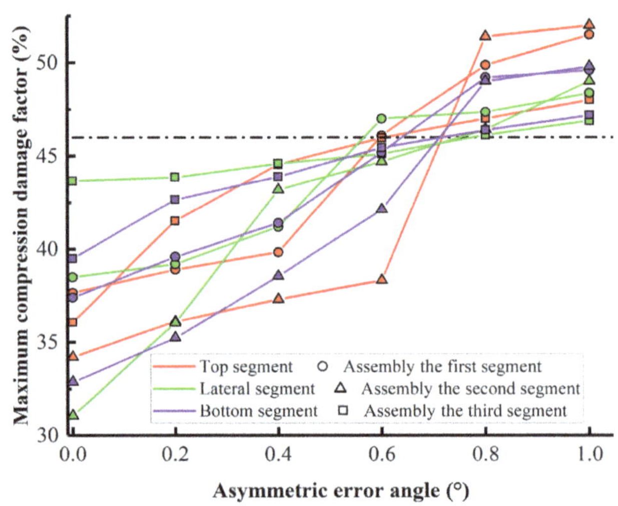

Figure 14. Maximum compression damage factor in the presence of inward deflection.

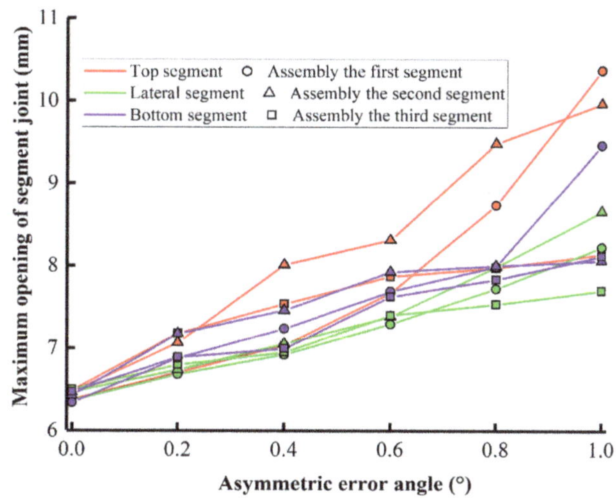

Figure 15. Maximum opening of segment joints in the presence of inward deflection.

6.2. Outward Radial Deflection Error

In the synchronous propulsion and assembly mode, the normal assembly of the segment will cause the segment to exhibit radial deflection, and the outward radial deflection error of the segment can reduce the possibility of segment damage to a certain extent. When a segment is deflected radially outward, the stress concentration effect on the packer is reduced, so the compression damage factor of the segment will not change abruptly. With the increase of the error, the compression damage factor increases with a more moderate trend. The maximum damage to the segment is concentrated on both sides of the middle segment, mainly showing the extrusion damage of the two sides and the compression damage of the segment gasket (Figure 16 and Table 5). When the initial segment is at the bottom, the outward radial deflection has the greatest influence on the damage to the segment. When the deflection is 0.41°, the segment will be damaged, as shown in Figure 17. The main reason for this is that the newly assembled segment is more prone to outward deflection when it is at the bottom than other initial positions. There are no other supports between the segment rings to provide additional support when radially deflected outwards, so the sides of the gasket are more prone to damage. When outward deflection occurs, the maximum tensile damage factor of the segment is 92.54%. Since the thrust is unchanged, the tensile damage of the segment is not greatly affected.

Table 5. Damage to the segments.

Location	Assembly Process	Damage Angle (°)	Destruction Location
Top	Assembly of the first segment	1.20	Both sides of segment
	Assembly of the second segment	1.00	Extrusion corner
	Assembly of the third segment	0.58	Extrusion corner
Lateral	Assembly of the first segment	1.10	Both sides of segment
	Assembly of the second segment	0.76	Extrusion corner
	Assembly of the third segment	0.58	Extrusion corner
Bottom	Assembly of the first segment	0.58	Both sides of segment
	Assembly of the second segment	0.76	Extrusion corner
	Assembly of the third segment	0.40	Extrusion corner and middle of outer arc

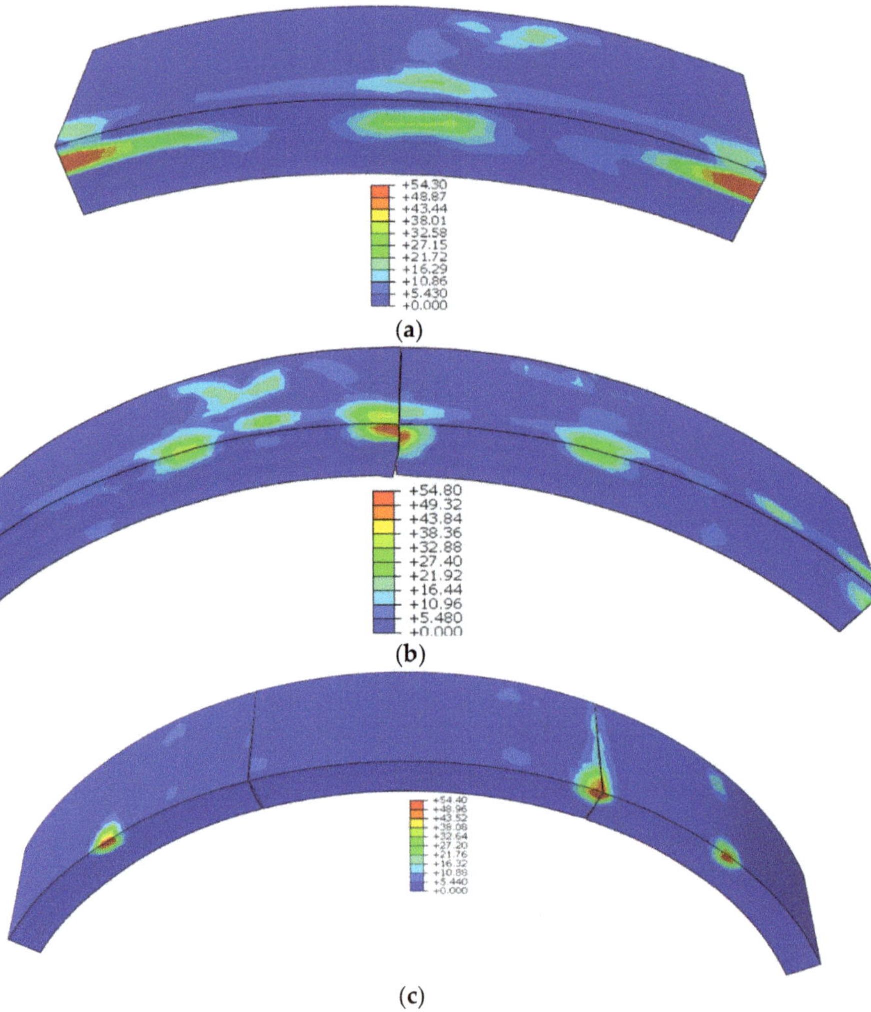

Figure 16. Damage to segment in the presence of outward deflection when the asymmetric error angle is 1.4°: (**a**) first segment assembly (%); (**b**) second segment assembly (%); (**c**) third segment assembly (%). (Take the segment at the bottom as an example.)

The outer joint width of the segment gradually decreases with the increase of the outward radial assembly error, as shown in Figure 18. During the segment assembly process, the joint quality of the segment is gradually improved, which means that the assembly process reduces the deflection freedom of the segment. The action position of the jack is distributed along the center of the bearing surface, and the gasket of the segment is closer to the outer side of the segment. Although the component force caused by the outward deflection error causes the segment to deflect to the outside, the component force is too small compared to the jack thrust in the synchronous propulsion and assembly mode. Therefore, under the outward radial deflection error of the segment, the influence on the maximum joint volume of the segment is small, and the maximum reduction is only 1 mm.

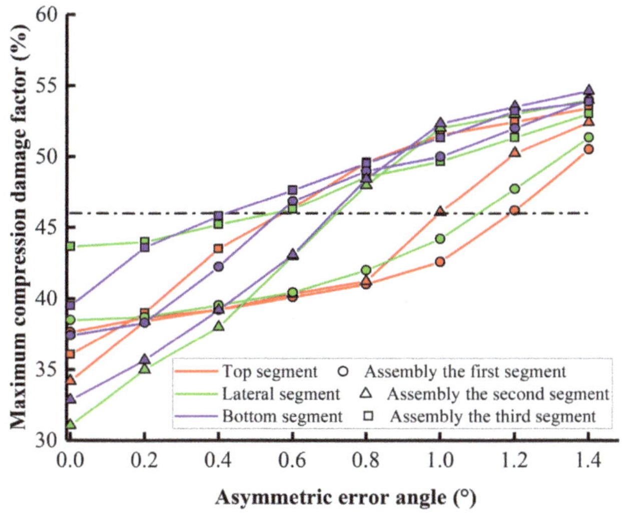

Figure 17. Maximum compression damage factor in the presence of outward deflection.

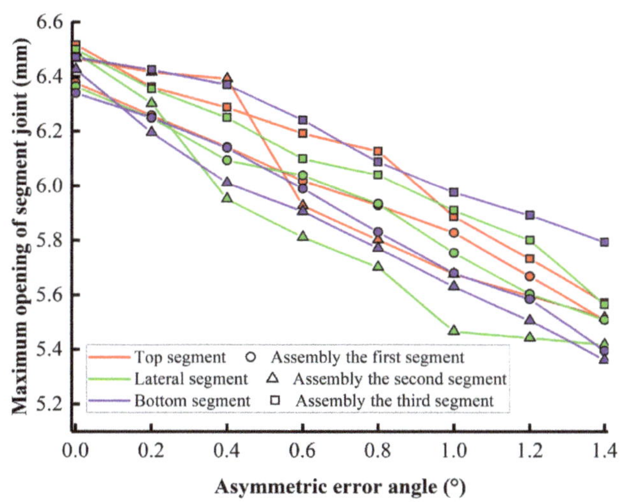

Figure 18. Maximum opening of segment joint in the presence of outward deflection.

6.3. Lateral Deflection Error

Lateral deflection is a deflection error that often occurs in shield construction. The larger the tunnel diameter, the greater the lateral deflection range. It can be seen from Figure 19 that, compared with radial deflection, as lateral deflection error increases, the compression damage factor of the segment increases slowly. The segment can tolerate lateral deflection well. Unlike radial deflection error, when the segment is subjected to the lateral deflection error, the damage to the segment is asymmetric. The maximum possible damage position is at the bottom end of the segment's lateral inclination (Table 6). In the process of segment assembly, the transformation of the rotation axis is an important reason for the change of the damage position. In the presence of lateral deflection error, the assembling position of the segment has little effect on the compression damage factor of the segment. When the segment is joined to the third segment, damage may occur at 1.4° (Figure 20). Although there are bolts between the segments to constrain the segment during the assembly process, the segment as a whole will show the same trend of inclination, and the extrusion effect will be superimposed. At this time, the segment will be damaged at a smaller error angle. The tensile damage factor of the segment also did not change greatly with the change of working conditions, and the peak value was 91.56%.

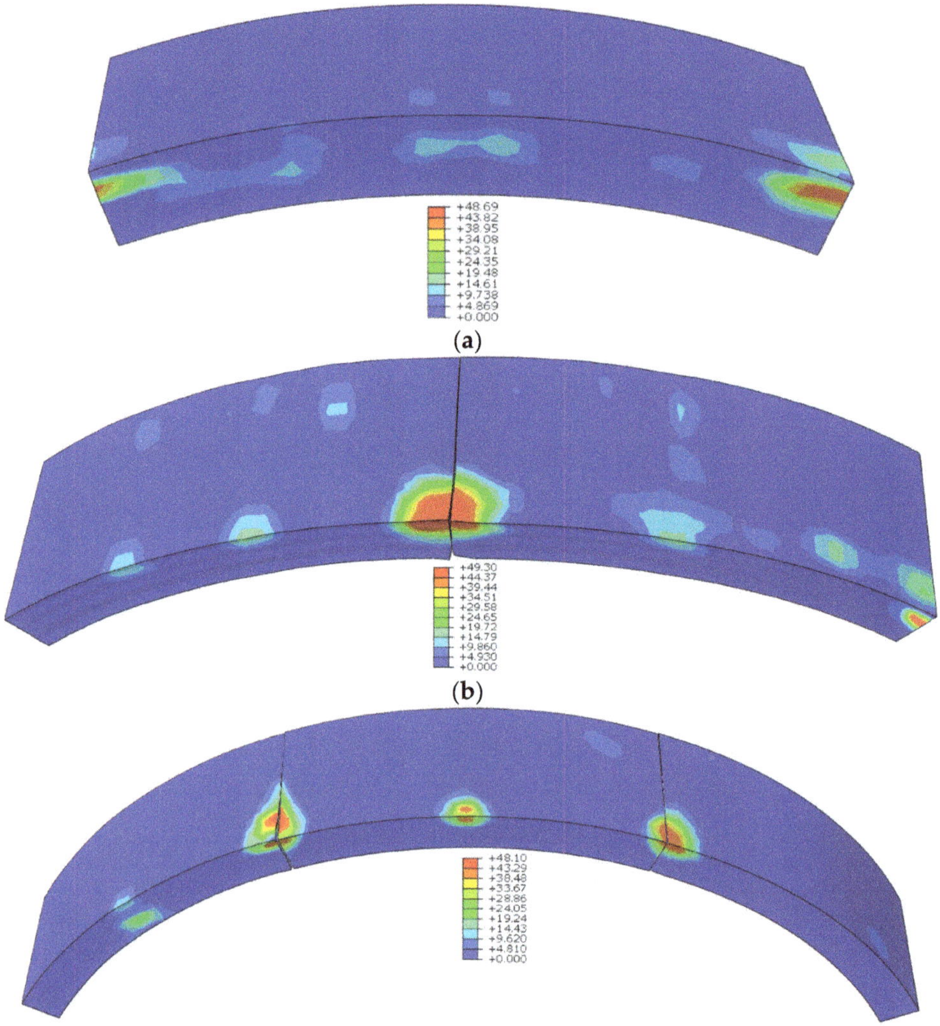

(c)

Figure 19. Damage to segment in the presence of lateral deflection when the asymmetric error angle is 2.4°: (**a**) first segment assembly (%); (**b**) second segment assembly (%); (**c**) third segment assembly (%). (Take the lateral segment as an example).

Table 6. Damage to the segments.

Location	Assembly Process	Damage Angle (°)	Destruction Location
Top	Assembly of the first segment	1.40	Sloping side of segment
	Assembly of the second segment	1.81	Sloping side of segment and extrusion corner
	Assembly of the third segment	1.38	Extrusion corner
Lateral	Assembly of the first segment	1.41	Sloping side of segment
	Assembly of the second segment	1.82	Sloping side of segment and extrusion corner
	Assembly of the third segment	1.37	Extrusion corner
Bottom	Assembly of the first segment	1.39	Sloping side of segment
	Assembly of the second segment	1.89	Sloping side of segment and extrusion corner
	Assembly of the third segment	1.38	Extrusion corner

The outer joint width of the segment gradually decreases with the increase of the outward radial assembly error, as shown in Figure 18. During the segment assembly process, the joint quality of the segment is gradually improved, which means that the assembly process reduces the deflection freedom of the segment. The action position of the jack is distributed along the center of the bearing surface, and the gasket of the segment is closer to the outer side of the segment. Although the component force caused by the outward deflection error causes the segment to deflect to the outside, the component force is too small compared to the jack thrust in the synchronous propulsion and assembly mode. Therefore, under the outward radial deflection error of the segment, the influence on the maximum joint volume of the segment is small, and the maximum reduction is only 1 mm.

Obviously, in the synchronous push-and-assemble mode, when the segment is subject to assembly error, the segment will be damaged within the maximum deflection error range. In practical engineering, it is necessary to control the assembly error of the segment within the damage angle so as to ensure the smooth development of the synchronous push-and-pull mode.

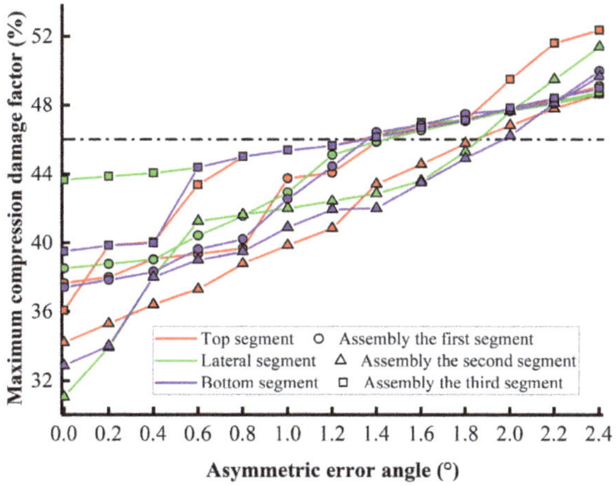

Figure 20. Maximum compression damage factor in the presence of lateral deflection.

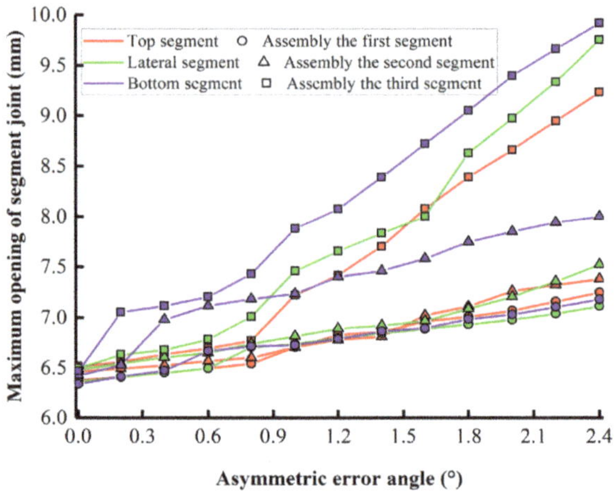

Figure 21. Maximum opening of segment joints in the presence of lateral deflection.

7. Conclusions

The influence of assembly error and the impact of jack thrust on the stress and structural damage of an unlooped segment is generally overlooked in the traditional shield construction process. In this paper, the interaction and damage of segments in the synchronous propulsion and assembly mode were investigated via a suite of multi-segment assembly finite element models. Through the research on the assembling position of the segment and the safety of the assembly process, the displacement, stress, and bending moment distribution of the newly assembled segment under different gravity directions were discussed. The goal of damage analysis is to explore the effects of different assembly errors, gravity directions, and the assembly process on segment damage and seam opening. The obtained results can be used to guide the smooth application of the synchronous propulsion and assembly method in practical engineering. It was found that:

1. In the synchronous propulsion and assembly mode, the segment will rotate radially inward during the assembly process. Under the influence of the propulsion force, the unlooped segment will firstly be squeezed on one side and then clamped on both sides. The rotational freedom of the segment will gradually decrease.
2. The direction of gravity affects the radial rotation of the segment, which in turn affects the interaction between the newly assembled segments. Inward rotation is greatest when the segment is at the top. The stress concentration of the segment will gradually decrease with the clamping effect, and the stress range is between 3.5–10.5 N/mm^2.
3. The bending moment of the segments that are initially assembled at different positions is different, which makes the segments present different bending conditions. With the assembly of adjacent segments, the bending of segments tends to be the same as the influence of the surrounding segments.
4. In the synchronous propulsion and assembly mode of the shield, the segment may be destroyed within the assembly error range. The longitudinal thrust of the jack will be redistributed due to the posture of the segment and the interaction between the segments. The stress concentration between the contact surfaces of the segments leads to the failure of the segments. The radial deflection error is more likely to damage the segment. In actual engineering, if the assembly error reaches the segment damage angle, the operation mode of the shield needs to be adjusted.
5. The maximum opening of the segment is directly related to the assembly error mode, and its peak value is strongly linearly related to the angle of the segment assembly error. Under the effect of assembly error, the maximum opening of the segment can

reach 10.4 mm, which may cause the segment gasket to bear excessive pressure and reduce the waterproof performance of the segment.

Author Contributions: Conceptualization, P.L.; Data curation, J.L. and J.F.; Funding acquisition, P.L.; Investigation, X.K.; Methodology, Z.D.; Project administration, J.F.; Resources, X.K.; Visualization, J.L.; Writing—original draft, Z.D.; Writing—review & editing, P.L. and X.W. All authors have read and agreed to the published version of the manuscript.

Funding: This work was supported by the "Social Development Project of Science and Technology Commission of Shanghai Municipality (21DZ1201105)", "The Fundamental Research Funds for the Central Universities (21D111320)", and "Shanghai Rising-Star Program of China (18QB1403800)".

Institutional Review Board Statement: Not applicable.

Informed Consent Statement: Not applicable.

Data Availability Statement: The data presented in this study are available on request from the corresponding author.

Acknowledgments: Thanks to research section and the members of research group for their great help in the engineering geological prospecting of the project.

Conflicts of Interest: The authors declare no conflict of interest.

References

1. Liu, X.; Yang, Z.; Men, Y. Study on time-varying law of longitudinal pressure between rings of shield tunnel. *J. Geotech. Eng.* **2021**, *43*, 188–193.
2. Yuan, Y.; Jiang, X.; Liu, X. Predictive maintenance of shield tunnels. *Tunn. Undergr. Space Technol.* **2013**, *38*, 69–86. [CrossRef]
3. Shi, P.; Li, P. Mechanism of soft ground tunnel defect generation and functional degradation. *Tunn. Undergr. Space Technol.* **2015**, *50*, 334–344. [CrossRef]
4. Zheng, G.; Cui, T.; Cheng, X.; Diao, Y.; Zhang, T.; Sun, J.; Ge, L. Study of the collapse mechanism of shield tunnels due to the failure of segments in sandy ground. *Eng. Fail. Anal.* **2017**, *79*, 464–490. [CrossRef]
5. Fang, J.; Zhang, Z.; Zhang, J. Application of artificial freezing to recovering a collapsed tunnel in Shanghai metro No.4 line. *China Civ. Eng. J.* **2009**, *42*, 124–128.
6. Jiang, X.; Zhang, X.; Wang, S.; Bai, Y.; Huang, B. Case Study of the Largest Concrete Earth Pressure Balance Pipe-Jacking Project in the World. *Transp. Res. Rec.* **2022**, *6*, 101–125. [CrossRef]
7. Chen, J.; Mo, H. Numerical study on crack problems in segments of shield tunnel using finite element method. *Tunn. Undergr. Space Technol.* **2009**, *24*, 91–102. [CrossRef]
8. Sugimoto, M. Causes of Shield Segment Damages During Construction. In Proceedings of the International Symposium on Underground Excavation and Tunnelling, Bangkok, Thailand, 2–4 February 2006.
9. Cavalaro, S.H.P.; Blom, C.B.M.; Walraven, J.C.; Aguado, A. Structural analysis of contact deficiencies in segmented lining. *Tunn. Undergr. Space Technol.* **2011**, *26*, 734–749. [CrossRef]
10. Lorenzo, S.G. In situ behaviour of an instrumented ring subjected to incipient TBM steering around a curve. *Eng. Struct.* **2021**, *249*, 113–124. [CrossRef]
11. Koyama, Y. Present status and technology of shield tunnelling method in Japan. *Tunn. Undergr. Space Technol.* **2003**, *18*, 145–159. [CrossRef]
12. Gemi, L.; Aksoylu, C.; Yazman. S.; Ozkilic. Y.O.; Arslan, M.H. Experimental investigation of shear capacity and damage analysis of thinned end prefabricated concrete purlins strengthened by CFRP composite. *Compos. Struct.* **2019**, *229*, 111399. [CrossRef]
13. Xu, M.; Zhang, Z. Cause Analysis and Treatment of Segment Damage of Large Diameter Shield Tummel. *Chin. J. Undergr. Space Eng.* **2013**, *9*, 1705–1712.
14. Krahl, P.A.; Palomo, I.I.; Almeida, S.J.C.; Henrique, S.G.; Pinto, J.N.O.; Vieira, J.L.C.M. Tolerances for TBM thrust load based on crack opening performance of fiber-reinforced precast tunnel segments. *Tunn. Undergr. Space Technol.* **2021**, *111*, 103847. [CrossRef]
15. Xu, G.; He, C.; Lu, D.; Wang, S. The influence of longitudinal crack on mechanical behavior of shield tunnel lining in soft-hard composite strata. *Thin-Walled Struct.* **2019**, *144*, 106282. [CrossRef]
16. Lorenzo, S.G. Structural response of concrete segmental linings in transverse interaction with the TBM. Part 2: Non-axisymmetric conditions. *Tunn. Undergr. Space Technol.* **2021**, *116*, 104024. [CrossRef]
17. Lorenzo, S.G. Structural response of concrete segmental linings in transverse interaction with the TBM. Part 1: Axisymmetric conditions. *Tunn. Undergr. Space Technol.* **2021**, *116*, 104023. [CrossRef]
18. Gruebl, F. Segmental rings (critical loads and damage prevention). In Proceedings of the International Symposium on Underground Excavation and Tunnelling, Bangkok, Thailand, 2–4 February 2006.

19. Maidl, B.; Herrenknecht, M.; Maidl, U.; Wehrmeyer, G. *Mechanised Shield Tunnelling*; Ernst & Sohn: Berlin, Germany, 2012.
20. Lorenzo, S.G. The role of temporary spear bolts in gasketed longitudinal joints of concrete segmental linings. *Tunn. Undergr. Space Technol.* **2020**, *105*, 103576. [CrossRef]
21. Liang, X.; Guan, L.; Wen, Z.; Sun, W.; Liu, X. Experimental Study of Structural Performance of Segmental Lining of Rectangular Shield Tunnel during Construction: A Case Study of Underground Connection Gallery in Hongqiao District of Shanghai. *Tunn. Constr.* **2016**, *36*, 1456–1464.
22. Wang, W.; Liu, J. Study on the influence of bolt hole assembly error on longitudinal joint misalignment. In *Waterproof and Drainage Professional Committee of Tunnel and Underground Engineering Branch of China Civil Engineering Society*; Shanghai Tunnel Engineering Rail Transit Design and Research Institute: Shanghai, China, 2015.
23. Li, X.; Yan, Z.; Wang, Z.; Zhu, H. Experimental and analytical study on longitudinal joint opening of concrete segmental lining. *Tunn. Undergr. Space Technol.* **2015**, *46*, 52–63. [CrossRef]
24. Zhu, B.; Li, X.; Fang, E.; Zhu, H. *Parametric Analysis on Joint Opening Width of Concrete Segmental Lining*; Geo-Shanghai: Shanghai, China, 2014.
25. Zhu, H.; Huang, B.; Li, X.; Hashimoto, T. Unified model for internal force and deformation of shield segment joints and experimental analysis. *Chin. J. Geotech. Eng.* **2014**, *36*, 2153–2160.
26. Li, Z.; Soga, K.; Wang, F.; Wright, P.; Tsuno, K. Behaviour of cast-iron tunnel segmental joint from the 3D FE analyses and development of a new bolt-spring model. *Tunn. Undergr. Space Technol.* **2014**, *41*, 176–192. [CrossRef]
27. Zhang, X.; Zhuang, X.; Zhu, H. Study on Boundary Conditions of Three-dimensional Numerical Models for Segment Joints. *Tunn. Constr.* **2014**, *34*, 161–166.
28. Gong, C.; Ding, W.; Mosalam, K.M.; Günay, S.; Soga, K. Comparison of the structural behavior of reinforced concrete and steel fiber reinforced concrete tunnel segmental joints. *Tunn. Undergr. Space Technol.* **2017**, *68*, 38–57. [CrossRef]
29. Liu, M.; Liao, S.; Men, Y.; Xu, J. Analytical solutions and in-situ measurements on the internal forces of segmental lining produced in the assembling process. *Transp. Geotech.* **2020**, *27*, 100478. [CrossRef]
30. Ding, W.; Gong, C.; Mosalam, K.M.; Soga, K. Development and application of the integrated sealant test apparatus for sealing gaskets in tunnel segmental joints. *Tunn. Undergr. Space Technol.* **2017**, *63*, 54–68. [CrossRef]
31. Gong, C.; Ding, W.; Xie, D. Parametric investigation on the sealant behavior of tunnel segmental joints under water pressurization. *Tunn. Undergr. Space Technol.* **2020**, *97*, 103231. [CrossRef]
32. Cao, W.; Shen, W. *Design of Super Large and Extra Long Shield Tunnel: Case Study of Shanghai Yangtze River Crossing Tunnel*; China Architecture & Building Press: Beijing, China, 2010.
33. *CB50010, 2010*; Code for Design of Concrete Structures. China Architecture and Building Press: Beijing, China, 2010. (In Chinese)
34. Weinberg, K.; Khosravani, M.R. On the tensile resistance of UHPC at impact. *Eur. Phys. J. Spec. Top.* **2018**, *227*, 167–177. [CrossRef]
35. Mazars, J. Application de la Mecanique de 1' Endommangement au au Comportement Non Lineaire et a la Rupture du Beton de Structure. Ph.D. Thesis, Université Paris VI, Paris, France, 1984.
36. Lubliner, J.; Oliver, J.; Oller, S.; Onate, E. A Plastic-Damage model for concrete. *Int. J. Solids Struct.* **1989**, *25*, 299–326. [CrossRef]
37. Li, M. Study on Concrete Damage. Master's Thesis, Henan University, Henan, China, 2005.
38. Li, C. Research on the Failure Behavior of Concrete Based on the Theory of Damage and Fracture. Ph.D. Thesis, Southwest Jiaotong University, Chengdu, China, 2012.
39. Sundara, R.; Yogananda, C.V. A three-dimensional stress distribution problem in the anchorage zone of a post-tensioned concrete beam. *Mag. Concr. Res.* **1966**, *18*, 75–84. [CrossRef]
40. Liao, L.; Albert, D.; Cavalaro, S.; Aguado, A.; Carbonari, G. Experimental and analytical study of concrete blocks subjected to concentrated loads with an application to TBM-constructed tunnels. *Tunn. Undergr. Space Technol.* **2015**, *49*, 295–306. [CrossRef]
41. Lei, M.; Zhu, B.; Gong, C.; Ding, W.; Liu, L. Sealing performance of a precast tunnel gasketed joint under high hydrostatic pressures: Site investigation and detailed numerical modeling. *Tunn. Undergr. Space Technol.* **2021**, *115*, 104082. [CrossRef]
42. Gong, C.; Ding, W.; Soga, K.; Mosalam, K.M. Failure mechanism of joint waterproofing in precast segmental tunnel linings. *Tunn. Undergr. Space Technol.* **2019**, *84*, 334–352. [CrossRef]

applied sciences

MDPI

Article

Experimental Study on the Influence of Slurry Concentration and Standing Time on the Friction Characteristics of a Steel Pipe-Soil Interface

Cong Zeng [1], Anfeng Xiao [1], Kaixin Liu [1], Hui Ai [2], Zhihan Chen [1] and Peng Zhang [1,*]

1 Faculty of Engineering, China University of Geosciences (Wuhan), Wuhan 430074, China; zengcong@126.com (C.Z.); xaf18385399786@163.com (A.X.); 15207162405@163.com (K.L.); cugzhchen@163.com (Z.C.)
2 Nuclear Industry JingXiang Construction Group Co., Ltd., Huzhou 313000, China; aihui19810603@163.com
* Correspondence: cugpengzhang@163.com; Tel.: +86-135-5400-5323

Abstract: In pipe jacking, the continued growth of friction resistance around the pipe may cause problems, such as an insufficient jacking force, soil collapse, and even land subsidence, which seriously endangers the structural safety of the pipe and surrounding structures. Bentonite slurry is often used as a lubrication material, but the friction resistance increases due to the inappropriate slurry concentration, and this may cause construction safety problems. In addition, the slurry standing time increases the difficulty of re-jacking construction. To further study the above problems, the friction characteristics of a steel pipe-soil interface under different slurry concentrations and slurry standing times were studied using direct shear tests. The test results show that the peak shear stress and friction coefficient of the interface first decrease and then increase with the increase in the concentration, which is less than or equal to 20%. The peak shear stress and friction coefficient increase with the increase in the concentration, which is greater than 20%, and the position of shear failure changes from between the slurry soil mixture and pipe wall to between the slurry and pipe wall, and finally to the slurry interior. The influence of the slurry standing time on the friction characteristics of the interface is as follows: For the same slurry concentration, the peak shear stress and friction coefficient of the interface increase with the standing time, approximately increasing first and then stabilizing; for different slurry concentrations, the amplification of the peak shear stress and friction coefficient increase with the increase in the concentration.

Keywords: pipe jacking; direct shear test; friction characteristics; slurry concentration; slurry standing time

Citation: Zeng, C.; Xiao, A.; Liu, K.; Ai, H.; Chen, Z.; Zhang, P. Experimental Study on the Influence of Slurry Concentration and Standing Time on the Friction Characteristics of a Steel Pipe-Soil Interface. *Appl. Sci.* **2022**, *12*, 3576. https://doi.org/10.3390/app12073576

Academic Editor: Francesca Scargiali

Received: 2 March 2022
Accepted: 29 March 2022
Published: 31 March 2022

1. Introduction

Pipe jacking is widely used in long-distance natural resource transportation projects because of its high construction accuracy, good safety, and little impact on the surrounding environment [1,2]. During pipe jacking construction, the pipe is continuously pushed into the stratum under the action of jacking force. There is sliding friction between the pipe and the surrounding soil, and the resulting friction resistance of the pipe soil increases with the increase in the jacking distance, which may lead to the problem of insufficient jacking force. However, blindly increasing the jacking force causes structural damage to the working shaft and endangers the construction safety. In addition, the friction resistance has a disturbing effect on the soil, and the increase in the friction resistance results in an enhanced disturbance action on the soil. This may lead to the soil body collapsing, resulting in land subsidence, which seriously endangers the structural stability of surrounding buildings and the safety of pipe jacking construction. Therefore, reducing the friction resistance of pipe soil has become an important link to ensure the smooth construction and construction safety of pipe jacking engineering.

After injecting bentonite slurry into the annular space between the pipe and the surrounding soil, the dry sliding friction between the pipe and the surrounding soil can be transformed into wet sliding friction, which can effectively reduce the friction coefficient of pipe soil to reduce the friction resistance of pipe soil and act as a lubricant. In addition, under the action of grouting pressure, the slurry penetrates and diffuses into the stratum, and combines with the surrounding soil to form a relatively dense mud cake. The mud cake transfers the grouting pressure, which exceeds the formation water pressure, to the soil particles, resulting in soil compaction and support of the stratum [3–5]. Traditional bentonite slurry is generally composed of bentonite, water, soda ash, and CMC according to a certain mass ratio, in which the mass of bentonite usually does not exceed 10% of the mass of water.

Many scholars have carried out a series of research work on pipe jacking slurry. Stein (1989) summarized the friction coefficient between the pipe and the soil in pipe jacking engineering under different stratum conditions, and believed that under the condition of grouting, the friction coefficient between the pipe and soil was in the range of 0.1~0.3 [6]. Jeffferis and Grouts (1992) believed that the ideal drag reduction effect can be achieved only when the slurry is wrapped on the outer pipe wall and the pipe is under the action of buoyancy [7]. Zhou et al. (2009) studied the influence of slurry injection on the jacking force and land subsidence during jacking in the silt stratum through a model test. The test showed that traditional thixotropic slurry has a better drag reduction effect and HL compound slurry is better suited to controlling subsidence [8]. Zhang et al. (2016) optimized lubrication slurry with PAC-HV (polyanionic cellulose high viscosity) as a treatment agent using the orthogonal test for the curved pipe jacking pipe roof project in the Gongbei tunnel [9]. Wang et al. (2018) found that highly viscous bentonite slurry mixed with polymer is more suitable for the lubrication of long-distance steel jacking by comparing the drag reduction effect and grouting amount with traditional bentonite slurry [10]. Chen et al. (2019) studied the influence of bentonite, caustic soda, sodium carboxymethyl cellulose, and hydrolyzed polyacrylamide on the performance parameters of drag reduction slurry, and proposed a formula for drag reduction slurry that is suitable for large-section rectangular pipe jacking [11].

Laboratory tests are a common method used to study the friction characteristics of the pipe–soil interface and calculate the friction coefficient of the interface. Shou et al. (2010) studied the friction characteristics of different kinds of lubricants using a simple laboratory test. The test showed that when "bentonite" is used as the lubricant, the friction coefficient of the interface is reduced by about 23%; and when "bentonite + plasticizer" is used as the lubricant, the friction coefficient of the interface is reduced by about 63% [12]. Namli and Guler (2016) studied the influence of the slurry pressure on the lubricating effect using a laboratory test. The test results showed that when bentonite slurry is continuously injected, even under very low pressure, a lubrication interface is formed between the slurry and the pipe, and the friction coefficient of the interface is reduced by almost 90% [13]. Li et al. (2019) tested the friction coefficient between sandstone and a concrete pipe under seven different contact conditions using the direct shear test. The test showed that the friction coefficient of the pipe–soil under the contact condition of pure bentonite slurry is the smallest [14]. Feng et al. (2022) performed a direct shear test to study the friction characteristics of seven kinds of soil and concrete pipes with or without lubrication slurry. The results showed that under slurry-lubricated conditions, the friction coefficient of the concrete–sand interface could be reduced by 60~80%, and the friction coefficient of the concrete–silty clay interface and the concrete–clay interface could be reduced by about 90% [15].

In summary, the existing research has mainly focused on the friction coefficient of the pipe–soil interface under different contact conditions, drag reduction slurry formula, and traditional performance parameters. Less research has focused on high-concentration bentonite slurry and the addition of a treatment agent to achieve an improvement, especially regarding the influence of the slurry concentration and standing time on the friction characteristics of the pipe–soil interface. However, research on the mechanical characteristic of the interface has mainly concentrated on geotechnical engineering, such as pile foundations, retaining walls, and the soil behind the wall. The influence law of different normal loads, mud skin moisture contents, and mud skin thicknesses on the mechanical properties of soil and the structural surface have mainly been studied through direct shear and single shear tests [16–20].

However, during downgrade jacking of a curved pipe in a water resources transportation project, due to the good fluidity and low viscosity of the traditional slurry, the slurry flows into the tunnel boring machine. In the front pipe, under the action of gravity, and the suction effect of the slag discharge pump of the machine, the slurry is then pumped into the slag discharge pipe, which cannot form a dense mud cake around the pipe wall, resulting in a poor lubrication effect. As such, the compound bentonite is used to prepare a liquid drag reduction slurry and "concentrated slurry" on site. The "concentrated slurry" is injected into the outer pipe wall of two pipes behind the tunnel boring machine, and the liquid drag reduction slurry is injected into the outer pipe wall of a subsequent pipe to alleviate the flow of traditional bentonite slurry into the tunnel boring machine due to gravity to achieve a better drag reduction effect. After the above grouting method is adopted, although the flow of liquid slurry into the tunnel boring machine is effectively alleviated, the jacking force surges in the normal jacking stage and rejacking stage. The jacking force curve is shown in Figure 1. The above problems may be due to unreasonable control of the slurry concentration during slurry configuration and the long interval of the rejacking construction.

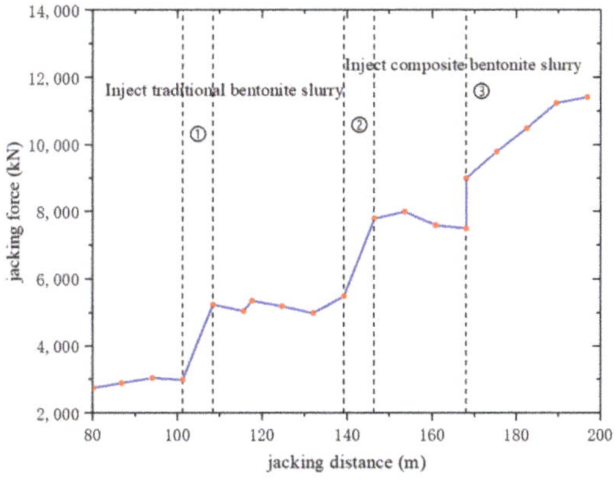

Figure 1. Jacking force–jacking distance curve.

Therefore, this study carried out indoor direct shear tests to deeply study the influence of the compound bentonite slurry on the friction characteristics of the pipe–soil interface from two aspects: the slurry concentration (the mass ratio of the compound bentonite to water, "ω") and the slurry standing time. By analyzing the test data and phenomena, combined with the physical properties of the slurry, the influencing mechanisms of the concentration and standing time on the friction characteristics of the interface were revealed.

2. Friction Characteristic Test

2.1. Test Scheme

The direct shear apparatus was used to define the friction characteristics of the pipe–soil interface under the conditions of different slurry concentrations and standing times. For testing purposes, the shape of the "pipe" was simplified during the test. The steel plate of the raw pipe material used on site was used, and one side of the steel plate used for the preservative treatment was used as the interface. This ensured that the interface between the soil and the outer pipe wall was consistent in practice.

During the test, a set of normal loads were processed by controlling the shear rate to be 5 mm/min (consistent with the jacking rate on site). The test was divided into two stages. In the first stage, the influence law of the concentration on the friction characteristics of the interface was studied. The concentration interval was set to 10~40% (determined according to the concentration interval of on-site slurry preparation), and the peak shear stress and friction coefficient of the pipe–soil interface under the condition of different slurry concentrations were tested.

On the basis of the first stage of the test, the slurry with a smaller friction coefficient of the interface was selected to carry out the second stage of the test, which studied the influence law of the slurry standing time on the friction characteristics of the interface. The standing time was set as 0~36 h, and the peak shear stress and friction coefficient of the interface were tested under different standing time conditions. The effects of the standing time on the peak shear stress and friction coefficient of the pipe–soil interface with different slurry concentrations were compared.

2.2. Test Apparatus and Material

Regarding the direct shear apparatus used in this test, a schematic diagram and physical diagram of the test are shown in Figures 2 and 3, respectively. The apparatus was used to measure the shear stress between the soil and structural plane under different shear rates and different normal loads.

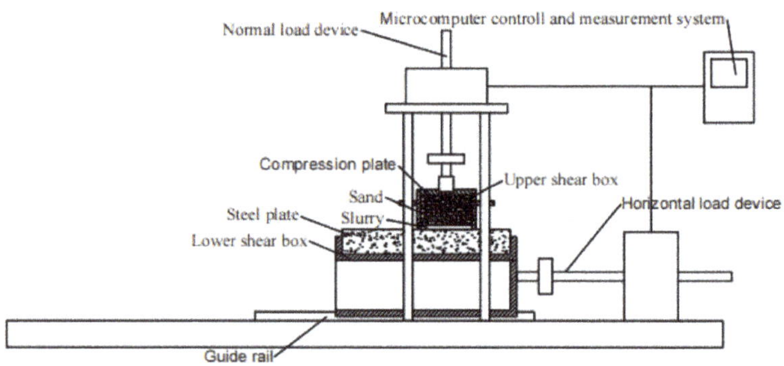

Figure 2. Schematic diagram of the direct shear apparatus.

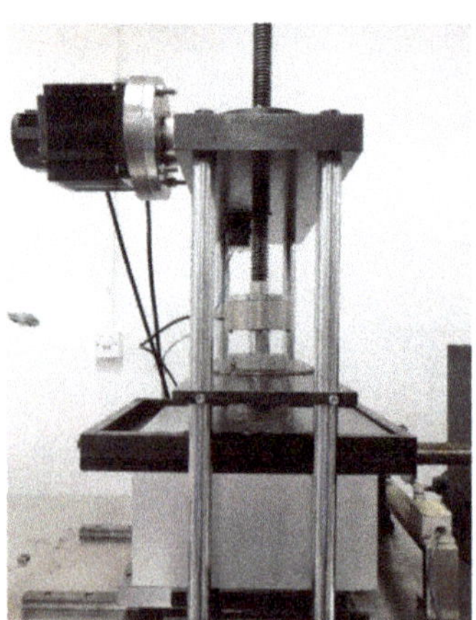

Figure 3. Physical drawing of the direct shear test.

The sand, steel plate, and slurry-making materials used in the indoor test were consistent with those used on site. The size of the steel plate was 350 mm × 250 mm × 20 mm. The slurry-making material used was the special compound bentonite from Trenchless (mainly composed of sodium bentonite and sodium carboxymethyl cellulose). The natural density of sand is 1.73 g/cm^3, the dry density is 1.60 g/cm^3, the natural moisture content is 8.1%, the internal friction angle is 35.5°, and the cohesion is 1.5 kPa. The particle size distribution is shown in Figure 4.

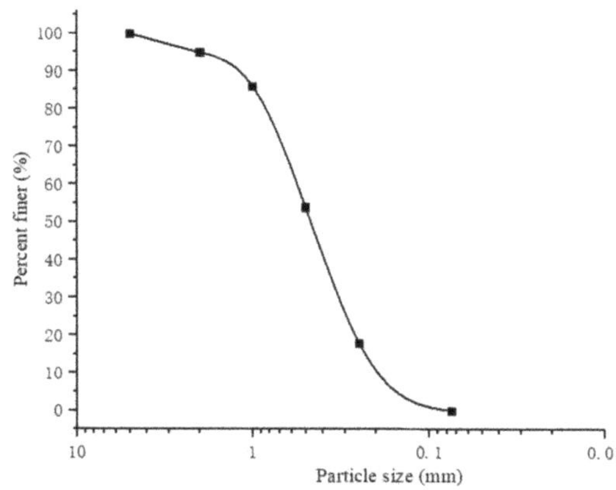

Figure 4. Particle size distribution curve.

2.3. Test Procedures

2.3.1. Slurry Configuration

During each slurry configuration, the quality of water was controlled to 1 kg. Bentonite and water were mixed according to a certain mass ratio and stirred with a mixer for 30 min. After stirring, it was left to stand for 24 h to pre-hydrate. Before each test, it was stirred with a mixer for 10 min. The performance parameters of the slurry with different concentrations are shown in Table 1.

Table 1. Slurry performance parameters at different concentrations.

Slurry Concentration	Slurry Performance Parameters				
	Plastic Viscosity (mPa·s)	Yield Point (Pa)	Initial Gel Strength/10-min Gel Strength (Pa)	Funnel Viscosity (Second)	Filtration Rate (mL/30 min)
ω = 10%	41	27.08	16.35/21.46	136 s	7.2
ω = 12%	44	30.10	17.50/22.50	150 s	7.0
ω = 14%	48	32.54	19.23/24.85	172 s	6.5
ω = 16%	55	36.70	23.30/28.06	190 s	5.8
ω = 18%	59	40.00	26.50/32.40	dropwise	5.2
ω = 20%	65	43.85	30.84/36.46	dropwise	4.5
ω = 25%	-	-	-	hardly drop	hardly filtrate
ω = 30%	-	-	-	hardly drop	hardly filtrate
ω = 35%	-	-	-	hardly drop	hardly filtrate
ω = 40%	-	-	-	hardly drop	hardly filtrate

2.3.2. Sample Filling

The compression plate (diameter 87 mm, height 5 mm) was placed in the upper shear box (outer diameter 98 mm, inner diameter 88 mm, height 100 mm), and then sand was used to fill the shear box. The sand was compacted to a filling height of 85 mm, and finally, the configurated slurry was used to fill the shear box and scraped with a scraper. When the friction characteristics of the interface between the steel plate and sand were tested, it was only necessary to fill the shear box with sand and compact it.

2.3.3. Shear Test

During the test, the upper shear box was controlled to be fixed, shear was generated by the movement of the lower shear box, and a certain normal load was applied to the upper shear box. After loading, the slurry that precipitated from the upper shear box was cleaned with a scraper, and the horizontal motor was controlled to drive the screw to push the lower shear box at a speed of 5 mm/min. The friction characteristics of the pipe–soil interface under the conditions of different concentrations and standing times were tested.

3. Test Result

3.1. Influence of the Slurry Concentration on the Friction Characteristics of the Interface

The normal load has no effect on the variation trend of the shear stress ("τ")–shear displacement ("s") curve of the interface and the friction coefficient of the interface, but only for the value of the shear stress of the interface. Therefore, taking the normal load of 100 kPa as an example, the variation trend of the shear stress–shear displacement curve of the pipe–soil interface under the normal load was analyzed (to facilitate the analysis of the friction characteristics of the interface, the pipe–soil interface without slurry was recorded as "interface I", and the pipe–soil interface with slurry was recorded as "interface II"), and the curve is shown in Figure 5. The following test results are shown in Figure 5.

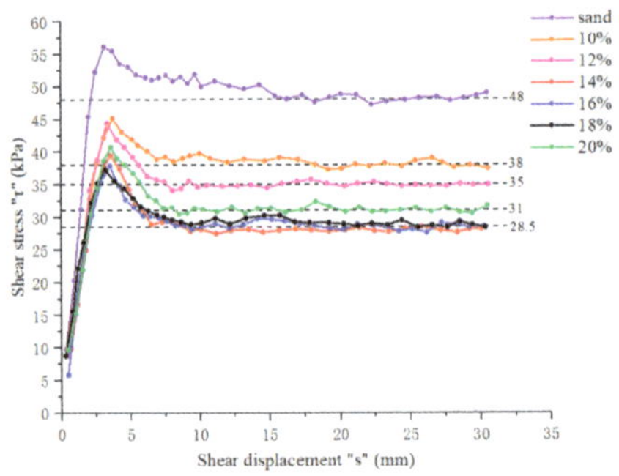

(a)

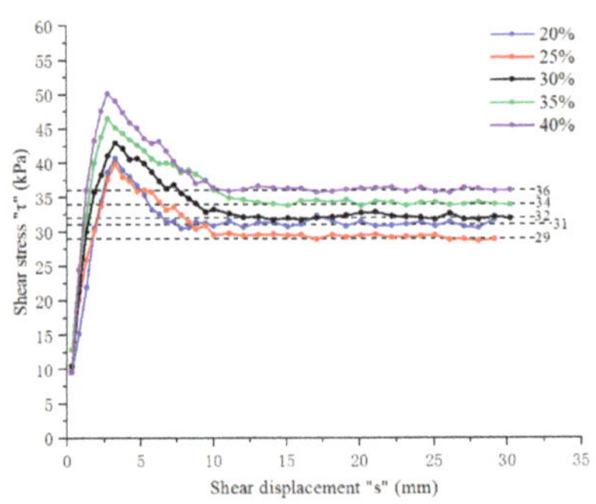

(b)

Figure 5. Friction characteristics of the pipe-soil interface with different slurry concentrations. (**a**) "τ–s"curve of the pipe-soil interface without slurry and with slurry with a con-centration range of 10~20%. (**b**) "τ–s" curve of the pipe-soil interface with slurry with a concentration range of 20~40%.

The peak value and stable value of the "τ–s" curve of "interface I" were greater than those of the "interface II" curve. This shows that the slurry reduced the shear stress on the interface between the pipe and soil, which means that the slurry reduced the friction when the pipe wall slid relative to the sand, and the slurry had a lubrication effect.

The "τ–s" curve of "interface I" had an obvious peak value. When the shear displacement continued to increase, the shear stress of the interface gradually declined and tended to a constant value, and the curve showed obvious strain softening characteristics. This occurred as the compacted sand particles were closely arranged, and the sand particles underwent physical changes, such as tumbling and dislocation, during the shear process, resulting in relative rolling between the particles. With the continuous shear, the particle positions were rearranged.

The "τ–s" curve of "interface II" also showed that was a strain softening type. This was because the slurry is thixotropic, which under the action of shear, caused the viscosity and static shear stress of the slurry to decrease while after standing, the viscosity and static shear stress increased. Under the action of slurry thixotropy, when the pipe wall and sand slid relatively, the shear stress on the interface rose rapidly and reached the peak value. With the further development of shear, the viscosity and static shear stress of the slurry under shear declined. At this time, the shear stress decreased from the peak value and tended to a stable value [21].

3.1.1. Concentration Less Than or Equal to 20%

The test data listed in Table 2 show that the peak shear stress ("τ_{max}") first decreased and then increased with the increase in the concentration, which is less than or equal to 20%. This was because the slurry with a concentration less than or equal to 12% did not support the soil well, resulting in contact between the slurry soil mixture and the pipe wall during shearing, and the occurrence of shear failure between the slurry soil mixture and the pipe wall (as shown in Figure 6a). With the increase in the slurry concentration, the contact degree between the sand and pipe wall was reduced and the supporting effect of the soil improved. The shear failure occurred between the pipe wall and slurry (as shown in Figure 6b,c), and the shear stress required to produce relative sliding decreased, so ${\tau_{max}}^{\omega=10\%} > {\tau_{max}}^{\omega=12\%} > {\tau_{max}}^{\omega=14\%} > {\tau_{max}}^{\omega=16\%} > {\tau_{max}}^{\omega=18\%}$. However, with the increase in the concentration, the viscosity of the slurry increased and the degree of adhesion to the pipe wall increased, so ${\tau_{max}}^{\omega=20\%} > {\tau_{max}}^{\omega=16\%} > {\tau_{max}}^{\omega=18\%}$.

Table 2. Friction test data of the pipe–soil interface with different concentrations.

	Normal Load "σ" (kPa)	Shear Displacement "s" (mm)	Peak Shear Stress "τ_{max}" (kPa)	Friction Coefficient "μ"
soil	100	3.075	56.155	0.48
ω = 10%	100	3.701	45.149	0.38
ω = 12%	100	3.295	44.413	0.35
ω = 14%	100	3.562	39.467	0.28
ω = 16%	100	3.509	37.896	0.285
ω = 18%	100	3.175	37.231	0.29
ω = 20%	100	3.578	40.770	0.31
ω = 25%	100	3.515	39.962	0.29
ω = 30%	100	3.488	42.962	0.32
ω = 35%	100	3.003	46.546	0.34
ω = 40%	100	2.804	50.218	0.36

Note: the shear displacement in the table is the shear displacement corresponding to the peak shear stress in the "τ–s" curve, and the friction coefficient is the ratio of the stable value of shear stress to the normal load.

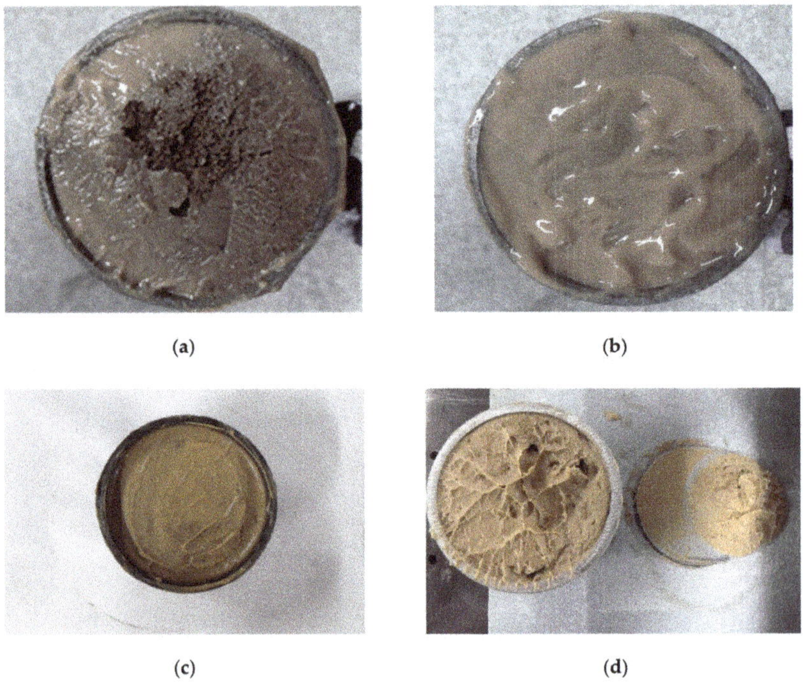

Figure 6. Friction characteristics of the pipe–soil interface with different slurry concentrations. (**a**) Shear failure surface of slurry with a concentration equal to 12%. (**b**) Shear failure surface of slurry with a concentration equal to 14%. (**c**) Shear failure surface of slurry with a concentration equal to 20%. (**d**) Shear failure surface of slurry with a concentration equal to 25%.

The friction coefficient ("μ") first decreased and then increased with the increase in the concentration, which is less than or equal to 20%. This occurred because the lower-concentration slurry did not support the soil and did not have a good lubrication effect. With the increase in the slurry concentration, the "barrier-lubrication" effect of the slurry improved, so $\mu^{\omega=14\%} < \mu^{\omega=12\%} < \mu^{\omega=10\%}$. However, the further increase in the slurry concentration increased the viscosity of the slurry and reduced its lubricity. Therefore, for slurry to have a good lubrication effect, it must have a good supporting role.

3.1.2. Concentration Greater Than 20%

Table 2 shows that the peak shear stress and friction coefficient increased with the increase in the concentration, which is greater than 20%. This is occurred because the slurry with a concentration greater than 20% was semi-solid, its "support-barrier" soil effect was remarkable, and the adhesion degree with the pipe wall was greatly improved. When the sand slid relative to the pipe wall, shear failure occurred in the slurry interior (as shown in Figure 6d). Therefore, when the pipe soil slid relatively, it had to overcome the spatial structure strength of the slurry, which is the gel strength. With the further development of shear, the relative sliding of the pipe soil was subjected to the cohesion of particles in the slurry, which is internal friction. As the gel strength (or static shear stress) of the slurry increased with the increase in the solid particle content, and the internal friction increased with the increase in the solid particle content, the greater the slurry concentration, the greater the peak shear stress and friction coefficient.

After comparing the test data of the slurry with a concentration of 20% and 25%, it was found that $\mu^{\omega=25\%} < \mu^{\omega=20\%}$ ${\tau_{max}}^{\omega=25\%} < {\tau_{max}}^{\omega=20\%}$. This is because shear failure of the former occurred between the pipe wall and the slurry while the shear failure of the latter occurred in the slurry interior, and the slurry with a concentration equal to 25% slurry had a lower solid particle content than the other higher-concentration slurry, so the lubricity was better.

In summary, the concentration has a critical value, which is 20%. When the concentration does not exceed the critical concentration, the slurry state is liquid; otherwise, it is semi-solid. Liquid slurry is mainly used for lubrication, and semi-solid slurry is mainly used to support soil. Different concentrations of slurry have different effects on the peak shear stress and friction coefficient of the pipe–soil interface. For liquid slurry, the peak shear stress of the interface depends on the structural shear strength of the interface, and the friction coefficient depends on the degree of friction between the slurry and the pipe wall. For semi-solid slurry, the peak shear stress depends on the spatial structure strength of the slurry, which is the static shear stress, and the friction coefficient depends on the internal friction of the slurry.

3.2. Influence of the Standing Time on the Friction Characteristics of the Interface

By analyzing the test data of the last stage and considering the slurry performance parameters, 3 groups of slurries with an improved lubrication performance of 14%, 16%, and 18% were selected for the test of this stage.

As shown in Figure 7, the standing time had little effect on the "τ–s" curve of the interface, and the curve was still the strain softening type. Under the same concentration condition, the "τ–s" curve of the interface of the slurry standing times of 24 and 36 h approximately coincided. According to the data shown in Table 3, under the same concentration, the peak shear stress and friction coefficient increased first and then stabilized with the increase in the slurry standing time.

Table 3. Friction test data of the pipe–soil interface at different standing times.

Concentration "ω"	Standing Time (h)	Normal Load "σ" (kPa)	Shear Displacement "s" (mm)	Peak Shear Stress "τ_{max}" (kPa)	Friction Coefficient "μ"
14%	0	100	3.562	39.467	0.28
	6	100	3.441	41.430	0.29
	12	100	3.192	46.301	0.31
	24	100	3.142	53.050	0.34
	36	100	3.138	53.810	0.34
16%	0	100	3.509	37.900	0.285
	6	100	3.401	44.706	0.305
	12	100	3.170	48.500	0.33
	24	100	2.939	53.699	0.35
	36	100	3.061	54.383	0.35
18%	0	100	3.175	37.231	0.29
	6	100	3.009	46.368	0.32
	12	100	2.881	50.707	0.34
	24	100	2.680	53.913	0.36
	36	100	2.646	54.439	0.36

Note: the shear displacement in the table is the shear displacement corresponding to the peak shear stress in the "τ–s"curve, and the friction coefficient is the ratio of the stable value of shear stress to the normal load.

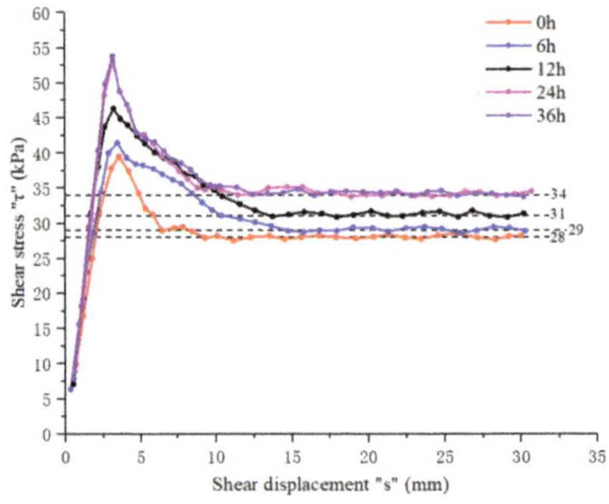

(a)

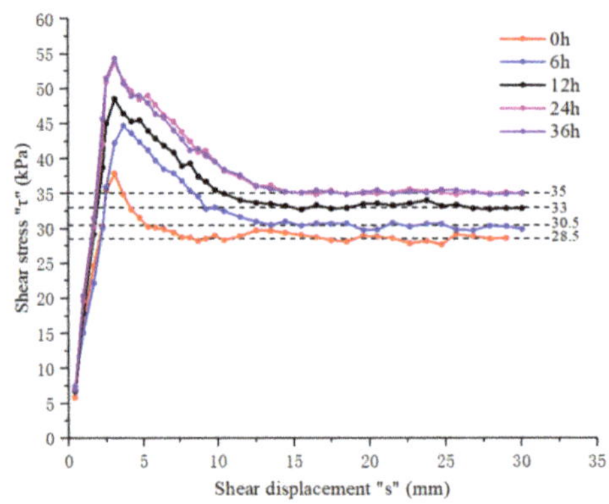

(b)

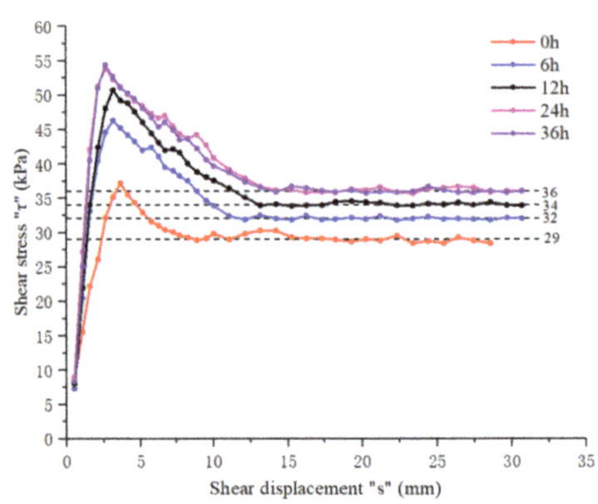

(c)

Figure 7. Friction characteristics of the pipe–soil interface with different slurry standing times. (**a**) "τ–s" curve of the pipe–soil interface with slurry with a concentration range of 14%. (**b**) "τ–s" curve of the pipe–soil interface with slurry with a concentration range of 16%. (**c**) "τ–s" curve of the pipe–soil interface with slurry with a concentration range of 18%.

The reason for this change trend was that, on the one hand, the slurry is thixotropic. After standing for a period of time, the viscosity and static shear stress of the slurry increased. For the slurry with a concentration range of 14~18%, shear failure occurred between the pipe wall and slurry. With the further development of shear, part of the slurry damaged by shear overflowed from the upper shear box. The other part was mixed with the slurry that was not damaged by shear and continued to contact with the pipe wall and produce shear. When the shear rate was smaller and the shear displacement was limited, the peak shear stress and friction coefficient increased with time. On the other hand, the thixotropy of the slurry is related to the type and content of the solid particles and whether treatment agent is added. When the contents of bentonite and treatment agent are certain, the viscosity and static shear stress of the slurry will not increase indefinitely with time [21], so the peak shear stress and friction coefficient will tend to be stable.

Under the same standing time, the friction coefficient increased with the increase in the concentration. This occurred because the slurry with a concentration range of 14~18% had a better "barrier-lubrication" effect, and its lubricity decreased with the increase in the concentration. However, when the standing time was greater than or equal to 6 h, the peak shear stress increased with the increase in the concentration, which is the opposite to what occurred when the standing time was 0 h. This occurred because after the slurry had stood for a period of time, its solid particles were overlapped with each other and formed a network structure, which made the slurry flocculate into colloid, and its viscosity and shear stress increased, which had a significant role in supporting the soil. At this time, the increase in the concentration increased the adhesion between the slurry and the pipe wall, resulting in the increase in the peak shear stress.

After comparing the data in Table 4, it can be seen that under the same standing time, the peak shear stress and friction coefficient of the pipe–soil interface with higher concentrations of slurry increased greatly. However, with the increase in the standing time, the amplification gap of the peak shear stress and friction coefficient on the interface with different concentrations of slurry gradually declined.

Table 4. Amplification of the peak shear stress and friction coefficient of the pipe–soil interface with different standing times.

Standing Time (h)	Amplification of Peak Shear Stress			Amplification of Friction Coefficient		
	ω = 14%	ω = 16%	ω = 18%	ω = 14%	ω = 16%	ω = 18%
6 h	5%	18.0%	28.3%	3.6%	7.0%	10.3%
12 h	17.3%	28.0%	36.2%	10.7%	15.8%	17.2%
24 h	34.4%	41.6%	44.8%	21.4%	22.8%	24.1%
36 h	36.34%	43.5%	46.2%	21.4%	22.8%	24.1%

4. Discussion

Considering that the time interval of normal re-jacking on site generally does not exceed 6 h, by comparing the data in Table 4, it was found that the amplification of the peak shear stress and friction coefficient of the interface with a slurry concentration equal to 18% was the greatest. However, excessive amplification of the friction coefficient and peak value of the shear stress will lead to a significant increase in the jacking force required for re-jacking, which is highly likely to lead to surging of the jacking force and increase the difficulty of re-jacking. Combined with the influence law of the concentration and standing time on the friction characteristics of the pipe–soil interface, the slurry with a concentration range of 14~16% showed a better lubrication effect and was less affected by the standing time.

According to the test data, the friction coefficient of the pipe–soil interface first decreased and then increased with the increase in the slurry concentration. This is because when the slurry concentration was less than or equal to 12%, the slurry did not support the soil well, resulting in contact between the slurry soil mixture and the pipe wall during shearing, and the friction coefficient of the interface was larger. When the slurry concentration increased, the "barrier-lubrication" effect of the slurry was improved, so the interface friction coefficient decreased. With the further increase in the slurry concentration, the adhesion between the slurry and pipe wall increased, so the friction coefficient increased. Within a certain standing time range, the interface friction coefficient increased with the increase in the standing time. This is because slurry is thixotropic. The standing time changes the state of the slurry, changing the slurry from a flowing state to a flocculation state. The viscosity of the slurry increases and the adhesion degree with the pipe wall increases, so the friction coefficient also increases. This is also almost consistent with Li's conclusion regarding the influence of the pipe jacking stagnation time on the pipe–soil interface [22].

Staheli (2006) proposed an equation to calculate the jacking force as follows [23]:

$$JF_{frict} = \mu_{\text{int}} \frac{\gamma \cdot r \cdot \cos\left(45 + \frac{\phi_r}{2}\right)}{\tan \phi_r} \cdot \pi \cdot d \cdot l \quad (1)$$

where JF_{frict} is the frictional component of the jacking force, μ_{int} is the soil pipe residual interface frictional coefficient, γ is the unit weight of the soil, ϕ_r is the residual friction angle of the soil, d is the outer diameter of the pipe, r is the pipe radius, and l is the length of the pipe.

According to the above formula, the frictional component of the jacking force is mainly related to the friction coefficient of the interface. The factors of the slurry concentration and slurry standing time studied in this paper have a significant effect on the friction coefficient of the interface, thus affecting the jacking force. Therefore, in future engineering practice, experiments should be carried out to study the effects of the slurry concentration and standing time on the friction characteristics of the pipe–soil interface, and to test the friction coefficient of the interface under the conditions of different slurry concentrations and standing times. This will provide an important basis for the calculation of the jacking force and provide guiding significance for the configuration of drag reduction slurry in engineering practice.

5. Conclusions

In this study, through the indoor direct shear test, the influence law of the slurry concentration and standing time on the friction characteristics of the pipe–soil interface was explored, and the following five conclusions were obtained.

The concentration changes the location of shear failure. When the concentration is greater than or equal to 25%, the shear failure occurs in the slurry interior; when the concentration is less than or equal to 12%, the shear failure occurs between the slurry soil mixture and the pipe wall; and when the concentration is more than 12% and less than 25%, the shear failure occurs between the pipe wall and the slurry.

Different concentrations of slurry have different effects on the peak shear stress and friction coefficient of the pipe–soil interface. For the pipe–soil interface with a slurry concentration of less than or equal to 20%, the peak shear stress depends on the structural shear strength of the interface, and the friction coefficient depends on the degree of friction between the slurry and the pipe wall. For the pipe–soil interface with a slurry concentration greater than 20%, the peak shear stress depends on the gel strength of the slurry, which is the static shear stress, and the friction coefficient depends on the internal friction of the slurry.

When the concentration is less than or equal to 20%, the peak shear stress and friction coefficient first decrease and then increase with the increase in the concentration. When the concentration is more than 20%, the peak shear stress and friction coefficient increase with the increase in the concentration.

When the concentration is unchanged, within a certain standing time range, the peak shear stress and friction coefficient of the interface increase with the increase in the standing time, and then are unchanged. When the standing time is greater than or equal to 6 h, the peak shear stress and friction coefficient increase with the increase in the concentration. Under the same standing time, the amplification of the peak shear stress and friction coefficient of the interface with different slurry concentrations increase with the increase in the concentration.

Unreasonable control of the slurry concentration and long standing times will increase the risk of the jacking force surging and adversely affect pipe jacking construction. Slurry with a concentration range of 14~16% has a better lubrication effect and is less affected by the standing time. Although the friction coefficient of the pipe–soil interface of the slurry within this concentration range belongs to the range of 0.1~0.3, the friction coefficient is slightly larger. Therefore, it is suggested that the slurry concentration should be within 16% and lubricants, such as polyacrylamide and graphite powder, should be added to further reduce the friction coefficient of the pipe–soil interface.

Author Contributions: Conceptualization, C.Z. and P.Z.; methodology, C.Z. and P.Z.; software, A.X.; validation, A.X., K.L. and Z.C.; formal analysis, A.X. and K.L.; investigation, Z.C.; resources, H.A.; data curation, K.L.; writing—original draft preparation, A.X.; writing—review and editing, A.X.; visualization, Z.C.; supervision, Z.C.; project administration, P.Z. All authors have read and agreed to the published version of the manuscript.

Funding: This research received no external funding.

Institutional Review Board Statement: Not applicable.

Informed Consent Statement: Not applicable.

Data Availability Statement: Not applicable.

Conflicts of Interest: The authors declare no conflict of interest.

References

1. Ma, B.S.; Najafi, M. Development and applications of trenchless technology in China. *Tunn. Undergr. Space Technol.* **2008**, *23*, 476–480. [CrossRef]
2. Shou, K.J.; Chang, F.W. Analysis of pipe-soil interaction for a miniature pipe jacking. *J. Mech.* **2006**, *22*, 213–220. [CrossRef]
3. Wei, G.; Xu, R.Q.; Shao, J.M.; Luo, M.H.; Jin, Z.L. Research on mechanism of reducing friction through injecting slurry in pipe jacking. *Rock Soil Mech.* **2004**, *25*, 930–934.
4. Khazaei, S.; Shimada, H.; Matsui, K. Analysis and prediction of thrust in using slurry pipe jacking method. *Tunn. Undergr. Space Technol.* **2004**, *19*, 356.
5. Cheng, W.C.; Wang, L.; Xue, Z.F.; Ni, J.C.; Rahman, M.M.; Arulrajah, A. Lubrication performance of pipejacking in soft alluvial deposits. *Tunn. Undergr. Space Technol.* **2019**, *91*, 102991. [CrossRef]
6. Dietrich, S. *Microtunnelling: Installation and Renewal of Nonman-Size Supply and Sewage Lines by the Trenchless*; Ernst & Sohn Verlag Fur Architektur Und Technische Wissenschaften: Berlin, Germany, 1989.
7. Jefferis, S.A.; Grouts, S. *Chapter 48 of Construction Materials Reference Book*; Doran, D.K., Ed.; Butterworth-Heineman: Oxford, UK, 1992.
8. Zhou, S.; Wang, Y.; Huang, X. Experimental study on the effect of injecting slurry inside a jacking pipe tunnel in silt stratum. *Tunn. Undergr. Space Technol.* **2009**, *24*, 466–471. [CrossRef]
9. Zhang, P.; Ma, B.; Zeng, C.; Xie, H.; Li, X.; Wang, D. Key techniques for the largest curved pipe jacking roof to date: A case study of Gongbei tunnel. *Tunn. Undergr. Space Technol.* **2016**, *59*, 134–145. [CrossRef]
10. Wang, J.; Wang, K.; Zhang, T.; Wang, S. Key aspects of a DN4000 steel pipe jacking project in China: A case study of a water pipeline in the Shanghai Huangpu River. *Tunn. Undergr. Space Technol.* **2018**, *72*, 323–332. [CrossRef]
11. Chen, X.; Ma, B.; Najafi, M.; Zhang, P. Long Rectangular Box Jacking Project:A Case Study. *Undergr. Space* **2021**, *6*, 101–125. [CrossRef]
12. Shou, K.; Yen, J.; Liu, M. On the frictional property of lubricants and its impact on jacking force and soil-pipe interaction of pipe-jacking. *Tunn. Undergr. Space Technol.* **2010**, *25*, 469–477. [CrossRef]
13. Namli, M.; Guler, E. Effect of Bentonite Slurry Pressure on Interface Friction of Pipe Jacking. *J. Pipeline Syst. Eng. Pract.* **2016**, *8*, 04016016.1–04016016.9. [CrossRef]
14. Li, C.; Zhong, Z.; Liu, X.; Tu, Y.; He, G. Numerical simulation for an estimation of the jacking force of ultra-long-distance pipe jacking with frictional property testing at the rock mass-pipe interface. *Tunn. Undergr. Space Technol.* **2019**, *89*, 205–221. [CrossRef]
15. Feng, X.; Zhang, Y.; Yuan, X.; Zhang, P. An Experimental Study on Frictional Properties of Concrete Pipe-Soil Interface. *J. Pipeline Syst. Eng. Pract.* **2022**, *13*, 04022004. [CrossRef]

16. Chen, C.; Wu-ming, L.; Qi, Y.; Zi-hao, J.; Ru-song, N.; Jun, Q. Experimental study of mechanical properties of concrete pile-slurry-sand interface. *Rock Soil Mech.* **2018**, *39*, 2461–2472.
17. Peng, K.; Zhu, J.G.; Zhang, D.; Wu, X.Y. Study of mechanical behaviors of interface between coarse-grained soil and concrete by simple shear test. *Chin. J. Rock Mech. Eng.* **2010**, *29*, 1893–1900.
18. Zhang, Z.J.; Rao, X.B.; Wang, Z.J.; Ding, H.S. Experimental study on influence of slurry thickness on mechanical behavior of interface between gravel and concrete. *Rock Soil Mech.* **2008**, *29*, 2433–2438.
19. Peng, K.; Zhu, J.G.; Wu, X.Y. Experiments on influence of slurry kinds on mechanical behavior of interface between gravel and concret. *J. Chongqing Univ.* **2011**, *34*, 110–115.
20. Zhao, W.; Li, T.L.; Han, J.Y.; Cheng, C. Experiment of shear properties on the interface between gravelly sand and concrete pipe. *J. Northeast. Univ. (Nat. Sci.)* **2020**, *41*, 424–429, 456.
21. Zhang, P.; Tan, L.X.; Ma, B.S. Formulae for frictional resistance considering mud thixotropy and pipe-soil contact characteristics. *Chin. J. Geotech. Eng.* **2017**, *39*, 2043–2049.
22. Li, T.; Zhao, W.; Liu, R.; Han, J.; Cheng, C. Experimental Study on the Pipe-Soil Interface under the Influence of Pipe Jacking Stagnation Time. *KSCE J. Civ. Eng.* **2021**, *26*, 1428–1438. [CrossRef]
23. Staheli, K. *Jacking Force Prediction: An Interface Friction Approach Based on Pipe Surface Roughness*; Georgia Institute of Technology: Atlanta, GA, USA, 2006.

applied sciences

MDPI

Article

Design Optimization Method of Feet-Lock Steel Pipe for Soft-Rock Tunnel Based on Load-Deformation Coordination

Yimin Wu [1], Changqing Tian [1], Peng Xu [1,*], Zhizhong Zhao [2], Jiawei Zhang [1] and Sijie Wang [3]

1 School of Civil Engineering, Central South University, Changsha 410075, China; wuyimin531@csu.edu.cn (Y.W.); 204811052@csu.edu.cn (C.T.); csuzhangjw@csu.edu.cn (J.Z.)
2 Gansu Provincial Highway Aviation Tourism Investment Group Co. Ltd., Lanzhou 730031, China; zzz30060232@163.com
3 Shanxi Road and Bridge Construction Group Co., Ltd., Taiyuan 030006, China; 15733170607@163.com
* Correspondence: xupeng1033@csu.edu.cn

Abstract: Feet-lock steel pipes have been widely applied in soft-rock tunnels, but their installation quantity and parameters depend on engineering experience. In this study, by simplifying the force between the feet-lock steel pipe and the primary lining, the load-deformation characteristics were analyzed separately, and inconsistent deformation was revealed. Next, combined with the passive and active bearing scheme, a design optimization method for feet-lock steel pipes was proposed under two working conditions, including the predicted large deformation and observed large deformation before the primary lining construction. Finally, the design method was applied to the Yulinzi Tunnel, and the results show that the original design of the feet-lock steel pipes could not meet the requirements of the controlling settlement. After the optimized design, both the passive bearing scheme installing two ϕ108 × 3-millimeter steel pipes and the active bearing scheme installing one ϕ108 × 4-millimeter steel pipe could meet the requirements of the controlling settlement. The active bearing design can significantly reduce the amount of steel required and the engineering cost. The scientificity and feasibility of the design optimization method were proven by the on-site monitoring and measurement.

Keywords: feet-lock steel pipe; design optimization method; load-deformation coordination; active bearing; soft-rock tunnel; engineering application

Citation: Wu, Y.; Tian, C.; Xu, P.; Zhao, Z.; Zhang, J.; Wang, S. Design Optimization Method of Feet-Lock Steel Pipe for Soft-Rock Tunnel Based on Load-Deformation Coordination. *Appl. Sci.* **2022**, *12*, 3866. https://doi.org/10.3390/app12083866

Academic Editor: Daniel Dias

Received: 15 March 2022
Accepted: 10 April 2022
Published: 11 April 2022

1. Introduction

The large settlement of tunnel vaults often occurs in soft-rock tunnels, mainly due to primary lining arch foot subsidence. The stability of the tunnel is a critical issue in the construction process [1,2]. If the settlement of the vault is not controlled, it may cause a large-scale collapse. At present, the most commonly adopted methods to control the subsidence of the arch foot include enlarging the arch foot, connecting the longitudinal beams, and installing feet-lock steel pipes (FLSPs) [3]. Among these methods, FLSP has the advantages of simple construction and a remarkable ability to control the settlement [4] and has been extensively utilized to control the settlement of tunnel vault in soft-rock areas [5–7]. The TFP is an inclined downward steel pipe with one end inserted into the surrounding stratum, and the other end welded to the steel arch [8], as shown in Figure 1. To explore and fully exploit the function of FLSP, it is necessary to carry out an in-depth study on the mechanical characteristics of FLSP and the selection of the relevant parameters.

At present, domestic and foreign scholars have carried out a series of studies on the mechanical characteristics of FLSP, utilizing theoretical analysis [9–11], numerical simulation [12–14], and field-test methods [15,16]. Chen et al. [11] proposed an analytical method to predict the vertical load and settlement at the feet of steel ribs with the support of the feet-lock pipe and validated this method with the results of field measurements. Cui et al. [14] explored the mechanism of action of foot-side reinforcement piles (FRSP) through model tests and numerical simulations and determined reasonable parameters. The results show that when the FRSP is long enough to cross the shear line, it can play a role in shear reinforcement, load redistribution, and internal pressure. FRSP effectively prevents the settlement of the tunnel and its surrounding foundations. Deng et al. [17] compared the effect of the feet-lock bolt and the systematic bolt under the condition of a cracked surrounding rock tunnel through a field test and a numerical simulation. The results show that applying feet-lock bolts is a feasible option to replace systematic bolts in cracked surrounding rock tunnels. Luo et al. [15] obtained the distribution law of the axial strain and the bending moment of the FLSP through electrical measurement and verified the accuracy of their test method by comparing the deformation of the pipe body before and after loading.

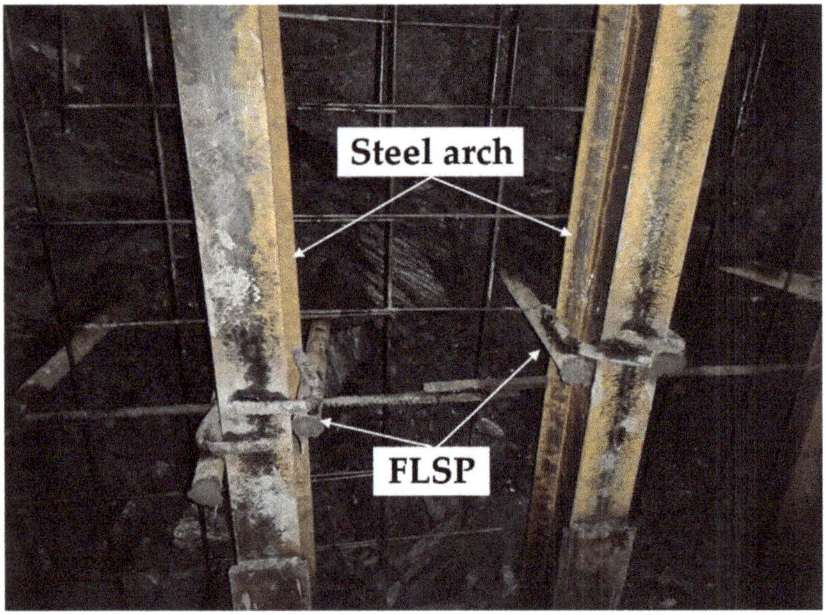

Figure 1. Feet-lock steel pipe.

In summary, it can be seen that the theoretical research on FLSPs has been relatively comprehensive. However, there is a lack of research on the parameter selection and installation quantity of FLSPs in actual projects. FLSPs need to undergo a certain level of deformation before providing the corresponding bearing capacity. Based on the concept of prestressed steel bars, the active load-bearing of FLSP is worth exploring. In order to guide the construction and design of FLSPs, a design optimization method is proposed in this paper. Firstly, the problem of the load-deformation inconsistency of FLSPs in controlling the vault settlement is raised. Next, the process of the design optimization method is explained. Finally, the design method is applied to the Yulinzi Tunnel. The change in the vault settlement before and after the design optimization was analyzed by monitoring and measurement and is presented here.

2. Inconsistent Load-Deformation Problem of FLSPs

2.1. Action Mechanism and Load-Deformation Characteristics of FLSPs

2.1.1. The Action Mechanism and Simplified Model of a FLSP

The force of FLSPs comprises two parts: on the one hand, the primary lining load (the axial force, the shear force, and the bending moment), and, on the other hand, the interaction between the FLSP and the surrounding rock (the friction along the direction of the pipe wall and the elastic stiffness perpendicular to the direction of the pipe wall). Under the primary lining axial load, FLSP exhibits displacement trends relative to the soil. The friction resistance between the soil and FLSP prevents the axial displacement of the FLSP. The FLSP bends under the shear and bending moment, and the soil resists deformation thanks to the elastic stiffness. Engineering practice shows that FLSPs mainly bear shear force, and the action between the FLSP and the primary lining is complicated. To simplify the calculation, the force between the FLSP and the primary lining is simplified as a pair of shear forces.

The mechanical analysis models of the interaction between the FLSP and the soil under shear force includes the Winkler, elastic continuum, and two-parameter foundation models. The Winkler model is shown in Figure 2. The model assumes that the foundation reaction force received at any point of the FLSP is proportional to the deflection:

$$\mathrm{P}(x) = KDy(x) \tag{1}$$

where P(x) is the foundation reaction force per unit length of the steel pipe, K is the elastic coefficient of the soil (MPa/m), D is the outer diameter of the FLSP (m), and $y(x)$ is the deflection of the FLSP (m).

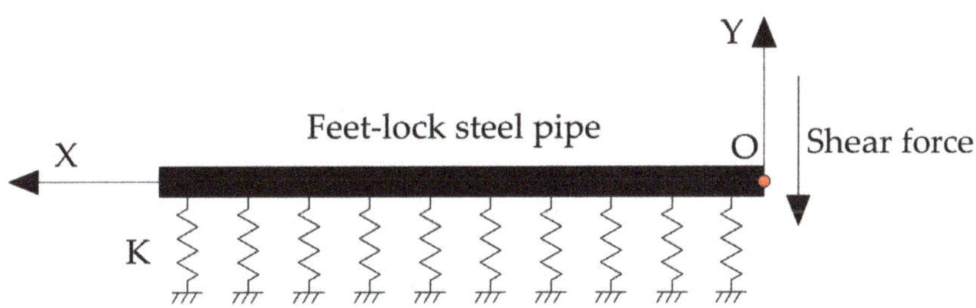

Figure 2. Winkler model of FLSP.

When the FLSP is subjected to shear load at the end, the differential equation of its deflection curve is:

$$\frac{\mathrm{d}^4 y}{\mathrm{d}x^4} + 4\alpha^4 y = 0 \tag{2}$$

where $\alpha = \sqrt[4]{\frac{KD}{4\mathrm{E}I_z}}$ is the deformation coefficient of the FLSP, $1/\alpha$ is the characteristic length (m), E is the elastic modulus of the FLSP (N/m^2), and I_z is the cross-sectional moment of inertia of the FLSP (m^4).

The general solution of the above formula is:

$$y = \mathrm{e}^{\alpha x}(\mathrm{A}_1\cos(\alpha x) + \mathrm{B}_1\sin(\alpha x)) + \mathrm{e}^{-\alpha x}(\mathrm{C}_1\cos(\alpha x) + \mathrm{D}_1\sin(\alpha x)) \tag{3}$$

where A_1, B_1, C_1, and D_1 are constants determined by boundary conditions.

In the Winkler model, an infinite beam when $\alpha L \geq \pi$ can be considered. According to the force and displacement boundary conditions, the deflection curve equation of the FLSP under the action of concentrated force can be calculated by:

$$y = \frac{2\alpha}{KD}\varphi_1 F_0 \tag{4}$$

where $\varphi_1 = \mathrm{e}^{-\alpha x}\cos(\alpha x)$.

At the end of the FLSP (x = 0):

$$y_0 = \frac{2\alpha}{KD}F_0 \tag{5}$$

To more intuitively express the influencing factors of the bearing characteristics of the FLSP, $\alpha = \sqrt[4]{\frac{KD}{4EI_z}}$ and $I_Z = \frac{\pi(D^4-d^4)}{64}$ can be substituted into Formula (5):

$$y_0 = F_0 \frac{2\alpha}{KD} = \sqrt[4]{\frac{256}{E\pi(D^4 - d^4)(KD)^3}} F_0 \tag{6}$$

According to Formula (6), the lateral stiffness of the FLSP is:

$$k = \frac{KD}{2\alpha} = \sqrt[4]{\frac{E\pi(D^4 - d^4)(KD)^3}{256}} \tag{7}$$

2.1.2. Load-Deformation Characteristics of FLSP

A ϕ 42 × 4-millimeter (diameter of 42 mm and thickness of 4 mm) steel pipe 3.5 m in length is taken as an example to analyze the load-deformation characteristics. Figure 3 shows the load-deformation curve of the FLSP according to Formula (6). The greater the deformation of the FLSP end, the greater the bearing capacity that the FLSP can provide. The deformation and bearing capacity change linearly. When the deformation of the FLSP end is 8.8 cm, the FLSP can provide a bearing capacity of 90 kN.

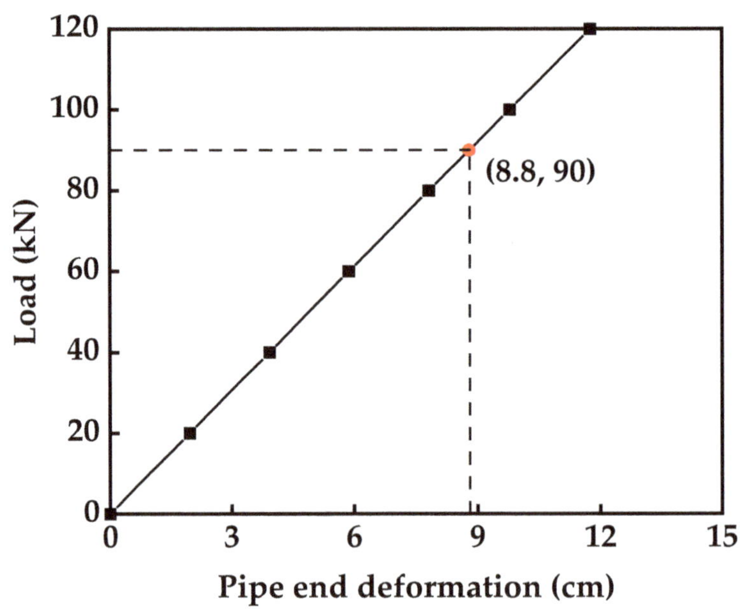

Figure 3. Load–pipe-end deformation curve.

2.1.3. Influencing Factors and Laws of Load-Deformation Characteristics

Combined with Formula (6), the load–pipe-end deformation laws of the FLSP under different parameter conditions were analyzed.

(1) Diameter of FLSP

Taking the FLSP with a length of 3.5 m and a wall thickness of 4 mm as an example, the pipe-end deformation–load characteristics of the FLSP with different diameters were analyzed. The foundation reaction coefficient was taken as 100 MPa/m. According to Formula (6), the deformation corresponding to the multi-level load was calculated. The deformation–load relationship curves of different diameters were drawn, as shown in

Figure 4. According to the steel-pipe cross-sectional area, length, and density, the weight of the steel pipe was calculated, as shown in Table 1. By combining Figure 4 and Table 1, it can be concluded that the bearing capacity of the ϕ89 steel pipe is three times that of the ϕ42 steel pipe under the same deformation, while the weight of the ϕ89 steel pipe is only twice that of the ϕ42 steel pipe. FLSPs of a large diameter can provide a greater bearing capacity per unit of weight under the same pipe-end deformation conditions. However, it is also necessary to consider the balanced selection of the drilling diameter range of the drilling equipment.

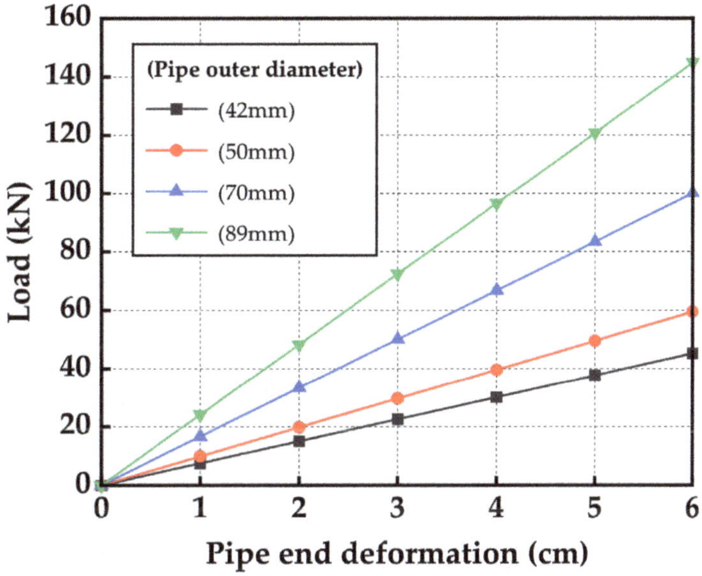

Figure 4. Deformation-load characteristics of FLSPs with different diameters.

Table 1. Weight of single FLSPs of different models (unit: kg).

	Diameter			
Thickness	**42 mm**	**50 mm**	**89 mm**	**108 mm**
2 mm	6.9	8.3	15.0	\
3 mm	10.1	12.2	22.3	27.2
4 mm	13.1	15.9	29.3	35.9
5 mm	\	19.3	36.0	44.2

\ indicates that this type of FLSP is not commonly used.

(2) Wall thickness of FLSP

Taking the FLSP with a length of 3.5 m and a diameter of 42 mm as an example, the pipe-end deformation–load-characteristics of the FLSP with different wall thicknesses were analyzed. The foundation reaction coefficient was taken as 100 MPa/m. The pipe-end-deformation–load-relationship curves with varying FLSP wall thicknesses were drawn, as shown in Figure 5. It can be seen from the figure that under the same deformation conditions, the difference in the bearing capacity of the FLSP with a 2.0-millimeter wall thickness and 4.0-millimeter wall thickness is less than 15%. The wall thickness has little effect on the pipe-end deformation–load characteristics of the FLSP. It can be seen from Table 1 that the weight of the steel pipe with the 4.0-millimeter wall thickness is twice that of the steel pipe with the 2.0-millimeter wall thickness. Under the same deformation conditions, the unit weights of thin-walled steel pipes can provide more bearing capacity.

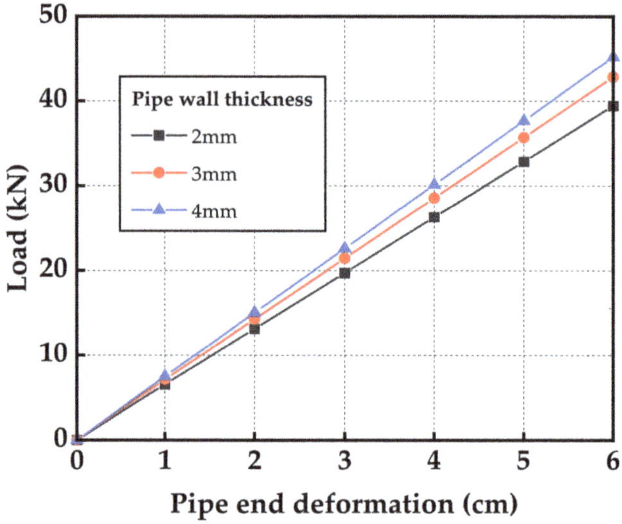

Figure 5. Deformation-load characteristics of FLSPs with different thicknesses.

(3) Foundation reaction coefficient

Taking the FLSP with a length of 3.5 m, a diameter of 42 mm, and a wall thickness of 4 mm as an example, the pipe-end deformation–load characteristics of the FLSP under the conditions of different foundation reaction coefficients were analyzed. The deformation-load curves under the conditions of varying foundation reaction coefficients were drawn. As shown in Figure 6, the larger the reaction coefficient of the surrounding rock, the stronger the capacity to constrain the deformation of the FLSP. Therefore, the grouting method can be adopted to increase the reaction coefficient when encountering soft surrounding rock, which can provide a large bearing capacity when the FLSP has a small displacement and reduce the settlement of the vault.

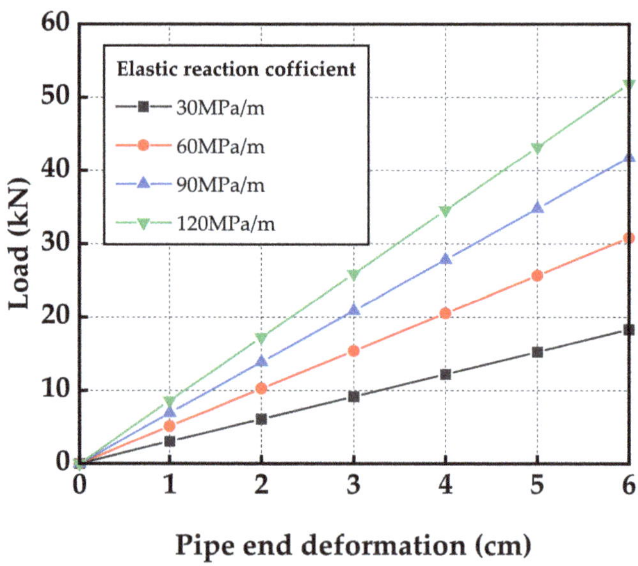

Figure 6. Deformation-load characteristics of FLSP with different foundation reaction coefficients.

2.2. Deformation Characteristics of Primary Lining under the Action of Arch Foot Bearing Capacity

The primary lining transfers the surrounding rock pressure generated by the tunnel excavation to the foundation and the FLSP. The primary lining relies on the bearing capacity provided by the arch foot to limit the deformation of the surrounding rock. The foundation bearing capacity supplied by the soft stratum is minimal, and can be ignored. Here, only the bearing capacity provided by the FLSP is considered.

Taking the Yulinzi Tunnel as an example to analyze the deformation characteristics of the primary lining under the action of various arch foot bearing capacities, the analysis model was established, as shown in Figure 7. The elastic–plastic model based on the Mohr–Coulomb strength criterion was adopted for the soil mass, and the soil parameters were obtained by using a direct shear test. The material parameters are shown in Table 2. The foundation bearing capacity provided by the soft foundation is ignored in the model. The change curve of the vault settlement and arch foot subsidence of the primary lining with various arch foot bearing capacities was obtained through a numerical analysis, as shown in Figure 8. The vault settlement and the arch foot subsidence of the primary lining decrease with the increase in the arch foot bearing capacity, and the overall change is linear. When the primary lining vault settlement is 16.5 cm, the required arch foot bearing capacity is 475 kN, and the corresponding arch foot subsidence is 6.2 cm.

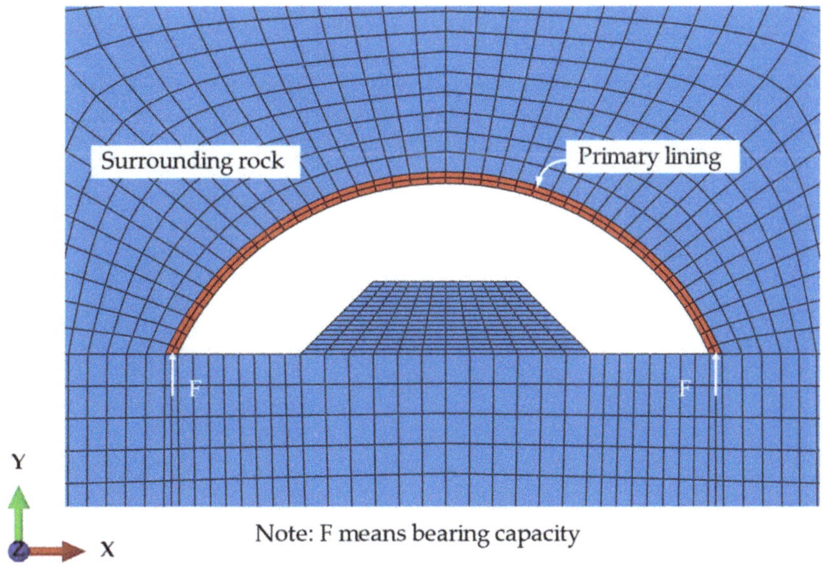

Figure 7. Finite element local model.

Table 2. Model calculation parameters.

Material	Surrounding Rock	C25 Shotcrete	I20b Steel Arch	Primary Lining
γ (heavy) ($kN \cdot m^{-3}$)	18	24	78.5	24.9
E (Young's modulus) (MPa)	8	28,000	206,000	31,863
υ (Poisson's ratio)	0.3	0.2	0.3	0.2
φ (friction angle) (°)	22	\	\	\
c (cohesion) (kPa)	40	\	\	\

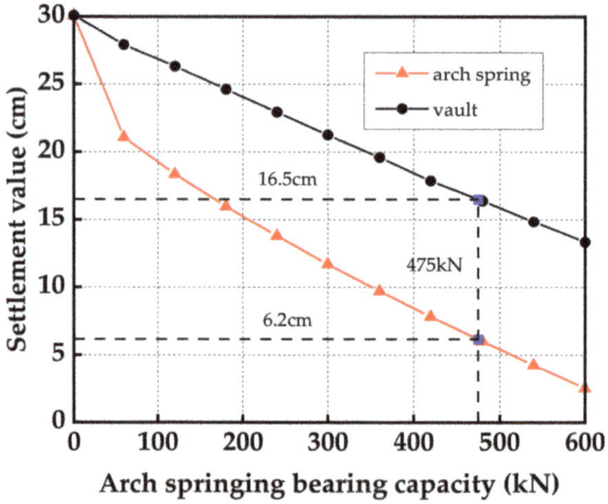

Figure 8. Vault and arch foot settlement–arch foot bearing capacity curve.

2.3. The Load-Deformation Inconsistency Problem of the FLSP

Taking the Yulinzi Tunnel as an example, we illustrate the load-deformation incoordination problem of the FLSP. According to the deformation characteristics of the primary lining in Section 2.2, it was concluded that when the vault settlement is controlled at 16.5 cm, the bearing capacity provided by the arch foot is 475 kN, and the arch foot subsidence of the primary lining is 6.2 cm. Because the numerical analysis adopts a plane model and the distance between the steel arches is 0.75 m, the bearing capacity required from the FLSP is 356 kN.

In the original design, four ϕ42 × 4 mm FLSPs 3.5 m in length are installed on the upper bench, as shown in Figure 9. The required bearing capacity of a single ϕ42 × 4-millimeter FLSP is approximately 90 kN. When the deformation value of a single ϕ42 × 4-millimeter FLSP has an arch foot subsidence value of 6.2 cm, it can provide a bearing capacity of 63 kN, but cannot reach the 90 kN required to control the settlement. The original FLSP cannot control the vault settlement within the range of 16.5 cm. Combined with the load-deformation characteristics of the FLSP in Section 2.1.2, it can be concluded that the FLSP provides a bearing capacity of 90 kN, and that the FLSP end needs to be deformed by 8.8 cm. At this point, the deformation value of the FLSP is greater than the primary lining arch foot subsidence value of 6.2 cm. There is a problem of deformation inconsistency between the FLSP and the primary lining arch foot.

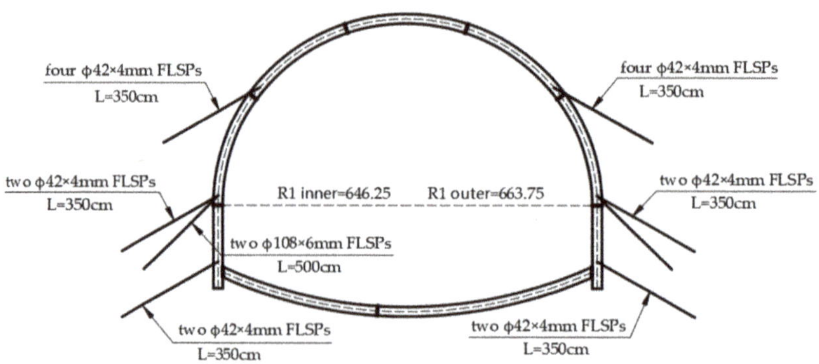

Figure 9. Design drawing of FLSP.

3. Design Optimization Method of FLSP Based on Load-Deformation Coordination

At present, the installation quantity and the parameters of FLSPs in tunnel design and construction mostly rely on relevant engineering experience and lack theoretical support. To solve the problem of the load-deformation inconsistency between the FLSP and the primary lining, a design optimization method for FLSPs is proposed for two working conditions: predicted large deformation and occurring large deformation.

From the pipe-end deformation-load characteristics of the FLSP, it can be concluded that the FLSP needs to be deformed to provide the corresponding bearing capacity. The FLSP is deformed under the primary lining load and then provides the bearing capacity, referred to as passive bearing. To rapidly and fully exert the bearing capacity of the FLSP, the method of pre-compressing the FLSP before connecting it with the primary lining is proposed. The FLSP can actively provide the bearing capacity, referred to as active bearing. A comparison between active and passive bearing is shown in Figure 10. The following describes the design optimization method of the FLSP.

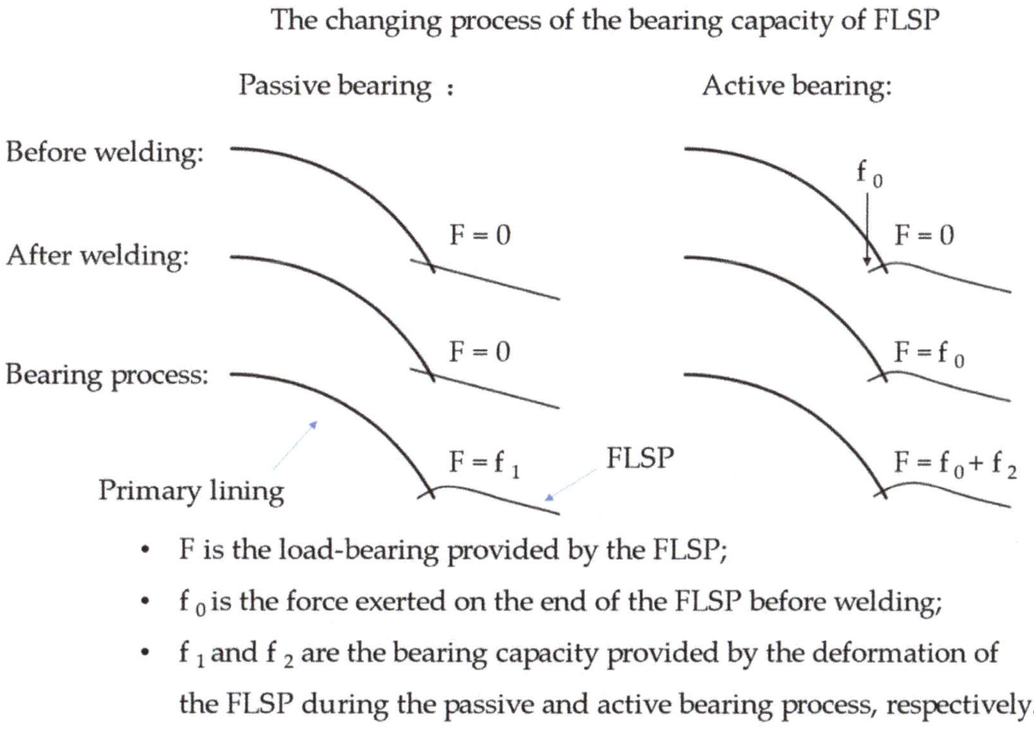

Figure 10. The comparison between the active bearing and passive bearing of the FLSP.

3.1. Predicted Large Deformation

When the numerical simulation method is applied, and it is judged that a large deformation will occur after tunnel excavation, the design optimization process of the FLSP shown in Figure 11 is applied. The design stage consists of four parts: the extraction of the vault settlement control index, the simplification and stiffness calculation of the FLSP, the finite element analysis of the primary lining, and the selection of the FLSP.

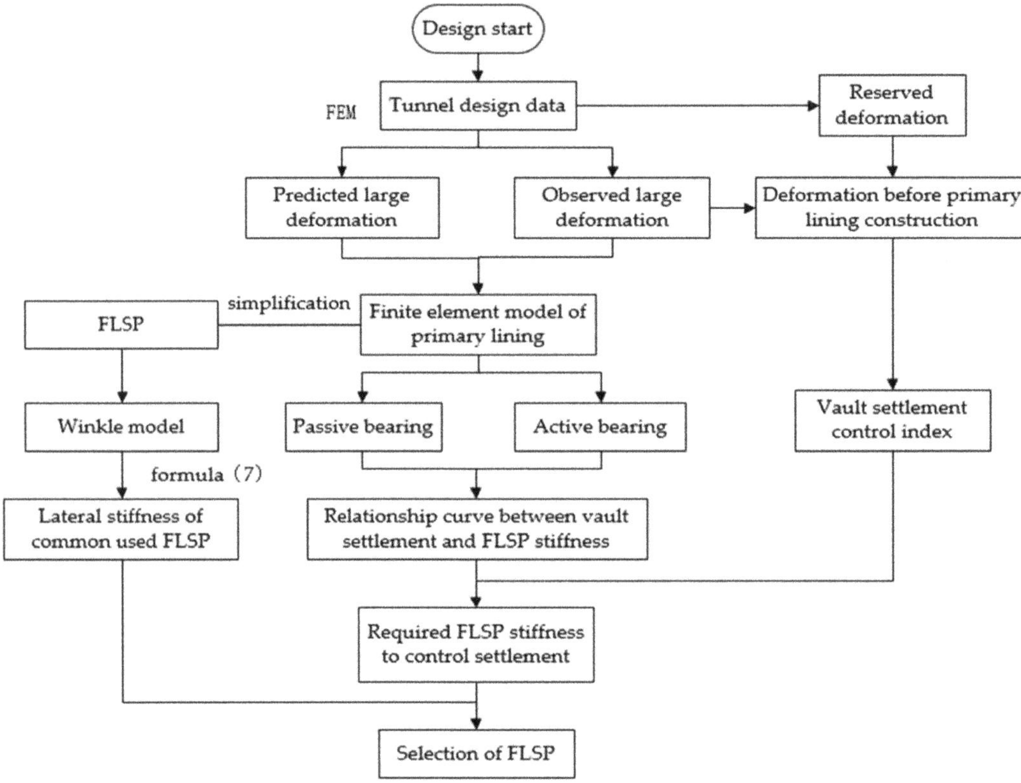

Figure 11. Flow chart of design optimization of FLSP.

3.1.1. Extraction of Vault Settlement Control Index

First of all, the reserved deformation of the tunnel is obtained from the design data. The reserved deformation is an essential index in the design of a tunnel. If the reserved deformation is too large, it increases the concrete volume of the secondary lining and the project cost. By contrast, if the reserved deformation is too small, the primary lining encroaches on the limit, resulting in the dismantling of the arch. For soft surrounding rock, Section 6.2.5 of the "Technical Specifications for Railway Squeezed Surrounding Rock Tunnels" (Q/CR 9512-2019), gives the reference reserved deformation, as shown in Table 3.

Table 3. Reference reserved deformation for extruded surrounding rock tunnel (unit: mm).

Deformation Level	One	Two	Three
Small, mid-span ($B \leq 12$ m)	100~200	200~300	300~400
Large-span and above	150~250	250~350	350~450

3.1.2. The Simplification and Stiffness Calculation of the FLSP

The connection between the FLSP and the primary lining is complicated. The following assumptions were made to simplify the analysis:

(1) Without considering the foundation bearing capacity, it was assumed that the FLSP bore all of the surrounding rock loads, and that the design was biased towards safety;
(2) The bending moment between the FLSP and the primary lining was ignored;
(3) The role of the FLSP is to limit the sinking of the arch foot, so the axial force of the FLSP was ignored, and only the lateral bearing effect was considered;

Therefore, only the shear force between the FLSP and the primary lining was considered. Several standard diameters and wall thicknesses of FLSPs applied in engineering practice were selected, and combined with Formula (7) to obtain the lateral stiffness of the FLSP.

3.1.3. Finite Element Analysis of Primary Lining

According to the engineering geology and construction drawing data, the finite element analysis model of the surrounding rock and the primary lining was established. The finite element analysis model simplified the FLSP as a spring connected to the primary lining and simultaneously realized the deformation coordination between the FLSP and the primary lining. The spring stiffness indicated the lateral stiffness of the FLSP. In the active load-bearing design analysis model, the active load-bearing effect is achieved by pre-compressing the spring before releasing the surrounding rock load. The curve between the spring stiffness and the settlement of the arch foot and vault was obtained by changing the spring stiffness. Finally, the required lateral stiffness of the FLSP was obtained according to the vault settlement control index.

3.1.4. Type Selection of FLSP

On the basis that the combined lateral stiffness of the FLSP was not less than the required lateral stiffness, the selection of the FLSP was carried out according to the principle of the least amount of steel.

3.2. *Observed Large Deformation*

When the surrounding rock has undergone a large deformation before the primary lining construction, the design optimization process of the FLSP shown in Figure 9 is applied. The overall design process is the same as the prediction of large deformation, differing only in the following two aspects:

(1) Because the large deformation occurs before the primary lining construction, the vault settlement control index is no longer the reserved deformation of the tunnel; it should instead be the reserved deformation minus the vault settlement before the primary lining construction. The vault settlement can be obtained through monitoring and measurement.

(2) During the finite element analysis of the primary lining, the stress should be released to the large deformation state. Next, the remaining stress is released after the primary lining construction. Similarly, it can be divided into two schemes: passive and active bearing. Because of the small allowable vault settlement, it is recommended to adopt the active bearing scheme of the FLSP to ensure the full exertion of the bearing capacity of the FLSP.

4. Case Study

Taking ZK279 + 840–860 of the Yulinzi Tunnel's left line in Zhengning County, Qingyang City, as an example, the optimized design of the FLSP was carried out. The surrounding rock was paleosol loess, the parameters of which are shown in Table 2. The buried depth of the tunnel, which is constructed by the three-bench seven-step method, was 50 m, and the excavation height of the upper bench was 3.8 m. The steel arch frame adopted an I20b beam, and the steel arch frame spacing was 75 cm. The shotcrete adopted C25 concrete, with a thickness of 24 cm, and the reserved deformation of the tunnel was 30 cm.

According to the available research, the vault settlement and peripheral convergence have the fastest deformation rate in the upper bench excavation stage [18]. The FLSPs installed in the upper bench play a crucial role in controlling the vault settlement. This paper takes the upper bench excavation as an example to select and design the FLSP. According to the measured settlement data of the on-site tunnel excavation, the vault settlement after upper bench excavation by the three-bench method was 55% of the maximum settlement value, so the control deformation of the upper bench was 16.5 cm. According to the concept of NATM, the secondary lining does not bear the load in theory and acts only as a safety reserve. In fact, it needs to bear 30% to 40% of the load in the soft surrounding rock. Here, the secondary lining bore 30%, the primary lining bore 70%, and the upper bench excavation bore 30%.

4.1. Numerical Calculation Model

The numerical simulation adopted the finite element software Abaqus. The excavation height of the upper bench was 3.8 m, the radius of the arc segment was 6.3 m, and the primary lining thickness was 0.24 m. In order to reduce the influence of the boundary on the calculation results, the overall model width (x-direction) was 60 m, the height (y-direction) was 80 m, and the top of the tunnel was 50 m from the free surface. The left and right of the model constrained the horizontal displacement, and the bottom constrained the horizontal and vertical displacement. The numerical model size and meshing are shown in Figure 12.

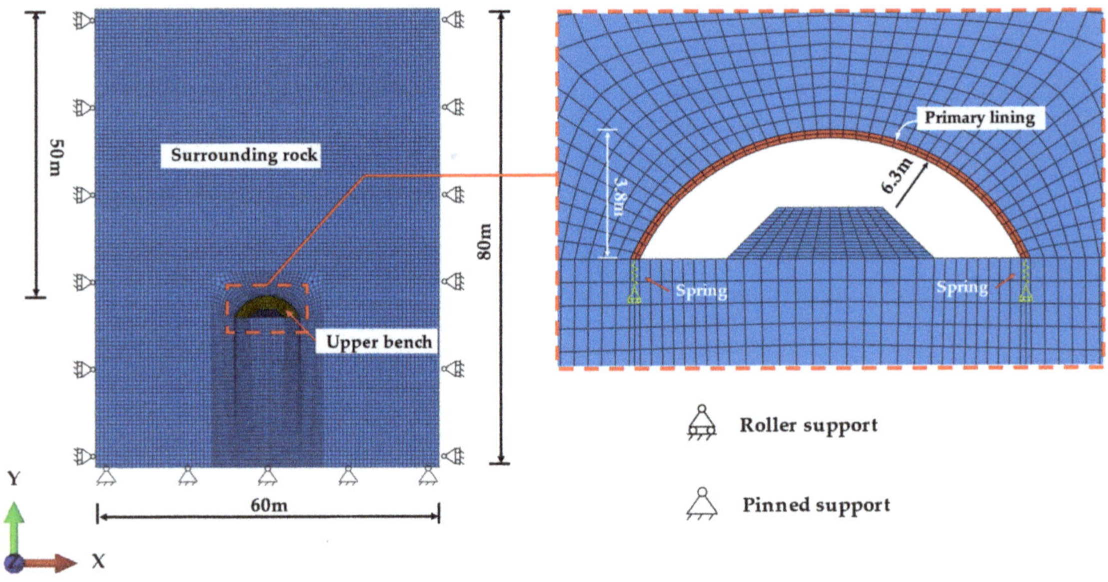

Figure 12. Finite element calculation model.

The elastic–plastic model based on the Mohr–Coulomb strength criterion was adopted for the soil mass. The primary lining included C25 concrete, an I20b steel arch, and FLSP, which adopted the elastic model. The steel arch's stiffness was transferred into the primary lining through stiffness conversion [19]. The plane strain element CPE4 was adopted for the primary lining and surrounding rock. The spring element simulated the FLSP. The spring and the primary lining were connected by node coupling. Common nodes were adopted to connect the primary lining and surrounding rock. The material parameters are shown in Table 2.

Equations (6) and (7) give the load–displacement relationship and the lateral stiffness of the connection part (point O in Figure 2) between the FLSP and the initial support, respectively. In order to facilitate the selection design and the realization of active bearing, the FLSP was simplified as a spring in the numerical simulation. The bottom of the spring was fixed, the upper part of the spring was coupled and connected to the lower end of the primary lining, and the active bearing effect of the foot-locking steel pipe could be achieved by pre-compressing the spring or setting a nonlinear spring.

4.2. Optimal Selection Analysis of FLSP

The numerical simulation results used to obtain the relationship curve between the vault settlement and the spring stiffness are summarized in Figure 13. It can be seen from the figure that the vault settlement decreased gradually with the increase in spring stiffness, indicating that the bearing force provided by the FLSP increased gradually. Under the no-spring condition, the vault settlement was greater than the control settlement value of 16.5 cm, so it was necessary to restrict and control the settlement of the arch vault with the FLSPs. The original design adopted four ϕ42 × 4 mm FLSPs. Because the corresponding spring stiffness was 4088 kN/m, and the steel arch spacing was 75 cm, the corresponding stiffness in the plane model was 5450 kN/m. The corresponding vault settlement calculated by the numerical simulation was 18.7 cm more than the vault settlement control value. Therefore, it was necessary to optimize the design of the FLSP. As can be seen in Figure 11, when the vault settlement was 16.5 cm, the corresponding spring stiffness was approximately 9860 KN/m, and the required stiffness of FLSP for each steel arch frame was 7400 kN/m.

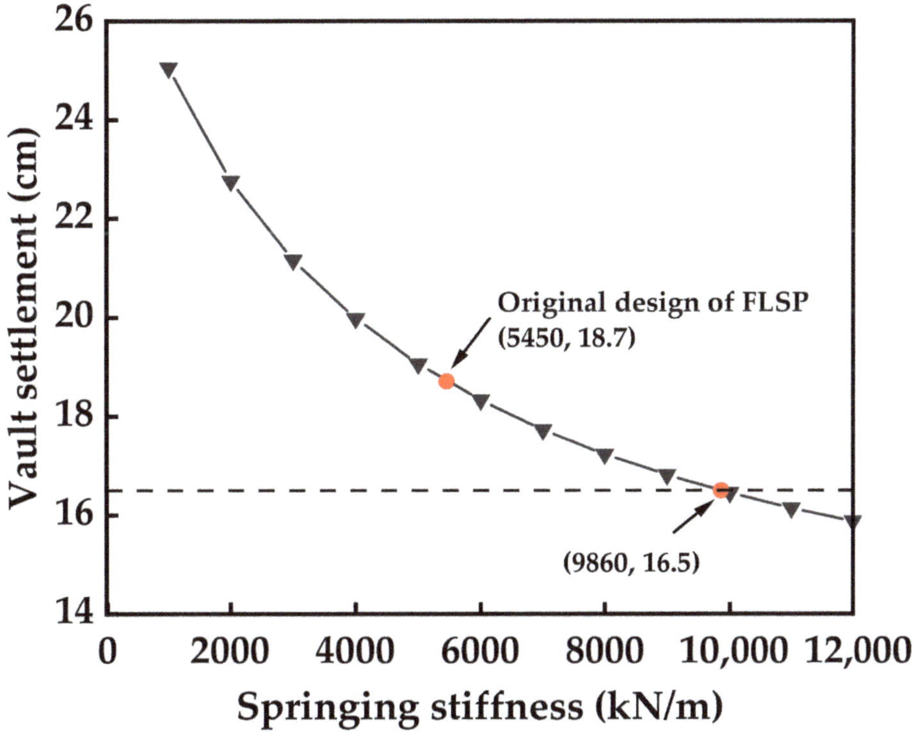

Figure 13. The relationship curve of vault settlement–spring stiffness.

FLSPs with diameters of 42 mm, 50 mm, 89 mm and 108 mm, and wall thicknesses of 2 mm, 3 mm, 4 mm and 5 mm were taken as examples for selection. The reaction coefficient of the surrounding rock foundation was 150 MPa/m. The corresponding lateral stiffness was obtained from Formula (7) and the material parameter calculation, as shown in Table 4. The design was based on the requirement that the total lateral stiffness of the FLSP should be not less than 7400 kN/m. At the same time, in combination with the steel consumption, the passive bearing scheme of the design optimization method selected two Φ108 × 3 mm FLSPs.

Table 4. Lateral stiffness of each type of FLSP (unit: kN/m).

Thickness	Diameter			
	42 mm	50 mm	89 mm	108 mm
2 mm	891	1164	2801	\
3 mm	968	1269	3074	4127
4 mm	1022	1343	3275	4404
5 mm	\	1398	3433	4624

4.3. Analysis of Active Bearing Effect

The vault settlements with different lateral stiffnesses were calculated when the spring was preloaded by 5 cm. The curve is shown in Figure 14. The effects of the active and passive bearing schemes were compared and analyzed. The change in the two schemes was the same, but the vault settlement was significantly reduced after preloading. With the increase in stiffness, the effect of the controlling settlement was more apparent. When the preload was 5 cm, the spring stiffness corresponding to the vault settlement of 16.5 cm was approximately 5570 kN/m, so the required stiffness of the FLSP was 4180 kN/m. Installing one ϕ108 × 4-millimeter FLSP can meet the settlement control requirements. The passive bearing scheme needs to install two ϕ108 × 3-millimeter FLSPs. According to Table 1, the steel consumption of the passive bearing scheme is 54.4 kg, and the steel consumption of the active bearing scheme is only 35.9 kg, which reduces the steel consumption by 34%. As a result, the active bearing can significantly reduce the amount of steel and reduce the project cost.

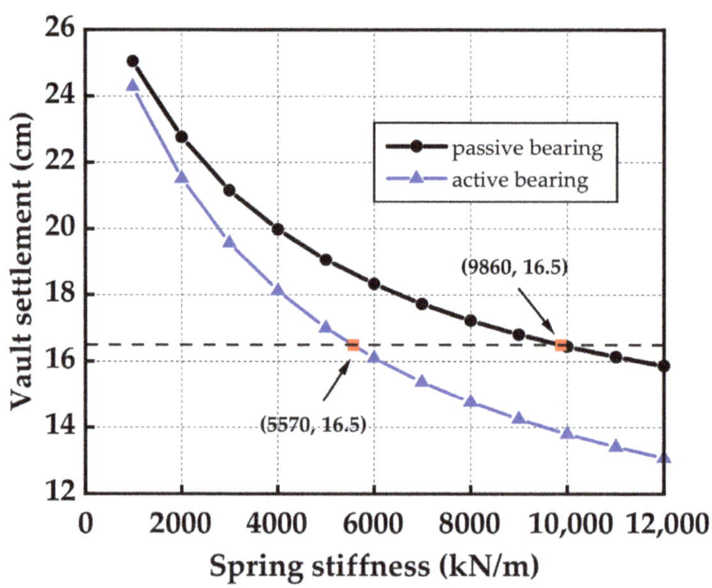

Figure 14. Comparison of the settlement of active and passive load-bearing vaults.

4.4. Monitoring and Measurement Results

According to the monitoring and measurement, the vault settlement results before and after the design were compared to verify the feasibility of the FLSP design method. In Figure 15, the time-history curve of the vault settlement is given, in which the section of mileage stake number ZK279 + 840 adopts the original design of four ϕ42 × 4 millimeter FLSPs, and the ZK279 + 850 section adopts two Φ108 × 3-millimeter FLSPs.

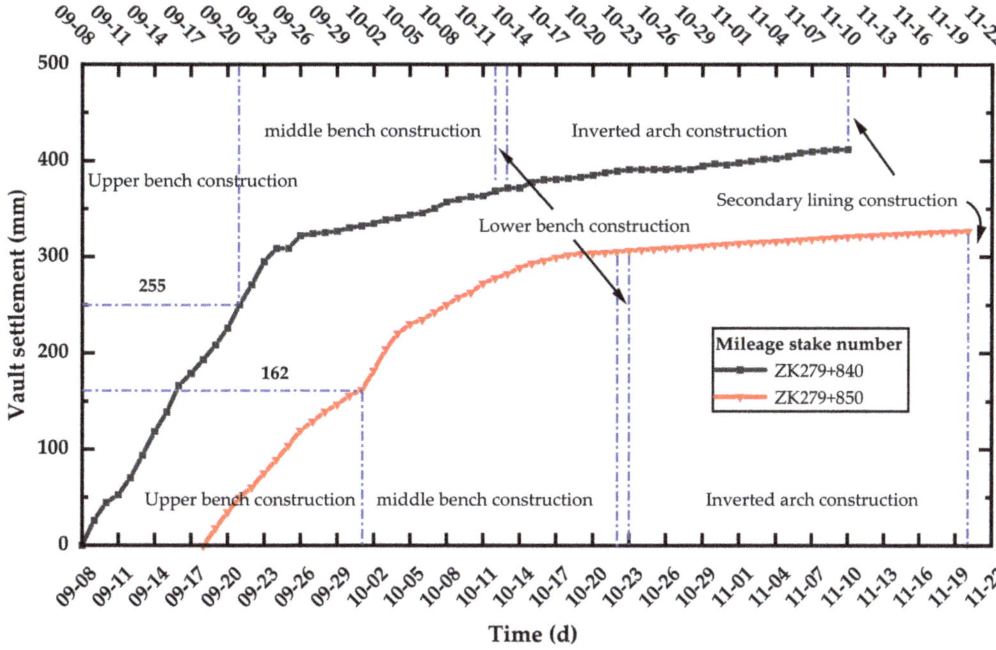

Figure 15. Vault settlement–time history curve.

The vault settlement of the ZK279 + 840 section was 25.5 cm after the upper bench construction and increased to 41.2 cm before the construction of secondary lining, exceeding the reserved deformation of the tunnel. Through the optimized design of the FLSP, the vault settlement of the ZK279 + 840 section was 16.2 cm after the upper bench construction. It increased to 32.6 cm before the construction of the secondary lining, which was roughly the same as the reserved deformation.

The numerical simulation results show that the vault settlement after the upper bench excavation on the ZK279 + 840 section was 18.7 cm, and that of the ZK279 + 850 section was 16.1 cm. The actual settlement of the ZK279 + 840 section was 25.5 cm, which was larger than the numerical simulation results. The actual settlement of the ZK279 + 850 section was 16.2 cm, which was roughly the same as the numerical simulation. On the one hand, the stress release rates of different sections are different; on the other hand, when the deformation of the loess tunnel is too large, the soil layer loses its bearing capacity, resulting in unpredictable loosening pressure.

From the analysis of the monitoring and measurement results, it can be seen that through the optimized design, the FLSP can control the settlement of the vault within the specified range, showing the feasibility of this method of design optimization.

5. Conclusions

(1) The interaction between the FLSP and the primary lining was simplified as a pair of shear forces. The Winkler model was used to analyze the load-deformation characteristics of the FLSP under the condition of shear force. Analyzing the deformation of the FLSP under specific load conditions and the deformation of the primary lining under specific bearing capacity conditions revealed the problem that the deformation of FLSP and the primary lining was inconsistent under particular load and bearing capacities.

(2) Under the two working conditions of predicted large deformation and observed large deformation, a design and optimization method for FLSPs based on the load-deformation coordination was proposed. The design method includes four parts:

the extraction of the vault settlement control index, the simplification and stiffness calculation of FLSP, the finite element analysis of the primary lining, and the selection of FLSP.

(3) For the Yulinzi Tunnel, after the design optimization method, using the passive bearing scheme, two 3.5-m-long ϕ108 × 3-millimeter FLSPs were installed, whereas, for the active bearing scheme, only one ϕ108 × 4-millimeter FLSP needed to be installed when preloaded by 5 cm. The active bearing of the FLSP can significantly reduce the amount of steel required and reduce the engineering cost.

In this study, an optimal design method for FLSPs was proposed. The design method adopts the Winkler model to consider the interaction between FLSP and the soil. Because the small deformation assumption is used in the derivation, only the elastic phase of the FLSP is considered. Next, the design method can be further improved by considering the plasticity of FLSP. The elastic–plastic solution of the deformation of FLSP can be obtained through the theoretical model, or the plasticity of FLSP can be considered through a numerical simulation. On the other hand, this study proposes the active bearing design of FLSPs, but it has not been applied in actual engineering works.

Author Contributions: Conceptualization, Y.W.; methodology, Y.W.; validation, C.T.; investigation, J.Z.; resources, Z.Z.; data curation, J.Z.; writing—original draft preparation, C.T.; writing—review and editing, P.X.; visualization, S.W.; project administration, Z.Z., and S.W.; funding acquisition, Z.Z. All authors have read and agreed to the published version of the manuscript.

Funding: This research was funded by the China Railway Corporation Science and Technology Research and Development Program, grant number 2018K019.

Institutional Review Board Statement: Not applicable.

Informed Consent Statement: Not applicable.

Data Availability Statement: All data are available from the author.

Conflicts of Interest: The authors declare no conflict of interest.

References

1. Keawsawasvong, S.; Ukritchon, B. Design equations for the stability of shallow unlined circular tunnels in the Hoek-Brown rock mass. *Bull. Eng. Geol. Environ.* **2020**, *79*, 4167–4190. [CrossRef]
2. Yodsomjai, W.; Keawsawasvong, S.; Likitlersuang, S. Stability of unsupported conical slopes in the Hoek-Brown rock mass. *Trans. Infrastructure. Geotechnical.* **2021**, *8*, 279–295. [CrossRef]
3. Wu, Y.; Lu, K.C.; Xu, Y. Bearing behaviors of steel foot pipes for tunnels in soft foundation. *Chin. J. Geotech. Eng.* **2009**, *31*, 1825–1832. (In Chinese)
4. Luo, Y.B.; Chen, J.X. Mechanical characteristics and mechanical calculation model of tunnel feet-lock bolt in soft surrounding rock. *Chin. J. Geotech. Eng.* **2013**, *35*, 1519–1525. (In Chinese)
5. Li, P.F.; Zhao, Y.; Zhou, X.J. Displacement characteristics of high-speed railway tunnel construction in loess ground by using multi-step excavation method. *Tunn. Undergr. Space Technol.* **2016**, *51*, 41–55. [CrossRef]
6. Fang, Q.; Zhang, D.L.; Wong, L.N.Y. Shallow tunnelling method (STM) for subway station construction in soft ground. *Tunn. Undergr. Space Technol.* **2012**, *29*, 10–30. [CrossRef]
7. Luo, Y.B.; Chen, J.X.; Huang, P.; Tang, M.Q.; Qiao, X.; Liu, Q. Deformation and mechanical model of temporary support sidewall in tunnel cutting partial section. *Tunn. Undergr. Space Technol.* **2017**, *61*, 40–49. [CrossRef]
8. Chen, L.J.; Chen, J.X.; Li, Y.; Luo, Y.B.; Mu, Y.J.; Hu, T.T.; Wang, C.W. Performance of Tunnel Feet-Lock Pipe (TFP) in Sharing Vertical Foundation Load. *KSCE J. Civ. Eng.* **2021**, *25*, 1086–1094. [CrossRef]
9. Chen, L.J.; Zhang, Y.L.; Ma, Z.Y. Analytical approach for support mechanism of feet-lock pipe combined with steel frame in weak rock tunnels. *KSCE J. Civ. Eng.* **2016**, *20*, 2965–2980. [CrossRef]
10. Liang, H.; Wang, Y.; Yan, X.; Liu, J.; Peng, H.; Guo, X. Research on mechanical stability of feet-lock pipe and steel arch in soft rock tunnel. *J. Beijing Jiaotong Univ.* **2020**, *44*, 34–41. (In Chinese)
11. Chen, L.J.; Chen, J.X.; Luo, Y.B.; Li, Y.; Hu, T.T. Vertical load and settlement at the foot of steel rib with the support of feet-lock pipe in soft ground tunnel. *Adv. Civ. Eng.* **2019**, *2019*, 6186748. [CrossRef]
12. Cui, Y.; Kishida, K.; Kimura, M. Analytical Study on the Control of Land Subsidence Accompanying Subsidence Phenomena with Foot Reinforced Side Pile. In Proceedings of the GeoShanghai International Conference 2010, Shanghai, China, 3 June 2010.
13. Deng, G.; Shao, S.; Tao, H.; Chen, C. A study of the effects of locking tremies on the internal force of tunnel support and the deformation of surrounding soil. *Chin. Civil. Eng. J.* **2010**, *43*, 108–113. (In Chinese)

14. Cui, Y.; Kishida, K.; Kimura, M. Prevention of the ground subsidence by using the foot reinforcement side pile during the shallow overburden tunnel excavation in unconsolidated ground. *Tunn. Undergr. Space Technol.* **2017**, *63*, 194–204. [CrossRef]
15. Luo, Y.B.; Diao, P.S.; Chen, J.X.; Li, D.; Gou, Y.L.; Qiao, X. New Method of Monitoring Tunnel Feet-Lock Pipe (TFP) Mechanics Using Fiber Bragg Grating (FBG). *J. Test Eval.* **2018**, *48*, 3125–3142. [CrossRef]
16. Shi, Z.; Luo, Y.; Chen, J.; Liu, W.; Yu, H. Field simulation test on mechanical characteristics of feet-lock anchor pipes for soft surrounding rock tunnel. *J. Highw. Transp. Res. Dev.* **2021**, *38*, 96–106. (In Chinese)
17. Deng, X.H.; Xia, D.H.; Wang, R.; Liu, Y.Y. Feet-lock bolt application in cracked surrounding rock tunnels. *Geotech. Geol. Eng.* **2019**, *37*, 3423–3434. [CrossRef]
18. Meng, D.; Tan, Z. Deformation control technology and supporting structure stress of large section loess tunnel. *Chin. Civil. Eng. J.* **2015**, *48*, 383–387. (In Chinese)
19. Gkikas, V.I.; Nomikos, P.P. Primary support design for sequentially excavated tunnel junctions in strain-softening hoek–brown rock mass. *Geotech. Geol. Eng.* **2021**, *39*, 1997–2018. [CrossRef]

applied sciences

MDPI

Article

Study on the Restraint Effect of Isolation Pile on Surface Settlement Trough Induced by Shield Tunnelling

Kan Huang [1,2], Yiwei Sun [1,3,*], Xilong Kuang [2,*], Xianqiang Huang [1], Runing Liu [1] and Qijiang Wu [1]

1 School of Civil Engineering, Changsha University of Science & Technology, Changsha 410114, China; hk_616@csust.edu.cn (K.H.); huangxianqiang99@163.com (X.H.); lrn345230@163.com (R.L.); w118176@163.com (Q.W.)
2 School of Civil Engineering, Changsha University, Changsha 410022, China
3 Shanghai Harbour Foundations Construction Group Ltd., Shanghai 200092, China
* Correspondence: yiweisun@stu.csust.edu.cn (Y.S.); kxl@ccsu.edu.cn (X.K.); Tel.: +86-18616631601 (X.K.)

Abstract: Isolation piles are widely used to control the influence of shield tunnelling on adjacent buildings as an effective protective measure. However, the restraint effect of isolation pile on surface settlement trough is rarely explored from the internal mechanism. Firstly, the restraint mechanism of isolation piles is investigated from the pile-soil-tunnel interaction mechanism. Secondly, based on the Melan solution and the Loganathan formula, the analytical solution of surface settlement trough under the influence of adjacent isolation piles is derived. Thirdly, in order to satisfy the engineering analysis scale and reflect the friction characteristics between isolation pile and soil particles simultaneously, the FDM-DEM coupling technique is introduced to establish a numerical model including discrete medium and continuum medium. Finally, the applicability and reliability of the analytical solution and the FDM-DEM coupling numerical solution are verified by comparing field measured data. The results indicate that the surface settlement trough under the influence of isolation piles will have an asymmetric distribution. Surface settlement tends to develop more to the opposite side of the tunnel when isolation piles are pre-installed on one side. The findings of the study have substantial theoretical significance and engineering reference value.

Keywords: shield tunnel; isolation pile; surface settlement trough; analytical solution; FDM-DEM coupling

Citation: Huang, K.; Sun, Y.; Kuang, X.; Huang, X.; Liu, R.; Wu, Q. Study on the Restraint Effect of Isolation Pile on Surface Settlement Trough Induced by Shield Tunnelling. *Appl. Sci.* **2022**, *12*, 4845. https://doi.org/10.3390/app12104845

Academic Editors: Mingfeng Lei, Chenjie Gong and Xianda Shen

Received: 20 April 2022
Accepted: 9 May 2022
Published: 11 May 2022

1. Introduction

In recent years, the increasing urban population density and residents' travel demand have brought huge traffic load to the limited ground space. The underground railway constructed by shield tunnelling has advantages of large unit transport, fast transport speed, high punctuality, and almost no influence from external weather, which can effectively mitigate the ground traffic congestion and gradually become the first choice for residents to travel in the urban area. However, shield tunnelling often needs to pass through a huge number of existing pile foundations at a close distance due to the large quantity of infrastructure in metropolitan areas [1–5]. Tunnelling-induced ground surface settlement, inclination, and discontinuous displacement could affect or even damage adjacent superstructures. The isolation pile between the existing pile and the tunnel is widely proven as an efficient protective measure for controlling the displacement of building's pile foundation induced by shield tunnelling [6–9].

In general, the surface settlement trough induced by shield tunnelling is approximately a Gaussian distribution in transverse direction [10]. However, when the isolation pile is pre-installed, the stiffness of the isolation pile is substantially larger than that of the surrounding soil, resulting in an asymmetrical distribution of surface settlement trough. Currently, various scholars have focused on this issue and conducted several studies. Fantera et al. [11] first presented a plane strain finite element analysis, in which a continuous diaphragm made by adjacent panels was established to evaluate the effects of various geometrical and mechanical parameters. Ding et al. [12] held the opinion that surface settlement trough based on Peck formula was not reasonable. Surface settlement trough will present cork distribution, skewed distribution, and normal distribution according to the scope of the disturbance when shield tunnelling passes through adjacent buildings. Huang et al. [13] highlighted that the surface settlement trough curve presented a 'hanging phenomenon' near the adjacent pile foundation. The existence of pile foundation will restraint the displacement of soil around pile induced by shield tunnelling. Rampello et al. [14] presented a numerical study to evaluate the efficiency of different schemes of embedded barriers. The results illustrated that the diaphragm wall could reduce both surface settlements and curvature of surface settlement trough beyond its location. Masini et al. [15] conducted a series of soil-structure interaction finite element analyses to evaluate the efficiency of a barrier made of a line of bored piles, comparing with field monitoring data from a test site in Rome and simple empirical equations. Franza et al. [16] proposed a three-dimensional linear elastic prediction method to evaluate the protective action of pile walls against surface and subsurface ground movements due to tunnel excavation. The barrier efficiency in reducing settlements is explored by comparing pile walls and diaphragm walls. Sun et al. [17] established three-dimensional finite element models to analyze the effects of different pile diameters and different distance between pile and tunnel on the surface settlement trough. An asymmetric skewness distribution curve was proposed by fitting the simulated data. Cao et al. [18] proposed an elastic solution for evaluating tunneling-induced vertical ground displacements due to the restraint of embedded isolation piles. Parametric analysis indicates that parameters associated with the isolation pile and the soil have significant influence on the surface settlement troughs and the barrier efficiency. To sum up, the above theoretical studies, numerical simulations, centrifugal tests, and field measured data adequately describe the restraint effect of isolation piles. However, most scholars performed parametric analyses, but did not investigate the underlying reasons for impact on the surface settlement trough induced by the pile-soil-tunnel interaction mechanism.

In view of the aforementioned issues, it is of great significance to study the restraint mechanism of isolation pile on soil displacement induced by shield tunnelling, which can help us better predict the surface settlement trough and accurately evaluate the protection effect of isolation pile on the adjacent buildings.

The outline of this paper is as follows: firstly, the Loganathan formula [19] based on virtual image principle is adopted to calculate the surface settlement trough without isolation piles. On this basis, Melan solution [20,21] in the semi-infinite plane based on elasticity is introduced to derive the ultimate surface settlement trough from the perspective of negative friction of pile edge. Secondly, an engineering case is found in the Changsha Metro Line 5 from East Laodong Road Station to Huaya Station, where shield tunnelling passes through Huaya International Hotel at a close distance. Thirdly, a two-dimensional numerical model is established based on FDM-DEM coupling, with the constraint effect of the isolation pile on the surface settlement trough reflected through the friction effect between particles and the wall. Finally, the applicability and reliability of the analytical and numerical solutions are verified by comparing field measured data, which has significant engineering guiding reference in analyzing such projects.

2. Restraint Mechanism of Isolation Pile

2.1. Pile-Soil-Tunnel Interaction Mechanism

Ground loss is an inevitable consequence of shield tunnelling. The ground loss in the disturbed zone will cause discontinuous settlement due to stress redistribution, resulting in soil particle wedge-caulking. According to the affecting degree of stratum disturbance induced by shield tunnelling, the strata perpendicular to the direction of shield tunnelling can be separated into three zones. As shown in Figure 1, they are, respectively, affected zone I (strong disturbance zone), affected zone II (medium disturbance zone), and affected zone III (weak disturbance zone). The larger the disturbance, the greater the soil displacement toward the excavation area.

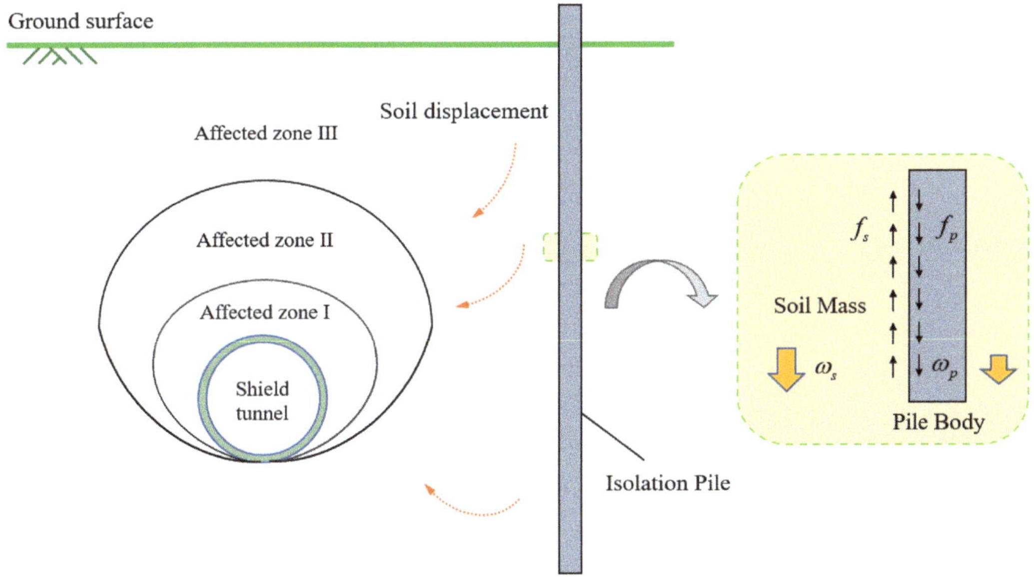

Figure 1. Pile-soil-tunnel interaction mechanism.

In Figure 1, ω_s is the vertical displacement of soil, ω_p is the vertical displacement of isolation pile, f_s is the restraint force of soil on pile side, f_p is the negative friction of isolation pile. As the vertical displacement of soil induced by tunnelling is larger than that of adjacent isolation pile, the surface of isolation pile will be affected by negative friction downward. According to the law of action and reaction, the soil around the isolation pile is subjected to upward by restraint force, resulting in the inconsistent surface settlement on both sides of the pile body.

2.2. Melan Solution

Melan first presented an elastic mechanics solution for a vertical line loading beneath the surface of a semi-infinite plane [20,21]. However, Melan's initial solution is merely a stress solution and does not satisfy the compatibility equation of elastic mechanics, which makes Melan solution less commonly utilized to guide engineering practice. On the basis of Melan solution, Verruijt and Booker [22] further derived the analytical solution of vertical displacement, in which a vertical line loading beneath the surface of a semi-infinite plane. In this study, Melan solution is introduced to analyze the restraint effect of adjacent isolation pile on the surface settlement trough through coordinate transformation. The calculation diagram of Melan solution is shown in Figure 2.

2.3. Analytical Solution

In this study, the Loganathan formula [19] is adopted to calculate the surface settlement trough without considering the influence of isolation piles. The vertical displacement of soil can be obtained by:

$$\omega_L = R^2\left\{-\frac{y-h}{x^2+(y-h)^2}+\frac{(3-4\mu)(y+h)}{x^2+(y+h)^2}-\frac{2y\left[x^2-(y+h)^2\right]}{\left[x^2+(y+h)^2\right]^2}\right\}\frac{4gR+g^2}{4R^2}\cdot\exp\left[-\frac{1.38x^2}{(h+R)^2}-\frac{0.69y^2}{h^2}\right] \quad (1)$$

where ω_L is the vertical displacement of soil induced by shield tunnelling without considering the effect of isolation pile (m), h is the depth of the tunnel axis from the ground surface (m), R is the radius of the tunnel (m), g is the gap parameter, x is the lateral distance from the tunnel axis (m), and y is the depth measured from the ground surface (m).

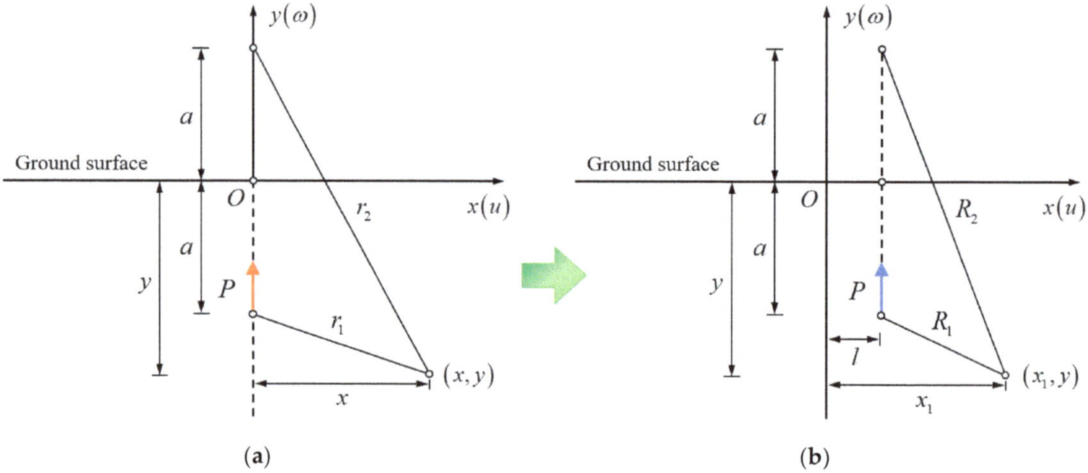

Figure 2. Calculation diagram of Melan solution: (**a**) Melan solution and (**b**) this study.

The vertical displacement of soil for a vertical line loading upward beneath the surface of a semi-infinite plane can be calculated by:

$$\omega_M = \frac{P}{8\pi G(1-\mu)}\left[(3-4\mu)\ln\frac{r_1}{r_2}+\frac{x^2}{r_1^2}-\frac{x^2}{r_2^2}\right]-\frac{P(1-\mu)}{\pi G}\ln r_1+\frac{P(y-a)^2}{2\pi G r_1^2}+\frac{Pya\left[x^2-(y-a)^2\right]}{4\pi G(1-\mu)r_1^4}+u_0 \tag{2}$$

where ω_M is the vertical displacement of soil caused by the vertical line load (m), P is the vertical line load per unit length (N/m), G is the shear modulus (N/m^2), a is the distance from load application point to ground surface (m), u_0 is the constant related to rigid body displacement (m), μ is the Poisson's ratio, r_1 is the distance from the observation point to load application point (m), and r_2 is the distance from the observation point to load application image point (m).

r_1 and r_2 can be expressed as:

$$r_1 = \left[x^2+(y-a)^2\right]^{\frac{1}{2}} \tag{3}$$

$$r_2 = \left[x^2+(y+a)^2\right]^{\frac{1}{2}} \tag{4}$$

Equation (2) can be expressed as Equation (5) after coordinate transformation.

$$\begin{aligned}\omega_M{}' = \frac{P}{8\pi G(1-\mu)}\left[(3-4\mu)\ln\frac{R_1}{R_2}+\frac{x_1^2}{R_1^2}-\frac{x_1^2}{R_2^2}\right]-\frac{P(1-\mu)}{\pi G}\ln R_1\\ +\frac{P(y-a)^2}{2\pi G R_1^2}+\frac{Pya\left[x_1^2-(y-a)^2\right]}{4\pi G(1-\mu)R_1^4}+u_0\end{aligned} \tag{5}$$

where $\omega_M{}'$ is the vertical displacement of soil caused by restraint effect of isolation pile (m), x_1 is the lateral distance from the tunnel axis after coordinate transformation (m), $x_1 = x - l$, l is the distance from isolation pile to tunnel axis (m), R_1 is the distance from the observation point to load application point after coordinate transformation (m), and R_2 is the distance from the observation point to load application image point after coordinate transformation (m).

R_1 and R_2 can be expressed as:

$$R_1 = \left[(x_1 - l)^2 + (y - a)^2\right]^{\frac{1}{2}} \quad (6)$$

$$R_2 = \left[(x_1 - l)^2 + (y + a)^2\right]^{\frac{1}{2}} \quad (7)$$

The vertical displacement of soil induced by shield tunnelling considering the restraint effect of adjacent isolation pile can be expressed as Equation (8). When $y \to 0$, the results degenerate into surface settlement trough.

$$\omega = \omega_L + \omega_M{}' \quad (8)$$

3. Case Study

As shown in Figure 3, a case study is found in the Changsha Metro Line 5 from East Laodong Road Station to Huaya Station. The interval right tunnel passes through Huaya International Hotel at a close distance between mileage YDK27 + 680 m and YDK27 + 750 m. The building is a frame-shear wall structure of 3–22 stories, with a basement of 1 story. The foundations are made of artificial bored piles with a diameter of 1000 mm. The pile toe enters a 1.5 m thick bearing layer. The minimum horizontal distance between the pile foundation of adjacent building and the tunnel edge is approximately 6.01 m. The thickness of soil in this section is about 16.0 m, and the stratum is mainly weathered conglomerate.

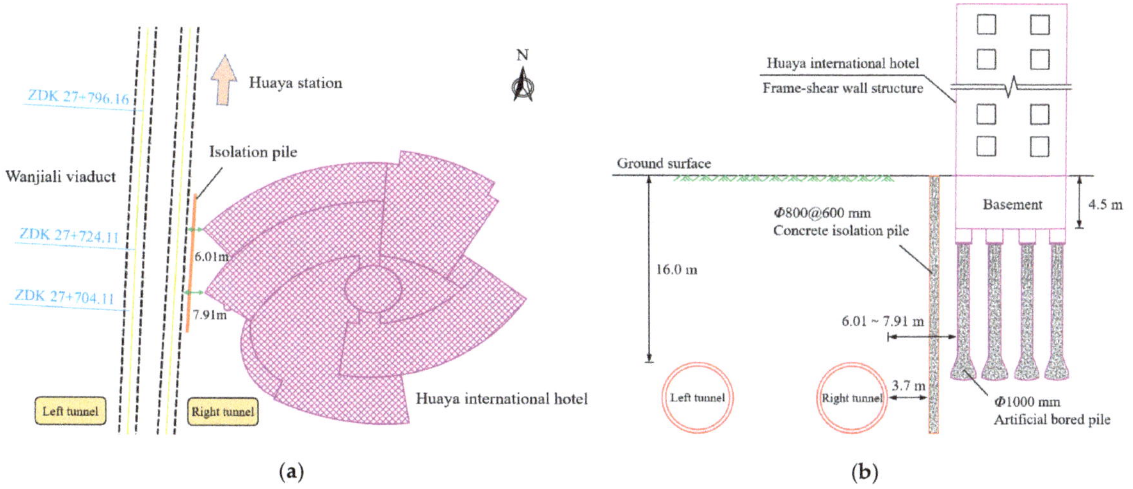

Figure 3. Relationship among building, isolation pile, and tunnel: (**a**) top view and (**b**) sectional view.

Before shield tunnelling to the Huaya International Hotel, the concrete piles with 800 mm diameter and 600 mm spacing are pre-installed in a single row to isolate and protect the building near the right tunnel side. Isolation piles made of C30 underwater concrete (strength is equivalent to ordinary C35 concrete) are installed from the ground surface to the bottom of the tunnel. At this time, the shortest distance between the edge of right tunnel and the isolation pile is merely 3.7 m.

In order to study the restraint effect of isolation pile on surface settlement trough induced by shield tunnelling, the surface settlement observation points were monitored by precision leveling instrument and indium steel ruler. The monitoring holes were drilled by 140 mm crystal drill.

4. FDM-DEM Coupling Model

The discrete element method (DEM) decomposes the soil mass into a series of particles in space, enabling for the calculation of friction between particles and walls. This study focuses on the restraint effect caused by friction between isolation piles and soil, which is suitable for simulation by discrete element method. However, the computational efficiency of the discrete element method is far lower than that of the continuum approaches, such as the finite element method (FEM) and the finite difference method (FDM). Therefore, the modeling size of DEM will be severely constrained. In order to satisfy the requirement of engineering analysis scale and reflect micro mechanical properties, this study introduces the FDM-DEM coupling technique [23–26] to realize the efficient coupling analysis between discrete medium and continuum medium.

4.1. FDM-DEM Coupling Technique

The FDM-DEM coupling mechanism based on FLAC^{2D} and PFC^{2D} is shown in Figure 4 for the plane strain problem. The force-displacement criterion is applied to each contact in each PFC^{2D} cycle. Particle motion is controlled according to Newton's second law. The positions of particles and walls are constantly updated. The unbalanced force on coupled wall-zone will be transmitted to FLAC^{2D} through the embedded Socket I/O interface. After obtaining the new stress and force, the new velocity and displacement will be calculated by balance equation. The node displacement in the coupled area will be transformed into a new displacement boundary condition through the Socket I/O interface again, so that the particles in the discrete domain will generate displacement. The cycle calculations will be carried out in turn.

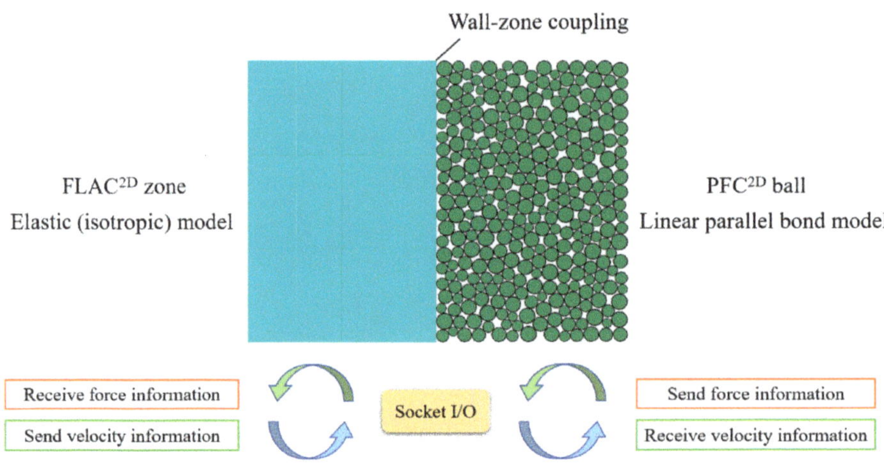

Figure 4. Coupling Mechanism Based on FLAC^{2D} and PFC^{2D}.

4.2. Numerical Model

According to the case mentioned above, the FDM-DEM coupling model is shown in Figure 5. The model has a length of 60 m and a height of 40 m. The discrete domain is 20 m in length and 18 m in height. The radius of shield tunnel is 3.0 m. The distance between the isolation pile and the right tunnel is 3.7 m. The length of the isolation pile is 22 m, and the width is 0.8 m. Ground loss induced by shield tunnelling is simulated by controlling walls shrinkage. The ground loss rate is estimated to be 2.5% based on feedback from field construction parameters. Isotropic elastic constitutive model is adopted in FLAC^{2D} zone. In order to accurately simulate the characteristics of rock and soil, linear parallel bond constitutive model is adopted in PFC^{2D} ball. Macroscopic and microscopic physical and mechanical parameters of soil are shown in Table 1.

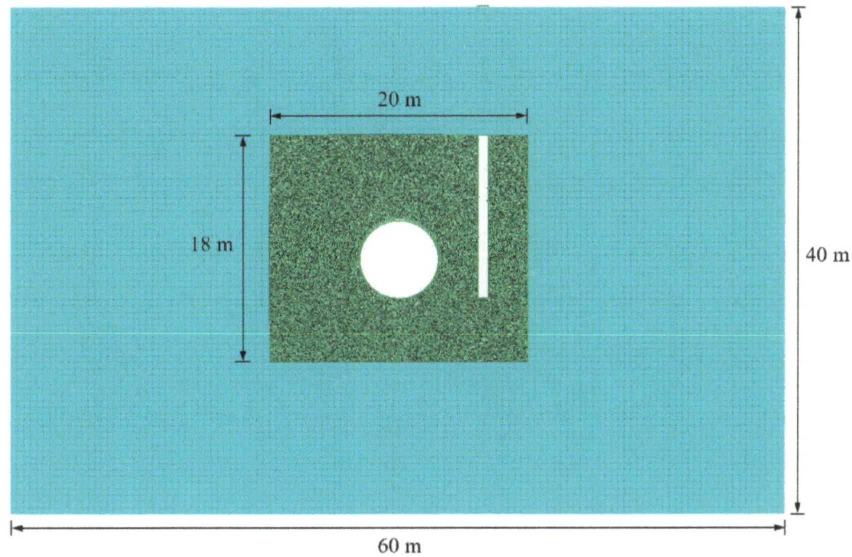

Figure 5. FDM-DEM Coupling Numerical Model.

Table 1. Macroscopic and microscopic physical and mechanical parameters of soil.

Macroscopic Parameters	Young's Modulus (MPa)	Density (kg/m^3)	Poisson's Ratio	Friction Angle (°)	Cohesion (kPa)
Soil mass	75	2250	0.25	30	40
Microscopic Parameters	**Effective Modulus (MPa)**	**Normal Stiffness (N·m^{-1})**	**Tangential Stiffness (N·m^{-1})**	**Normal Bond (kPa)**	**Shear Bond (kPa)**
Soil mass	18.5	7.5×10^6	4.5×10^6	40	45

5. Analysis and Verification

5.1. Results of Analytical Solution

The surface settlement trough calculated by the analytical solution proposed in this study is shown in Figure 6. The friction of adjacent isolation piles on the surrounding soil restricts the surface settlement trough to some extent, resulting in an asymmetric distribution of the ultimate surface settlement trough. According to the results of the Melan solution, it can be found that the restraint effect becomes more obvious with the shortening of the distance from the isolation pile. The surface settlement decreases remarkably near the isolation pile. If the isolation pile is located on the right side of the tunnel, the surface settlement trough shows the characteristics of left low and right high. The analytical solution proposed in this study reveals the reasons of asymmetric surface settlement trough deformation resulting from the pile-soil-tunnel interaction mechanism.

5.2. Results of FDM-DEM Coupling

Figure 7 presents the contours of soil displacement around the tunnel induced by shield tunnelling with and without isolation piles. It can be seen that the affected zone of soil displacement above the top of the tunnel is wider and the displacement variation is more uniform without isolation pile. The affected zone of soil displacement above the tunnel top is restricted under the influence of isolation piles. The soil around the pile is restrained to some extent under the impact of friction, which blocks soil displacement to progress to the other side of the isolation pile.

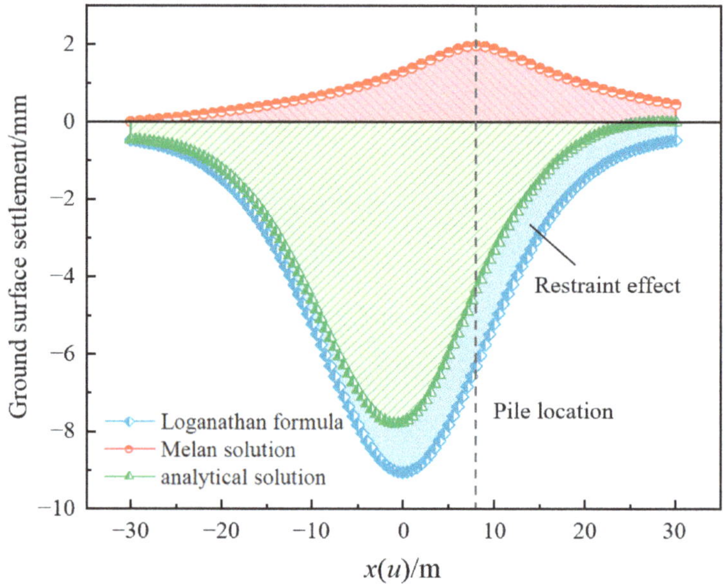

Figure 6. Results of analytical solution.

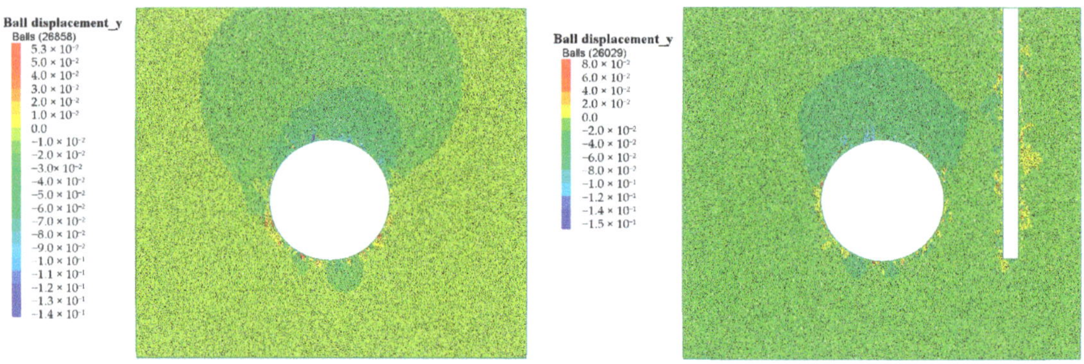

(**a**) (**b**)

Figure 7. Contours of soil displacement: (**a**) without isolation pile and (**b**) with isolation pile.

The comparisons of surface settlement trough calculated by FDM-DEM coupling is shown in Figure 8. There is an asymmetric phenomenon in the surface settlement trough without isolation piles due to the randomness of particle generation and particle distribution after initial stress balance. The overall results are highly similar to those calculated by Loganathan formula. The maximum surface settlement with isolation piles is slightly less than that without isolation piles. In this case, the maximum surface settlement without isolation piles is 8.937 mm, and the maximum surface settlement with isolation piles is 8.225 mm, with a decrease of 7.96%. The numerical results indicate that if isolation piles are pre-installed on the right side of the tunnel, the soil displacement will be restrained on the right side, but the soil displacement will transfer to the side without pile.

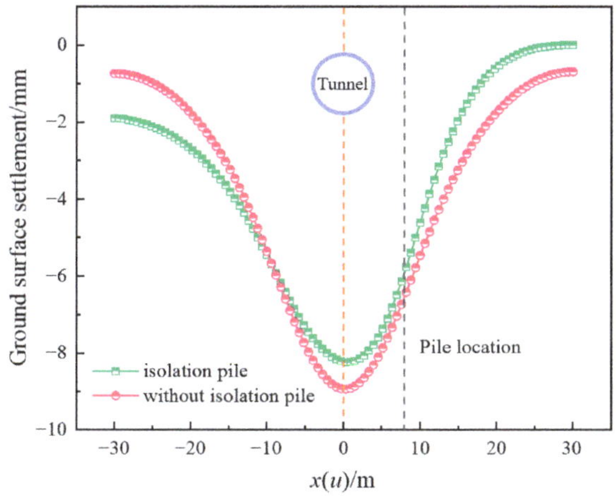

Figure 8. Results of FDM-DEM Coupling.

5.3. Comparison with Field Measured Ddata

The comparison results of the analytical solution, FDM-DEM coupling, and field measured data are shown in Figure 9. The results indicate that the analytical solution proposed in this study and the numerical solution based on FDM-DEM coupling are both in good agreement with the field measured data. The numerical solution based on FDM-DEM coupling is slightly larger than the measured data when analyzing the surface settlement on the left side of the tunnel. The displacement of surface settlement trough at x = −30 m is −1.893 mm, which is larger than −0.459 mm calculated by analytical solution. The reason for this phenomenon is that in the FDM-DEM coupling calculation, the calculation mode of the continuum medium needs to be adjusted to the 'model largestrain' mode. The results of soil displacement are more significantly affected by the parameters, and the calculation results may be larger than the measured data. This finding clearly indicates that when isolation piles are pre-installed on one side of the tunnel, surface settlement tends to develop more on the opposite side. In the meantime, the surface settlement trough could be discontinuous while passing through the position of isolation pile in the actual situation.

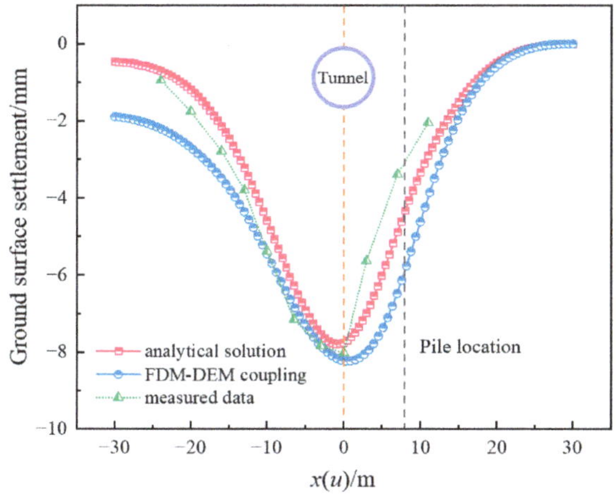

Figure 9. Comparison with field measured data.

6. Conclusions

(1) The restraint mechanism of isolation piles on the surface settlement trough induced by shield tunnelling is analyzed based on pile-soil-tunnel interaction mechanism. The friction between pile and surrounding soil is the principal reason of restraint effect. Based on the Melan solution and Loganathan formula, the analytical solution is derived. The analytical solution could reflect the restraint effect of isolation pile on the surface settlement trough from the internal mechanism;

(2) Taking the Changsha Metro Line 5 from East Laodong Road Station to Huaya Station, where shield tunnelling passes through Huaya International Hotel at a close distance as an engineering background. The FDM-DEM coupling technique is introduced to establish the numerical model. The coupling numerical model could simultaneously satisfy the engineering analysis scale and reflect the friction characteristics between isolation pile and soil particles;

(3) The FDM-DEM coupling model can simulate the impact of different ground loss rates on surface settlement trough by deleting and activating preset circle walls. The affected zone above the tunnel top has a wider range and a more uniform variation after shield tunnelling without isolation pile, while the range of affected zone decreases under the influence of isolation pile. The existence of isolation pile blocks the continuous development of soil displacement;

(4) Comparing with field measured data, the applicability and reliability of the analytical solution and the FDM-DEM coupling model proposed in this study are verified. The ultimate surface settlement trough induced by shield tunnelling presents an asymmetric distribution due to the restraint effect of isolation pile on surrounding soil. The maximum surface settlement with isolation piles is less than that without isolation piles. Meanwhile, when isolation piles are pre-installed on one side of the tunnel, surface settlement tends to develop more on the side without isolation piles.

Author Contributions: Conceptualization, Y.S.; methodology, Y.S.; validation, K.H., Y.S. and X.K.; investigation, Y.S. and X.H.; resources, Y.S.; data curation, Y.S. and R.L.; writing—original draft preparation, Y.S.; writing—review and editing, K.H.; visualization, Y.S. and Q.W.; supervision, K.H.; funding acquisition, K.H. All authors have read and agreed to the published version of the manuscript.

Funding: This research was funded by the National Natural Science Foundation of China, grant number 52078060, the National Science Foundation of Hunan Province, grant number 2020JJ4606, the International Cooperation and Development Project of Double-First-Class Scientific Research in Changsha University of Science & Technology, grant number 2018IC19 and the Innovative Program of Key Disciplines with Advantages and Characteristics of Civil Engineering of Changsha University of Science & Technology, grant number 18ZDXK05.

Institutional Review Board Statement: Not applicable.

Informed Consent Statement: Not applicable.

Data Availability Statement: Some or all data and models from the study are available from the corresponding author by request.

Conflicts of Interest: The authors declare no conflict of interest.

References

1. Huang, K.; Sun, Y.W.; Zhou, D.Q.; Li, Y.J.; Jiang, M.; Huang, X.Q. Influence of water-rich tunnel by shield tunneling on existing bridge pile foundation in layered soils. *J. Cent. South Univ.* **2021**, *28*, 2574–2588. [CrossRef]
2. Ding, W.Q.; Gong, C.J.; Mosalam, K.M.; Soga, K. Development and application of the integrated sealant test apparatus for sealing gaskets in tunnel segmental joints. *Tunn. Undergr. Space Technol.* **2017**, *63*, 54–68. [CrossRef]
3. Gong, C.J.; Ding, W.Q.; Mosalam, K.M.; Günay, S.; Soga, K. Comparison of the structural behavior of reinforced concrete and steel fiber reinforced concrete tunnel segmental joints. *Tunn. Undergr. Space Technol.* **2017**, *68*, 38–57. [CrossRef]
4. Lei, M.F.; Zhu, B.B.; Gong, C.J.; Ding, W.Q.; Liu, L.H. Sealing performance of a precast tunnel gasketed joint under high hydrostatic pressures: Site investigation and detailed numerical modeling. *Tunn. Undergr. Space Technol.* **2021**, *115*, 104082. [CrossRef]

5. Zhou, Z.; Zhang, J.J.; Gong, C.J. Automatic detection method of tunnel lining multi-defects via an enhanced You Only Look Once network. *Comput. Aided Civ. Infrastruct. Eng.* **2022**, *37*, 762–780. [CrossRef]
6. Huang, K.; Sun, Y.W.; Yang, J.S.; Li, Y.J.; Jiang, M.; Huang, X.Q. Three-dimensional displacement characteristics of adjacent pile induced by shield tunneling under the influence of multiple factors. *J. Cent. South Univ.* **2022**, *29*. [CrossRef]
7. Bai, Y.; Yang, Z.H.; Jiang, Z.W. Key protection techniques adopted and analysis of influence on adjacent buildings due to the Bund Tunnel construction. *Tunn. Undergr. Space Technol.* **2014**, *41*, 24–34. [CrossRef]
8. Fu, J.Y.; Yang, J.S.; Zhu, S.T.; Shi, Y.F. Performance of jet-grouted partition walls in mitigating the effects of shield-tunnel construction on adjacent piled structures. *J. Perform. Constr. Facil.* **2017**, *31*, 04016096. [CrossRef]
9. Huang, K.; Sun, Y.W.; He, J.; Huang, X.Q.; Jiang, M.; Li, Y.J. Comparative study on grouting protection schemes for shield tunneling to adjacent viaduct piles. *Adv. Mater. Sci. Eng.* **2021**, *2021*, 5546970. [CrossRef]
10. Peck, R.B. Deep excavations and tunneling in soft ground. In Proceedings of the 7th International Conference on Soil Mechanics and Foundation Engineering, Mexico City, Mexico, 29 August 1969.
11. Fantera, L.; Rampello, S.; Masini, L. A mitigation technique to reduce ground settlements induced by tunnelling using diaphragm walls. *Procedia Eng.* **2016**, *158*, 254–259. [CrossRef]
12. Ding, Z.; Wei, X.J.; Wei, G. Prediction methods on tunnel-excavation induced surface settlement around adjacent building. *Geomech. Eng.* **2017**, *12*, 185–195. [CrossRef]
13. Huang, K.; Sun, Y.W.; Huang, X.Q.; Li, Y.J.; Jiang, M.; Liu, R.N. Effects of different construction sequences on ground surface settlement and displacement of single long pile due to twin paralleled shield tunneling. *Adv. Civ. Eng.* **2021**, *2021*, 5559233. [CrossRef]
14. Rampello, S.; Fantera, L.; Masini, L. Efficiency of embedded barriers to mitigate tunnelling effects. *Tunn. Undergr. Space Technol.* **2019**, *89*, 109–124. [CrossRef]
15. Masini, L.; Rampello, S. Predicted and observed behaviour of pre-installed barriers for the mitigation of tunnelling effects. *Tunn. Undergr. Space Technol.* **2021**, *118*, 104200. [CrossRef]
16. Franza, A.; Losacco, N.; Ledesma, A.; Viggiani, G.M.B.; Jimenez, R. Protecting surface and buried structures from tunnelling using pile walls: A prediction model. *Can. Geotech. J.* **2021**, *58*, 1590–1602. [CrossRef]
17. Sun, Y.W.; Huang, K.; Li, Y.J. Influence of existing bridge pile foundation on the deformation of surface settlement trough induced by shield tunneling. *J. Transp. Sci. Eng.* **2022**, *38*, 79–87. (In Chinese)
18. Cao, L.Q.; Chen, X.S.; Shen, X.; Zhang, D.L.; Su, D.; Fang, H.C. Theoretical analysis of the barrier effect of embedded isolation piles on tunneling-induced vertical ground displacements. *Comput. Geotech.* **2022**, *144*, 104609. [CrossRef]
19. Loganathan, N.; Poulos, H.G. Analytical prediction for tunneling-induced ground movements in clays. *J. Geotech. Geoenviron. Eng.* **1998**, *124*, 846–856. [CrossRef]
20. Melan, E. Der Spannungszustand der durch eine Einzelkraft im Innern beanspruchten Halbscheibe. *J. Appl. Math. Mech.* **1932**, *12*, 343–346. (In German) [CrossRef]
21. Poulos, H.G.; Davis, E.H. *Elastic Solutions for Soil and Rock Mechanics*, 1st ed.; John Wiley and Sons: Hoboken, NJ, USA, 1974; pp. 1–28.
22. Verruijt, A.; Booker, J.R. Complex variable analysis of Mindlin's tunnel problem. In Proceedings of the Developments in Theoretical Geomechanics, Sydney, Australia, 16–17 November 2000.
23. Cai, M.; Kaiser, P.K.; Morioka, H.; Minami, M.; Maejima, M.; Tasaka, Y.; Kurose, H. FLAC/PFC coupled numerical simulation of AE in large-scale underground excavations. *Int. J. Rock Mech. Min. Sci.* **2007**, *44*, 550–564. [CrossRef]
24. Trivino, L.F.; Mohanty, B. Assessment of crack initiation and propagation in rock from explosion-induced stress waves and gas expansion by cross-hole seismometry and FEM–DEM method. *Int. J. Rock Mech. Min. Sci.* **2015**, *77*, 287–299. [CrossRef]
25. Qu, T.M.; Wang, S.Y.; Fu, J.Y.; Hu, Q.X. Numerical examination of EPB shield tunneling-induced responses at various discharge ratios. *J. Perform. Constr. Facil.* **2019**, *33*, 04019035. [CrossRef]
26. Yin, Z.Y.; Wang, P.; Zhang, F.S. Effect of particle shape on the progressive failure of shield tunnel face in granular soils by coupled FDM-DEM method. *Tunn. Undergr. Space Technol.* **2020**, *100*, 103394. [CrossRef]

applied sciences

MDPI

Article

Effect of Jack Thrust Angle Change on Mechanical Characteristics of Shield Tunnel Segmental Linings Considering Additional Constrained Boundaries

Lichao He [1], Yu Jiang [1,2] and Wenjun Zhang [3,4,*]

[1] Beijing First Expressway Construction Management Co., Ltd., Beijing 100165, China; helichao@bchd.com.cn (L.H.); jiangyu@bchd.com.cn (Y.J.)
[2] Department of Industrial Engineering, Tsinghua University, Beijing 100084, China
[3] School of Civil Engineering, Tianjin University, Tianjin 300350, China
[4] School of Civil Engineering and Architecture, Guangxi University, Nanning 530004, China
* Correspondence: wjzhang@tju.edu.cn

Abstract: With the attitude adjustment of shield machine, the jack thrust angle will change and may lead to great damage to segmental lining structures. In this paper, a three-dimensional refined finite element method is proposed to study the action mechanism and influence of jack thrust angle change on the mechanical characteristics of shield tunnel, considering additional constrained boundaries formed by various construction loads. Firstly, the finite element method is established and verified. Then, a series of analyses is carried out to investigate the mechanical performance of a certain tunnel lining with eleven rings under conditions of different jack thrust angles. An in-depth discussion is conducted on the obtained displacement, deformation, and stress distribution. A quantitative standard is finally gained for controlling the jack thrust angle. The results show that the maximum segment displacement occurs at the lower part of the arch waist of the second ring out of the shield tail, which has little relationship with the attitude adjustment. With the increase of the jack thrust angle, the maximum offset between the rings increases linearly, which is not conducive to the safety of the lining structures. The development direction of the tunnel displacement is opposite to that of the jack thrust deflection. The displacement of the tunnel is most affected by the jack thrust deflection when the shield machine is upward, which is also the most unfavorable situation for the segmental linings. The results in this paper provide novel insights into the effect of jack thrust on shield tunnel.

Keywords: shield tunnel; jack thrust angle; mechanical properties; numerical analysis; construction load

Citation: He, L.; Jiang, Y.; Zhang, W. Effect of Jack Thrust Angle Change on Mechanical Characteristics of Shield Tunnel Segmental Linings Considering Additional Constrained Boundaries. *Appl. Sci.* **2022**, *12*, 4855. https://doi.org/10.3390/app12104855

Academic Editors: Mingfeng Lei, Chenjie Gong and Xianda Shen

Received: 25 March 2022
Accepted: 10 May 2022
Published: 11 May 2022

Publisher's Note: MDPI stays neutral with regard to jurisdictional claims in published maps and institutional affiliations.

1. Introduction

During the shield tunneling construction, different thrust has to be assigned to distinct hydro-cylinders to control the attitude of shield machine, so that its axis can be the same as that of the designed tunnel [1]. Under such conditions, the resultant jack force generally provides eccentric load for the segmental lining structures, which may lead to local damage to the segments [2]. Therefore, it is very critical to maintain good control of the jack thrust and its angle. Moreover, the segmental lining structures are under complex action of various types of loads in the construction stage, e.g., water and soil pressure, grouting pressure, reaction force of shield tail brush, and jack thrust. However, the aforementioned loads were commonly studied as separate factors in the past studies. To obtain more reliable guidance for practical engineering, it is essential to investigate the influence of the jack thrust on shield tunnel with consideration of the additional constrained boundaries formed by other construction loads.

Based on the reports published by Lorenzo [3], there is increasing evidence implying that the jack thrust forces are transient load cases with respect to the reasonable design of the reinforced concrete segments, which means that the jack thrust forces and angle have a great influence on the safety of the whole segmental linings. Looking at this problem, a great number of efforts have been made to reveal the impact of the jack thrust on the mechanical properties of the segments. Lee et al. [4] established a fluid-structure interaction numerical model to examine the stability of the shield machine and the nearby segments by considering jack thrust at different levels. Qiao et al. [5] simplified the segmental lining structures as the continuous circular beam to analyze the control methods of the tunneling parameters by governing the appropriate jack thrust forces under different soil stratum. From their conclusions, the deviation of the horizontal axis for the constructed shield tunnel greatly grows with the increasing jack thrust. Additionally, Zhou et al. [6] conducted exploration of the causes of damage to the tunnel segments in the construction stage and found that the uneven jack thrust caused a great proportion of adverse effects. Similar results were gained during the study on the splitting and crushing behavior of the lining structures under TBM hydraulic jack thrust [7]. Generally, great concentrated jack thrust forces are imposed on the linings by the shield machine, resulting in high compressive stresses under the thrust shoes during the lining construction process [8]. These high compressive stresses may further develop into crush damage and lead to unexpected leakage failure of the segments. Although it is obvious that the jack thrust has fundamental importance in controlling the mechanical properties of the segmental lining structures, there is still a great challenge to determine the reasonable governing parameters, especially for tunnels constructed in special environments [9]. To obtain the appropriate values of the jack thrust, Peng et al. [10] fully analyzed the force-arranged characteristics of the key segment with action of jack thrust force experimentally, theoretically, and numerically, during which various influence factors were considered, including the arrangement and distance of the jack as well as the loading of thrust force. Considering the fact that the conventional reinforced concrete segments are vulnerable under jack thrust force, Avanaki [11] studied the mechanical characteristics of hybrid steel fiber reinforced composite segments subjected to thrust. It was found that fiber-reinforced concrete segments may be a good choice for tunnels whose construction is in need of frequent adjustment of jack thrust [12,13]. Another method to accurately control the jack thrust is to introduce the intellectualized technology, with which the real-time updating can be realized by the prediction from the analysis of the relationship between the historical data of the thrust and geological condition [14]. The aforesaid publications make great contributions to a better understanding of the effect of jack thrust, and some qualitative conclusions are available for certain conditions [15]. However, further investigation is still necessary to get more detailed knowledge of the mechanical influence of the jack thrust on the segmental linings.

Distinct methods have been adopted in analyzing the impact of jack thrust, among which the most popular one may be the numerical approach [16–19]. Although some researchers also conducted experiments on the safety analysis of the segmental linings subjected to jack thrust [20], the numerical method shows more attraction due to its significant advantages in a parametric study [21]. It should be noted that most of the existing research applied simplified models and structures, which may cause limitations to the calculation results. In this paper, a refined three-dimensional model considering the bolts and other construction loads is proposed to study the effect of jack thrust angle change on mechanical characteristics of shield tunnel segmental linings. Discussions are carried out on the displacement, deformation, and stress of the tunnel segments and the connecting bolts. According to the obtained results, final recommendations will be provided to control the jack thrust angle.

2. Project Overview

The total length of a shield tunnel project is 16.75 km, while the buried depth of the tunnel slab bottom is 8.3~30 m. The project is a marine alluvial plain with a simple fill on the surface. Soft soil is widely distributed along the tunnel, while sandy clay, silt, and silty sand are occasionally found. The ground elevation is 2.5~4 m, and the design water level is 2 m. The shield tunnel adopts a general pipe ring design. The outer diameter of the tunnel is 6.6 m, whereas the segment thickness is 0.35 m, and the wedge-shaped amount of the universal pipe ring is 0.04 m. Each segmental ring can be divided into six segments, including three standard segments, two temporary connection segments, and one K segment, as shown in Figure 1. The segment joints are connected by two M30 bending bolts, and the rings are connected by sixteen M30 bending bolts at 22.5° equal spacing. The specific joint structure is shown in Figure 2.

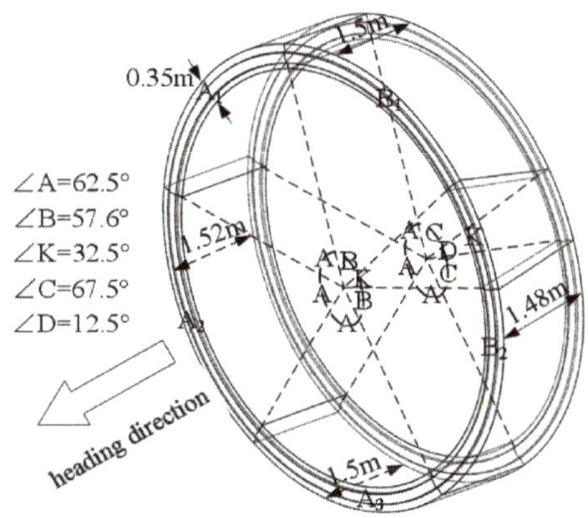

Figure 1. Segmental lining ring.

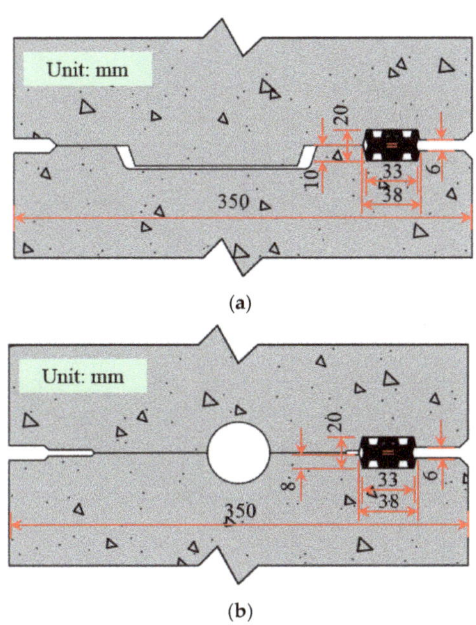

Figure 2. Circumferential and longitudinal joint. (**a**) Circumferential joint. (**b**) Longitudinal joint.

3. Numerical Modelling

3.1. D Finite Element Model

The structure-load model is established to study the mechanical properties of the shield tunnel under the action of the jack thrust angle. The joint of the shield segment is divided into ring joint and segment joint. Joint is the weak link of shield tunnel, where local damage of segment is easy to occur, and then cause leakage. ABAQUS finite element software is used to establish the standard segment, temporary connection segment, and K segment, considering the circumferential groove. All the segments are assembled together with the bolts to form eleven rings in total, as shown in Figure 3. In this model, the weakness of the joint, the distribution of construction loads along the tunnel axis, and the longitudinal difference of the tunnel are considered. The segment part is simulated by the C3D8 element, while the bolt is simulated using B31 element. The mesh division is shown in Figure 4. The joint is composed of joints and connecting bolts. The interaction between segments is simulated by setting a surface-to-surface contact relationship, and the contact attribute is defined as a hard attribute. The connecting bolts connect the segment components to each other through the embedded constraint relationship.

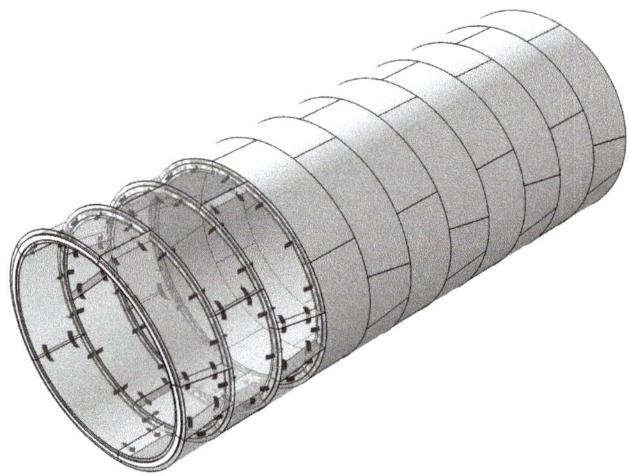

Figure 3. Tunnel model.

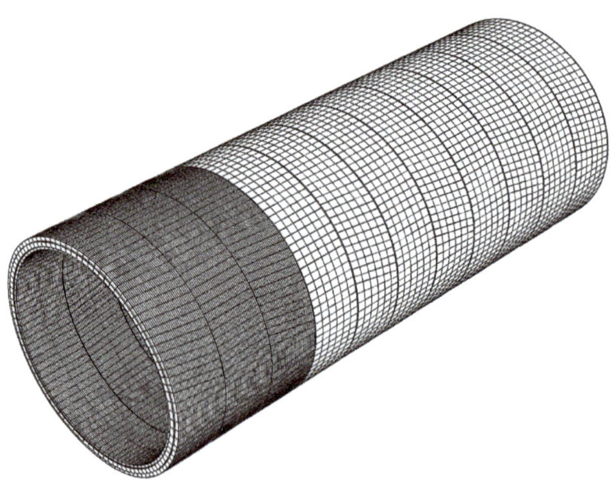

Figure 4. Mesh of numerical model.

The axial direction constraint of the tunnel is set from the current position of the jack, and the ground springs are arranged outside the tunnel to simulate the soil constraint. Additionally, springs are arranged outside the third ring segment from the jack shoes to simulate the constraint effect of the shield machine tail, as shown in Figure 5.

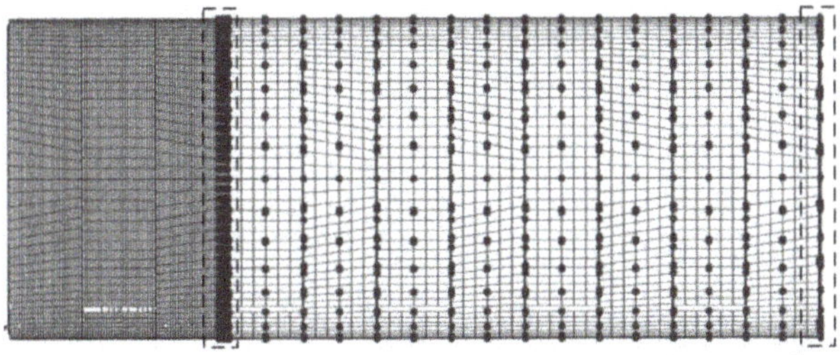

Figure 5. Boundary conditions.

3.2. Material Properties

The strength grade of segment concrete is C60, and the damage to plastic constitutive relation of CDP concrete is adopted. The elastic modulus is 3.45×10^4 MPa. Poisson's ratio is 0.2. Density is 2420 kg/m^3. Shear expansion angle is 38°. Eccentricity is 0.1. The ratio of biaxial ultimate compressive strength to uniaxial ultimate compressive strength is 1.16, and the flow potential offset value is 0.67. The cohesion coefficient is 0.001, while the damage factor can be seen in Table 1.

Table 1. Compressive and tensile properties of C50 concrete.

Uniaxial Compressive Stress/(MPa)	Uniaxial Inelastic Compressive Strain	Damage Parameter	Uniaxial Tensile Stress/(MPa)	Tensile Strain Due to Uniaxial Cracking	Tensile Damage Factor
5.769	0	0	2.640	0	0
21.107	5.96×10^{-5}	0.037	2.168	7.93×10^{-5}	0.275
30.056	3.03×10^{-4}	0.122	1.677	1.26×10^{-4}	0.438
32.454	7.37×10^{-4}	0.239	1.344	1.69×10^{-4}	0.565
27.726	1.54×10^{-3}	0.435	1.120	2.08×10^{-4}	0.658
21.198	2.40×10^{-3}	0.610	0.962	2.45×10^{-4}	0.725
16.413	3.21×10^{-3}	0.730	0.845	2.82×10^{-4}	0.775
13.142	3.97×10^{-3}	0.807	0.756	3.17×10^{-4}	0.813
10.858	4.71×10^{-3}	0.857	0.685	3.52×10^{-4}	0.842
9.205	5.43×10^{-3}	0.891	0.628	3.86×10^{-4}	0.864
7.966	6.13×10^{-3}	0.914	0.581	4.21×10^{-4}	0.882
7.009	6.82×10^{-3}	0.931	0.541	4.54×10^{-4}	0.897
5.636	8.20×10^{-3}	0.953	0.508	4.88×10^{-4}	0.909

M30 bending bolt with a strength grade of 8.8 is used as the joint bolt, and the ideal elastic-plastic constitutive relation is set. The elastic modulus is 210 GPa, while Poisson's ratio is 0.167, and the yield stress is 660 MPa.

3.3. Water and Soil Pressure

The loading state of the shield tunnel in the stratum is simulated by applying water and soil pressure and foundation spring. The foundation spring is a nonlinear spring under uniaxial compression in the radius direction of the tunnel. The calculation method of spring stiffness is as follows:

$$K_R = \begin{cases} k \cdot n \cdot S_i & , R^+ \\ 0 & , R^- \end{cases} \quad (1)$$

In the above formula, K_R is the foundation spring stiffness, k is the foundation bed coefficient, n is the number of grids represented by the foundation spring, S_i is the unit area divided by the outer segment surface, R^+ denotes tunnel radius direction is positive, while R^- means tunnel radius direction is reversed, and the coordinate system is shown in Figure 6. The foundation spring stiffness in the Z direction and T direction is taken as 1/3 of R positive spring stiffness in the radius direction of the tunnel.

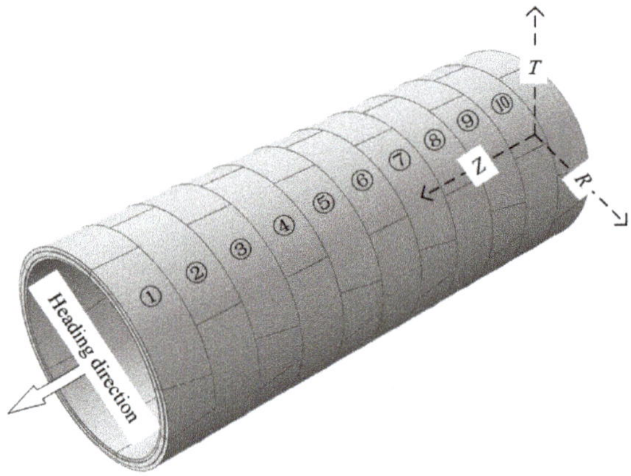

Figure 6. Tunnel coordinates.

The typical stratum section of the shield tunnel is selected to calculate the water and soil pressure. The specific soil layer parameters are shown in Table 2. The buried depth of the tunnel is set at 9.9 m. The water and soil pressure are calculated by the soil and water integrated overburden earth pressure, and the distribution model is shown in Figure 7.

Table 2. Soil parameter.

Soil Layer Number	The Thickness of the Soil Layer/m	Saturation Gravity/(kN/m^3)	Coefficient of Subgrade Bed/(MPa/m)
plain fill soil ① 2	1	16	-
silty clay ④ 1	2.1	18.57	11.38
muddy clay ⑥ 21	5.4	17.47	9.07
Silty clay ⑥ 4	10.3	19.04	14.31

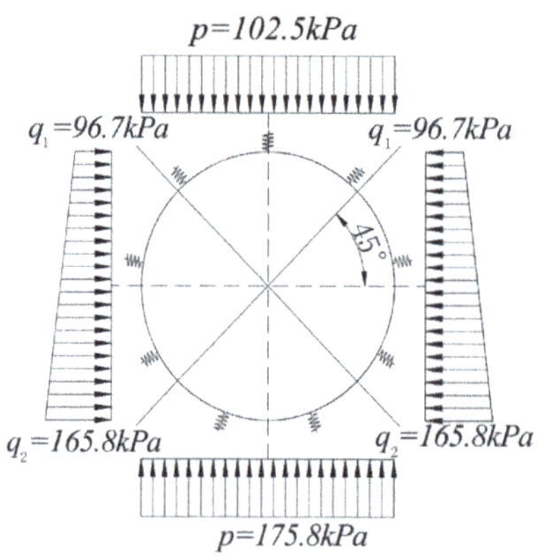

Figure 7. Water and soil pressure.

3.4. Grouting Pressure

The shield tail synchronous grouting method is adopted in shield tunneling. The circular distribution mode of grouting pressure is selected as a fan-shaped filling diffusion model, and Bingham fluid is selected as grouting slurry. The annular distribution formula of grouting pressure is as follows:

$$P = P_i + A(\cos\alpha_i - \cos\alpha) \pm B(\alpha_i - \alpha) \tag{2}$$

$$A = \rho g R \tag{3}$$

In the above formula, P is the grouting pressure, P_i is the grouting hole pressure, which is generally 0.1 MPa~0.2 MPa higher than the water pressure at the grouting hole position. A is the slurry weight parameter, α is the diffusion angle, α_i is the grouting hole angle, B is the slurry shear force, which is related to the fluid property, generally taken as 0.84 kPa, ρ is the slurry fluid density, g is the gravity acceleration, R is the outer tunnel diameter. The fourth ring after synchronous grouting pipe ring attenuates to 0 linearly, as shown in Figure 8.

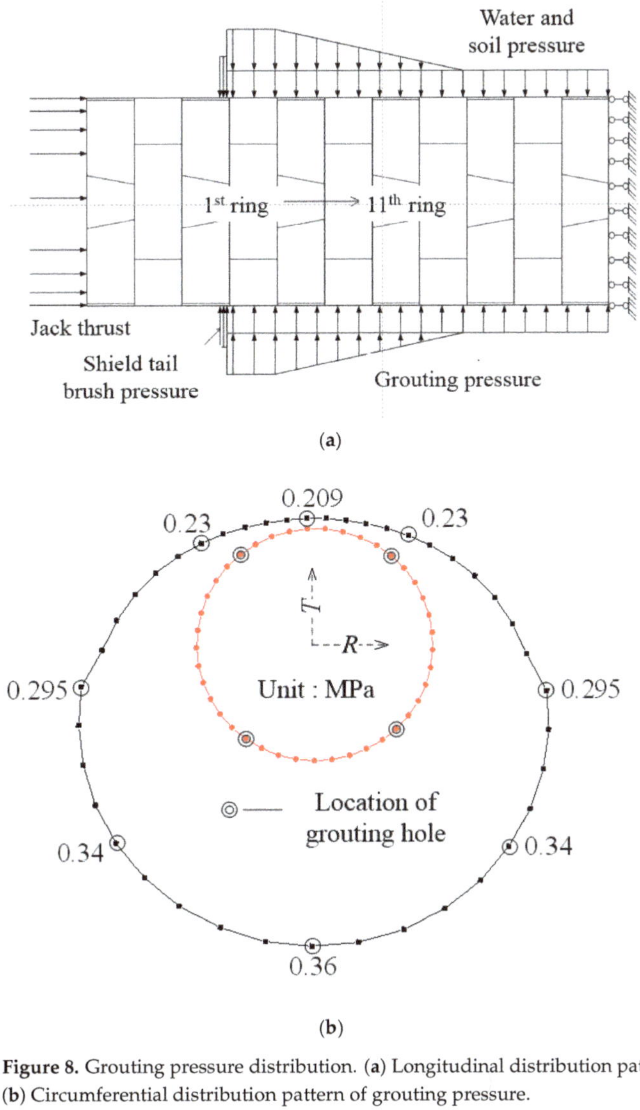

Figure 8. Grouting pressure distribution. (**a**) Longitudinal distribution pattern of grouting pressure. (**b**) Circumferential distribution pattern of grouting pressure.

3.5. Tail Reaction

Generally, there is a gap of 0.8~1% in the outer diameter of the pipe ring between the shield tail and the segment. Under this condition, the extrusion force between the shield tail brush and the segment is minimal. However, under the joint influence of the posture adjustment and synchronous grouting at the shield tail, the shield tail gap is unevenly distributed along the circumferential direction. The grouting slurry is easy to be immersed into the shield tail brush. The stiffness of the shield tail brush increases obviously, and the reaction force tends to be non-uniform distribution. Under the coupling effect of the tail grease cavity pressure, the shield tail brush reaction force, the grouting pressure, and the segment which is about to come out of the shield tail can be easily damaged.

The nonlinear spring has the characteristics of automatic adjustment, distribution, and rebalancing load mode according to the deformation of the shield tunnel. The real stress state of the shield tail segment in shield tunneling is simulated by using a nonlinear shield tail spring and initial position shield tail brush reaction force. The curve of the relationship between the reaction force of the shield tail brush and the compression amount is shown in Figure 9.

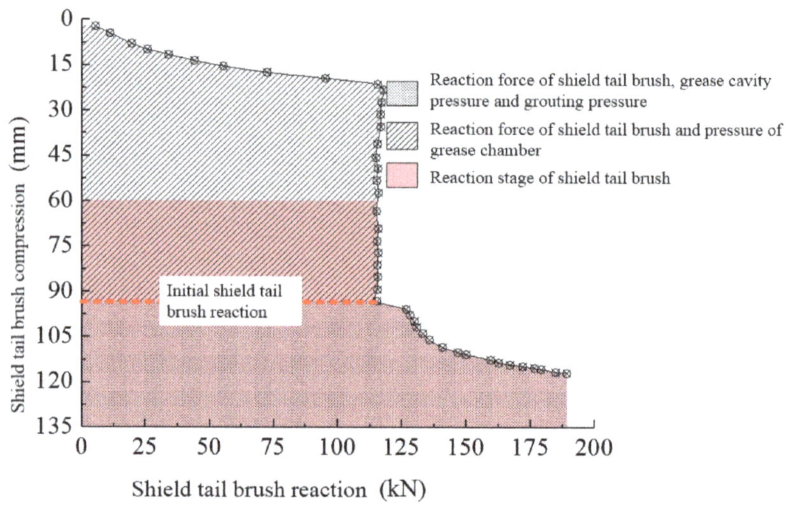

Figure 9. Reaction of shield tail.

3.6. Jack Thrust

The total thrust of the jack includes six parts: Friction between shield machine periphery and soil mass, earth pressure at tunnel excavation face, friction resistance between the inner segment of shield tail and shield shell, traction resistance of rear trolley, penetration resistance generated in front of cutting ring edge during shield tunneling, and variable direction resistance of tunnel turning correction. Through the calculation of the selected typical section, the total thrust of the jack is about 14,000 kN. The jack thrust distribution is shown in Table 3, while the layout of the jack thrust is shown in Figure 10.

Table 3. Table of the jack thrust distribution.

Shield Posture	Single Cylinder Thrust (kN)				Additional Bending Moment/(kN·m)	
	A	B	C	D	MX	MY
upstream	−626	−752	−1222	−752	7402	0
downstream	−894	−738	−1117	−738	3608	0
left	−983	−1080	−931	−588	29	−5812

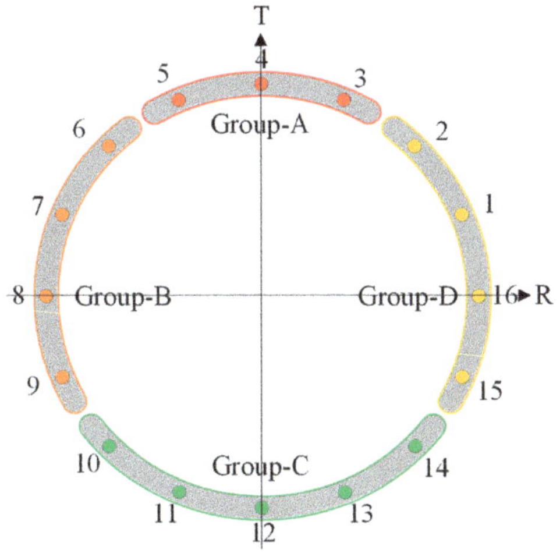

Figure 10. Layout of Jack Thrust.

3.7. Verification

To verify the reliability and accuracy of the established finite element, a smaller model with totally identical modeling criteria was set up according to the experiments conducted by He et al. [22]. The validated experiments include three segments and four bolts, which can properly reflect the overall characteristics of the proposed numerical method in this paper.

The length, width, and depth of the segments on the sides are 350, 1600, and 500 mm, while those of the middle segment are 2700, 1600, and 500 mm, respectively, as shown in Figure 11. Besides, the material properties of the segments and bolts are the same as those described in Section 3.2. The two circumferential joints are subjected to shear force during the experiments, which is simulated in the verified model. More detailed information about the verified experiments can be directly found in [22].

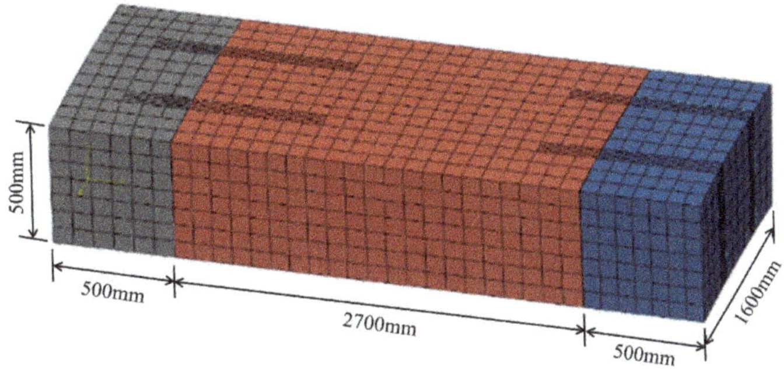

Figure 11. Verified numerical model.

Additionally, a convergence study is also carried out by comparing calculation results obtained from numerical models with different mesh criteria, i.e., large mesh, medium mesh, and small mesh. Figure 12 illustrates the results of the numerical analysis against those of the experiments. It can be seen that the overall behavior of the results from the two different approaches shows good agreement, no matter which mesh size is utilized. However, the numerical results with large mesh size (nearly 40 cm) are not accurate enough, whereas those with medium mesh size (nearly 20 cm) and small mesh size (nearly 5 cm) both exhibit satisfactory reliability compared with the corresponding experimental results.

The aforementioned verification study provides confidence for the reliability and accuracy of the proposed numerical method so that it can be used to further investigate the effect of the jack thrust on the mechanical characteristics of the segmental lining structures. Besides, the mesh of the first three segmental rings adopts a small size, while the other ones uses a medium size. In this way, both the accuracy and computational efficiency can be considered during the numerical analysis.

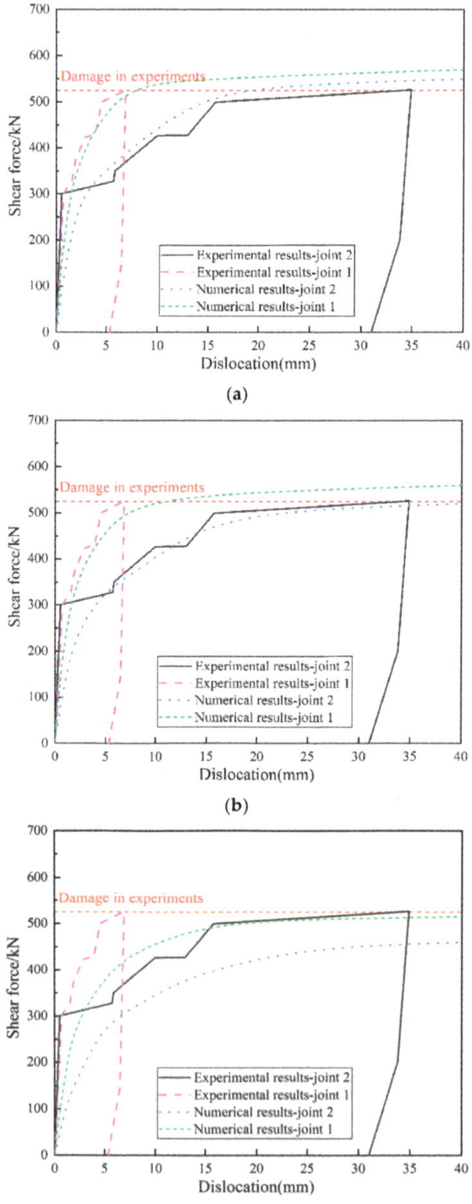

Figure 12. Comparison between numerical and experimental results. (**a**) Small mesh size. (**b**) Medium mesh size. (**c**) Large mesh size.

4. Analysis of Calculation Results

4.1. Segment Staggering

It can be seen from Figure 13a that the first ring and the second ring are only affected by the longitudinal Jack thrust force in the Z direction of the tunnel axis combined with the numerical simulation calculation structure. They are all less than 10^{-1} mm. That is, no comparative analysis is conducted.

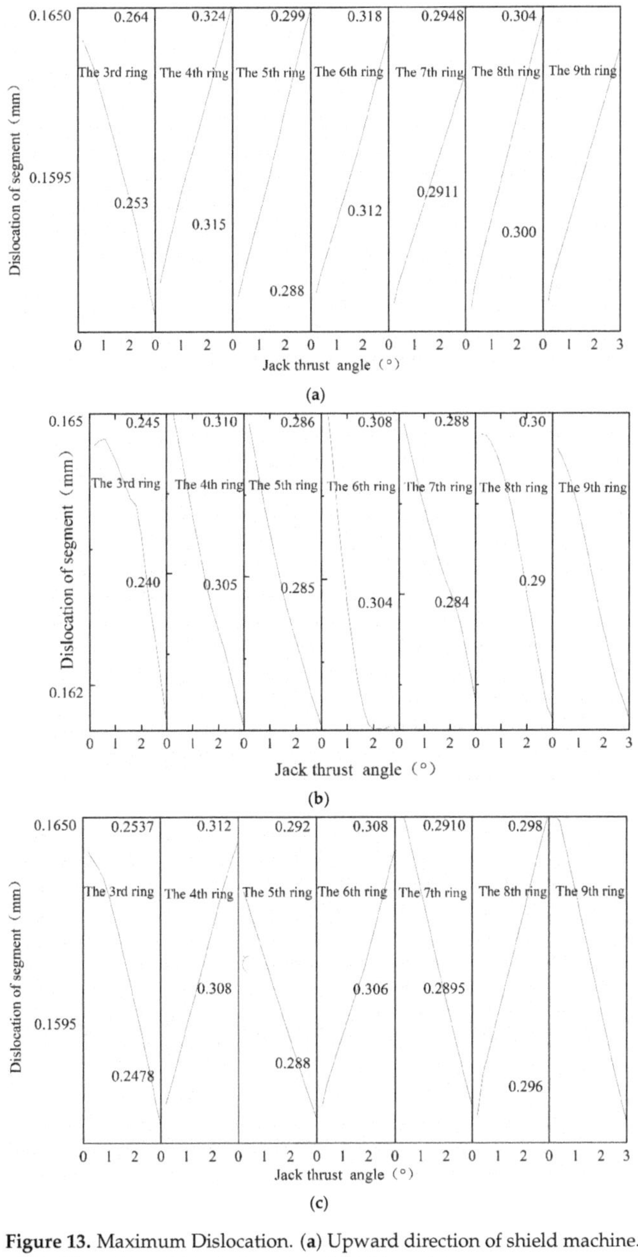

Figure 13. Maximum Dislocation. (a) Upward direction of shield machine. (b) Downward direction of shield machine. (c) Left direction of shield machine.

Under the action of jack thrust angle, the statistical results of the maximum segment staggering position of the 3rd~9th rings are shown in Table 4.

Table 4. Statistical table for the location of maximum segment dislocation.

Ring Number	Upstream	Downstream	Left
3	A2–A3 (213.75°)	A2–A3 (213.75°)	A2–A3 (213.75°)
4	A1–A2 (326.25°)	A1–A2 (326.25°)	A1–A2 (326.25°)
5	A2–A3 (213.75°)	A2–A3 (213.75°)	A2–A3 (213.75°)
6	A1–A2 (326.25°)	A1–A2 (326.25°)	A1–A2 (326.25°)
7	A2–A3 (213.75°)	A2–A3 (213.75°)	A2–A3 (213.75°)
8	A1–A2 (326.25°)	A1–A2 (326.25°)	A1–A2 (326.25°)
9	A2–A3 (213.75°)	A2–A3 (213.75°)	A2–A3 (213.75°)

According to the statistical results, the maximum staggering number of segments in two adjacent rings occurs at the joints of A2–A3 and A1–A2 segments, respectively. The adjacent faces are lower than the arch waist of the K segment, and the vertical section of the tunnel is symmetrically distributed.

When the shield machine is upward, the amount of segment staggering in the third ring decreases linearly with the increase of the jack thrust angle, while the segment staggering amount of the fourth to ninth rings increases linearly with the increase of the jack thrust angle. Under the condition of the jack thrust angle of 3°, the segment staggering amount of the fifth ring in the third to ninth rings is the largest, which is 0.324. When the shield machine is downward, the staggering amount of segments in rings 3~9 decreases with the increase of jack thrust angle. When the jack thrust angle is 3°, the segment staggering amount of the fifth ring is the largest, which is 0.31. When the shield machine runs on the left, the segment staggering amount of odd number ring (K segment is on the left side of shield tunneling prevention) decreases linearly with the increase of jack thrust deflection angle, while that of eighth number ring (K segment is on the right side of shield machine tunneling prevention) increases linearly with the increase of jack thrust deflection angle. When the jack thrust deflection angle is 0.2°, the segment staggering amount of the fifth ring is the largest, which is 0.31 mm.

According to the statistical results of the maximum displacement of segments and the variation law with the jack thrust angle, it can be seen that the amount of segment staggering is jointly controlled by the *RT* plane transverse load and the *Z*-axis longitudinal load. The location of the maximum segment staggering amount in each ring is the maximum position of the tunnel transverse shear force, which is consistent with the actual situation. The adjustment of the shield machine attitude has no effect on the position of the maximum segment stagger but only changes the development law of segment staggering amount with jack's thrust deflection angle. No matter how the shield machine posture is adjusted, the segment staggering amount of the second ring, which is detached from the shield tail, is the largest.

4.2. Displacement and Dislocation between Rings

In the process of shield machine attitude adjustment, when the shield machine is pitching, the jack thrust produces vertical component force due to the existence of a deflection angle, which only affects the vertical displacement of the tunnel. In the same way, when the shield machine yaws, the horizontal component force of jack thrust is produced due to the existence of a deflection angle, and the horizontal displacement of the entering tunnel will be affected. Consequently, the vertical displacement and deformation of the shield machine are analyzed, whereas the horizontal displacement and deformation of the left row of the shield machine are neglected.

It can be seen from Figure 14a that when the shield machine goes up, the displacement of the tunnel vault and arch bottom develops vertically and downward (the opposite direction of the *Z*-axis) is consistent with the stress direction. The influence degree of the jack thrust angle decreases with the increase of the distance from the shield machine and finally tends to be stable. The vertical displacement of the tunnel increases by 51.4%, and the rising slope is 1.82 mm/°. The vertical deformation increases with the increase of the distance from the shield machine and finally approaches 6.83 mm, and it is not affected by the jack thrust angle. From Figure 14b, when the shield machine goes down, the displacement of the tunnel vault and arch bottom vertically upward (*Z*-axis downward) is consistent with the stress direction. The variation law of the influence degree of the jack thrust angle with the distance from the shield machine is the same as that of the upward direction of the shield machine. The results show that the displacement of the arch crown and arch bottom decreases with the increase of the jack thrust deflection angle, and the vertical displacement of the tunnel on jack's action surface decreases by 43.9%, and the decline slope is 1.08 mm/°. The vertical deformation increases with the increase of the distance from the shield machine and finally approaches 7.03 mm, which is not affected by the jack thrust deflection angle. From Figure 14c, the displacement of the left and right sides of the tunnel moves horizontally to the right (in the opposite direction of the *R*-axis), which is consistent with the stress direction. The influence of the jack thrust angle decreases with the distance from the shield machine and finally tends to be stable. The displacement of the left and right sides of the tunnel increases with the increase of the thrust deflection angle of the jack. The horizontal displacement of the working surface of the jack increases by 1750.3%, and the rising slope is 1.31 mm/°.

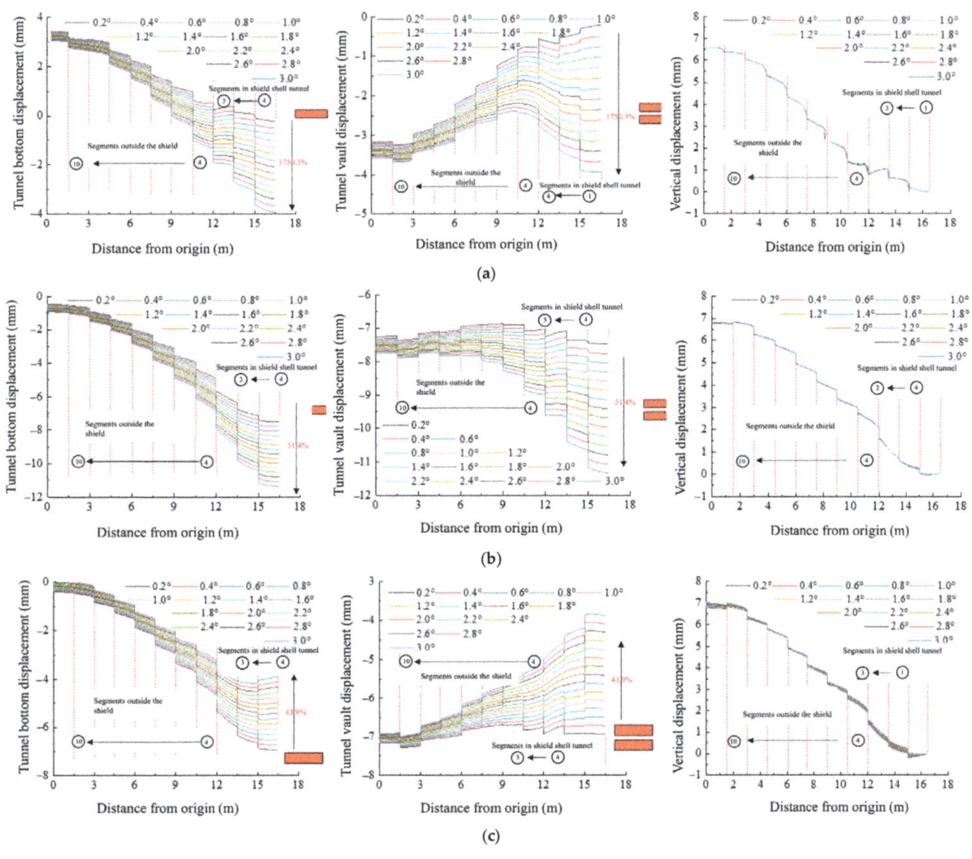

Figure 14. Tunnel displacement and transverse deformation. (**a**) Upward direction of shield machine. (**b**) Downward direction of shield machine. (**c**) Left direction of shield machine.

By comparing the amount of staggered platform between the four positions of the tunnel vault, arch bottom, and both sides of the arch waist, it is finally determined that the maximum amount of staggered abutment is found between the arch bottom ring and the arch bottom ring. Compared with the adjacent pipe rings, the amount of staggering between the third and fourth rings is the largest. It can be seen from Figure 15 that when the shield machine goes up, the maximum amount of staggering between rings increases linearly with the increase of jack thrust deflection angle, and the rising slope is 0.127 mm/°. When the shield machine goes down and left, the maximum amount of staggering between rings decreases linearly with the increase of Jack thrust deflection angle, and the decline rate is −0.105 mm/° and −0.005 mm/°.

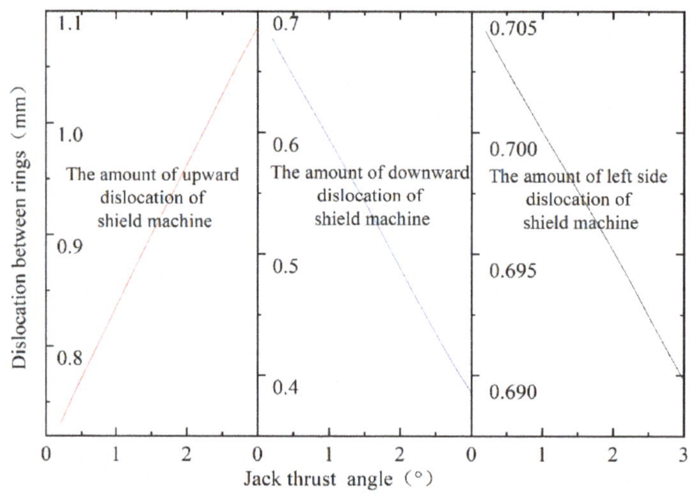

Figure 15. Maximum dislocation of circumferential joint.

According to the above analysis of tunnel displacement and staggered platform amount between rings, it can be seen that the adjustment of the shield machine attitude does not influence the lateral deformation of the tunnel. The tunnel displacement is sensitive to the jack thrust deflection angle, and the tunnel displacement direction is opposite to the shield machine posture adjustment direction, which is consistent with the jack thrust *RT* plane component direction and has a linear change. When the shield goes down, with the increase of jack thrust deflection angle, the tunnel displacement direction is more sensitive to the jack thrust deflection angle. After the deformation of the tunnel, the axis of the tunnel is gradually parallel to the *Z*-axis, and the stress of the structure tends to be reasonable. The angle between the tunnel axis and the *Z*-axis will increase with the increase of jack thrust when the shield machine goes up and left. The structure will have excessive regional deformation and easy segment damage.

4.3. Stress of Segment and Bolt

It can be seen from Figure 16 that the maximum principal stress is distributed at the joints between rings 1st–2nd and 2nd–3rd, the outside of the arch waist segments of the rings 8, 9, and 10, and the inside of the segments at the top and bottom of the arch. Due to the influence of the jack thrust deflection angle, the vertical shear stress of the joints between the rings increases gradually. The maximum principal stress of the grooves at the bottom of the arch of the second and third rings in the 1st–2nd and 2nd–3rd rings is the largest, which is the thinness of the structure. When the shield machine goes down, the maximum principal stress is distributed on the outer side of the segment at the eighth, ninth, and tenth ring, which are the central positions of the segment at the bottom of the arch. Besides, the joint between the 1st–2nd and 2nd–3rd rings is only at the spandrel position, which is smaller than that in the upward direction of the shield machine. This further verifies that the shield machine in Section 4.2 is under relatively reasonable stress. On the left side of the shield machine, the maximum principal stress is distributed on the outer side of the arch waist segment of the eighth, ninth, and tenth rings, the outer groove of the arch waist position of the joint between the 1st–2nd and 2nd–3rd rings and the inner groove at the arch bottom. Similarly, due to the influence of the jack pushing angle, the horizontal shear force of the joint between the rings increases gradually, which leads to the maximum principal stress on the right side of the arch waist of the 1st–2nd and 2nd–3rd ring joints. It is larger than the maximum principal stress of the left spandrel.

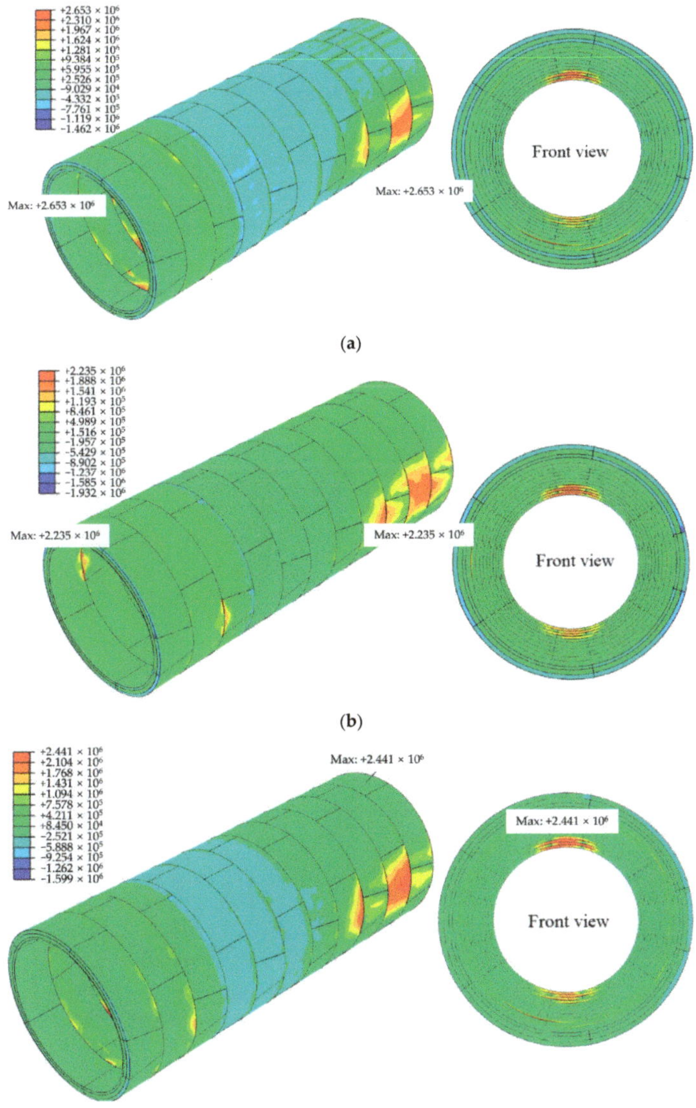

Figure 16. Maximum stress of segmental lining structures. (**a**) Upward direction of shield machine. (**b**) Downward direction of shield machine. (**c**) Left direction of shield machine.

According to the comparison between the left and right pictures in Figure 17, no matter how the shield machine posture is adjusted, the bolt stress between the rings is higher than that between segments. The maximum stress of the bolts between the rings is distributed at the arch bottom of the joint between the third and fourth rings, which is also the most significant position of the amount of staggering between rings. The actual maximum stress of bolts between segments is distributed near the lower position of arch waist, which is consistent with the maximum segment staggering. The measurement positions correspond to each other.

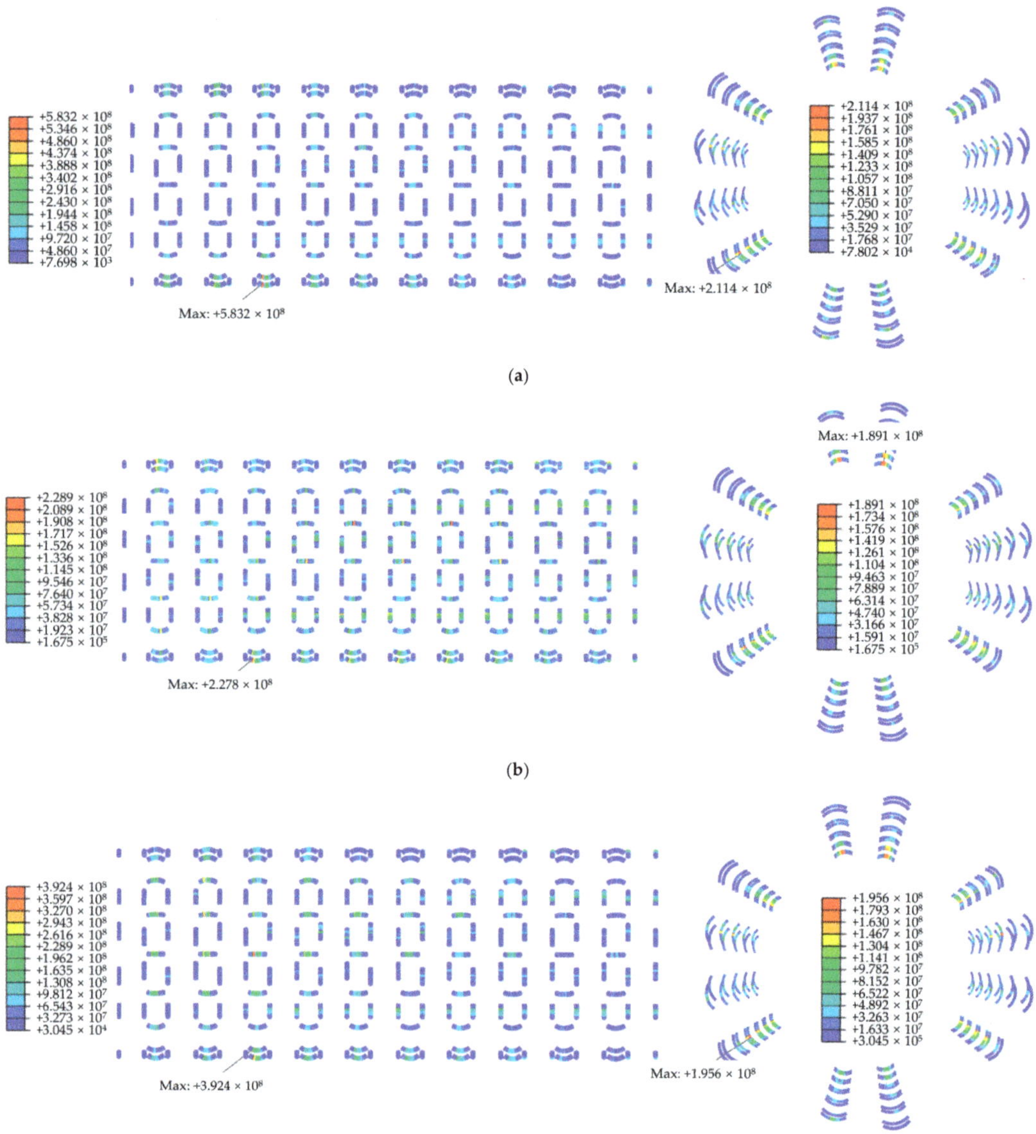

Figure 17. Stress of Bolts. (**a**) Upward direction of shield machine. (**b**) Downward direction of shield machine. (**c**) Left direction of shield machine.

It can be seen from Figure 18a that the maximum von Mises stress of all bolts in the tunnel increases linearly with the increase of the jack thrust angle when the shield machine goes up, and the increase slope is 53.28 MPa/°. When the shield machine goes down and left, the maximum von Mises stress of bolts decreases linearly with the increase of jack thrust angle, and the descending ramps are −47.83 MPa/° and −1.6 MPa/°, respectively. From Figure 18b–d, when the shield machine goes up and down, the position where the maximum principal stress of the segment appears is continuously changing and finally stays at the 2nd–3rd ring. When the shield machine runs on the left, the maximum principal stress of the segment appears on the inner side of the 1tenth ring vault. When the shield machine goes up, the jack thrust deflection angle is within 2.2° and does not exceed the ultimate material strength. When the shield machine goes up, the arch waist position on the right side of the joint between the second and third rings is on the right side. When the shield machine goes down, the maximum principal stress of the segment decreases rapidly within 1.2° of jack thrust deflection angle, and it is basically stable at 2.23 MPa after exceeding 1.2°, which is less than the ultimate strength of the material. When the shield machine goes left, the maximum principal stress of the segment decreases linearly with the increase of jack thrust angle.

Based on the above analysis, the bolt stress of the inter ring joint is higher than that of the segment joint, and the position of the maximum bolt stress is the same as that of the maximum staggering amount. When the shield machine goes up, the jack thrust deflection angle on the structural stress and the tunnel structure under the conventional external loads (water and soil pressure, grouting pressure, shield tail brush reaction force) are superimposed as the most unfavorable combination for the structure.

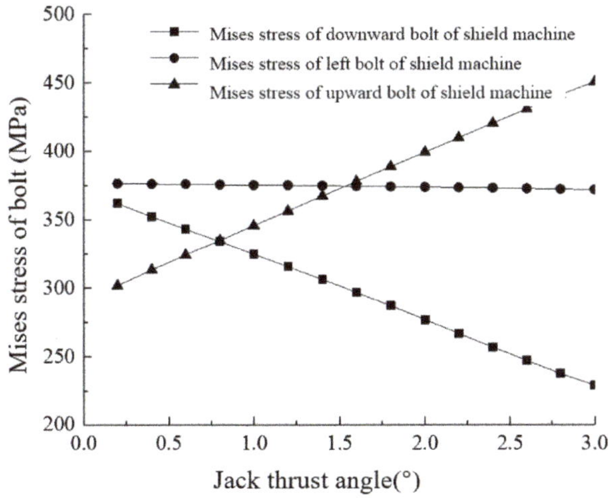

(**a**)

Figure 18. *Cont.*

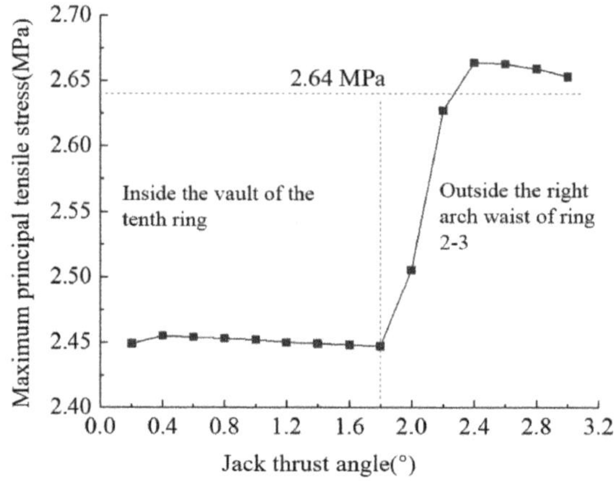

(**b**)

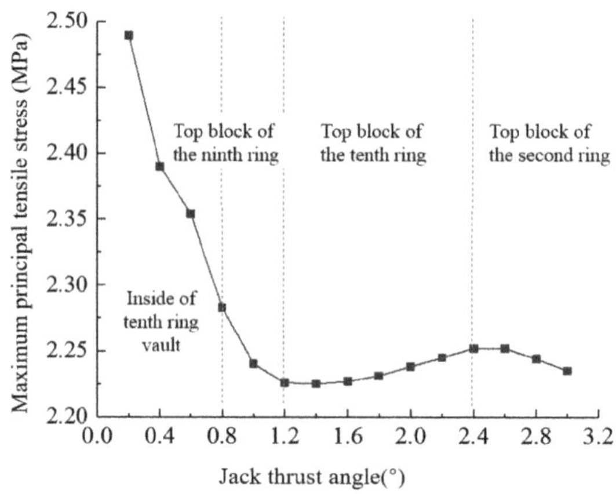

(**c**)

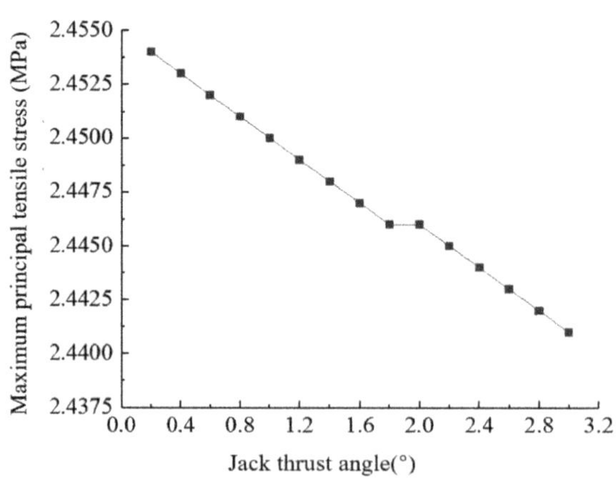

(**d**)

Figure 18. Stress of tunnel and bolt. (**a**) Stress of bolt. (**b**) Stress of tunnel in upward direction. (**c**) Stress of tunnel in downward direction. (**d**) Stress of tunnel in left direction.

5. Conclusions

This paper focuses on the mechanical influence of jack thrust angle on the tunnel structure caused by the shield machine posture adjustment. Through the analysis of the calculation results, the quantitative change rules of tunnel joint deformation, displacement, and stress with jack thrust angle are obtained. The control range of the deflection angle is given, which has particular guiding significance for practical engineering. The specific conclusions are as follows:

(1) No matter how the shield machine attitude is adjusted, the maximum amount of segment staggering occurs at the lower side of the arch waist, and the second ring segment out of the shield tail has the most substantial amount.

(2) When the shield machine goes up, down, and left, the maximum amount of staggering between rings occurs at the joint position of the 3rd–4th ring at the bottom of the arch. When the shield machine goes up, the joint between the rings increases linearly by 0.127 mm/° with the increase of the jack thrust deflection angle.

(3) The development direction of tunnel displacement is opposite to that of jack's thrust angle, and the lateral deformation of the tunnel is not affected by the posture adjustment of the shield machine.

(4) When the shield machine goes down, the structural stress tends to be reasonable. When the shield machine goes up, the structure has been damaged, which is an unfavorable combination of tunnel load, and the deflection angle should be controlled within 2.2°. When the shield machine goes left, it only affects the variation of deformation and stress with the deviation angle of jack thrust force.

Author Contributions: Formal analysis, Y.J.; methodology, W.Z.; software, L.H.; visualization, Y.J.; writing—original draft, L.H.; writing—review & editing, W.Z. All authors have read and agreed to the published version of the manuscript.

Funding: This research was supported by the National Natural Science Foundation of China (Grant No. 51978460) and the Scientific Research Project Investment Plan of Beijing Capital Highway Development Group Co., Ltd. (Grant No. 2021-4).

Acknowledgments: This research was funded by the National Natural Science Foundation of China (Grant No. 51978460) and the Scientific Research Project Investment Plan of Beijing Capital Highway Development Group Co., Ltd. (Grant No. 2021-4), which are gratefully acknowledged.

Conflicts of Interest: The authors declare no conflict of interest.

References

1. Shi, C.H.; Wang, Z.X.; Gong, C.J.; Liu, J.W.; Peng, Z.; Cao, C.Y. Prediction of the additional structural response of segmental tunnel linings induced by asymmetric jack thrusts. *Tunn. Undergr. Space Technol.* **2022**, *124*, 104471. [CrossRef]
2. Shen, X.; Yuan, D.J.; Jin, D.L. Influence of shield attitude change on shield–soil interaction. *Appl. Sci.* **2019**, *9*, 1812. [CrossRef]
3. Lorenzo, S.G. Design guidelines for concrete segmental linings in transverse interaction with the TBM. *Tunn. Undergr. Space Technol.* **2022**, *122*, 104388. [CrossRef]
4. Lee, H.S.; Na, Y.M.; Jang, H.S.; Suk, I.H.; Kang, S.H.; Kim, H.T.; Park, J.K. A Study on the Stability of Shield TBM Thrust Jack in the Behavior of Operating Fluid According to Thrust Force. *J. Korean Soc. Manuf. Process. Eng.* **2019**, *18*, 38–45. [CrossRef]
5. Qiao, S.; Xu, P.; Liu, R.; Wang, G. Study on the Horizontal Axis Deviation of a Small Radius TBM Tunnel Based on Winkler Foundation Model. *Appl. Sci.* **2020**, *10*, 784. [CrossRef]
6. Zhou, J.H.; Wu, D.; Zhou, S.H.; Cui, Y.X. Analysis on damage cause of shield tunnel segments during construction. *Adv. Mat. Res.* **2012**, *594*, 1308–1313. [CrossRef]
7. Medeghini, F.; Trabucchi, I.; Conforti, A.; Tiberti, G.; Plizzari, G.A. Local splitting and crushing behavior under TBM hydraulic jacks. *Galle. Grand. Opere. Sotte.* **2018**, *127*, 37–44.
8. Conforti, A.; Tiberti, G.; Plizzari, G.A. Splitting and crushing failure in FRC elements subjected to a high concentrated load. *Compos. Part B Eng.* **2016**, *105*, 82–92. [CrossRef]
9. Zhe, G.; Li, S.C.; Zhao, S.S.; Zhang, J.Y. Nonlinear dynamic analysis and application of thrust force for slurry balance shield. *Rock Soil Mech.* **2018**, *39*, 469–476.
10. Peng, Z.Y.; Liu, W.N.; Ding, D.Y.; Yang, X.R.; Yan, M.; Lu, W.D. Arrangement of thrust force about partitioned key segment. *J. Cent. Sou. Univ. Sci. Technol.* **2014**, *45*, 4018–4025.

11. Jamshidi Avanaki, M. Effects of hybrid steel fiber reinforced composites on structural performance of segmental linings subjected to TBM jacks. *Struct. Concr.* **2019**, *20*, 1909–1925. [CrossRef]
12. Yao, Y.; Bakhshi, M.; Nasri, V.; Mobasher, B. Interaction diagrams for design of hybrid fiber-reinforced tunnel segments. *Mater. Struct.* **2018**, *51*, 35. [CrossRef]
13. Neu, G.E.; Edler, P.; Freitag, S.; Gudžulić, V.; Meschke, G. Reliability based optimization of steel-fibre segmental tunnel linings subjected to thrust jack loadings. *Eng. Struct.* **2022**, *254*, 113752. [CrossRef]
14. Zhou, Q.; Chen, J.; He, Z.; Zhang, H. Design of intellectualized controller of shield machine. *J. Tongji Univ. Nat. Sci.* **2008**, *36*, 76–80.
15. Cho, S.H.; Kim, J.; Won, J.; Kim, M.K. Effects of jack force and construction steps on the change of lining stresses in a TBM tunnel. *KSCE J. Civ. Eng.* **2017**, *21*, 1135–1146. [CrossRef]
16. Chaipanna, P.; Jongpradist, P. 3D-FEM analysis of shield tunnel construction with ground-spring model. *Geomate J.* **2017**, *12*, 58–62. [CrossRef]
17. Deng, K.S.; Li, Y.Y.; Yin, Z.R. Thrust distribution characteristics of thrust systems of shield machines based on spatial force ellipse model in mixed ground. *J. Mech. Sci. Technol.* **2016**, *30*, 279–286. [CrossRef]
18. Chaipanna, P.; Jongpradist, P. 3D response analysis of a shield tunnel segmental lining during construction and a parametric study using the ground-spring model. *Tunn. Undergr. Space Technol.* **2019**, *90*, 369–382. [CrossRef]
19. Dastjerdy, B.; Hasanpour, R.; Samimi Namin, F.; Chakeri, H. 3D computational study of tunnel segment behaviour under loading by TBM thrust jacks. *Geomech. Geoengin.* **2021**, *16*, 182–197. [CrossRef]
20. Jamshidi, A.M.; Abedi, M.; Hoseini, A. Experimental and numerical-based design of hybrid steel fibre-reinforced concrete tunnels. *Mag. Concr. Res.* **2020**, *72*, 720–733. [CrossRef]
21. Trabucchi, I.; Tiberti, G.; Plizzari, G.A. A parametric numerical study on the behavior of large precast tunnel segments during TBM thrust phase. *Eng. Struct.* **2021**, *241*, 112253. [CrossRef]
22. He, Y.; Yang, Z.H.; Liu, X.; Ding, W.Q. Theoretical solution for failure mechanism of circumferential joint of shield tunnel segment connected using diagonal bolts. *Tunn. Constr.* **2021**, *41*, 933. [CrossRef]

applied sciences

MDPI

Article

Effects of Sidewall Brightness on LED Lighting Environment and Visual Performance in Road Tunnels

Xuefeng Chen [1,2], Shouzhong Feng [1,3], Jie Li [4], Long Zhou [1,*], Weixing Mao [3] and Hehua Zhu [1]

1 Department of Geotechnical Engineering, College of Civil Engineering, Tongji University, Shanghai 200092, China; 1811419@tongji.edu.cn (X.C.); fsz63@vip.163.com (S.F.); zhuhehua@tongji.edu.cn (H.Z.)
2 Guizhou Highway Engineering Group Co., Ltd., Guiyang 550008, China
3 Anhui Zhongyi New Material Science and Technology Co., Ltd., Chuzhou 239500, China; satellitemao939@163.com
4 Wuhan Guangyi Transportation Science and Technology Incorporated Company, Wuhan 430073, China; lijiechuzhou@163.com
* Correspondence: zhoulonglmn@126.com; Tel./Fax: +86-021-6598-5140

Abstract: The lighting design of a road tunnel focuses on the setting of pavement luminance. As for the tunnel sidewall luminance, it simply follows the principle of no less than 60% of the pavement luminance. In fact, the sidewall is one of the important components of the tunnel lighting environment; however, the impact of the improvement of sidewall brightness on the LED lighting environment in a tunnel has not attracted enough attention. In this study, the impact of the improvement of tunnel sidewall brightness on the lighting environment and visual characteristics of human eyes were analyzed based on the lighting experiments when the tunnel sidewalls were decorated with two different types of materials and illuminated by LED lamps with five different color temperatures. The test results show that the tunnel sidewall luminance will increase if the energy-storage and luminescent coating with high reflectance is decorated on tunnel sidewalls, and the pavement luminance increases with the increasing sidewall luminance. After the improvement of the sidewall and pavement luminance, the visible distance for a small object in the tunnel will increase to a certain extent. Furthermore, there is no discomfort for drivers during driving, and the changing of eye pupil diameter is relatively stable, which indicates that the increase in tunnel sidewall luminance has a positive effect on the visual performance of drivers.

Keywords: tunnel lighting; sidewall brightness; lighting environment; visual characteristics; experiment

Citation: Chen, X.; Feng, S.; Li, J.; Zhou, L.; Mao, W.; Zhu, H. Effects of Sidewall Brightness on LED Lighting Environment and Visual Performance in Road Tunnels. *Appl. Sci.* **2022**, *12*, 4919. https://doi.org/10.3390/app12104919

Academic Editor: Nicola Bosso

Received: 30 March 2022
Accepted: 11 May 2022
Published: 12 May 2022

1. Introduction

Road tunnel lighting systems consume a lot of energy, and the power consumption of 24 h continuous lighting in tunnels has become a heavy burden during tunnel operation [1–3]. According to the statistics, the power consumption of tunnel lighting accounts for more than 50% of the total energy consumption for a tunnel [4–6]. In China, there were 21,316 road tunnels with a total length of 21,999 km by the end of 2020, and the total length of road tunnels still increases at a rate of more than 10% each year [7,8]. With the continuous growth of total road tunnel length, tunnel lighting faces a serious problem of energy consumption [9,10].

In recent years, LED lamps have become the first choice for tunnel energy-saving lighting [11–13]. Fruitful studies have been performed covering the energy-saving effects and layout optimization of LED lamps in tunnels. Sato and Hagio [14] and Renzler et al. [15] investigated the influence of lamp layout on lighting energy saving in road tunnels for the symmetric lighting, counter-beam lighting and pro-beam lighting situations. Fan and Yang [16] established a parameter optimization model to acquire the most energy-saving luminaire installation parameters including luminaire installation height and angle, longitudinal and transverse installation spacing, and luminaire power in the interior zone of road tunnels. Shi et al. [17] analyzed the road surface illumination and evenness obtained from different lighting simulations based on DIALux software to get the best scheme for tunnel lamp installation. Domenichini et al. [18] revealed the influence of the lighting system on the driver's behavior in road tunnels by means of driving simulator experiments. Zhang et al. [19], Zhao et al. [20] and He et al. [21] studied the effects of the spatial distribution of lighting environment parameters on drivers' visual performance. Li et al. [22] proposed an optimal design model for determining light sources, powers, layout modes and equipment spacing of lamps in tunnels, which can not only effectively reduce the overall installation and operation cost, but also ease the complexity of manual calculations during the design process of tunnel lighting systems.

The use of LED lighting can save energy consumption on the basis of ensuring the tunnel lighting requirements, and the existing statistical data show that the application of LED lamps in the tunnel can achieve energy saving of more than 30% compared with other types of illumination lamps [23–25]. For the LED lighting environment, the lighting specifications of the "*Guide for the Lighting of Road Tunnels and Underpasses* (CIE 88 2004)" [26] and "*Guidelines for Design of Lighting of Highway Tunnels* (JTG/T D70/2-01-2014)" [27] specify the road surface luminance in tunnels and demand that the average luminance of the tunnel sidewalls, up to at least a height of 2 m, must be at least 60% of the average road surface luminance at the relevant location. In the existing tunnel lighting design, the lighting calculation and analysis focus on the setting of road surface luminance. As for the sidewall luminance, it simply follows the principle of no less than 60% of the road luminance. In fact, the tunnel sidewall is one of the important components in tunnel lighting environment; however, the impact of the improvement of sidewall brightness on the LED lighting environment in the tunnel has not attracted enough attention. For example, when the sidewall luminance is 80% of the road surface luminance, or even greater than the road luminance, what impact will it have on the lighting environment in the tunnel and the visual characteristics of human eyes? In this study, experimental research was carried out in a tunnel when the tunnel sidewalls were decorated with two different types of materials and illuminated by LED lamps with five different color temperatures, to analyze the impact of improvement of tunnel sidewall brightness on the lighting environment and visual characteristics of human eyes.

2. Test Arrangement

2.1. Tunnel and LED Lamps

The test tunnel is 160 m long, with a clear width of 10.4 m, a clear height of 7.5 m and a wall thickness of 0.25 m. The cross-section dimensions are shown in Figure 1. In the area of 4.6–5.4 m away from the road surface, five rows of LED lamps were installed, and the vertical interval between each row was 0.2 m. In each row, the LED lamps were symmetrically arranged on both sides of the tunnel. The longitudinal interval between two adjacent LED lamps was 10.5 m, and the distance from the first lamp to the tunnel entrance is 12.5 m. The color temperatures of the five rows of lamps from low position to high position are 2500 K, 3500 K, 4500 K, 5500 K and 6500 K, respectively. Two different decorative materials were decorated on both sides of the tunnel sidewalls within the range of 3 m, as shown in Figures 1 and 2. Firstly, the energy-storage and luminescent coating with high reflectance was painted on the tunnel sidewalls, and the indexes of lighting environment and visual characteristics of human eyes in the tunnel were tested. The energy-storage and luminescent coating is a modified rare-earth aluminate material and using energy-storage and luminescent coating for auxiliary lighting in tunnel can increase the brightness because of its high reflection; moreover, the energy-storage and luminescent coating can release the stored energy in the form of light for emergency lighting [28,29]. Then, the energy-storage and luminescent coating was covered with gray cloth, and the same indexes were tested. The appearance and interior decoration of the tunnel decorated with energy-storage and luminescent coating and gray cloth are shown in Figure 2.

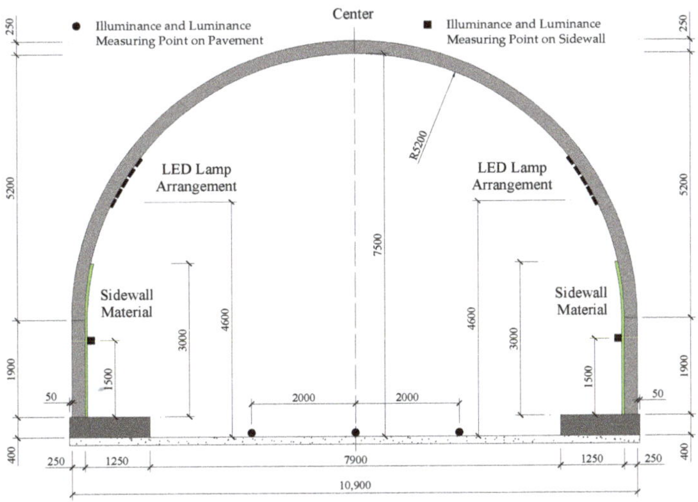

Figure 1. Dimensions of tunnel cross-section (Unit: mm).

(a) (b)

Figure 2. Tunnel appearance and decoration. (**a**) Sidewalls with energy-storage and luminescent coating, (**b**) sidewalls with gray cloth.

2.2. Lighting Environment Setup

The electric power of LED lamps in the tunnel was taken as the control index, and two different electric powers were selected. The luminous efficiency of the LED lamps is 108 lm/W, and the light distribution curve is shown in Figure 3. Combined with the lighting conditions of the test tunnel, two lighting states were set. For the tunnel in which the sidewalls were decorated with energy-storage and luminescent coating, the LED lamps with five different color temperatures were turned on at electric power of 10 W sequentially; then, the LED lamps with five different color temperatures were turned on at electric power of 18 W sequentially, i.e., a total of ten lighting environments. For the tunnel in which the sidewalls were decorated with gray cloth, the power control of LED lamps was the same as that decorated with energy-storage and luminescent coating, which will not be repeated here. The luminance, illuminance, spectral distribution, S/P value and human visual characteristics were tested in the twenty lighting environments to provide a basis for the analysis in this study.

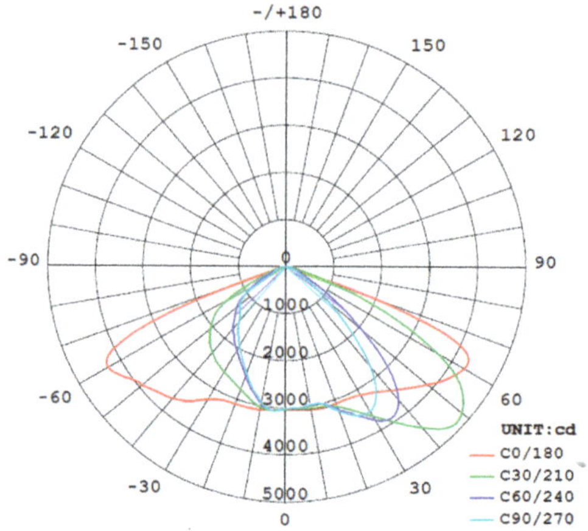

Figure 3. Light distribution curve of LED lamps.

2.3. Experimental Scheme

After the energy-storage and luminescent coating and gray cloth were decorated on the tunnel sidewalls, the material reflectance was measured to be 0.9 and 0.3 for the coating and cloth, respectively, by the C84-III reflectance tester (see Figure 4a). The test was carried out after 8 p.m. to avoid the influence of natural light on the lighting environment in the tunnel. The tested indexes of the lighting environment mainly include the illuminance and luminance in tunnel, and the spectral distribution and S/P value of LED light sources. The illuminance and luminance in tunnel were measured by the CL-500A illuminance meter and LM-3 luminance meter, respectively (see Figure 4b,c), and three cross-section positions of 55 m, 65 m and 75 m away from the tunnel entrance were selected for the measurement. In each cross-section, the measuring points for the illuminance and luminance of road surface were the center of the pavement and the left and right sides which are 2 m from the center (see Figure 1), and there were nine measuring points for the pavement. The sidewall measuring points were at the height of 1.5 m on the left and right sidewalls (see Figure 1) with a total of six measuring points of the three cross-sections. In the testing of sidewall and pavement illuminance, the optical probe of the illuminance meter was along the vertical direction of the sidewall and pavement, respectively. The spectral distribution and S/P value of the LED light sources were measured by the OHSP-350Z multifunctional spectrometer (see Figure 4d), and the measured position was selected at the cross-section of 65 m away from the tunnel entrance.

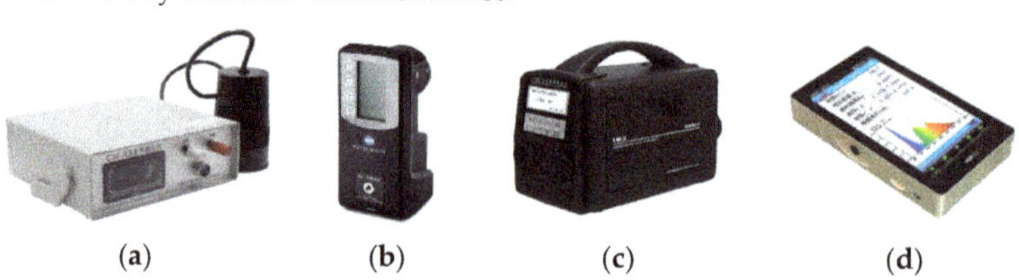

Figure 4. Testing devices for lighting environment parameters in tunnel. (**a**) Reflectance tester, (**b**) illuminance meter, (**c**) luminance meter, (**d**) multifunctional spectrometer.

The test for human visual characteristics includes two parts: the visible distance test for a small object in the tunnel and the dynamic change of the driver's eye pupil diameter during driving in the tunnel. When testing the visible distance for the small object, three observers with normal sight and without color blindness, color weakness, or other eye diseases stood outside the tunnel with a distance of 38 m away from the tunnel entrance and observed the small object randomly placed in the tunnel, as shown in Figure 5. The small object is a gray cube with a size of 10 cm × 10 cm × 10 cm and a surface reflectivity of 0.2 [19]. The observer firstly stood outside the tunnel and a black cloth was placed in front of him; then, the black cloth was removed, and the observer was requested to recognize the small object randomly placed on the pavement of the tunnel for 0.1 s [19,30]. The observer was asked the following three questions: (1) Is there any object? (2) Where is the position of the object on the pavement? and (3) What is the shape and color of the object? If the three questions were answered correctly, it was considered that the observer can recognize the small object [19]. During the test, the distance between the small object and the observer was repeatedly adjusted until the observer cannot recognize the small object, and then recorded the maximum visible distance between the observer and the small object.

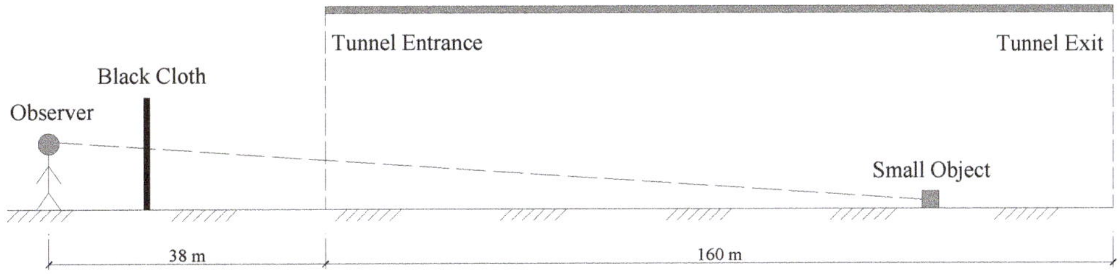

Figure 5. Visible distance test for a small object in tunnel.

When testing the dynamic change of driver's eye pupil diameter during driving, one of the three observers who had the longest driving age and was the most skilled at driving was selected as the test object. The Tobii Pro Glasses 3 eye tracker (see Figure 6a), which can realize high-quality eye tracking and ensure accurate and stable eye tracking data, was used. Before the test, the driver parked the vehicle at the entrance of the tunnel, then wore the eye tracker and calibrated it, as shown in Figure 6b. Subsequently, the driver started the vehicle and drove at a constant speed in the tunnel. Driving to the position about 40 m from the tunnel exit, the driver began to brake until the vehicle stopped. Because the test tunnel is only 160 m, the constant speed for the driving was determined as 20 km/h and 40 km/h in the test. During the test, a test assistant sat at the co-driver's position to be responsible for the data storage of the eye tracker and gave the command to the driver to brake when the vehicle was about 40 m away from the tunnel exit, so as to ensure driving safety (see Figure 6b).

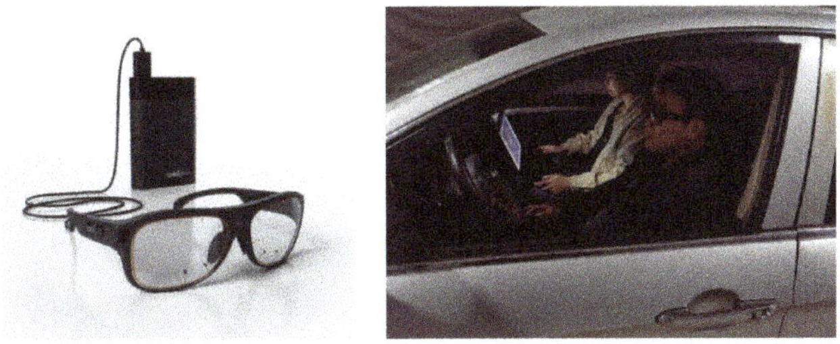

(a) (b)

Figure 6. Acquisition of a driver's pupil diameter data. (**a**) Eye tracker, (**b**) the driver wearing eye tracker.

3. Tunnel Lighting Environment Parameters

3.1. Luminance and Illuminance in Tunnel

The measured luminance and illuminance in tunnel when the sidewalls decorated with energy-storage and luminescent coating and gray cloth are shown in Tables 1 and 2, respectively. It can be seen from Table 1 that when the LED lamps turned on at the electric power of 10 W, in the tunnel decorated with energy-storage and luminescent coating, the average luminance values of the nine measuring points on pavement corresponding to the LED lamps with color temperatures of 2500 K, 3500 K, 4500 K, 5500 K and 6500 K were 2.7 cd/m^2, 2.7 cd/m^2, 2.8 cd/m^2, 2.7 cd/m^2 and 2.9 cd/m^2, respectively, and the average luminance values of the six measuring points on sidewalls corresponding to the LED lamps with color temperatures of 2500 K, 3500 K, 4500 K, 5500 K and 6500 K were 3.6 cd/m^2, 3.6 cd/m^2, 3.8 cd/m^2, 3.7 cd/m^2 and 3.9 cd/m^2, respectively. It can be seen from Table 2 that in the tunnel decorated with gray cloth when the LED lamps turned on at the same power, the average luminance values of the nine measuring points on pavement illuminated by the LED lamps with five different color temperatures were 2.1 cd/m^2, 2.2 cd/m^2, 2.3 cd/m^2, 2.1 cd/m^2 and 2.2 cd/m^2, and the average luminance values of the six measuring points on sidewalls illuminated by the LED lamps with five different color temperatures were 1.3 cd/m^2, 1.3 cd/m^2, 1.4 cd/m^2, 1.3 cd/m^2 and 1.4 cd/m^2. For the tunnel illuminated by the LED lamps with five different color temperatures, the luminance and illuminance distribution in the tunnel is basically the same. The luminance and illuminance distribution in tunnel illuminated by the LED lamps with the electric power of 18 W is similar to the above situation and will not be repeated here. As for the relationship between illuminance and luminance of pavement in the tunnel, the lighting specification of "*Guidelines for Design of Lighting of Highway Tunnels* (JTG/T D70/2-01-2014)" [27] specifies that the conversion relationships between the illuminance and luminance are 10 and 15 on concrete and asphalt pavements, respectively. The pavement of the test tunnel is concrete pavement, and the relationships between pavement illuminance and luminance was 8.4–10.3 (see Tables 1 and 2). These values were close to the value specified in the specification, which verified the rationality of the test.

Table 1. Luminance and illuminance in tunnel decorated with energy-storage and luminescent coating.

Color Temperature (K)	Single Lamp Power of 10 W					Single Lamp Power of 18 W				
	Sidewall Luminance (cd/m^2)	Pavement Luminance (cd/m^2)	Pavement Illuminance (lx)	Luminance Ratio of Sidewall to Pavement	Ratio of Illuminance to Luminance of Pavement	Sidewall Luminance (cd/m^2)	Pavement Luminance (cd/m^2)	Pavement Illuminance (lx)	Luminance Ratio of Sidewall to Pavement	Ratio of Illuminance to Luminance of Pavement
2500	3.6	2.7	23.0	1.3	8.5	6.7	5.0	41.8	1.3	8.4
3500	3.6	2.7	24.3	1.3	9.0	6.6	5.1	43.8	1.3	8.6
4500	3.8	2.8	24.7	1.4	8.8	6.8	5.1	43.8	1.3	8.6
5500	3.7	2.7	24.2	1.4	9.0	6.6	5.0	43.0	1.3	8.6
6500	3.9	2.9	25.6	1.3	8.8	6.9	5.2	47.3	1.3	9.1

Table 2. Luminance and illuminance in tunnel decorated with gray cloth.

Color Temperature (K)	Single Lamp Power of 10 W					Single Lamp Power of 18 W				
	Sidewall Luminance (cd/m^2)	Pavement Luminance (cd/m^2)	Pavement Illuminance (lx)	Luminance Ratio of Sidewall to Pavement	Ratio of Illuminance to Luminance of Pavement	Sidewall Luminance (cd/m^2)	Pavement Luminance (cd/m^2)	Pavement Illuminance (lx)	Luminance Ratio of Sidewall to Pavement	Ratio of Illuminance to Luminance of Pavement
2500	1.3	2.1	17.8	0.6	8.5	2.1	3.5	34.3	0.6	9.8
3500	1.3	2.2	20.5	0.6	9.3	2.2	3.4	34.5	0.6	10.1
4500	1.4	2.3	20.8	0.6	9.0	2.3	3.5	34.8	0.7	9.9
5500	1.3	2.1	20.2	0.6	9.6	2.3	3.4	33.2	0.7	9.8
6500	1.4	2.2	22.6	0.6	10.3	2.3	3.7	37.1	0.6	10.0

Brightness is an important index for the evaluation of tunnel lighting quality. The lighting specifications of "*Guide for the Lighting of Road Tunnels and Underpasses* (CIE 88 2004)" [26] and

"*Guidelines for Design of Lighting of Highway Tunnels* (JTG/T D70/2-01-2014)" [27] specify that the average luminance of the tunnel sidewalls, up to at least a height of 2 m, must be at least 60% of the average road surface luminance at the relevant location. The ratios of sidewall luminance to pavement luminance in the tunnel decorated with energy-storage and luminescent coating and gray cloth measured in the test were 1.3–1.4 and 0.6–0.7, respectively (see Tables 1 and 2), which meets the above specification requirements. When the pavement luminance in the tunnel having the sidewalls with gray cloth was in the range of 2.1–2.3 cd/m^2, the pavement luminance in the tunnel having the sidewalls with energy-storage and luminescent coating increased by 20–30% on the premise of the electric power of LED lamps keeps unchanged (see Table 1). On the same premise, when the pavement luminance in the tunnel with gray cloth was in the range of 3.4–3.7 cd/m^2, the pavement luminance in the tunnel with energy-storage and luminescent coating increased by 40–50% (see Table 2). The application of energy-storage and luminescent coating for auxiliary lighting can increase the brightness in tunnel [28,29]. The tunnel sidewall luminance will increase if the energy-storage and luminescent coating is decorated on tunnel sidewalls, and the road surface luminance increases with the increasing sidewall luminance; however, the increase rate of luminance will be different and needs to be analyzed if the tunnel structural style, lighting system arrangement and other parameters are changed.

3.2. Spectral Distribution of LED Light Sources

The spectral distribution of LED light sources with different color temperatures is shown in Figure 7. According to the spectral distribution in Figure 7, the spectral irradiance of the LED light sources increased with the increasing brightness in the tunnel. The spectral irradiance distribution reflects the radiation energy of light at different wavelengths. When the electric power of LED lamps was the same, the spectral irradiance of the LED light sources in the tunnel decorated with energy-storage and luminescent coating was greater than that decorated with gray cloth. For the LED light sources, the spectrum presented the bimodal distribution characteristics, and there were obvious differences in the bimodal distribution characteristics of LED light sources with different color temperatures. The spectrum with relatively high peak on the left (about 450 nm, see Figure 7c–e) corresponded to high color temperature, and that on the right (between 540 nm and 590 nm, see Figure 7a,b) corresponded to low color temperature. For the LED light sources with low color temperatures of 2500 K and 3500 K, the blue light component with the wavelength of 430–480 nm was less, while the red and yellow light component with the wavelength of 580–780 nm was more. As the color temperature increased to more than 4500 K, the blue light in the range of 430–480 nm gradually increased and the red and yellow light in the range of 580–780 nm gradually decreased.

3.3. S/P Value

The S/P value is related to the spectral irradiance distribution of the LED light sources. In general, the S/P value of the LED light sources is $\geq$1, and the S/P value for LED light sources with more blue light components can be greater than 2 [31]. The S/P values in different lighting environment measured in the test are shown in Table 3. Previous study claimed that in the mesopic vision (between about 0.001 and 3 cd/m^2) for human eyes, the higher the S/P value of the LED light source, the driver can feel the lighting with higher brightness under the condition of consuming the same electric energy, and it is beneficial to driving safety in road tunnel lighting [31–33].

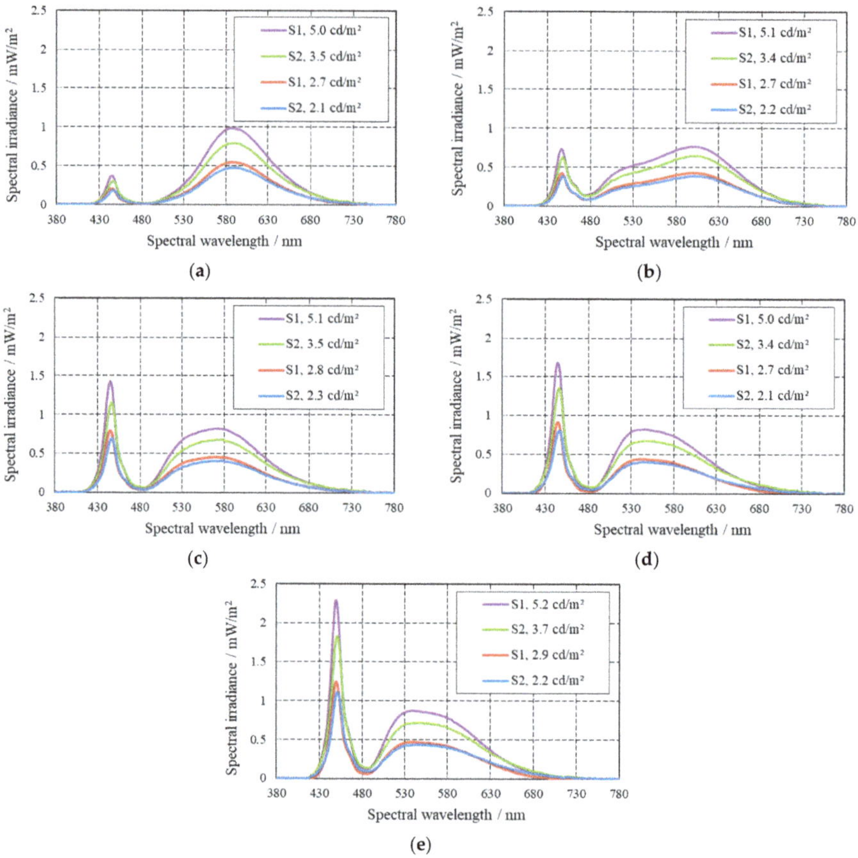

Figure 7. Spectral distribution of LED light sources (S1 and S2 represent sidewalls with energy-storage and luminescent coating and gray cloth, respectively). (**a**) 2500 K, (**b**) 3500 K, (**c**) 4500 K, (**d**) 5500 K, (**e**) 6500 K.

Table 3. S/P value.

Color Temperature (K)	Tunnel Sidewalls Decorated with Energy-Storage and Luminescent Coating		Tunnel Sidewalls Decorated with Gray Cloth	
	Single Lamp Power of 10 W	Single Lamp Power of 18 W	Single Lamp Power of 10 W	Single Lamp Power of 18 W
2500	0.811	0.808	0.844	0.836
3500	1.539	1.526	1.555	1.535
4500	1.572	1.566	1.584	1.582
5500	1.757	1.774	1.798	1.801
6500	2.034	2.036	2.070	2.058

4. Visual Characteristics of Human Eyes in Tunnel

4.1. Visible Distance for Small Object

Figure 8 shows the comparison of the lighting environment in the tunnel decorated with different materials (energy-storage and luminescent coating and gray cloth) illuminated by LED lamps with different color temperatures (only the situations for electric power of LED lamps of 10 W are listed). The color temperatures of the LED light sources can be divided into warm color (less than 3300 K, and majority of the light is red light), intermediate color (between 3300 K and 5300 K) and cold color (greater than 5300 K, and majority of the light is blue light) [31,32,34]. In general, the tunnel lighting environment was obviously dim when the color temperature was 2500 K, the light was relatively soft when the color temperature was 3500–5500 K, and the lighting environment gave the observers a high cold feeling when the color temperature was 6500 K.

(**a**)

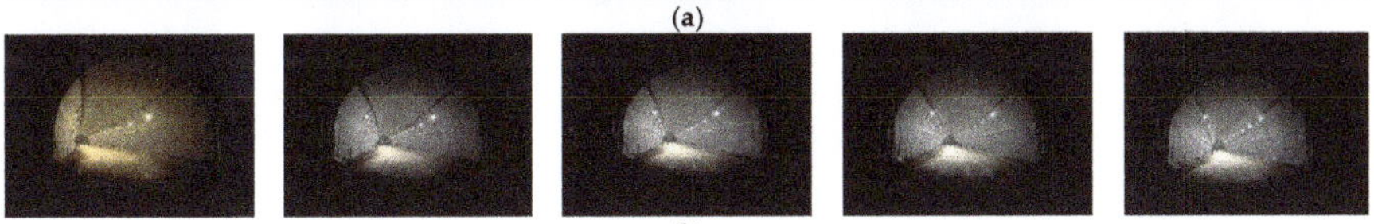

(**b**)

Figure 8. Comparison of lighting environment in tunnel (from left to right: color temperatures of 2500 K to 6500 K). (**a**) Sidewalls with energy-storage and luminescent coating, (**b**) sidewalls with gray cloth.

Figure 9 shows the visible distance for a small object with different color temperatures of LED light sources and pavement luminance (the data in the figure is the average value of three observers' visible distances). As mentioned in Section 3.3, it is beneficial to driving safety if the S/P value of the LED light source increases in the mesopic vision which is between about 0.001 and 3 cd/m^2; however, the data in Figure 9 illustrates that the increase in color temperature and S/P value of LED light had no obvious influence on the visible distance of human eyes. Compared with the color temperature and S/P value, the visibility of a small object was more sensitive to the brightness. The visible distance increased to a certain extent after the luminance of the sidewall and road surface in the tunnel improved by the energy-storage and luminescent coating. For the conditions of single lamp power of 10 W and 18 W, the visible distance of the small object increased by 6–10% and 3–10%, respectively.

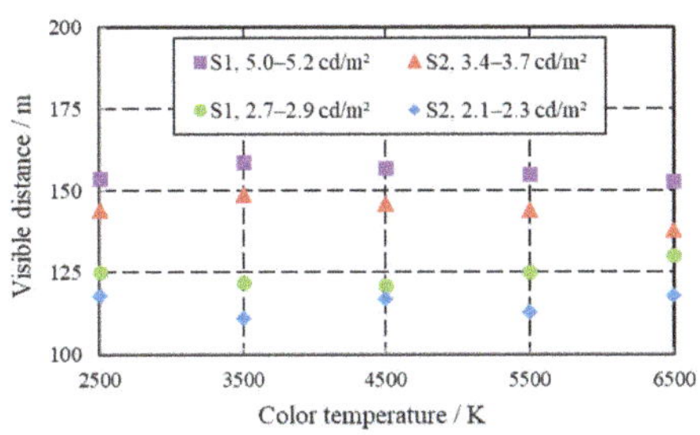

Figure 9. Visible distance for a small object in tunnel (S1 and S2 represent sidewalls with energy-storage and luminescent coating and gray cloth, respectively).

4.2. Dynamic Change of Driver's Pupil Diameter during Driving

Figure 10 shows the change of the driver's eye pupil diameter at an interval of 0.5 s while driving in the tunnel with different color temperatures of LED light sources and pavement luminance. The data in Figure 10 shows that for the tunnel with two different sidewalls, the pupil diameter of human eye was basically in the range of 3.8–5.0 mm, and the changing trend was relatively stable when driving in the tunnel at the speed of 20 km/h and 40 km/h. In addition, there was no discomfort for the driver during driving.

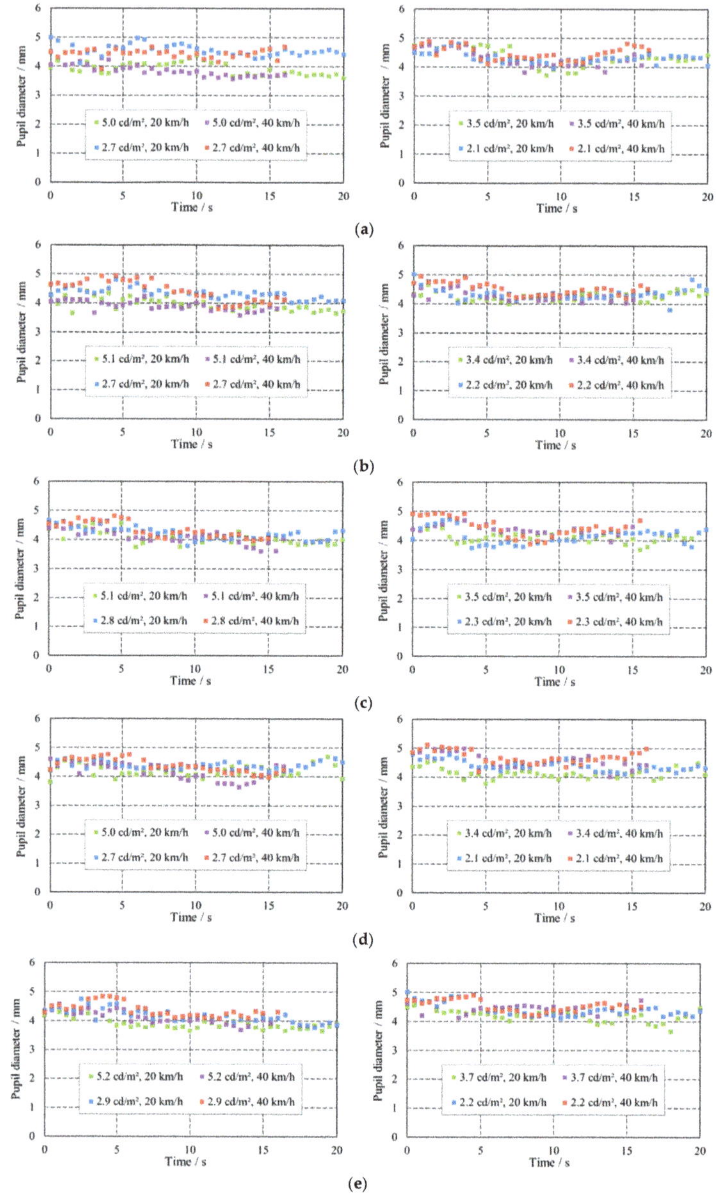

Figure 10. Driver's pupil diameter change (left: sidewalls with energy-storage and luminescent coating; right: sidewalls with gray cloth). (**a**) 2500 K, (**b**) 3500 K, (**c**) 4500 K, (**d**) 5500 K, (**e**) 6500 K.

Figure 11 shows the change rate of the driver's eye pupil diameter at an interval of 0.5 s while driving in the tunnel with different color temperatures of LED light sources and pavement luminance. The data shows that for the tunnel with two different sidewalls, the change rate of pupil diameter in 0.5 s interval was −15–15%. Within the range of this change rate, there was no discomfort for the driver. Combined with the distribution characteristics of tunnel luminance in Section 3.1 and the visible distance of the small object in Section 4.1, it can be seen that for LED light sources with different color temperatures, the pavement luminance increased with the increasing sidewall luminance; this improvement increased the driver's visible distance to a small object in the tunnel to a certain extent, and had no effect on the driver's pupil diameter change, which indicates that the increase in tunnel sidewall luminance has a positive effect on the visual performance of drivers.

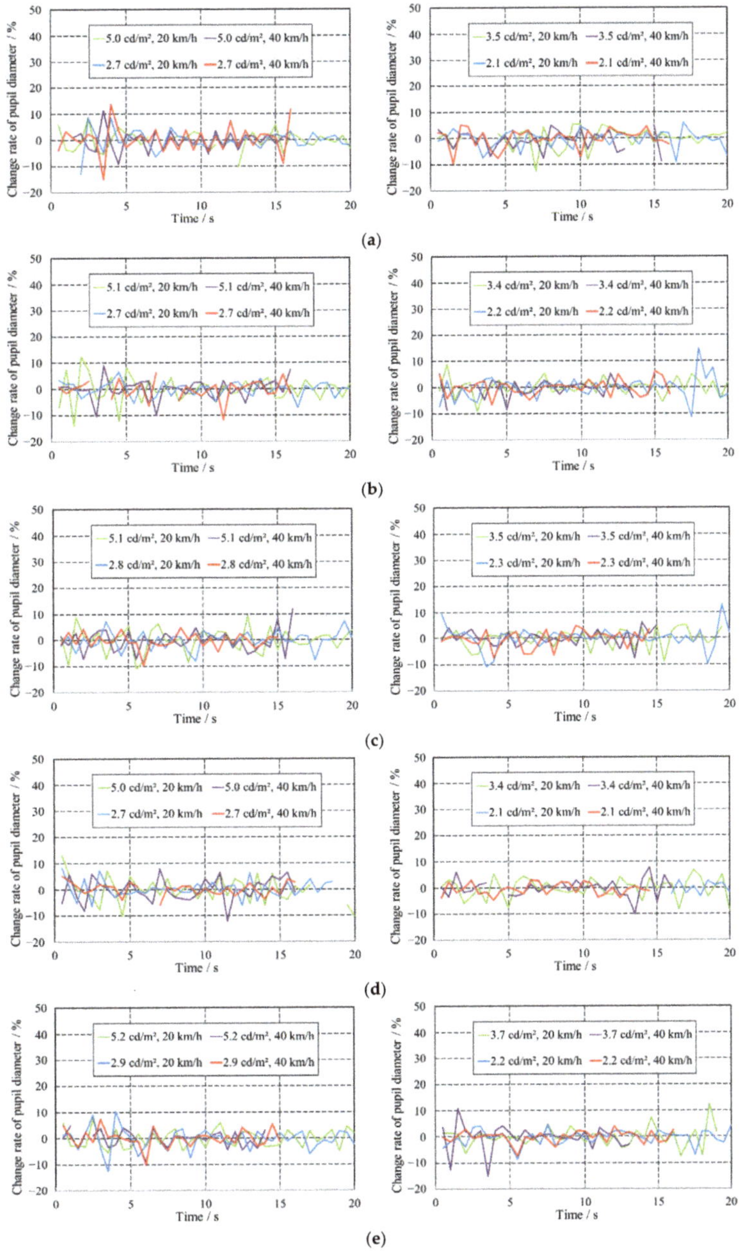

Figure 11. Change rate of driver's pupil diameter (left: sidewalls with energy-storage and luminescent coating; right: sidewalls with gray cloth). (**a**) 2500 K, (**b**) 3500 K, (**c**) 4500 K, (**d**) 5500 K, (**e**) 6500 K.

5. Conclusions

The impact of the improvement of tunnel sidewall brightness on the lighting environment and visual characteristics of human eyes are analyzed based on the lighting experiments in a tunnel when the tunnel sidewalls are decorated with two different types of materials and illuminated by LED lamps with five different color temperatures. According to the results, the main conclusions are as follows.

The tunnel sidewall luminance increases if the energy-storage and luminescent coating with high reflectance is decorated on tunnel sidewalls, and the pavement luminance increases with the increasing sidewall luminance. Taking the energy-storage and luminescent coating applied in this study as an example, the pavement luminance can be increased by more than 20% compared with the tunnel in which the sidewalls are decorated with gray cloth and illuminated by the LED lamps with the same electric power. In addition, the ratio of sidewall to pavement luminance in the tunnel decorated with energy-storage and luminescent coating is 1.3–1.4, which is far greater than the demand of lighting specifications.

The visibility of human eyes is more sensitive to the brightness than the color temperature and S/P value of the LED light sources. After the improvement of the tunnel sidewall and pavement luminance, the visible distance of a small object in the tunnel will increase to a certain extent. Furthermore, there is no discomfort for drivers during driving, and the changing of eye pupil diameter is relatively stable, which indicates that the increase in tunnel sidewall luminance has a positive effect on the visual performance of drivers; however, the possibility of a glare issue remains to be discussed in a future study.

Author Contributions: Conceptualization, S.F.; methodology, X.C.; validation, J.L.; formal analysis, X.C. and L.Z.; investigation, L.Z. and J.L.; resources, S.F. and H.Z.; data curation, W.M.; writing—original draft preparation, X.C.; writing—review and editing, L.Z.; visualization, L.Z. and J.L.; supervision, W.M.; project administration, H.Z.; funding acquisition, S.F. All authors have read and agreed to the published version of the manuscript.

Funding: This research was funded by the Science and Technology Project of Department of Communications of Guizhou Province, China (No. 2021-122-025), and the Science and Technology Project of Department of Communications of Hebei Province, China (No. TH-201909).

Institutional Review Board Statement: Not applicable.

Informed Consent Statement: Not applicable.

Data Availability Statement: Not applicable.

Conflicts of Interest: The authors declare no conflict of interest.

References

1. Gil-Martín, L.M.; Peña-García, A.; Jiménez, A.; Hernández-Montes, E. Study of light-pipes for the use of sunlight in road tunnels: From a scale model to real tunnels. *Tunn. Undergr. Space Technol.* **2014**, *41*, 82–87. [CrossRef]
2. Wang, Y.Q.; Cui, Y.W.; Chen, F.; Ren, R. An "Illumination Moving with the Vehicle" Intelligent Control System of Road Tunnel Lighting. *Sustainability* **2020**, *12*, 7314. [CrossRef]
3. Shen, Y.; Ling, J.X.; Li, T.C.; Zhou, L.; Feng, S.Z.; Zhu, H.H. Diffuse reflection-based lighting calculation model and particle swarm optimization algorithm for road tunnels. *Tunn. Undergr. Space Technol.* **2022**, *124*, 104457. [CrossRef]
4. Moretti, L.; Cantisani, G.; Carrarini, L.; Bezzi, F.; Cherubini, V.; Nicotra, S. Italian Road Tunnels: Economic and Environmental Effects of an On-Going Project to Reduce Lighting Consumption. *Sustainability* **2019**, *11*, 4631. [CrossRef]
5. Sun, D.; Athienitis, A.; D'Avignon, K. Application of semitransparent photovoltaics in transportation infrastructure for energy savings and solar electricity production: Toward novel net-zero energy tunnel design. *Prog. Photovolt. Res. Appl.* **2019**, *27*, 1034–1044. [CrossRef]
6. Qin, L.; Shi, X.H.; Leon, A.S.; Tong, C.D.; Ding, C. Dynamic luminance tuning method for tunnel lighting based on data mining of real-time traffic flow. *Build. Environ.* **2020**, *176*, 106844. [CrossRef]
7. Qin, L.; Peña-García, A.; Leon, A.S.; Yu, J.C. Comparative Study of Energy Savings for Various Control Strategies in the Tunnel Lighting System. *Appl. Sci.* **2021**, *11*, 6372. [CrossRef]
8. Zhou, Z.; Zhang, J.J.; Gong, C.J. Automatic detection method of tunnel lining multi-defects via an enhanced You Only Look Once network. *Comput.-Aided Civ. Infrastruct. Eng.* **2022**, *37*, 762–780. [CrossRef]

9. Qin, L.; Cao, Q.L.; Leon, A.S.; Weng, Y.N.; Shi, X.H. Use of Pupil Area and Fixation Maps to Evaluate Visual Behavior of Drivers inside Tunnels at Different Luminance Levels—A Pilot Study. *Appl. Sci.* **2021**, *11*, 5014. [CrossRef]
10. Shen, Y.; Deng, Y.; Li, T.C.; Zhou, L.; Feng, S.Z.; Zhu, H.H. Determining multidimensional diffuse reflection effects in city tunnel lighting environment. *Build. Environ.* **2022**, *212*, 108796. [CrossRef]
11. Shailesh, K.R.; Tanuja, S.; Kamath, M.V. Analysis of energy savings from replacing HPSV lighting with LED lighting in road lighting application. In Proceedings of the International Conference on Emerging Trends in Electrical Engineering and Energy Management, Chennai, India, 13–15 December 2012; pp. 473–477.
12. Li, X.J.; Ling, J.X.; Shen, Y.; Lu, T.; Feng, S.Z.; Zhu, H.H. The impact of CCT on driving safety in the normal and accident situation: A VR-based experimental study. *Adv. Eng. Inform.* **2021**, *50*, 101379. [CrossRef]
13. Lee, C.T.; Chen, L.B.; Chu, H.M.; Hsieh, C.J.; Liang, W.C. An Internet of Things (IoT)-Based Master-Slave Regionalized Intelligent LED-Light-Controlling System. *Appl. Sci.* **2022**, *12*, 420. [CrossRef]
14. Sato, M.; Hagio, T. Visibility Enhancement and Power Saving by Pro-beam LED Tunnel Lighting Method. *J. Light Vis. Environ.* **2014**, *38*, 89–93. [CrossRef]
15. Renzler, M.; Reithmaier, N.; Reinhardt, R.; Pohlb, W.; Ußmüllera, T. A road tunnel model for the systematic study of lighting situations. *Tunn. Undergr. Space Technol.* **2018**, *72*, 114–119. [CrossRef]
16. Fan, S.J.; Yang, C. Parameters Optimization and Energy-Saving of Highway Tunnel Stagger Luminaire Distribution Lighting with LED. In Proceedings of the International Conference on Civil, Transportation and Environment, Guangzhou, China, 30–31 January 2016; pp. 436–443.
17. Shi, N.; Dong, L.L.; Qin, L.; Xu, W.H. Study on lamp-layout scheme of highway tunnel lighting based on DIALux. In Proceedings of the International Conference on Civil, Transportation and Environment, Guangzhou, China, 30–31 January 2016; pp. 773–781.
18. Domenichini, L.; Torre, F.L.; Vangi, D.; Virga, A.; Branzi, V. Influence of the lighting system on the driver's behavior in road tunnels: A driving simulator study. *J. Transp. Saf. Secur.* **2016**, *9*, 216–238. [CrossRef]
19. Zhang, X.Q.; Hu, J.B.; Wang, R.H.; Gao, X.J.; He, L.C. The comprehensive efficiency analysis of tunnel lighting based on visual performance. *Adv. Mech. Eng.* **2017**, *9*, 1–9. [CrossRef]
20. Zhao, E.Z.; Dong, L.L.; Chen, Y.; Lou, Q.; Xu, W.H. The Impact of LED Color Rendering on the Dark Adaptation of Human Eyes at Tunnel Entrances. *Int. J. Environ. Res. Public Health* **2020**, *17*, 1566. [CrossRef]
21. He, S.Y.; Liang, B.; Tähkämöb, L.; Maksimainenb, M.; Halonenb, L. The Influences of Tunnel Lighting Environment on Drivers' Peripheral Visual Performance during Transient Adaptation. *Displays* **2020**, *64*, 101964. [CrossRef]
22. Li, S.G.; Tu, G.; Zhou, Q.H. An optimal design model for tunnel lighting systems. *Optik* **2021**, *226*, 165660. [CrossRef]
23. Sędziwy, A.; Kotulski, L. Towards highly energy-efficient roadway lighting. *Energies* **2016**, *9*, 263. [CrossRef]
24. Djuretic, A.; Kostic, M. Actual energy savings when replacing high-pressure sodium with LED luminaires in street lighting. *Energy* **2018**, *157*, 367–378. [CrossRef]
25. Zhao, L.; Qu, S.C.; Zhang, W.G.; Xiong, Z.L. An energy-saving fuzzy control system for highway tunnel lighting. *Optik* **2019**, *180*, 419–432. [CrossRef]
26. CIE. *Guide for the Lighting of Road Tunnels and Underpasses (CIE88-2004)*; Internal Commission on Illumination: Vienna, Austria, 2004.
27. Ministry of Transport of the People's Republic of China. *Guidelines for Design of Lighting of Highway Tunnels (JTG/T D70/2-01-2014)*; China Communications Press Co., Ltd.: Beijing, China, 2014.
28. He, S.Y.; Liang, B.; Pan, G.B.; Wang, F.; Cui, L.L. Influence of dynamic highway tunnel lighting environment on driving safety based on eye movement parameters of the driver. *Tunn. Undergr. Space Technol.* **2017**, *67*, 52–60. [CrossRef]
29. Feng, S.Z.; Chen, X.F.; Mao, W.X.; Zhou, L.; Zhu, H.H. Study on Light Environment in Tunnels Decorating with Multifunctional Energy-storage and Luminescent Material. *Chin. J. Undergr. Space Eng.* **2021**, *17*, 2030–2036.
30. Schreuder, D.A. *The Lighting of Vehicular Traffic Tunnels*; Centrex Publishing Company: Eindhoven, The Netherlands, 1964.
31. Ma, F.; Wu, M.J.; Xie, H.B. *Lighting of Tunnel*; Science Press: Beijing, China, 2018.
32. Liu, Y.Y.; Peng, L.; Lin, L.Y.; Chen, Z.L.; Weng, J.; Zhang, Q.W. The impact of LED spectrum and correlated color temperature on driving safety in long tunnel lighting. *Tunn. Undergr. Space Technol.* **2021**, *112*, 103867. [CrossRef]
33. CIE. *Light as a True Visual Quantity: Principles of Measurement*; Bureau Central de la CIE: Paris, France, 1978.
34. Yang, W.Y.; Jeon, J.Y. Effects of Correlated Colour Temperature of LED Light on Visual Sensation, Perception, and Cognitive Performance in a Classroom Lighting Environment. *Sustainability* **2020**, *12*, 4051. [CrossRef]

applied sciences

MDPI

Article

Elastic Modulus Prediction Model of Foamed Concrete Based on the Walsh Formula

Zhong Zhou, Jiangfeng Hu, Fan Li *, Junjie Zhang and Mingfeng Lei *

School of Civil Engineering, Central South University, Changsha 410075, China; dazhong78@csu.edu.cn (Z.Z.); jiangfenghu@csu.edu.cn (J.H.); 204811075@csu.edu.cn (J.Z.)
* Correspondence: 194811082@csu.edu.cn (F.L.); mingfenglei@csu.edu.cn (M.L.)

Abstract: Foamed concrete consists of cement matrix and air-foam and has more complicated structure characteristics than normal concrete. However, current research on the elastic modulus of foamed concrete is still limited to empirical equations. In this work, a new theoretical prediction model was proposed for calculating the elastic modulus of foamed concrete. The elastic modulus calculation model for the spherical shell element of foamed concrete is constructed based on the Walsh formula and the assumption of spherical pores. Moreover, the theoretical prediction model is established by introducing the two-layer embedded model for the elastic modulus of foamed concrete. Then, the compressive test is employed to verify the accuracy of the model. The results show that the elastic modulus of foamed concrete decreases with the increase in porosity and matrix Poisson's ratio, and increases with the increase in the matrix elastic modulus. The research results can improve the mechanical theories of foamed concrete materials and have good engineering application values.

Keywords: foamed concrete; elastic modulus; Walsh formula; two-layer embedded model; spherical pore assumption; mesomechanics

Citation: Zhou, Z.; Hu, J.; Li, F.; Zhang, J.; Lei, M. Elastic Modulus Prediction Model of Foamed Concrete Based on the Walsh Formula. *Appl. Sci.* **2022**, *12*, 5142. https://doi.org/10.3390/app12105142

Academic Editors: Stefano Invernizzi and Dario De Domenico

Received: 14 April 2022
Accepted: 16 May 2022
Published: 19 May 2022

1. Introduction

Foamed concrete is mainly composed of cement matrix and foam in proportion [1–3]. The production process of foamed concrete is shown in Figure 1. A porous structure with light bulk density and high strength can be formed after curing through the cast-in-place construction method. In addition, the density and strength properties of foamed concrete can be adjusted according to needs, which is very effective in controlling foundation settlement and reducing structural cracks [4–6]. Therefore, foamed concrete has good applicability in practical projects. Foamed concrete was first used in airport roadbed engineering in Japan and was later introduced into China by scholar Chen [7]. After more than 40 years of engineering practice and unremitting exploration by relevant scholars, foamed concrete has become increasingly extensive, and the research results have been significantly improved.

Foamed concrete has a large number of internal cellular structures, which is different from concrete. These microscopic features [8–10] and the properties of matrix materials [11,12] determine the mechanical properties of foamed concrete, and directly affect the elastic modulus or other parameters, thus affecting its deformation properties. Therefore, scholars have made useful attempts on the mechanical characteristics of foamed concrete. Ramamurthy [13,14] found that the volume, size, and spacing of pores have a great influence on the compressive strength and density of foamed concrete, and established a compressive strength model of foamed concrete. Their research shows that foamed concrete with uniform distribution of air-void sizes, circular air-voids, and optimal spacing between voids has better mechanical properties. Falliano et al. [15–17] studied the dependence of the compressive strength of foamed concrete on dry density, water content, curing conditions, cement type, and foaming agent, and found that compressive strength showed a linear trend with the increase in density. Panesar et al. [18] thoroughly studied the influence of the microstructure of foamed concrete on the mechanical parameters, and the study shows that the microstructure is remarkably important to the strength and elastic modulus. Based on these typical research results, therefore, the elastic modulus is the key parameter in which to study the mechanical properties of foamed concrete, which can be used to design roadbed padding and predict the subgrade under load mechanical response of a foamed concrete roadbed in subgrade engineering [19,20].

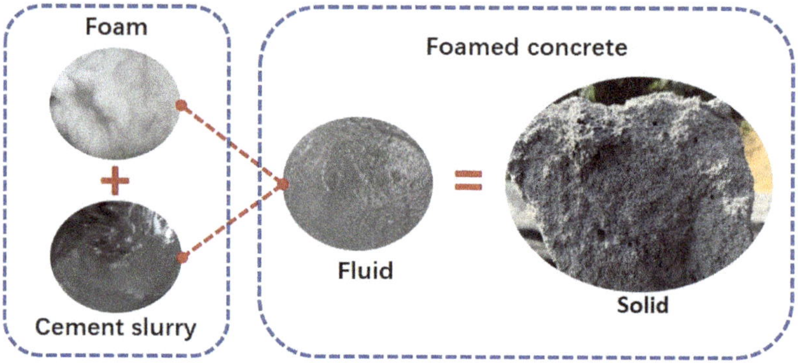

Figure 1. Production process of foamed concrete.

Numerical simulation and mechanical tests are effective methods to explore the elastic modulus of foamed concrete. Youssef et al. [21] conducted a 3D numerical simulation study on foamed concrete, which showed that the elastic modulus of foamed concrete mainly depends on its internal porosity. When the porosity is at a low level, 3D numerical simulations and mean field schemes provide consistent estimates of the overall stiffness of a biphasic composite material made of spherical pores embedded in a matrix. This study divides foamed concrete into matrix and embedded spherical pores, and preliminarily proves the rationality of this assumption. Therefore, the mechanical analysis of foamed concrete based on a two-layer embedded model is valuable and feasible. According to the study of Jones et al. [22], the elastic modulus of foamed concrete is related to the matrix material, and the elastic modulus of foamed concrete using fine sand is higher than that using fly ash as fine aggregate. Their research only qualitatively analyzed the influence of matrix materials on the elastic modulus of foamed concrete, but it is necessary to quantitatively analyze how matrix parameters specifically affect the elastic modulus. Ge et al. [23] performed mechanical performance tests on foamed concrete with wet densities of 700 and 800 kg/m^3. The test results showed that the elastic modulus meets the requirements of cast-in-place foamed concrete subgrades. They concluded that both air-void content and pore size distribution lead to the significant difference in mechanical performance of mixes. The analysis of the factors affecting the elastic modulus of foamed concrete shows that porosity and matrix material properties have a significant impact on the elastic modulus of foamed concrete.

In addition to the test method, the empirical formula method is also an effective method to determine the elastic modulus. Tan et al. [24] conducted a series of uniaxial compression tests of foamed concrete and summarized that the elastic modulus of foamed concrete with densities of 450, 600, and 800 kg/m^3 increased with the increase in height-to-diameter (H:D) ratio. This study proposed a prediction model for elastic modulus at different densities and temperatures on the basis of experimental data. Mydin [25] recommended a model for the relationship between elastic modulus and porosity of foamed concrete at ambient temperature. Their study confirmed that the mechanical performance model of ordinary strength concrete can predict the reduction in mechanical properties of foamed concrete, and verified the feasibility of using the Balshin equation to calculate the compressive elastic modulus at ambient temperature. Li and Purkiss [26] present a critical review of the currently available models for the mechanical behavior of concrete at elevated temperatures, and proposed a stress–strain–temperature model of concrete based on the existing models and test data, which provides a great reference foundation for the prediction of the elastic modulus of foamed concrete.

However, both the test methods and empirical formula methods have limitations. The test methods are time-consuming and can only be used to determine the elastic modulus after foamed concrete is manufactured. The scope of application for the empirical formula is conditional. If the actual situation is different from that of the empirical formula method, the prediction results of the empirical formula may have deviation.

Most of the existing studies on the elastic modulus of foamed concrete remain at the macromechanical level, mainly including mechanical tests [27–30] and empirical formula analysis [31–33]. However, theoretical research on foamed concrete from the perspective of mesomechanics is lacking. Thus, this paper considers the influence of the internal pore structure characteristics and matrix material properties of foamed concrete on the overall elastic modulus. Moreover, the Walsh formula and spherical pore assumption are introduced to construct an elastic modulus calculation model of the spherical shell element of foamed concrete. Combined with the two-layer embedded model, an inclusion theory model of the elastic modulus of foamed concrete is established. Then, the influence law of the elastic parameters of foamed concrete is thoroughly explored in accordance with the laboratory tests and the existing theoretical models to verify the theory. The research results can improve the mechanical theories of foamed concrete materials and have good engineering application values.

2. Micromechanical Analysis of Foamed Concrete Elements

2.1. Walsh Formula

Foamed concrete is a typical porous medium material, and the content of internal air bubbles is closely related to the macroscopic deformation parameters of foamed concrete. Therefore, this study introduces the Walsh formula [34] for the theoretical analysis of the elastic modulus of foamed concrete elements from the perspective of mesomechanics. In accordance with the reciprocity theorem of work, Walsh deduces the relationship between the volume flexibility of the pore matrix in the material and the volume flexibility of the solid matrix, as given in Equation (1) [34]:

$$\delta = \delta_s + \frac{\Delta V_c}{V \Delta \sigma} \tag{1}$$

where V is the unit volume of the porous medium, δ is the volume compliance of the overall unit, V_c is the pore volume, δ_s is the volume compliance of the soil particle matrix, and ΔV_c is the variation in pore volume under the action of volume stress $\Delta\sigma$.

Cao et al. [35] transformed the derivation of bulk flexibility into the solution of elastic modulus by introducing the intermediate parameter of bulk modulus. They obtained the relationship between the elastic modulus of the element matrix and the equivalent elastic modulus of spherical shell elements on the basis of the Walsh formula, as shown by Equation (2) [35]:

$$\frac{1}{E_2} = \frac{1}{E_1} + \frac{\Delta V_c}{3(1-2\nu) V \Delta\sigma} \tag{2}$$

where E_2 is the equivalent elastic modulus of the spherical shell element and E_1 is the matrix elastic modulus. The schematic of the spherical shell section of the medium element is shown in Figure 2.

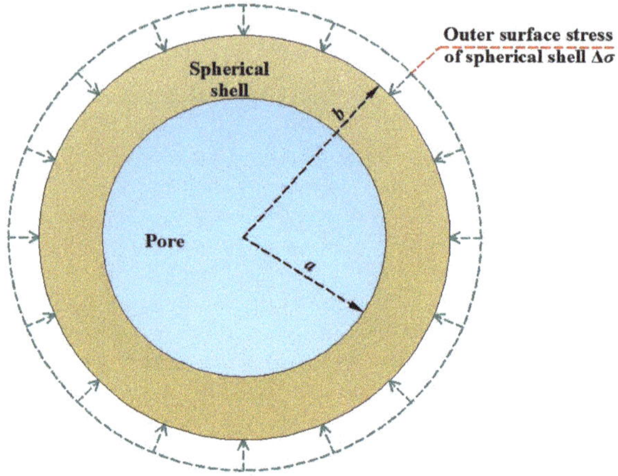

Figure 2. Schematic of the spherical shell section of the medium element.

2.2. Spherical Pore Assumption

In accordance with the elastic mechanics calculation formula for uniformly distributed pressure inside and outside the spherical shell, the following assumptions are made. (1) The representative material unit is a spherical shell, and its inner and outer diameters are a and b, respectively. The inner sphere represents the pore part of the material, and the spherical shell represents the skeleton of the material. (2) The spherical shell, that is, the material skeleton, is a linear elastic body. (3) The pores in the material are independent of one another. The schematic of the spherical shell of a foamed concrete unit is shown in Figure 3. In consideration of the volume stress on the outer surface, the radial displacement of the spherical shell can be obtained by elastic mechanics, as shown by Equation (3) [35]:

$$u_a = -\frac{3a(1-\nu)\Delta\sigma}{E_1(1-n)} \tag{3}$$

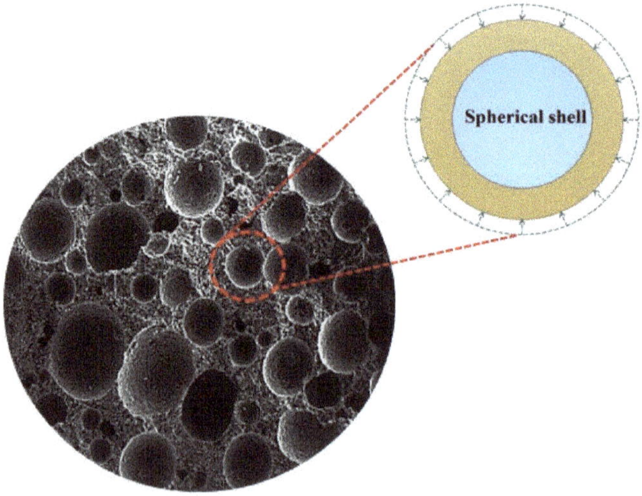

Figure 3. Schematic of the spherical shell of a foamed concrete unit.

2.3. Spherical Shell Elastic Modulus of Foamed Concrete

Owing to the existence of a large number of closed air bubbles in foamed concrete, mechanical analysis of foamed concrete microelements can be conducted on the basis of the assumption of spherical pores. First, a foamed concrete unit containing spherical pores is selected, which can be regarded as a spherical shell with inner and outer diameters a and b, respectively. When the internal pore connectivity is poor, it is assumed that no pore water pressure exists. The inner diameter variation for the spherical shell has the following relationship with volume variation, as shown by Equations (4)–(6).

$$\Delta V_c = 4\pi a^2 \Delta a = -4\pi a^2 u_a \tag{4}$$

$$V = \frac{4}{3}\pi b^3 \tag{5}$$

$$\frac{\Delta V_c}{V} = \frac{9n(1-\nu)\Delta\sigma}{E_1(1-n)} \tag{6}$$

When Equation (6) is substituted into Equation (2), the calculation formula of the elastic modulus of foamed concrete element with respect to porosity can be obtained, as shown by Equations (7) and (8).

$$E_2 = \frac{1-n}{1+\beta n}E_1 \tag{7}$$

$$\beta = \frac{1+v}{2(1-2v)} \tag{8}$$

The initial porosity of foamed concrete can be measured experimentally. Nevertheless, in the actual compression process, the porosity changes with the pore deformation of foamed concrete. Therefore, this study analyzes it from the perspective of stress. The derivation process is shown in Equations (9) and (10).

$$n_0 = \frac{V_c}{V} \tag{9}$$

$$n = \frac{V_c - \Delta V_c}{V - \Delta V_c} = \frac{V_c/V - \Delta V_c/V}{1 - \Delta V_c/V} \tag{10}$$

Equation (10) is substituted into Equation (6) to obtain Equation (11).

$$\frac{\Delta V_c}{V} = \frac{9(n_0 - \Delta V_c/V)(1-v)\Delta\sigma}{E_1(1-n_0)} \tag{11}$$

Based on Equation (11), Equation (12) can be obtained:

$$\frac{\Delta V_c}{V} = \frac{9n_0(1-v)\Delta\sigma}{E_1(1-n_0)+9(1-v)\Delta\sigma} \tag{12}$$

Equation (12) is substituted into Equation (2), and the calculation formula of the spherical shell elastic modulus of foamed concrete is obtained in Equation (13).

$$E_2 = E_1 - \frac{3n_0(1-v)E_1}{1-2v+(2-v)n_0+9(1-v)(1-2v)\Delta\sigma/E_1} \tag{13}$$

3. Two-Layer Embedded Model Analysis of Foamed Concrete

3.1. Two-Layer Embedded Model

Foamed concrete is a composite material, which mainly includes fine aggregate and a large number of dense circular holes. Owing to the random distribution of internal holes, its stress condition is extremely complicated. For simplicity, a two-layer embedded model in the porous material theory is proposed. The basic principle of the model is that foamed concrete is regarded as a two-phase composite material, and the overall elastic modulus is obtained through the homogenization process. From a microscopic analysis, the holes in foamed concrete can be regarded as spherical shells, so its mesoscopic structure can be regarded as the form of a matrix unit wrapping a spherical shell unit. The schematic of the calculation model is shown in Figure 4. From the figure, foamed concrete conforms to the two-layer embedded wrapping model [36–38] on the mesostructure.

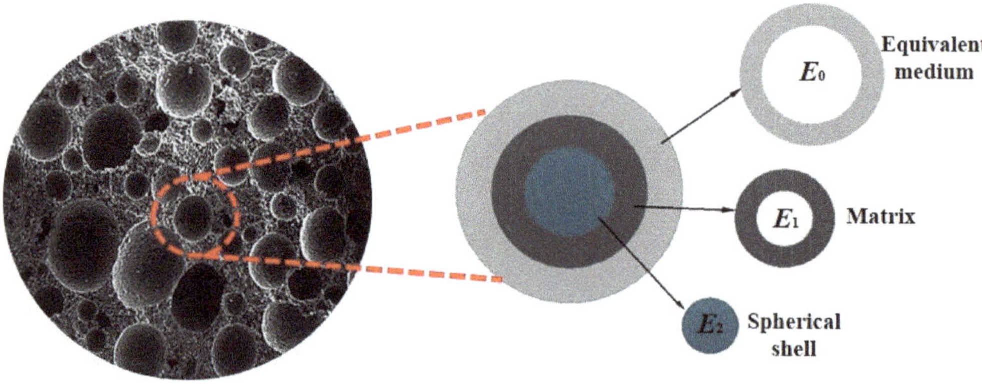

Figure 4. Two-layer embedded model of foamed concrete.

3.2. Prediction Model for the Elastic Modulus of Foamed Concrete

Li et al. [39] proposed an analytical solution for elastic theory for the two-layer embedded model. Accordingly, this study performs the theoretical analysis of the mesomechanics of foamed concrete by introducing the model. The prediction model for the elastic modulus of foamed concrete is shown in Figure 5, and the derivation process is shown in Equations (14)–(18) [39].

$$u_{0c} = \frac{1}{E_0}\left[(1+v_0)\frac{b^2c^2(P-P_0)}{(c^2-b^2)c} + (1-v_0)\frac{Pc^2-P_0b^2}{c^2-b^2}c\right] \tag{14}$$

$$u_{0b} = \frac{1}{E_0}\left[(1+v_0)\frac{b^2c^2(P-P_0)}{(c^2-b^2)b} + (1-v_0)\frac{Pc^2-P_0b^2}{c^2-b^2}b\right] \tag{15}$$

$$u_{1b} = \frac{1}{E_1}\left[(1+v_1)\frac{a^2b^2(P_0-P_1)}{(b^2-a^2)b} + (1-v_1)\frac{P_0b^2-P_1a^2}{b^2-a^2}b\right] \tag{16}$$

$$u_{1a} = \frac{1}{E_1}\left[(1+v_1)\frac{a^2b^2(P_0-P_1)}{(b^2-a^2)a} + (1-v_1)\frac{P_0b^2-P_1a^2}{b^2-a^2}a\right] \tag{17}$$

$$u_{2a} = \frac{1-v_2}{E_2}P_1a \tag{18}$$

where u is the displacement; u_{1a} and u_{2a} are the displacements of matrix, spherical shell at boundary $r = a$, respectively; u_{0b} and u_{1b} are the displacements of equivalent medium, matrix at boundary $r = b$, respectively; u_{0c} is the displacement of equivalent medium at boundary $r = c$; and v_0, v_1, and v_2 are the Poisson's ratio of the equivalent medium, foamed concrete matrix, and spherical shell, respectively. On the basis of the theory of elasticity, Li [39] further proposed formulas for the effective elastic modulus E_0 of a composite material in Equations (19)–(21) [39].

$$E_0 = \frac{E_1(1-f)(1-v_0)}{x_1 - \frac{4E_2f}{E_1(1-f)(1-v_0)+E_2x_2}} \tag{19}$$

$$x_1 = f(1+v_1) + (1-v_1) \tag{20}$$

$$x_2 = (1+v_1) + f(1-v_1) \tag{21}$$

where f is the volume percentage of inclusions. The volume ratio of the spherical shell inclusion can be regarded as the bubble porosity for foamed concrete, as shown in Equation (22). E_1 and v_1 can be obtained through the performance test of foamed concrete matrix material. For the convenience of analysis, it is assumed that $v_2 = v_1$, but v_0 is still unknown. Therefore, the 3D problem should be simplified to a 2D problem. The relationship between the size of the spherical shell and the volume ratio is shown in Equation (23).

$$f = n_0 \tag{22}$$

$$b = a/\sqrt{f} \tag{23}$$

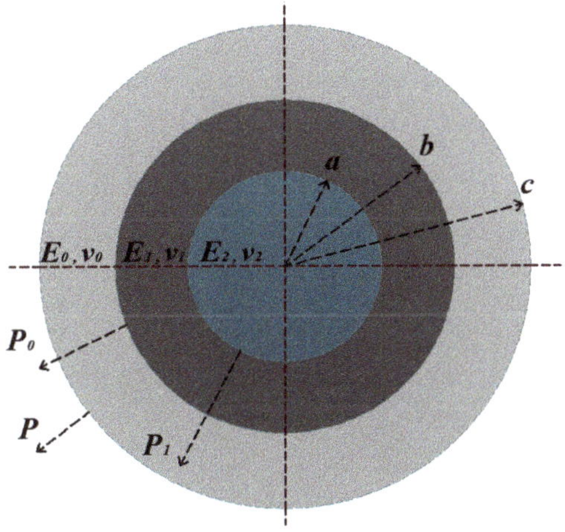

Figure 5. Prediction model for the elastic modulus of foamed concrete.

With reference to the formula for calculating the shear modulus of single-inclusion composites proposed by Christensen et al. [40], the effective elastic modulus E_0 of foamed concrete material can be completely determined, which is also based on the two-layer embedded model. The calculation formula for the effective shear modulus μ_0 can be derived, as shown in Equations (24)–(29) [40].

$$[\mu_0/\mu_1]^2 A + [\mu_0/\mu_1] B + D = 0 \tag{24}$$

$$\begin{aligned} A = {} & 3f(1-f)^2(\mu_2/\mu_1 - 1)(\mu_2/\mu_1 + \eta_2) + \\ & [(\mu_2/\mu_1)\eta_1 + \eta_2\eta_1 - ((\mu_2/\mu_1)\eta_1 - \eta_2)f^3] \times \\ & [f\eta_1(\mu_2/\mu_1 - 1) - ((\mu_2/\mu_1)\eta_1 + 1)] \end{aligned} \tag{25}$$

$$\begin{aligned} B = {} & -6f(1-f)^2(\mu_2/\mu_1 - 1)(\mu_2/\mu_1 + \eta_2) + \\ & [(\mu_2/\mu_1)\eta_1 + (\mu_2/\mu_1 - 1)f + 1] \times \\ & [(\mu_2/\mu_1 + \eta_2)(\eta_1 - 1) - 2((\mu_2/\mu_1)\eta_1 - \eta_2)f^3] + \\ & (\eta_1 + 1)f(\mu_2/\mu_1 - 1) \times \\ & [\mu_2/\mu_1 + \eta_2 + ((\mu_2/\mu_1)\eta_1 - \eta_2)f^3] \end{aligned} \tag{26}$$

$$\begin{aligned} D = {} & 3f(1-f)^2(\mu_2/\mu_1 - 1)(\mu_2/\mu_1 + \eta_2) + \\ & [(\mu_2/\mu_1)\eta_1 + (\mu_2/\mu_1 - 1)f + 1] \times \\ & [\mu_2/\mu_1 + \eta_2 + ((\mu_2/\mu_1)\eta_1 - \eta_2)f^3] \end{aligned} \tag{27}$$

$$\eta_1 = 3 - 4v_1 \tag{28}$$

$$\eta_2 = 3 - 4v_2 \tag{29}$$

where μ_0, μ_1, and μ_2 represent the shear elastic moduli of the equivalent medium, matrix, and spherical shell, respectively. In accordance with the principle of elastic mechanics, the relationship between elastic modulus, shear modulus, and Poisson's ratio is shown in Equation (30).

$$v_0 = \frac{E_0}{2\mu_0} - 1 \tag{30}$$

The elastic modulus prediction model of foamed concrete can be obtained by solving Equations (19), (24) and (30) simultaneously.

3.3. Calculation Process for the Elastic Modulus of Foamed Concrete

Foamed concrete is a two-layer embedded single-inclusion composite material. The main idea of its calculation follows. The inclusion spherical shell is placed into foamed concrete matrix material (assuming its elastic characteristic is *F*, which is determined by the material properties, such as elastic modulus *E* and Poisson's ratio *v*, and assuming its elastic characteristic be *S*). Then, the formula above is used to complete the homogenization process to obtain the macroscopic properties of foamed concrete (setting its elastic characteristics as *F-SM*). The calculation diagram is shown in Figure 6.

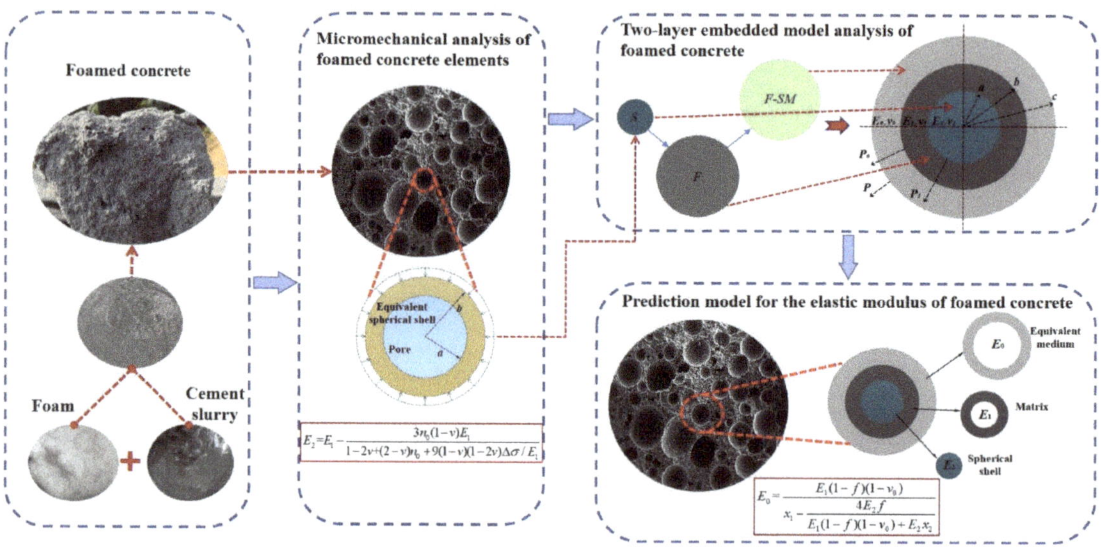

Figure 6. Schematic of the calculation of a single inclusion in foamed concrete.

From the mesomechanical analysis process described above, the calculation process for the elastic modulus of foamed concrete under normal conditions can be determined when combined with the basic physical performance parameters of foamed concrete, as shown in Figure 7.

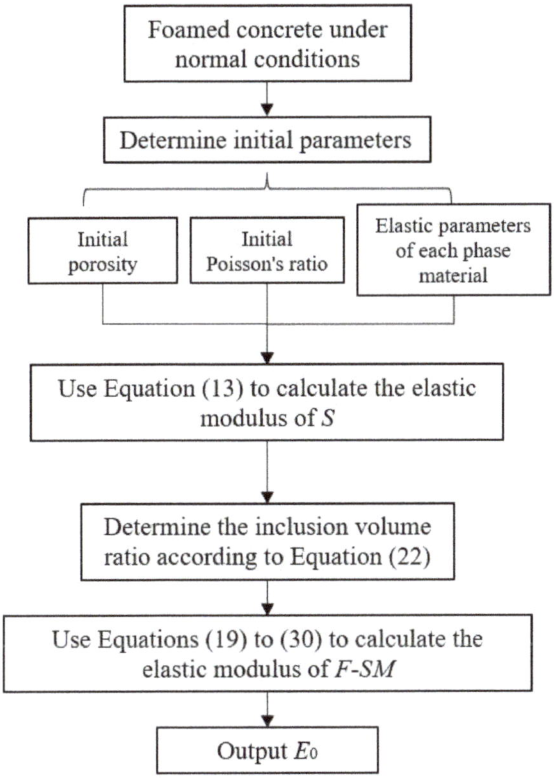

Figure 7. Flowchart of the calculation for the elastic modulus of foamed concrete.

4. Model Validation

This section mainly verifies the inclusion theoretical model of foamed concrete constructed above on the basis of the comparison of laboratory tests and existing research literature.

4.1. Comparison with Test Results

In this test, an instron1346 universal testing machine produced by the Instron company of Britain is used to carry out the indoor compression test of foamed concrete test block. The raw materials of foamed concrete are 42.5-grade ordinary Portland cement, class F fly ash, and QW-100-type foaming agent. The size of the test block is set to 100 mm × 100 mm × 100 mm. The curing conditions are implemented in a sealed plastic bag for constant-temperature curing, and the curing temperature is 20 ± 2 °C. The test process is shown in Figure 8.

In the test, a controlled strain rate of 50 mm/min is used to load the foamed concrete test block, and the stress–strain curve is immediately generated using the universal testing machine. Loading is conducted until the foamed concrete specimen fails, and the maximum failure load is recorded. The elastic modulus of the foamed concrete test block can be obtained by measuring the slope of the straight-line segment on the stress–strain curve.

In accordance with the compressive strength test results of foamed concrete cubes, the test data of the test blocks are listed in Table 1.

(a) Preparation of cement slurry (b) Preparation of foam clusters (c) Preparation of foamed concrete slurry (d) Maintenance of foamed concrete (e) Test piece demolding (f) Test piece under compression

Figure 8. Schematic of test.

Table 1. Test value of the compressive strength of foamed concrete test blocks.

Test Group Number	Water–Solid Ratio	Bubble Rate	Wet Severity/kN·m^{-3}	28-Day Unconfined Compressive Strength/MPa	Elastic Modulus/MPa
1	1:1.8	0.71	5.39	1.52	718
2	1:1.8	0.68	5.70	1.79	826
3	1:1.8	0.68	6.03	1.67	778
4	1:1.8	0.65	6.19	1.74	806
5	1:1.8	0.65	6.25	1.77	818
6	1:1.8	0.65	6.32	1.90	870
7	1:1.8	0.62	6.67	1.96	894
8	1:1.8	0.62	7.05	2.12	958
9	1:1.8	0.59	7.07	2.54	1126
10	1:1.8	0.59	7.53	2.80	1230

The porosity of foamed concrete and the mechanical constants of the matrix material are mainly involved in the analysis and calculation of this model. The former is mainly obtained by measuring the bubble rate in indoor tests, and the relevant parameters of the latter are listed in Table 2.

Table 2. Values of the relevant parameters of matrix materials.

Matrix Material Parameters	v_1	E_1/MPa	μ_1/MPa	ρ_s/kg·m^{-3}
value	0.16	2450	1060	1.9

From the experimental–theoretical comparison results given in Figure 9, the elastic modulus of foamed concrete calculated using the theoretical model reported in this paper is in good agreement with the laboratory test data. The overall error is within a reasonable range. In addition, the distribution law of the elastic modulus of foamed concrete with bulk density indicates that a certain linear quantitative relationship exists between its elastic modulus and its density, which can be described by an empirical formula.

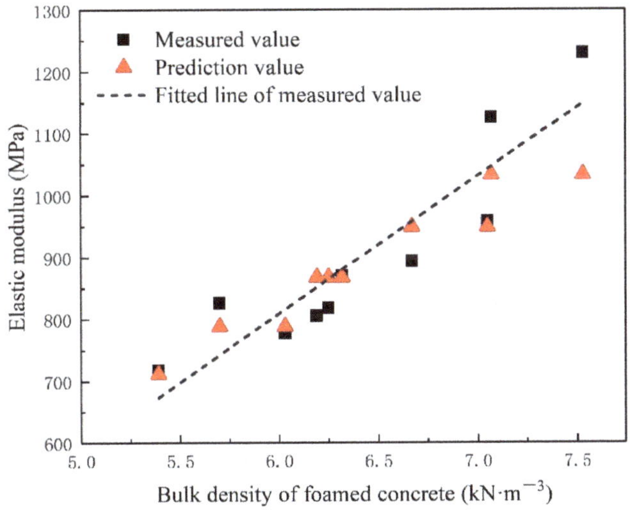

Figure 9. Comparison of elastic modulus test and theory of foamed concrete.

4.2. Comparison with Existing Research

Tan et al. [41] conducted a study on the stress–strain characteristics of foamed concrete with different densities; they provided a simple description formula of elastic modulus based on laboratory tests, as shown by Equation (31) [41]:

$$E = 1.9572\rho - 370.41 \tag{31}$$

Therefore, on the basis of the compressive strength test of foamed concrete, the relevant parameters of the indoor test block are substituted into the theoretical model described in this paper for comparative analysis with Equation (31). The results are shown in Figure 10.

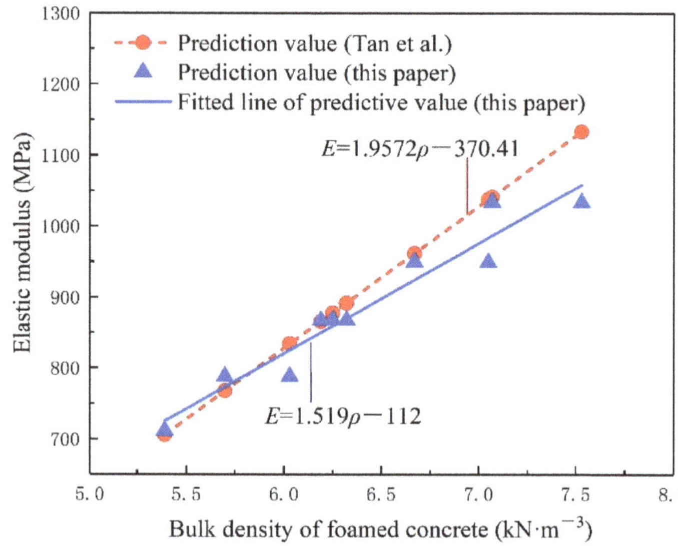

Figure 10. Comparison between the elastic modulus theory of foamed concrete and existing research (Tan et al. [41]).

From the comparison of the calculation results shown in Figure 10, the model described in this paper is relatively close to the calculation results of the existing theoretical results, and the data have a high degree of agreement. Moreover, in accordance with the results of this model, the linear description of the elastic modulus on the density of foamed concrete can be obtained, as shown in Equation (32). Comparison of the linear distribution laws of the two indicates that the overall slope deviation is not large.

$$E = 1.519\rho - 112 \tag{32}$$

McCormick et al. [31] conducted a series of mechanical tests on foamed porous concrete materials, and they obtained the relationship between the elastic modulus, wet bulk density, and compressive strength of foamed concrete based on a large number of test data, such as Equation (33); Jones et al. [22] gave the relationship between the elastic modulus and compressive strength of foamed concrete based on experimental studies, as shown in Equation (34).

$$E = 33W^{1.5}f_c^{0.5} \tag{33}$$

$$E = 0.42f_c^{1.18} \tag{34}$$

Referring to the test data of each group of test blocks in Table 1, the relevant parameters are substituted into the theoretical model of this paper for comparative analysis with Equation (33); the calculation results are shown in Figure 11. Equation (34) is compared with the model in this paper, and the calculation results are shown in Figure 12. In order to more intuitively reflect the rationality of the model described in this paper, the calculated values of McCormick's, Jones's, Tan's, and the model in this paper are compared with the experimental values, and the results are shown in Figure 13.

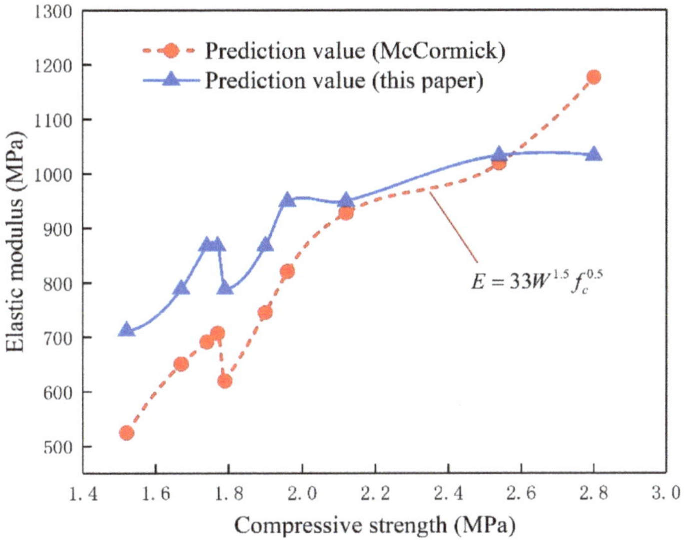

Figure 11. Comparison of elastic modulus theory of foamed concrete and existing research (McCormick [31]).

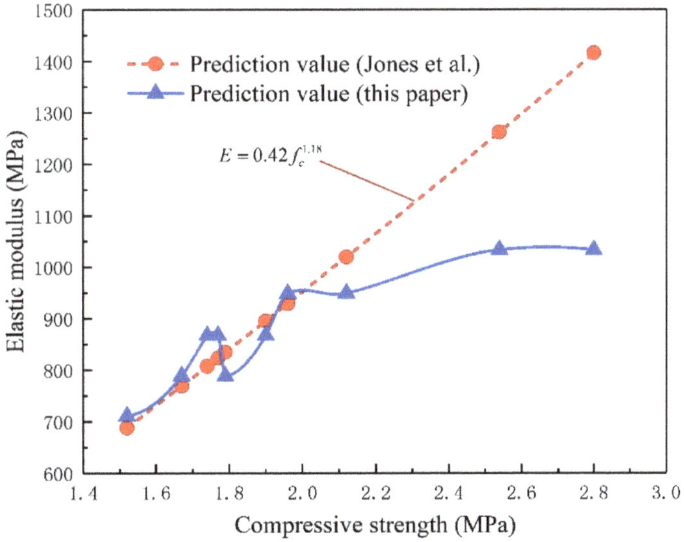

Figure 12. Comparison of elastic modulus theory of foamed concrete with existing research (Jones et al. [22]).

It can be seen from Figure 11 that the proposed model and McCormick's model have a certain similarity in the calculation results. It can be seen from Figure 12 that when the compressive strength is in the range 1.5–2.0 MPa, the calculation results of the proposed model and Jones's model are relatively close. When the compressive strength of foamed concrete is higher than 2.0 MPa, the difference between the two calculated values becomes increasingly larger.

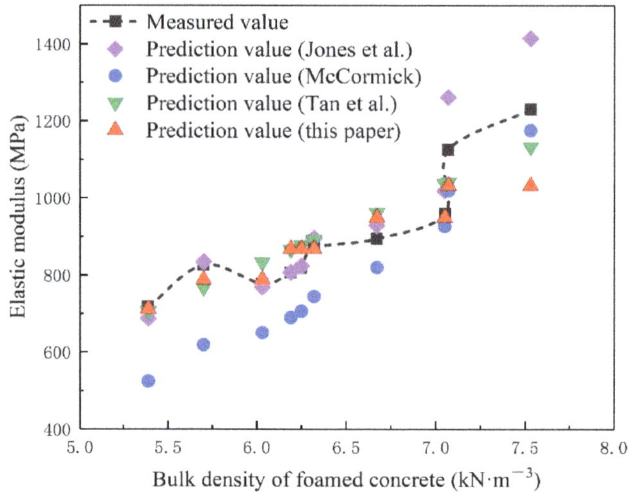

Figure 13. Comprehensive comparison of calculation results of elastic modulus of foamed concrete (Jones et al. [22], McCormick [31] and Tan et al. [41]).

According to the comparison of elastic modulus values shown in Figure 13, it can be seen that compared with the research results of McCormick, Jones, and Tan, the calculation results of the elastic modulus prediction model of foam concrete proposed in this paper are closer to the experimental data, indicating that the accuracy of this model is better than other models. Therefore, the theoretical model of the elastic modulus of foamed concrete based on the Walsh formula has good rationality.

5. Parameter Sensitivity Analysis

In accordance with the relevant parameters of foamed concrete used in the laboratory test, parameter sensitivity analysis is further performed for the theoretical model of a foamed concrete inclusion. The basic parameters of the model are shown in Table 2. From the previous analysis, the elastic modulus of foamed concrete is closely related to the density, and the material density is related to the porosity and the elastic parameters of the matrix material. Therefore, the influence of these main parameters on the elastic modulus of foamed concrete is explored in depth.

5.1. Porosity of Foamed Concrete

The elastic modulus results of foamed concrete with porosity 0.59, 0.62, 0.65, 0.68, and 0.71 are calculated. Figure 14 presents the relationship between the elastic modulus and porosity of foamed concrete. As shown in Figure 14, compared with foamed concrete with porosity 0.59, the elastic modulus of foamed concrete with porosity 0.62, 0.65, 0.68, and 0.71 decreased by 8.1%, 16%, 23.7%, and 31.1%, respectively. This shows that the calculation results of this model are negatively correlated with porosity; the higher the porosity of foamed concrete, the lower the elastic modulus. The main reason is that porosity is directly related to the overall stiffness of the test block, and an excessively high porosity greatly reduces the overall stiffness of foamed concrete.

This law can also be obtained from the laboratory test data, but the difference is that the test data do not show a strict linear correlation with porosity. Accordingly, in practice, the relationship between the elastic modulus and the porosity of foamed concrete is nonlinear, but on the whole, the model is in good agreement with the experimental data in the porosity range 0.62–0.71. This model can be used to explore the effect of porosity on the elastic modulus of foamed concrete.

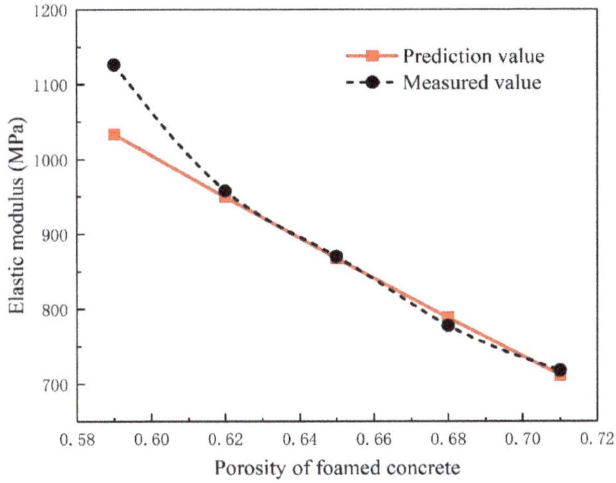

Figure 14. Variation in elastic modulus of foamed concrete with porosity.

5.2. Poisson's Ratio of Foamed Concrete Matrix

The elastic modulus of foamed concrete with Poisson's ratio of matrix material in the range 0.1~0.28 were calculated, and the corresponding curves drawn, as shown in Figure 15. It can be seen from Figure 15 that when the matrix Poisson's ratio increases from 0.1 to 0.28, the elastic modulus of foamed concrete with porosity 0.59, 0.62, 0.65, 0.68, and 0.71 decreased by 18.7%, 19.7%, 20.6%, 21.5%, and 22.4%, respectively. This shows that the elastic modulus decreases continuously with the increase in Poisson's ratio. This is mainly because the material is not prone to lateral deformation when Poisson's ratio is small; on the contrary, when the Poisson's ratio is large, the material is prone to lateral deformation, and the stiffness of the material is large. Therefore, the elastic modulus is negatively correlated with the matrix Poisson's ratio.

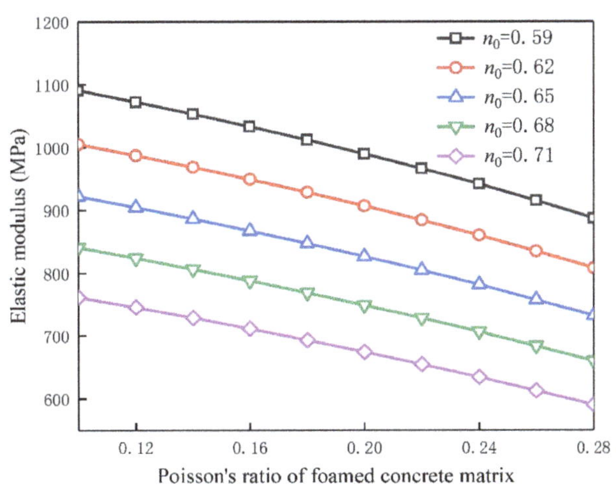

Figure 15. Variation in elastic modulus of foamed concrete with Poisson's ratio of matrix.

5.3. Elastic Modulus of Foamed Concrete Matrix

The elastic modulus of foamed concrete with the elastic modulus of matrix material in the range 1450~3450 MPa was calculated; Figure 16 shows the relationship between the elastic modulus of foamed concrete and the elastic modulus of the matrix. When the elastic modulus of the matrix increases from 1450 to 3450 MPa, the elastic modulus of foamed concrete with porosity 0.59, 0.62, 0.65, 0.68, and 0.71 increases by 118%, 116%, 113%, 111%, and 108%, respectively. This shows that there is a positive correlation between the elastic modulus of foamed concrete and that of the matrix material.

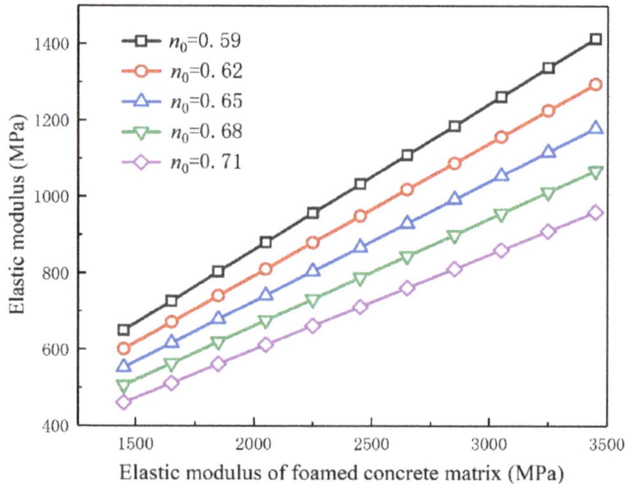

Figure 16. Variation law of the elastic modulus of foamed concrete with the elastic modulus of matrix.

In addition, comparison of the curve slopes under different porosity calculation conditions shows that the smaller the porosity, the higher the curve slope. That is, the elastic modulus of foamed concrete has a high correlation with the elastic properties of the matrix in the case of low porosity. The performance of foamed concrete in the case of high porosity is not only related to the matrix but is also restricted by the pore state.

Quantitative analysis based on theory shows that for foamed concrete, a two-phase medium material composed of matrix material and bubble pores, its elastic properties are controlled by two main factors: material porosity and matrix material properties. Adjusting the level of these two factors can effectively change the elastic properties of foamed concrete.

6. Conclusions

In this paper, an elastic modulus prediction model of foamed concrete is established based on the derivation of the Walsh formula and inclusion theory. The research can be mainly summarized as the following points:

(1) According to the internal structure characteristics of foamed concrete, the calculation formula is established for elastic modulus of spherical shell by introducing the Walsh formula and spherical pore hypothesis. Moreover, the elastic modulus prediction model of foamed concrete is constructed based on the two-layer embedded model in inclusion theory.

(2) The uniaxial compression test of foamed concrete is carried out, and the proposed model is compared with the test data and the existing theoretical models. The verification results show that the elastic modulus values calculated using the proposed model are consistent with the experimental data. The values are also in good agreement with the distribution law of the calculation results calculated by using other theoretical models.

(3) The effects of porosity, Poisson's ratio of matrix, and elastic modulus of matrix on the elastic modulus of foamed concrete are analyzed by the proposed model. The results show that the elastic modulus of foamed concrete decreases with the increase

in porosity and matrix Poisson's ratio, and increases with the increase in matrix elastic modulus.

Author Contributions: Conceptualization, F.L. and J.H.; methodology, F.L.; validation, F.L. and J.H.; formal analysis, Z.Z. and M.L.; investigation, J.Z.; data curation, F.L. and J.Z.; writing—original draft preparation, F.L.; writing—review and editing, J.H.; supervision, Z.Z. and M.L.; project administration, Z.Z.; funding acquisition, Z.Z. All authors have read and agreed to the published version of the manuscript.

Funding: This research was funded by the National Natural Science Foundation of China, grant number 50908234, and Hunan Provincial Natural Science Foundation of China, grant number 2020JJ4743.

Institutional Review Board Statement: Not applicable.

Informed Consent Statement: Not applicable.

Data Availability Statement: Data are contained within the article.

Conflicts of Interest: The authors declare no conflict of interest.

Abbreviations

Main symbols and notations used in the paper

Symbol	Description
a	inner diameter of spherical shell
b	outer diameter of spherical shell
c	outer diameter of equivalent medium
E	elastic modulus
E_0	elastic modulus of equivalent medium
E_1	elastic modulus of the matrix
E_2	elastic modulus of the spherical shell element
f	volume percentage of inclusions
f_c	compressive strength of foamed concrete
n	porosity
n_0	initial porosity of foamed concrete
P	radial effective stress at the outer diameter of the equivalent medium
P_0	radial effective stress at the outer diameter of spherical shell
P_1	radial effective stress at the inner radius of spherical shell
u_a	radial displacement of the spherical shell
u_{0b}	displacement of equivalent medium at boundary $r = b$
u_{0c}	displacement of equivalent medium at boundary $r = c$
u_{1a}	displacement of matrix at boundary $r = a$
u_{1b}	displacement of matrix at boundary $r = b$
u_{2a}	displacement of spherical shell at boundary $r = a$
V	volume of the porous medium
V_c	pore volume
W	wet bulk density of foamed concrete
δ	volume compliance of the overall unit
δ_s	volume compliance of the matrix
ΔV_c	change in pore volume
Δa	change of inner diameter of spherical shell
$\Delta\sigma$	the action of volume stress
ρ	density of foamed concrete
ρ_s	density of the matrix
v	Poisson's ratio
v_0	Poisson's ratio of foamed concrete
v_1	Poisson's ratio of the matrix
v_2	Poisson's ratio of the spherical shell

References

1. Othuman, M.A.; Wang, Y.C. Elevated-temperature thermal properties of lightweight foamed concrete. *Constr. Build. Mater.* **2011**, *25*, 705–716. [CrossRef]
2. Kearsley, E.P.; Wainwright, P.J. The effect of porosity on the strength of foamed concrete. *Cem. Concr. Res.* **2002**, *32*, 233–239. [CrossRef]
3. Kearsley, E.P.; Wainwright, P.J. The effect of high fly ash content on the compressive strength of foamed concrete. *Cem. Concr. Res.* **2001**, *31*, 105–112. [CrossRef]
4. Zhou, Z.; Zhang, J.J.; Gong, C.J. Automatic detection method of tunnel lining multi-defects via an enhanced You Only Look Once network. *Comput.-Aided Civil Infrastruct. Eng.* **2022**, *37*, 762–780. [CrossRef]
5. Huang, J.J.; Su, Q.; Zhao, W.H.; Li, T.; Zhang, X.X. Experimental study on use of lightweight foam concrete as subgrade bed filler of ballastless track. *Constr. Build. Mater.* **2017**, *149*, 911–920. [CrossRef]
6. Kim, T.H.; Kim, T.H.; Kang, G.C. Performance evaluation of road embankment constructed using lightweight soils on an unimproved soft soil layer. *Eng. Geol.* **2013**, *160*, 34–43. [CrossRef]
7. Chen, Z.P.; Wang, J.B.; Liu, J.F.; Zhao, W.H. Experimental study on dynamic engineering characteristics of foam lightweight soil. *Highway* **2019**, *64*, 77–80. (In Chinese)
8. Jiang, C.; Wu, Y.F.; Jiang, J.F. Effect of aggregate size onstress-strain behavior of concrete confined by fiber composites. *Compos. Struct.* **2017**, *168*, 851–862. [CrossRef]
9. Hilal, A.A.; Thom, N.H.; Dawson, A.R. The Use of Additives to Enhance Properties of Pre-Formed Foamed Concrete. *Int. J. Eng. Technol.* **2015**, *7*, 286–293. [CrossRef]
10. Hilal, A.A.; Thom, N.H.; Dawson, A.R. On void structure and strength of foamed concrete made without/with additives. *Constr. Build. Mater.* **2015**, *85*, 157–164. [CrossRef]
11. Amran, Y.H.M.; Farzadnia, N.; Ali, A.A.A. Properties and applications of foamed concrete; a review. *Constr. Build. Mater.* **2015**, *101*, 990–1005. [CrossRef]
12. Liu, C.; Luo, J.; Li, Q.; Gao, S.; Su, D.; Zhang, J.; Chen, S. Calcination of green highbelite sulphoaluminate cement (GHSC) and performance optimizations of GHSC-based foamed concrete. *Mater. Des.* **2019**, *182*, 107986. [CrossRef]
13. Ramamurthy, K.; Nambiar, E.K.K.; Ranjani, G.I.S. A classification of studies on properties of foam concrete. *Cem. Concr. Compos.* **2009**, *31*, 388–396. [CrossRef]
14. Ranjani, G.I.S.; Ramamurthy, K. Behaviour of foam concrete under sulphate environments. *Cem. Concr. Compos.* **2012**, *34*, 825–834. [CrossRef]
15. Falliano, D.; De Domenico, D.; Ricciardi, G.; Gugliandolo, E. Experimental investigation on the compressive strength of foamed concrete: Effect of curing conditions, cement type, foaming agent and dry density. *Constr. Build. Mater.* **2018**, *165*, 735–749. [CrossRef]
16. Falliano, D.; De Domenico, D.; Ricciardi, G.; Gugliandolo, E. 3D-printable lightweight foamed concrete and comparison with classical foamed concrete in terms of fresh state properties and mechanical strength. *Constr. Build. Mater.* **2020**, *254*, 119271. [CrossRef]
17. Falliano, D.; De Domenico, D.; Ricciardi, G.; Gugliandolo, E. Compressive and flexural strength of fiber-reinforced foamed concrete: Effect of fiber content, curing conditions and dry density. *Constr. Build. Mater.* **2019**, *198*, 479–493. [CrossRef]
18. Panesar, D.K. Cellular concrete properties and the effect of synthetic and protein foaming agents. *Constr. Build. Mater.* **2013**, *44*, 575–584. [CrossRef]
19. She, W.; Du, Y.; Zhao, G.; Feng, P.; Zhang, Y.; Cao, X. Influence of coarse fly ash on the performance of foam concrete and its application in high-speed railway roadbeds. *Constr. Build. Mater.* **2018**, *170*, 153–166. [CrossRef]
20. Zhang, Y.; Chen, D.; Liang, Y.; Qu, K.; Lu, K.; Chen, S.; Kong, M. Study on engineering properties of foam concrete containing waste seashell. *Constr. Build. Mater.* **2020**, *260*, 119896. [CrossRef]
21. Ben Youssef, M.; Lavergne, F.; Sab, K.; Miled, K.; Neji, J. Upscaling the elastic stiffness of foam concrete as a three-phase composite material. *Cem. Concr. Res.* **2018**, *110*, 13–23. [CrossRef]
22. Jones, M.R.; McCarthy, A. Preliminary views on the potential of foamed concrete as a structural material. *Mag. Constr. Res.* **2005**, *57*, 21–32. [CrossRef]
23. Ge, Z.; Guan, H.Q.; Sun, R.J.; Zhang, H.Z.; Wang, W.L.; Qi, H. Use of green calcium sulphoaluminate cement to prepare foamed concrete for road embankment: A feasibility study. *Constr. Build. Mater.* **2020**, *237*, 117791. [CrossRef]
24. Tan, X.J.; Chen, W.Z.; Yang, D.S.; Liu, H.Y.; Chan, A.H.C. Experimental and theoretical studies on effect of height-to-diameter ratios on failure forms and mechanical characteristics of foamed concrete. *J. Mater. Civ. Eng.* **2019**, *31*, 04018341. [CrossRef]
25. Mydin, M.A.O.; Wang, Y.C. Mechanical properties of foamed concrete exposed to high temperatures. *Constr. Build. Mater.* **2012**, *26*, 638–654. [CrossRef]
26. Li, L.; Purkiss, J. Stress-strain constitutive equations of concrete material at elevated temperatures. *Fire Saf. J.* **2005**, *40*, 669–686. [CrossRef]
27. Shi, X.N.; Huang, J.J.; Su, Q. Experimental and numerical analyses of lightweight foamed concrete as filler for widening embankment. *Constr. Build. Mater.* **2020**, *250*, 118897. [CrossRef]
28. Horpibulsuk, S.; Suddeepong, A.; Chinkulkijniwat, A.; Liu, M.D. Strength and compressibility of lightweight cemented clays. *Appl. Clay. Sci.* **2012**, *69*, 11–21. [CrossRef]

29. Cong, M.; Bing, C. Properties of a foamed concrete with soil as filler. *Constr. Build. Mater.* **2015**, *76*, 61–69. [CrossRef]
30. Nambiar, E.K.K.; Ramamurthy, K. Air-void characterisation of foam concrete. *Cem. Concr. Res.* **2007**, *37*, 221–230. [CrossRef]
31. McCormick, F.C. Rational proportioning of preformed foam cellular concrete. *ACI J. Proc.* **1967**, *64*, 104–110.
32. Saint-Jalmes, A.; Peugeot, M.L.; Ferraz, H.; Langevin, D. Differences between protein and surfactant foams: Microscopic properties, stability and coarsening. *Colloids Surf. A Physicochem. Eng. Asp.* **2005**, *263*, 219–225. [CrossRef]
33. Raj, A.; Sathyan, D.; Mini, K.M. Physical and functional characteristics of foam concrete: A review. *Constr. Build. Mater.* **2019**, *221*, 787–799. [CrossRef]
34. Walsh, J.B. The effect of cracks on the compressibility of rock. *J. Geophys. Res.* **1965**, *70*, 381–389. [CrossRef]
35. Cao, W.G.; Li, P.; Xu, X.; He, M. An analytical method of ground foundation settlement influenced by modulus change and thre e-dimensional stress. *J. Hunan Univ. (Nat. Sci.)* **2018**, *45*, 142–148. (In Chinese)
36. Yang, H.; Zhou, Z.; Wang, X.C.; Zhang, Q.F. Elastic modulus calculation model of a soil-rock mixture at normal or freezing temperature based on micromechanics approach. *Adv. Mater. Sci. Eng.* **2015**, *2015*, 576080. [CrossRef]
37. Zhou, Z.; Yang, H.; Xing, K.; Gao, W.Y. Prediction models of the shear modulus of normal or frozen soil-rock mixtures. *Geomech. Eng.* **2018**, *15*, 783–791. [CrossRef]
38. Zhou, Z.; Liu, Z.Z.; Yang, H.; Gao, W.Y.; Zhang, C.C. Freeze-thaw damage mechanism of elastic modulus of soil-rock mixtures at different confining pressures. *J. Cent. South Univ.* **2020**, *27*, 554–565. [CrossRef]
39. Li, G.; Li, Y.; Metcalf, J.B.; Pang, S.S. Elastic modulus prediction of asphalt concrete. *J. Mater. Civ. Eng.* **1999**, *11*, 236–241. [CrossRef]
40. Christensen, R.M.; Lo, K.H. Solutions for effective shear properties in three phase sphere and cylinder models. *J. Mech. Phys. Solids* **1979**, *27*, 315–330. [CrossRef]
41. Tan, X.J.; Chen, W.Z.; Liu, H.Y.; Chan, A.H.C. Stress-Strain Characteristics of Foamed Concrete Subjected to Large Deformation under Uniaxial and Triaxial Compressive Loading. *J. Mater. Civ. Eng.* **2018**, *30*, 04018095. [CrossRef]

applied sciences

MDPI

Article

Failure Mechanism Analysis and Optimization Analysis of Tunnel Joint Waterstop Considering Bonding and Extrusion

Yimin Wu [1], Haiping Wu [1,*], Dinghai Chu [2], Sheng Feng [3], Junjian Zhang [4] and Haoran Wu [1]

1 School of Civil Engineering, Central South University, Changsha 410075, China; wuyimin531@csu.edu.cn (Y.W.); wuhaoran@csu.edu.cn (H.W.)
2 China Railway 21st Bureau Group the 3rd Engineering Co., Ltd., Xianyang 712023, China; 15733170607@163.com
3 Hubei Provincial Communications Planning and Design Institute Co., Ltd., Wuhan 430050, China; whpeason@163.com
4 Hubei Provincial Communications Investment West Hubei Expressway Construction Management Co., Ltd., Wuhan 430000, China; songluyaocat@163.com
* Correspondence: 204812375@csu.edu.cn

Abstract: In waterproofing mountain tunnels, the tunnel joint is the weak link. To explore the waterproof failure mechanism of the tunnel joint waterstop belt and to propose an optimization method for the waterstop belt, this paper combines tests and numerical simulations, summarizes the waterproofing mechanism of the waterstop belt, establishes a finite element model of the waterstop belt considering bonding and extrusion, and studies the waterproofing ability and mechanical properties of the waterstop. The main conclusions are as follows: (1) The waterproofing capacity of the water stop belt depends on its surface contact pressure and bonding force. (2) Waterstop deformation will partially destroy the bonding between the waterstop and concrete, reducing the reliability of the waterproofing mechanism. (3) When the deformation of the waterstop belt reaches a certain degree, its stress value is too large to meet the requirements of its service life. (4) The design can be optimized from two aspects: the bond between the waterstop and concrete, and the size of the waterstop.

Keywords: mountain tunnel; joint waterproofing; waterstop; the waterproof mechanism; water seepage resistance; finite element analysis

Citation: Wu, Y.; Wu, H.; Chu, D.; Feng, S.; Zhang, J.; Wu, H. Failure Mechanism Analysis and Optimization Analysis of Tunnel Joint Waterstop Considering Bonding and Extrusion. *Appl. Sci.* **2022**, *12*, 5737. https://doi.org/10.3390/app12115737

Academic Editors: Mingfeng Lei, Chenjie Gong and Xianda Shen

Received: 13 May 2022
Accepted: 4 June 2022
Published: 5 June 2022

1. Introduction

With the continuous increase in traffic in China, the scale of tunnels continues to increase and a large number of technical problems have been encountered during the construction and operation of tunnels. Most tunnels have varying degrees of issues, among which tunnel leakage is one of the most common, affecting the normal service use of tunnels [1,2]. Although railway and highway mountain tunnels use a special anti-drainage system [3], quite a few tunnels face serious leakage soon after they are built, and in China, there is even a saying: "ten tunnels and nine leaks". An anti-drainage system may be damaged during construction and service, in which case, when the groundwater pressure is great, the water will penetrate the waterproof layer [4]. In this case, the waterproofing of the concrete lining becomes the key to the waterproofing of the entire tunnel project. Lining joints are undoubtedly the weak link in the waterproofing of lining structures [5]. To avoid water leakage through these joints, waterstops are generally used.

However, in actual projects, it is often difficult to achieve the desired effect through waterstop waterproofing. To meet the requirements of tunnel waterproofing, scholars at home and abroad have made a lot of optimizations. Some scholars [6–8] believe that the existing waterstop structure is not reasonable enough, based on which the structure of the waterstop has been optimized. Some studies [9–13] indicate that improving the adhesion of waterstops to concrete can effectively improve its waterproofing performance. On the basis of the contact between the waterstop and concrete, some scholars [14–16] have proposed inflatable waterstops, bituminous surface waterstops, and flexible waterstops as options. According to Kudritz [17], a proper waterstop profile is critical. The above analysis of the causes of failure of a waterstop is based on engineering experience, and there is no theoretical analysis of this, which makes it difficult to explain the rationality of its optimization measures.

Some scholars [18–21] have simulated the waterstop zone through finite element analysis, analyzing its stress and deformation characteristics. Lin [22] simulated the deformation of the waterstop of the deformed seam of the pipe gallery, analyzed the stress characteristics of the waterstop under different working conditions, and optimized the structure of the existing waterstop. Meng and Li [23,24] simulated the drawing of rubber strips and the deformation of the optimized waterstop, and analyzed the stress characteristics of the waterstop. They adopted the method of binding or applying tangential friction to the contact between the waterstop and the concrete, but did not consider the adhesion of the waterstop and the concrete, which is more limited. At the same time, the extrusion effect of concrete on the waterstop was considered, nor was the failure mechanism of the waterstop waterproofing studied. Only the deformation stress characteristics of the waterstop were analyzed.

In this paper, the buried rubber waterstop commonly used in the project is the research object, and the waterproofing principle of the waterstop, the bonding of the waterstop and concrete, the extrusion effect of the concrete pouring, etc., are taken into account to carry out the finite element analysis, study the failure characteristics, and propose the relevant optimization scheme to provide a reference for the optimization of the waterproof measures of the mountain tunnel joint.

2. Analysis of the Working Mechanism of the Buried Waterstop

2.1. The Working State of the Buried Waterstop

The ring joints of mountain tunnels are generally divided into two categories: deformation joints and construction joints, and the buried waterstop belt is set along the ring in the middle of the lining thickness, as shown in Figure 1. The following two kinds of deformation often occur in the tunnel: (1) longitudinal uneven settlement due to uneven distribution of the stratum and surrounding load changes and (2) longitudinal telescopic settlement due to the temperature difference, concrete shrinkage and creep, and vehicle load.

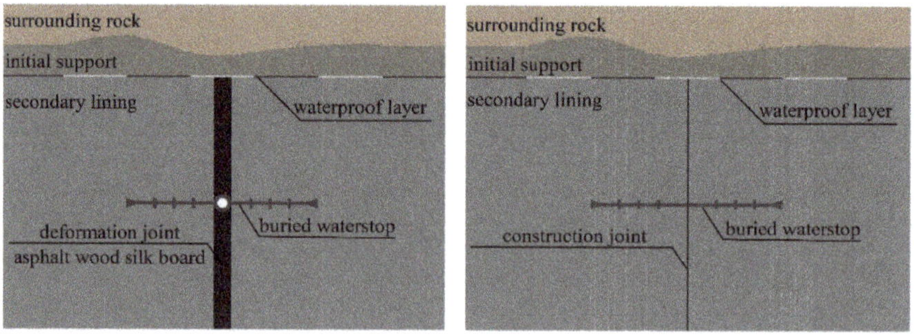

(a) (b)

Figure 1. Schematic diagram of the waterproof structure of the mountain tunnel joints. (**a**) Deformation joint, (**b**) Construction joint.

The deformation joint is set mainly to improve the uneven force deformation of the tunnel, which is generally divided into settlement joints and telescopic joints, and the width should be 20 to 30 mm. The specification stipulates that the maximum allowable settlement difference of the deformation joint for settlement shall not be greater than 30 mm [25].

There is no clear provision on the maximum allowable telescopic value of the deformation joint, which can be calculated by referring to the available literature [26,27] and taking into consideration unfavorable conditions and a certain safety factor. The maximum elongation amount of the deformation joint after rounding is 20 mm, and the maximum shortening amount is 10 mm.

In summary, when working normally, the buried waterstop at the deformation joint will be in an uneven settlement or telescopic state with the lining deformation. The maximum settlement value is 30 mm, the maximum elongation value is 20 mm, and the maximum compression value is 10 mm.

2.2. Waterproof Mechanism of a Buried Waterstop

The waterproofing principle of a buried waterstop mainly includes the following two aspects:

(1) Extrusion sealing: in a similar way to the sealing principle of the shield tunnel rubber gasket [28,29], the rubber sealing belt transfers pressure to its contact surface. When extrusion occurs between the waterstop and the concrete lining, contact pressure is generated. Without considering the bonding force between the waterstop belt and concrete, leakage occurs when the water pressure is greater than the contact pressure (Figure 2a).

(2) Bonding water plugging: There is a bond between the contact surface of the buried waterstop and the concrete lining. Without considering the sealing of the waterstop belt, when the water pressure is greater than the bond force, the bond area experiences "intact–damage–failure" and then leakage occurs (Figure 2b).

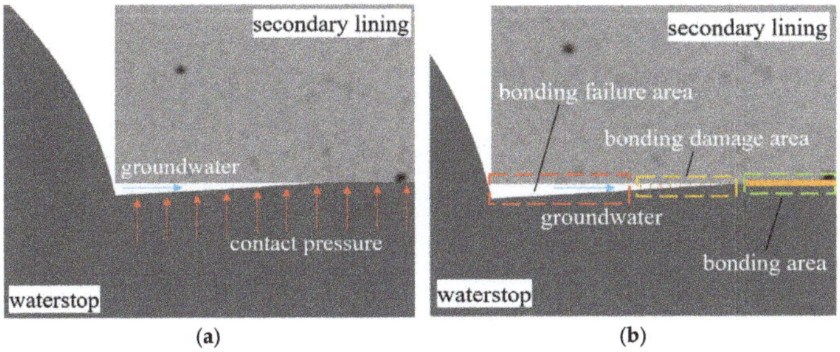

Figure 2. Waterproofing mechanism of a waterstop. (**a**) Extrusion sealing, (**b**) Bonding water plugging.

If the extrusion seal and bonding of the waterstop are simultaneously considered, when the water pressure is greater than the contact pressure, the contact pressure will fail. The bonded area then begins to be stressed, and the magnitude of the force is the water pressure minus the contact pressure. As the water pressure increases, the bonded area enters the damage stage, and then completely fails, at which point the waterstop belt fails to block the water, and the tunnel joints leak.

The force of the waterstop preventing groundwater from passing through the seepage channel is called water infiltration resistance (Figure 3). The waterproofing ability of the waterstop depends on its water infiltration resistance, involving contact pressure and adhesion force. When water infiltration resistance is greater than water pressure, the waterstop can play a waterproofing role, that is:

$$P_{wl} \geq \gamma P_w \tag{1}$$

$$P_{wl} = F + C \tag{2}$$

where P_{wl} is the water infiltration resistance, P_w is the theoretical water pressure, γ is the safety factor, F is the contact pressure, and C is the adhesion force. At the same time, the long seepage path formed by the complex cross-sectional form of the waterstop can improve the waterproof reliability of the joint, remove the head pressure, and increase the difficulty for groundwater to seep through the tunnel joint.

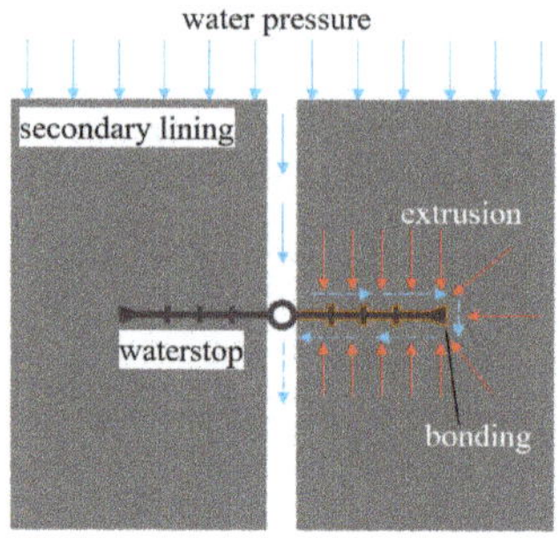

Figure 3. Water infiltration resistance of tunnel joints.

As can be seen from the preceding discussion, tunnel joint deformation constantly changes with temperature, load, etc., and with the development of deformation, the contact pressure between the waterstop and the concrete may change (Figure 4). The bonding of the waterstop to the concrete may also break with deformation, and the resistance of the waterstop to water seepage will also change. The water infiltration resistance after waterstop deformation can be regarded as short-term, P^{st}_{wl}, and the water infiltration resistance when the waterstop is not deformed can be regarded as long-term, P^{lt}_{wl}.

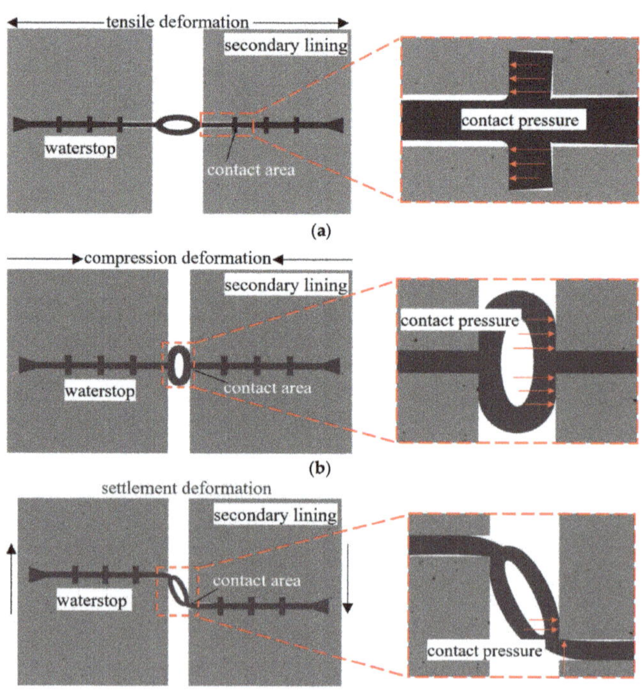

(a)

(b)

(c)

Figure 4. The deformation contact pressure of the waterstop. (**a**) Tensile contact pressure. (**b**) Compressive contact pressure. (**c**) Settling contact pressure.

At the same time, according to relevant research reports, the service life of artificial rubber and other polymers can reach more than 100 years when the stress level does not exceed 20% of their tensile strength at fracture [30]. According to the specification, the tensile strength of the rubber waterstop is not less than 10 MPa, so the dangerous stress value is 2 MPa. In the following section, using test and finite element analysis, the failure characteristics of a waterstop will be studied from two aspects: water infiltration resistance and forced deformation of the waterstop.

3. Analysis of Failure Characteristics of a Buried Waterstop

3.1. Bonding Strength Test and Initial Contact Pressure Calculation

The waterproofing ability of tunnel joints depends on their resistance to seepage, which is composed of bonding force and contact pressure. To obtain resistance of the waterstop to seepage and provide parameters for finite element analysis, the adhesive force of rubber and concrete is tested and the initial contact pressure of the waterstop belt is calculated when the joint is not deformed.

3.1.1. Adhesive Strength Test of Rubber and Concrete

Since the test specimens in this test were small, and the bond between concrete and rubber is mainly the bond between cement mortar and rubber, cement test blocks were used instead of concrete in this test. The cement used was ordinary Portland cement, with a compressive strength of 42.5 MPa (28 d). The tensile bond strength and the shear bond strength of rubber–cement were used as evaluation indexes.

The rubber was cut into rectangular blocks of 100 mm × 50 mm × 20 mm. The size of the cement test block was the same as that of the rubber block. The better of the two ends of the rubber block was selected for pouring. After pouring, the specimen was left for curing for 28 days at a temperature of 20 ± 2 °C and a relative humidity of more than 95%. The specimens were divided into tensile and shear groups, with three in each group. The tensile bond strength test surface was a rectangle surface of 50 mm × 20 mm, and the shear bond strength test surface was a square surface of 50 mm × 50 mm (Figure 5).

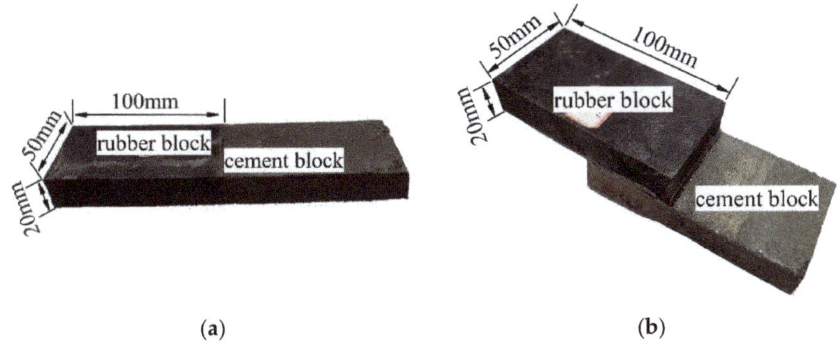

Figure 5. Waterproofing mechanism of the waterstop. (**a**) Tensile specimen. (**b**) Shear specimen.

A tension machine (0–1000 N) was used as test equipment (Figure 6). Since the specimen was rectangular, the bonding force between rubber and concrete was small, and the bonding interface brittle. So, the selected tensile speed was 0.1 mm/min.

The tensile or shear strength is the maximum stress that the test piece bears in the process of bond failure caused by an external force. The average strength of the three specimens in each group was taken as the test result (Table 1).

Figure 6. The tension machine.

Table 1. The bonding strength of rubber to concrete.

Conditions	1	2	3	Bonding Strength/MPa
Tensile specimen	0.091	0.101	0.097	0.096
Shear specimen	0.215	0.208	0.211	0.211

3.1.2. Initial Contact Pressure

When the tunnel joint is not deformed, the contact pressure between the waterstop belt and concrete is regarded as the initial contact pressure, and the initial contact pressure is formed by the extrusion of concrete pouring on the waterstop belt. The effect of concrete shrinkage on extrusion was ignored, extrusion simplification was regarded as the result of concrete self-weight, and the self-weight height was the vertical distance from the pouring surface to the initial setting of concrete. The tunnel lining of the project was analyzed. The casting time of the lining of a mold is 7 h, and the initial setting time of concrete is 2 h, so the extrusion pressure at each point of the tunnel can be known. The initial contact pressure at the vault was the smallest, with a size of 0.005 MPa (Figure 7). It increased continuously along the lining contour and then remained unchanged, with a maximum value of 0.048 MPa (Figure 7).

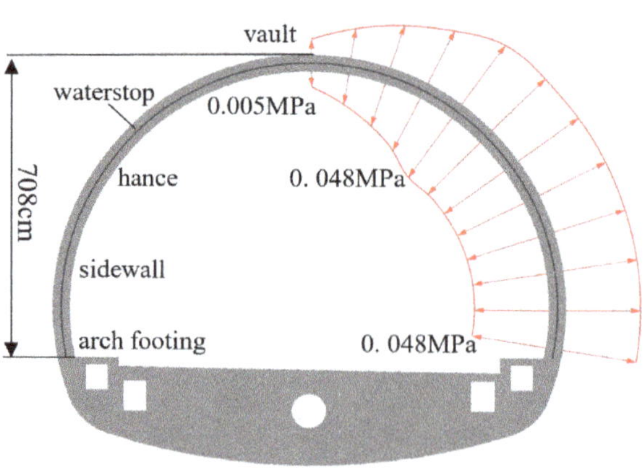

Figure 7. The initial contact pressure of the waterstop.

3.2. Finite Element Analysis of Force and Deformation of Waterstop

3.2.1. Finite Element Model

To analyze the impact of waterstop deformation on its waterproofing ability, ABAQUS/Standard is used to establish the plane strain finite element model of the buried waterstop in the concrete lining. The buried rubber waterstop is simulated by the Mooney–Rivlin model. The size is shown in Figure 8, and to distinguish it from the following, it is named waterstop I. The concrete damage-plastic model is used to simulate the tunnel lining, with a width of 300 mm and a height of 400 mm. The bonding of the waterstop and concrete is simulated by cohesive force contact. The construction quality of the waterstop is assumed to be good, and the finite element model is shown in Figure 9.

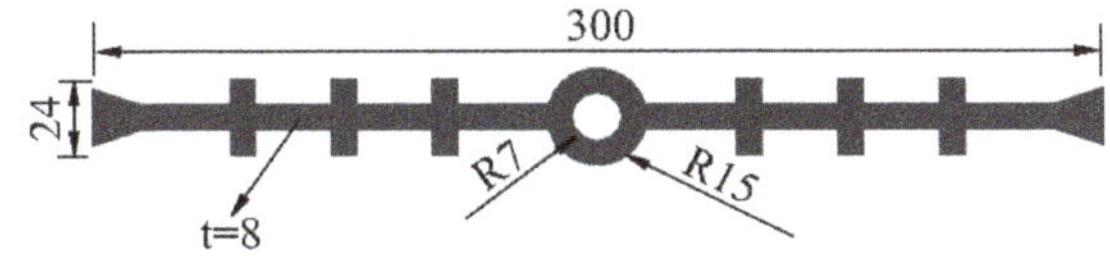

Figure 8. Waterstop I (unit: mm).

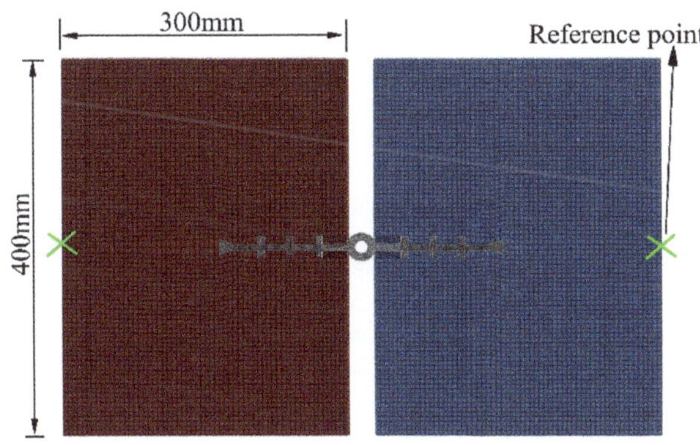

Figure 9. Finite element analysis model.

The strength grade of concrete lining is considered as C30, the elastic modulus E_c is 30 GPa, and Poisson's ratio ν_c is 0.2. Lining failure is simulated in ABAQUS using the concrete damage plastic model [31]. Among the damage plasticity parameters, the eccentricity e_f, the ratio of biaxial to uniaxial compressive strength f_{b0}/f_{c0}, and the coefficient k use default values of 0.1, 1.16, and 0.667, respectively; and the viscosity parameter μ_c is set to 0.005 based on ABAQUS implicit solution; the expansion angle ψ is set to 30°; and the stress–strain curves of concrete uniaxial compression and tension are selected for the concrete design code [32], as shown in Table 2, where σ_c and σ_t are compressive stress and tensile stress and ε_c^{in}, and ε_t^{in} are inelastic strain and cracking strain, respectively.

Table 2. Uniaxial stress–strain relationship of concrete.

σ_c	ε_c^{in}	σ_t	ε_t^{in}
14.070	0	2.030	0
20.100	0.000802	2.010	0.000028
14.637	0.002456	1.232	0.000149
10.073	0.004080	0.849	0.000257
7.501	0.005638	0.661	0.000359
5.931	0.007162	0.548	0.000458
4.890	0.008668	0.473	0.000556
4.153	0.010165	0.419	0.000653
3.607	0.011655	0.378	0.000749
3.186	0.013141	0.346	0.000846

In the study of rubber material, strain energy function can be used to simulate its constitutive model, such as the Mooney–Rivlin model, the Ogden model, the Yeoh model, and the Arruda–Boyce model. Among them, the Mooney–Rivlin model is widely used. The rubber waterstop is simulated by the dual coefficient Mooney–Rivlin in ABAQUS, and the strain energy function relationship is:

$$W = C_{10}(l_1 - 3) + C_{01}(l_2 - 3) \tag{3}$$

where W is strain energy density, l_1 and l_2 are invariants of the strain tensor, and C_{10} and C_{01} are mechanical properties constants of materials. The relationship between elastic modulus E_0 and material constant can be obtained from the incompressibility of rubber:

$$E_0 = 6(C_{10} + C_{01}) \tag{4}$$

Rubber elastic modulus E_0 can be determined by rubber hardness H_A, and its relationship is as follows:

$$E_0 = (15.75 + 2.15H_A)/(100 - H_A) \tag{5}$$

It can be seen from the specification that the hardness of the rubber waterstop of the tunnel joint is 60 $\pm$ 5. In this simulation, the median hardness value of 60 is used for calculation. From experience, it is known that C_{10}/C_{01} generally takes 0.25 for better fitting [33]. It can be calculated that C_{10} = 0.484 MPa and C_{01} = 0.121 MPa.

The contact between the waterstop and the concrete is set as the normal hard contact, the tangential direction adopts the penalty function method, and the friction coefficient is 0.3. The bond between the waterstop and the concrete is set as cohesive contact, the normal stiffness is 1.96 $\times$ 10^6 MPa, the tangential stiffness is 4.33 $\times$ 10^6 MPa, the normal bonding force is 0.095 MPa, the tangential bonding force is 0.211 MPa, and the plastic displacement is 0.001 mm. Self-contact is set in the middle hole of the waterstop, and the friction coefficient is 0.8.

To analyze the impact of deformation on the waterproof effect of the waterstop belt and study the difference of the waterstop under different initial contact pressure conditions, the type, magnitude, and initial contact pressure of deformation described above as well as the working condition were established as shown in Table 3.

Table 3. Working conditions of finite element analysis.

Working Conditions	Initial Contact Pressure (MPa)	Deformation Types	Deformation (mm)
1-1		Tensile	20
1-2	0.005	Compression	10
1-3		Settlement	30
2-1		Tensile	20
2-2	0.048	Compression	10
2-3		Settlement	30

3.2.2. Damage Failure Analysis of Concrete Lining

Taking working condition 1 as an example, Figure 10 shows concrete compression and tensile damage when lining tension, compression, and uneven settlement deformation reach the maximum. DAMAGEC is the concrete compression damage value, and DAMAGET is the concrete tensile damage value. The results show that the damage value of concrete lining in tensile and compressive deformation is 0, the compressive damage value of concrete lining in settlement deformation is 0, and the tensile damage value is approximately equal to 0. It can be considered that the concrete has no plastic damage, which will not be considered in the following.

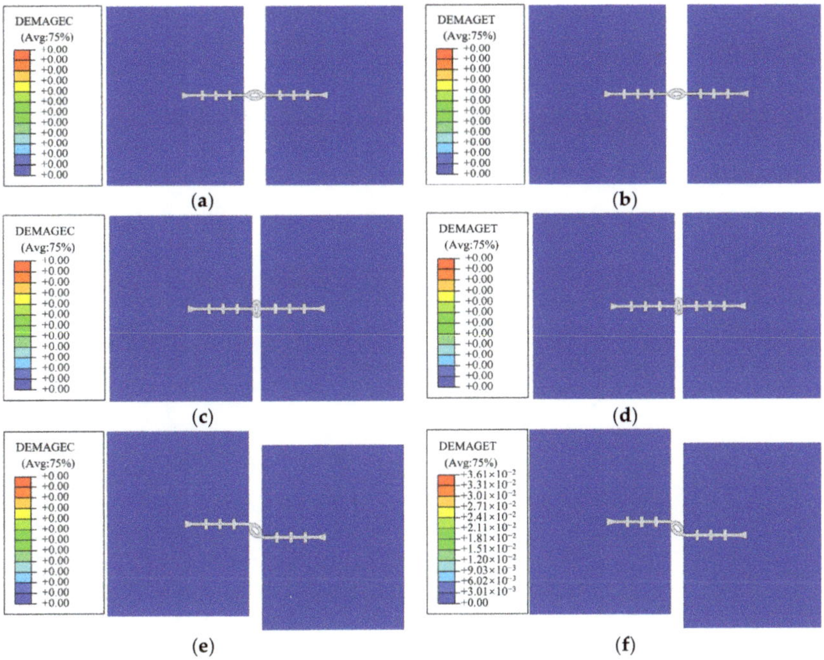

Figure 10. Damage cloud diagram of concrete stiffness in working condition 1. (**a**) 1-1 concrete compression damage. (**b**) 1-1 concrete tensile damage. (**c**) 1-2 concrete compression damage. (**d**) 1-2 concrete tensile damage. (**e**) 1-3 concrete compression damage. (**f**) 1-3 concrete tensile damage.

3.2.3. Deformation Stress Characteristics of the Waterstop

Taking operating condition 1 as an example, Figure 11 shows the von Mises stress distribution when the tensile, compressive, and uneven settlement deformation of the lining reaches the maximum. The stress on the waterstop is mainly concentrated between the first rib on the left and right sides during tensile deformation, and the stress on the waterstop during compression deformation and settlement deformation is mainly concentrated in the middle hole of the waterstop, whilst the farther side of the waterstop faces almost no force.

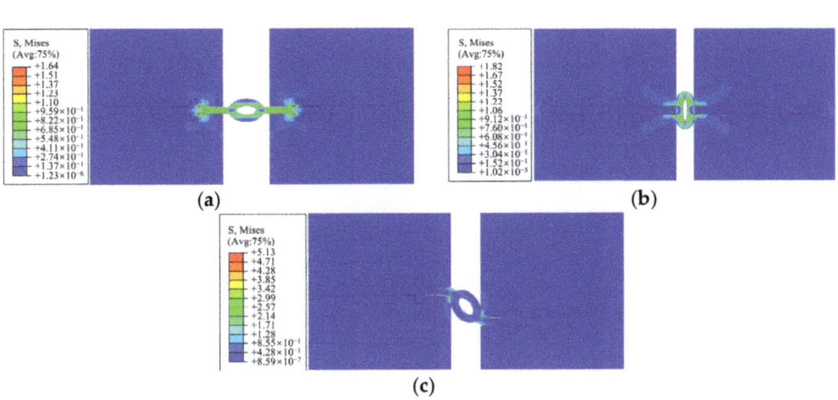

Figure 11. Condition 1 von Mises stress cloud diagram (unit: MPa). (**a**) 1-1 stress cloud diagram. (**b**) 1-2 stress cloud diagram. (**c**) 1-3 stress cloud diagram.

The bond damage between the waterstop and the concrete when the deformation of the lining reaches the maximum is shown in Figure 12, where CSDMG is the bond stiffness damage value. Tensile deformation and settlement deformation inflict greater damage on the bond between the waterstop and the concrete, and the bond failure area is between the second ribs on both sides of the waterstop. However, the damage to the bond by compression deformation is small, and only the bond near the middle hole fails.

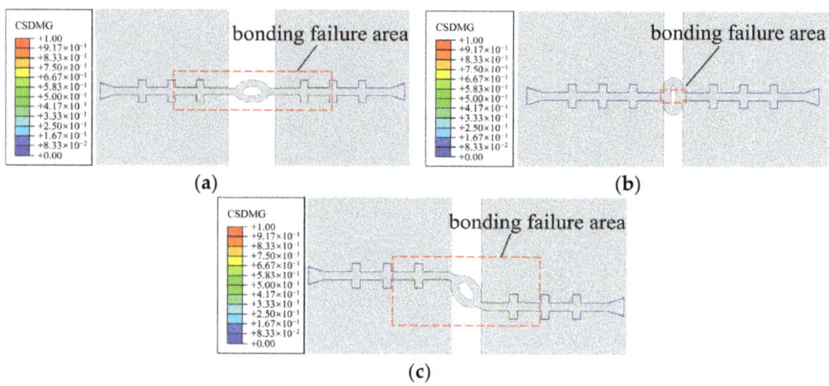

Figure 12. Damage cloud diagram of bonding stiffness in working condition 1. (**a**) 1-1 bonding stiffness damage. (**b**) 1-2 bonding stiffness damage. (**c**) 1-3 bonding stiffness damage.

It can be seen that: (1) due to the failure of the bonding part after deformation, the short-term resistance of the waterstop belt to seepage is only provided by the deformation contact pressure; (2) deformation will not completely destroy the bond between the waterstop belt and concrete, and joint deformation is uncertainty, so the waterproofing ability of the waterstop belt depends on its resistance to water seepage in the long term; (3) after deformation, the effective seepage path of the waterstop becomes shorter and the reliability of the waterstop as a barrier to leakage decreases. Once the joint is restored from a large deformation to a small deformation, compared with the joint without deformation, the waterproof is more likely to fail.

3.2.4. Deformation Stress on the Waterstop

Taking working condition 1 as an example, the development of the maximum von Mises stress on the waterstop with the increase in the deformation is plotted in Figure 13. It can be seen that the overall von Mises stress value on the waterstop increases with the increase in deformation, and is approximately linearly related to the amount of deformation. With different types of deformation, different parts of the waterstop belt become dangerous. Under tensile deformation, the joint between the first rib and the side of the waterstop belt is the most stressed part. Under compression deformation, the upper and lower positions of the inner holes of the waterstop belt are the most stressed parts. Under settlement deformation, the joint between the side wing of the waterstop and the middle hole, and the position not in contact with the lining is the most stressed part.

As can be seen from the figure, under compression deformation, the stress on the waterstop grows the fastest, followed by tensile deformation and settlement deformation. However, due to the difference in deformation, the waterstop is under the maximum stress when the settlement deformation reaches the maximum value. During tensile and compressive deformation, the von Mises stress value of the waterstop is always within the level of 2 MPa. When settlement deformation occurs, once the deformation becomes greater than 29.5 mm, the von Mises stress exceeds the dangerous level, and it is difficult to meet the service life requirements.

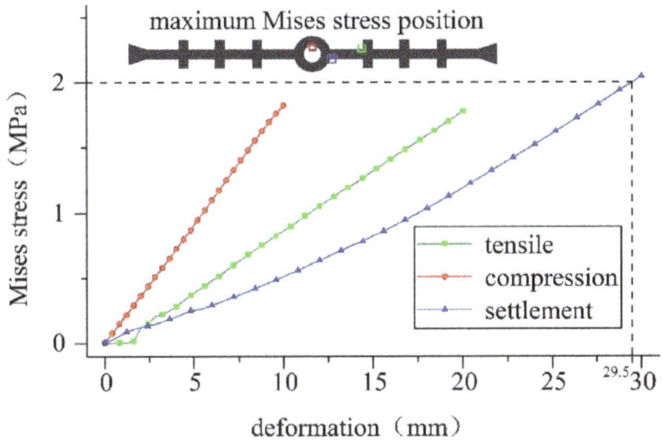

Figure 13. Maximum Mises stress on the waterstop in working condition 1.

3.3. *Waterproof Capacity Analysis of the Waterstop*

3.3.1. Long-Term Resistance of the Waterstop to Seepage

As shown in Figure 14, the maximum nominal stress criterion is regarded as the damage initiation criterion for the bonding between the waterstop belt and concrete, and the bonding stress is 0.096 MPa according to the test results.

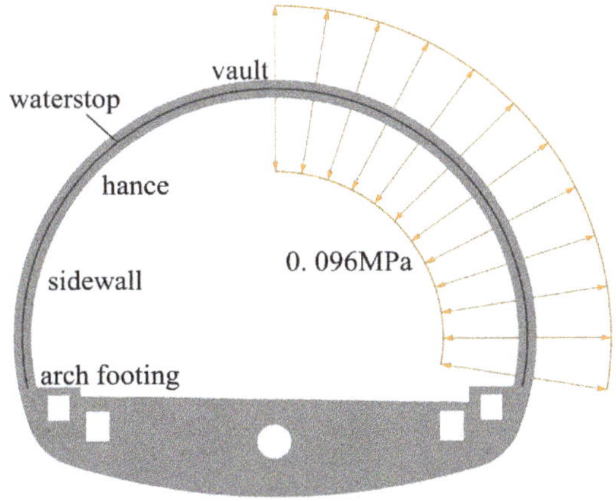

Figure 14. The adhesive force of the waterstop.

The initial contact pressure can be calculated according to the discussion in the preceding section. According to Equation (2), the long-term resistance of tunnel joints to water seepage is the minimum at the vault, which is 0.101 MPa. Initially, it increases along the lining contour and then remains unchanged, and the maximum value is 0.144 MPa (Figure 15).

Tunnel leakage occurs in the vault, hance, sidewall, and arch footing [34,35]. According to the distribution of the long-term resistance of the waterstop to seepage, the reasons for leakage are as follows: (1) the long-term resistance to seepage at the vault is the smallest in all parts of the tunnel lining. (2) The water pressure increases downward along the lining contour, and the long-term resistance of the waterstop to seepage is the same from the hance to the arch footing. When the water pressure after the lining increases, the water pressure at the sidewall and arch footing will reach the maximum water pressure that the joint can bear first, and then leakage occurs.

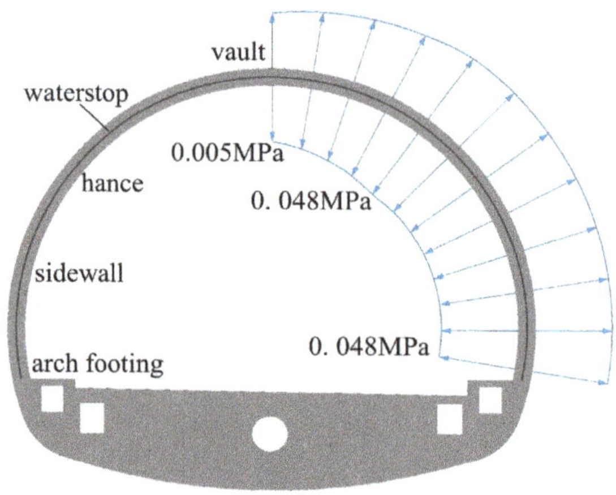

Figure 15. Long-term resistance of the waterstop to seepage.

3.3.2. Short-Term Resistance of the Waterstop to Seepage

Taking working condition 1 as an example, the development and change of the maximum value of contact pressure of the waterstop with the increase in the deformation are plotted in Figure 16.

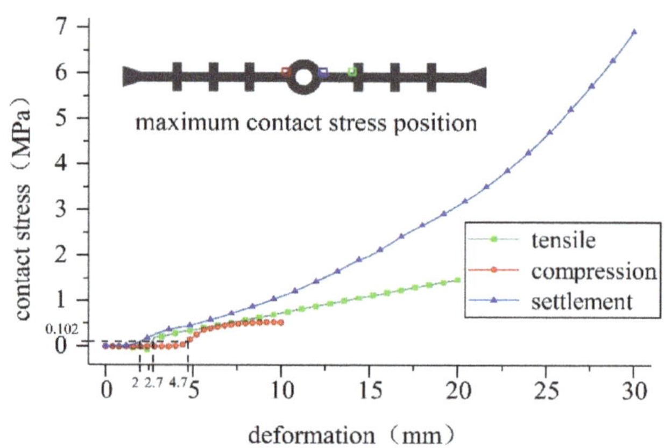

Figure 16. The maximum contact pressure of the waterstop in working condition 1.

Figure 16 shows that the overall contact pressure between the waterstop and the concrete increases with the increase in deformation. During tensile deformation, the connection between the first rib of the waterstop and the flank is the part with the highest contact pressure; during compression deformation, the contact position around the middle hole and the lining is the position with the highest contact pressure; when the settlement is deformed, the contact position between the flanks of the waterstop and the lining is the position with the highest contact pressure.

It can be seen from above that short-term resistance to seepage is offered by the deformation contact pressure. Figure 16 shows that when the tensile deformation is greater than 2.7 mm, the compression deformation is greater than 4.7 mm, and the settlement deformation is greater than 2 mm, the short-term resistance to water seepage is greater than the long-term resistance. Subsequently, with increasing deformation, the short-term resistance to water seepage also continues to increase.

The related results of working conditions 1 and 2 are shown in Table 4, where P^{st}_{wl} is the long-term seepage resistance, the critical deformation t_c is the deformation when the short-term seepage resistance is equal to the long-term seepage resistance, $Mises_{max}$ is the maximum von Mises stress on the waterstop, the allowable deformation t_a is the maximum deformation when the hose can meet the service life requirements, and P^{lt}_{wlmax} is the maximum short-term seepage resistance within the allowable deformation.

Table 4. Waterstop I waterproof capability parameters.

Conditions	P^{st}_{wl} (MPa)	t_c (mm)	$Mises_{max}$ (MPa)	t_a (mm)	P^{lt}_{wlmax} (MPa)
1-1		2.7	1.78	20.0	1.48
1-2	0.101	4.7	1.82	10.0	0.53
1-3		2.0	2.05	29.5	6.92
2-1		3.1	1.75	20.0	1.45
2-2	0.144	4.7	1.89	10.0	0.51
2-3		1.4	2.00	29.9	6.61

In summary, the following conclusions can be drawn:

(1) The stress on waterstop I at the maximum tensile and compressive deformation does not reach the dangerous level. After the settlement deformation reaches 29 mm, the stress exceeds the dangerous level, which may make it difficult for the tunnel joint to meet the service life requirements.

(2) When the deformation of the tunnel joint is greater than the critical deformation, the short-term resistance to water seepage is greater than the long-term resistance to water seepage. Therefore, if the waterstop can maintain a certain amount of deformation for a long time, it may be more conducive to the waterproofing of the tunnel joint.

(3) Except for the long-term water seepage resistance, the difference between working conditions 1 and 2 is small, which means that the initial contact pressure has little effect on the waterproofing ability of the waterstop after deformation.

4. Optimization Analysis of the Waterproof Performance of the Waterstop

According to the above analysis: (1) The deformation of the waterstop may lead to such a massive increase in stress that it becomes difficult for the tunnel joint to meet the service life requirements. (2) The deformation will cause local damage to the bond between the waterstop and the concrete, reducing its waterproof reliability. (3) The waterproofing of tunnel joints mainly depends on the long-term resistance of the waterstop to water seepage. Therefore, for the waterstop to be effectively waterproof for a long time, it is necessary to improve the stress and deformation state of the waterstop by ensuring good construction quality, and to improve the long-term resistance of the waterstop to water seepage.

4.1. Dimensional Optimization of the Waterstop

The deformation state of the waterstop can be improved by optimizing the size of the waterstop. At present, there are many specifications for buried waterstops used in tunnels. To compare the influence of the size of the middle hole of the waterstop and the thickness of the waterstop on its deformation, various inner diameters of the middle hole are used for waterstop I. At the same time, specifications and engineering cases are referred to in order to select two types of waterstops. As shown in Figures 17 and 18, waterstop II is thinner than waterstop I, the thickness of waterstop III is the same as that of waterstop II, and the size of the mesopore of waterstop III is larger than that of waterstops I and II. The plane strain finite element models are established.

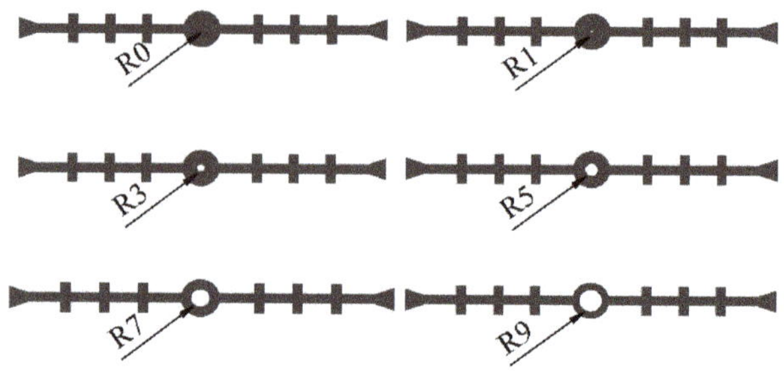

Figure 17. Waterstops I with different inner diameters of the middle hole (unit: mm).

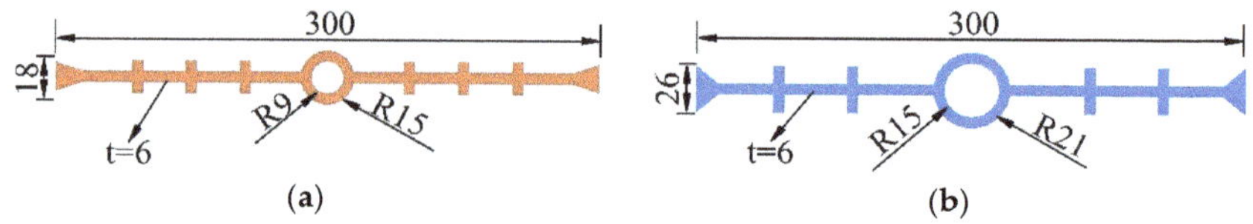

Figure 18. Dimensions of waterstops II and III (unit: mm). (**a**) waterstop II. (**b**) waterstop III.

As can be seen, the concrete has no plastic damage, so it can be treated as a rigid body, and the deformation applied is the same as above. The calculation results obtained are shown in Figures 19–21.

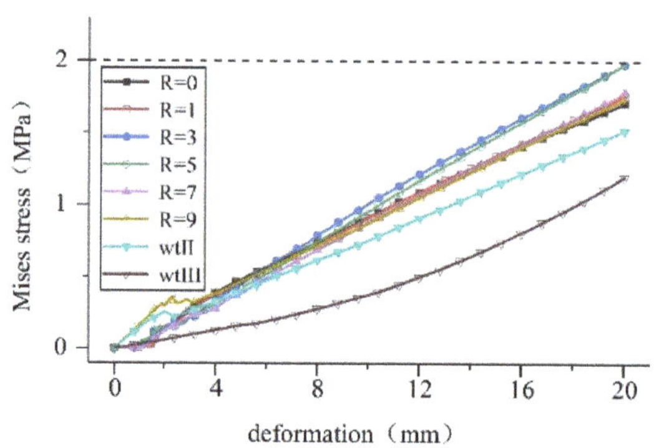

Figure 19. Tensile deformation stress on waterstops of various sizes.

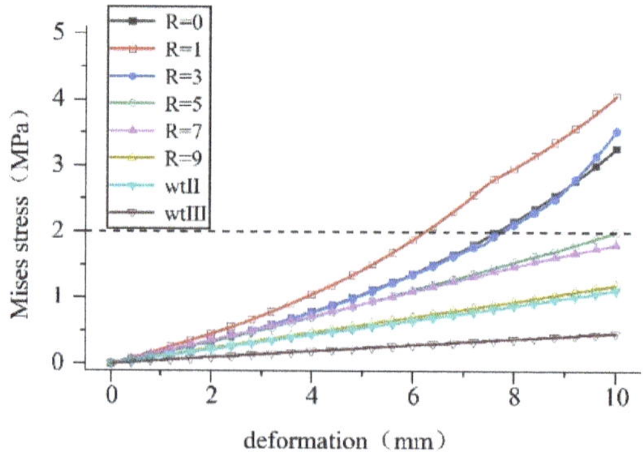

Figure 20. Compression deformation stress on waterstops of various size.

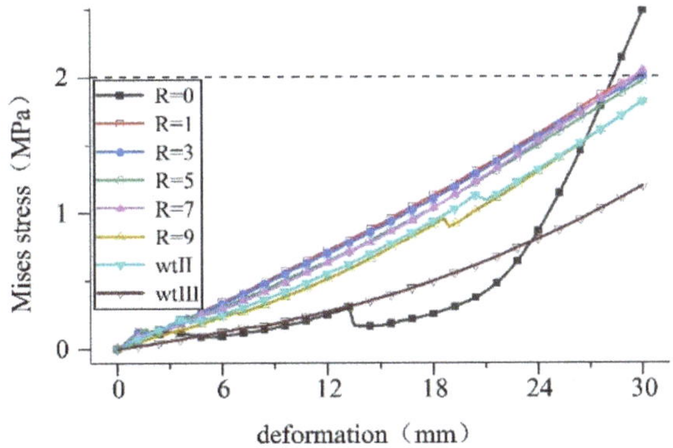

Figure 21. Settlement deformation stress on waterstops of various size.

Calculations show that: (1) When the middle hole has a specific outer diameter, increasing its inner diameter can improve the stress state during the deformation of the waterstop belt. The compression deformation is more obvious, and the maximum stress value is reduced by 71% when R = 9 compared to that when R = 1. (2) When the middle hole is a specific size, reducing the thickness of the waterstop belt can improve the stress state of the tensile deformation of the waterstop belt. At R = 9, the maximum stress value of waterstop II decreased by 13% compared with waterstop I. (3) Increasing the size of the hole in the waterstop belt can obviously improve the stress state during the deformation of the waterstop belt. Taking the settlement deformation as an example, the maximum stress value of waterstop III decreased by 34% compared with waterstop II.

A waterstop is made of rubber, and if it is too thin, it will increase the risk of damage such as tearing and puncturing. The tunnel deformation joint has as designated width. On the premise of satisfying these two conditions, the above analysis shows that stress deformation can be improved by reducing the thickness of the waterstop and increasing the inner diameter and size of the hole in the waterstop. This will also reduce the damage to the bond between the waterstop and the concrete due to deformation, and reduce the risk of waterstop failure caused by excessive deformation.

4.2. Optimal Analysis of the Adhesion between the Waterstop and Concrete

The waterproof capacity of the waterstop depends on its surface contact pressure and adhesive force. Because the contact pressure is difficult to change, the waterproof capacity can be optimized by increasing the adhesive force of the waterstop as follows: (1) the surface of the waterstop a double-sided adhesive layer can be affixed, (2) the surface of the waterstop can be coated with strong adhesive coatings, such as epoxy-quartz sand, and (3) by beating or setting grooves in the waterstop surface. There are also different ways to improve the adhesive force of the waterstop. To analyze the deformation state of the waterstop under different bonding forces, the finite element model was established by taking the waterstop II as an example. This can be accomplished by setting the bond force C = 1 between rubber and concrete obtained in the test, setting different bond forces, and applying the same deformation as above. The calculated results are shown in Figures 22–24.

The calculation results show that: (1) the size of the adhesive force does not affect the compression deformation stress on the waterstop, and (2) in the case of tensile and settlement deformation, the greater the bond force, the greater the deformation stress on the waterstop and when the bond force reaches a certain degree, the deformation stress exceeds the dangerous level.

To sum up, the bonding optimization of the waterstop and concrete should be limited to a certain range, combined with size optimization, to increase the long-term resistance of the waterstop to water seepage and meet the requirements of service life.

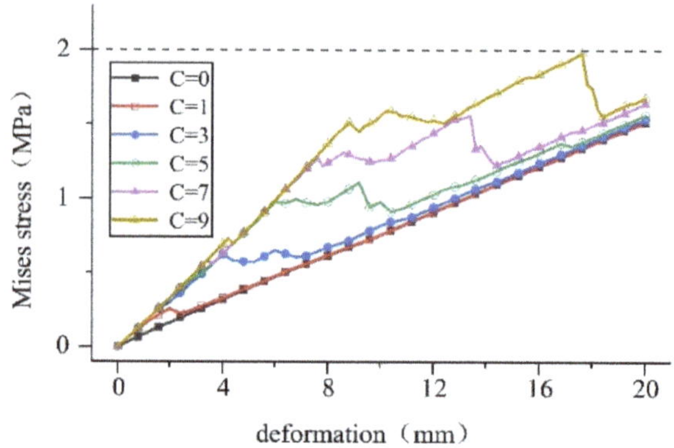

Figure 22. Tensile deformation stress on the waterstop under different bonding force conditions.

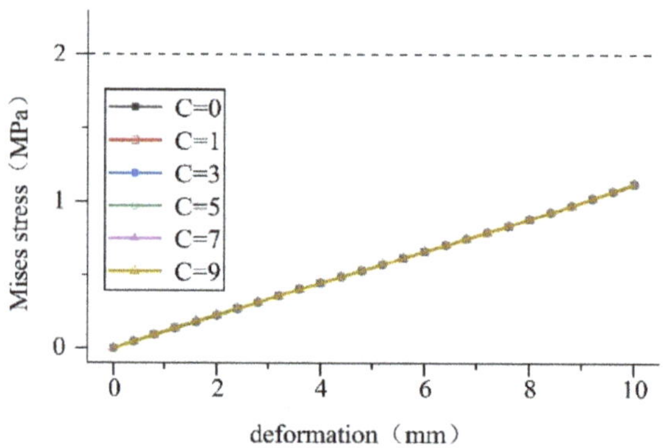

Figure 23. Compression deformation stress on the waterstop under different bonding force conditions.

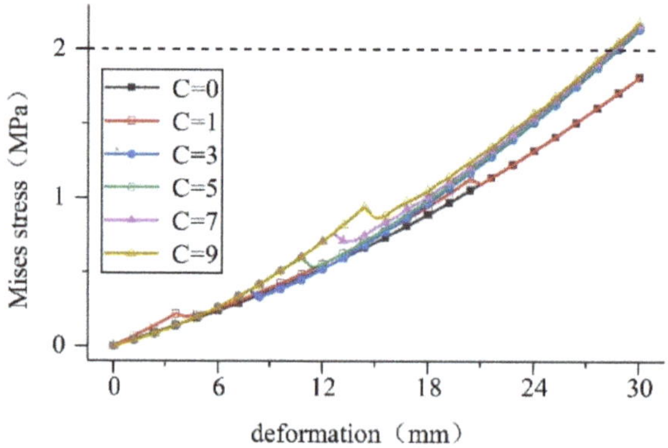

Figure 24. Settlement deformation stress on the waterstop under different bonding force conditions.

5. Conclusions

Given the current situation with frequent instances of water leakage in mountain tunnels, this paper summarizes the waterproofing mechanism of the middle-buried waterstop, establishes a finite element model considering the bonding and extrusion of the waterstop, and simulates the waterstop. Taking waterstops I, II, and III as examples, the influences of the size, specification, and deformation of the waterstop belt on the waterproof effect is analyzed, and optimization measures for the waterstop belt are put forward. The specific conclusions are as follows:

(1) The waterproofing ability of the buried waterstop mainly depends on its resistance to seepage, and the resistance involves contact pressure and bonding force between the waterstop and concrete. According to whether the tunnel joint is deformed or not, it can be classified as short-term or long-term seepage resistance.
(2) The deformation force of the waterstop is mainly concentrated between the first ribs on the left and right sides. The deformation does not completely destroy the bond between the waterstop belt and concrete, but only reduces the waterproofing reliability of the waterstop. Therefore, the waterproofing ability of the waterstop depends on its long-term resistance to water seepage.
(3) The stress on the waterstop may exceed the dangerous level when the deformation is large, making it difficult for the tunnel joints to meet the service life requirements. For example, the stress on the waterstop I in this paper will exceed the dangerous level after the settlement deformation reaches 29 mm.
(4) The long-term resistance of the waterstop to water seepage is the smallest at the vault, first increases continuously along with the lining profile and then remains unchanged. When the deformation of the tunnel joint is greater than the critical deformation, the short-term resistance to water seepage is greater than the long-term resistance to it.
(5) Reducing the thickness of the waterstop and increasing the size of the hole in the waterstop are beneficial to improving the deformation and stress state of the waterstop and to reducing the damage to the bond between the waterstop and the concrete. Increasing the adhesion between the waterstop and the concrete can significantly improve the long-term resistance of the waterstop to water seepage, but at the same time, the deformation stress on the waterstop will increase.

Author Contributions: Conceptualization, Y.W.; methodology, Y.W.; validation, H.W. (Haiping Wu); investigation, H.W. (Haoran Wu); resources, S.F.; data curation, D.C.; writing—original draft preparation, H.W. (Haiping Wu); writing—review and editing, S.F.; visualization, J.Z.; project administration, J.Z. and D.C. All authors have read and agreed to the published version of the manuscript.

Funding: This research was funded by the Department of Transport of Hubei Province Science and Technology Project, grant number 2020-186-2-5.

Institutional Review Board Statement: Not applicable.

Informed Consent Statement: Not applicable.

Data Availability Statement: All data are available from the author.

Conflicts of Interest: The authors declare no conflict of interest.

References

1. Muhammad, N.Z.; Keyvanfar, A.; Majid, M.Z.A.; Shafaghat, A.; Mirza, J. Waterproof Performance of Concrete: A Critical Review on Implemented Approaches. *Constr. Build. Mater.* **2015**, *101*, 80–90. [CrossRef]
2. Sañudo, R.; Miranda, M.; García, C.; García-Sanchez, D. Drainage in Railways. *Constr. Build. Mater.* **2019**, *210*, 391–412. [CrossRef]
3. Li, P.; Liu, H.; Zhao, Y.; Li, Z. A Bottom-to-up Drainage and Water Pressure Reduction System for Railway Tunnels. *Tunn. Undergr. Space Technol.* **2018**, *81*, 296–305. [CrossRef]
4. Hohmann, R. Fugenausbildung und -Abdichtung Bei Wasserundurchlässigen Bauwerken Aus Beton: Neue Regelwerke Setzen Maßstäbe. *Beton-Stahlbetonbau* **2004**, *99*, 938–948. [CrossRef]
5. Song, J.; Oh, K.; Kim, B.; Oh, S. Performance Evaluation of Waterproofing Membrane Systems Subject to the Concrete Joint Load Behavior of Below-Grade Concrete Structures. *Appl. Sci.* **2017**, *7*, 1147. [CrossRef]
6. Lv, K.; Ma, C.; Ji, Z.; Xu, P. Development of Butterfly Drainage Waterstop for Tunnel Lining Construction Joint. *Mod. Tunn. Technol.* **2011**, *48*, 23–26+37. [CrossRef]
7. Lv, K.; Ji, Z.; Ma, C. Tunnel Construction Joints Back Stick Type Waterstop Technology Several Improvements. *Mod. Tunn. Technol.* **2012**, *49*, 39–43. [CrossRef]

8. Liu, Q.; Tan, Z.; Ren, H.; Wang, X.; Xu, R. Study on New Water Stop of New Sea Tunnel. *China J. Highw. Transp.* **2016**, *29*, 116–123. [CrossRef]
9. Cho, B.H.; Nam, B.H.; Seo, S.; Kim, J.; An, J.; Youn, H. Waterproofing Performance of Waterstop with Adhesive Bonding Used at Joints of Underground Concrete Structures. *Constr. Build. Mater.* **2019**, *221*, 491–500. [CrossRef]
10. Cho, B.H.; Nam, B.H.; Seo, S.; Kim, J.; An, J.; Youn, H. Preliminary Study of Waterstop with Adhesive Bonding Technique for Diaphragm Wall Construction. In Proceedings of the Transportation Research Board 98th Annual Meeting, Washington, DC, USA, 13–17 January 2019.
11. Ma, W.; Guo, X.; Ma, C.; Guo, Y.; Jin, J.; Li, P. Structure Design and Performance Test of Self-Adhesive Waterstop. *China Railw.* **2015**, 30–33. [CrossRef]
12. Guo, X.; Ma, W.; Ma, C.; Guo, Y.; Jin, J.; Chen, S. Model Test Study on Comprehensive Drainage Effect of Self-Adhesive Waterstop. *Railw. Eng.* **2016**, 25–28.
13. Guo, X. Experimental Study on Anchorage Performance of Rubber Waterstop in Railway Tunnel. *Railw. Eng.* **2017**, *57*, 49–51.
14. Association of State Dam Safety Officials (Ed.) *Association of State Dam Safety Officials Annual Conference 2010, [ASDSO Annual Conference]: Seattle, WA, USA, 19–23 September 2010*; Curran: Red Hook, NY, USA, 2011; ISBN 978-1-61782-979-6.
15. Popchenko, S.N.; Stabnikov, N.V. Bituminous Surface Waterstops for Joints of Hydraulic Structures. *Gidrotekhnicheskoe Stroit.* **1973**, 15–18.
16. Laning, A. Flexible Waterstops. *Aberd. Concr. Constr.* **1993**, *38*.
17. Kudritz, J.; Adams, R.; Davis, C. Putting the STOP in Waterstops. In Proceedings of the 2017 ASDSO Annual Dam Safety Conference, San Antonio, TX, USA, 10–14 September 2017; Association of State Dam Safety Officials: San Antonio, TX, USA, 2017; pp. 493–501.
18. Wang, W.; Luo, Y.; Kong, L.; Zhang, Y.; Chen, Y. Numerical Simulation and Experimental Study on Mechanical Properties of GINA Waterstop. *Mod. Tunn. Technol.* **2021**, *58*, 237–243. [CrossRef]
19. Luo, Y.; Feng, Z.; Liu, J.; He, B.; Guo, H. Research on Structure and Performance of Internal Rubber Waterstop. *Railw. Eng.* **2013**, 93–96.
20. Liu, S.; Xu, Y.; Guo, L.; Wang, Y. Numerical Analysis of Corrugated Waterstop of High CFRDs. *Water Power* **2019**, *45*, 28–31.
21. Chen, E.; Chen, K.; Ma, C.; Xin, Y. Finite Element Analysis of Rubber Waterstop Belt under Stress State. *World Rubber Ind.* **2010**, *37*, 15–19.
22. Lin, P.; Zhao, Z.; Fan, B.; Han, S.; Liu, G. Analysis of the Deformation Capacity and Size Optimization of Rubber Waterstop in the Deformation Joint of Utility Tunnel. *Bull. Sci. Technol.* **2020**, *36*, 57–63. [CrossRef]
23. Meng, C.; Li, R.; Liu, J.; Xu, Y.; Li, J.; Hao, J. Research on Internally-Side Attached Rubber Waterstop Belt: Part I Pull-out Tests of Rubber Strip and Its Numerical Simulation. *Water Power* **2021**, *47*, 69–72+95.
24. Li, R.; Meng, C.; Zhou, J.; Xu, Y.; Li, J.; Hao, J. Research on Interally-Side Attached Rubber Waterstop Belt: Part II Numerical Simulation on Waterstop Belt and Its Model Test. *Water Power* **2021**, *47*, 42–45.
25. *TZ 331-2009*; Technical Guide for Construction of Railway Tunnel Drainage. China Railway Publishing Institute: Beijing, China, 2009.
26. Wang, T. *Control of Cracking in Engineering Structure*, 2nd ed.; China Building Industry Press: Beijing, China, 2017; ISBN 978-7-112-20690-2.
27. Pu, C.; Xia, C.; Li, Y.; Weng, X. Study of Thermo-Stress of Highway Tunnel and the Law of Line-f Issure Development Caused by the Thermo-Stress. *China J. Highw. Transp.* **2000**, 78–81. [CrossRef]
28. Xue, S. *Waterproof Technology of Underground Construction Engineering*; China Building Industry Press: Beijing, China, 2003; ISBN 7-112-05904-6.
29. Dong, L.; Jiang, Y.; Yang, Z.; Cheng, J.; Liu, C.; Zhang, J. Experimental Study and Water-Resistant Mechanism of Gaskets in Joints of Tunnel Segments. *Chin. J. Geotech. Eng.* **2017**, *39*, 469–474.
30. Gu, G. *Mechanical Analysis of Rubber Waterstop Belt and Plastic Waterstop Belt*; Chinese Society of Hydroelectric Engineering: Beijing, China, 2004; pp. 399–406.
31. Lee, J.; Fenves, G.L. Plastic-Damage Model for Cyclic Loading of Concrete Structures. *J. Eng. Mech.* **1998**, *124*, 892–900. [CrossRef]
32. *GB 50010-2010*; Code for Design of Concrete Structures. China Building Industry Press: Beijing, China, 2015.
33. Gent, A.N. (Ed.) *Engineering with Rubber: How to Design Rubber Components*, 3rd ed.; Hanser Publishers: Munich, Germany; Cincinnati, OH, USA, 2012; ISBN 978-3-446-42764-8.
34. Zou, Y.; He, C.; Zhou, Y.; Zhang, Z.; Fu, J. Statistics and Cause Analysis of Leakage Diseases in Operating Expressway Tunnels in Chongqing. *J. Highw. Transp. Res. Dev.* **2013**, *30*, 86–93+101.
35. Chen, J.; Liu, X.; Zhang, W.; Zhang, J. Investigation and Statistical Analysis of Water Leakage in Highway Tunnel in Guizhou Province. *Mod. Tunn. Technol.* **2011**, *48*, 7–11.

applied sciences

MDPI

Article

Seismic Response Analysis of Anchor Joint in Shield–Driven Tunnel Considering Soil–Structure Interaction

Gaole Zhang [1], Wenjun Zhang [1,2,*], Jianbing Qi [1], Rongjian Niu [1] and Chi Zhang [1]

1 School of Civil Engineering, Tianjin University, Tianjin 300350, China; zhanggaole100@tju.edu.cn (G.Z.); qi_jianbing123@tju.edu.cn (J.Q.); rigel07@tju.edu.cn (R.N.); carezhang@tju.edu.cn (C.Z.)
2 School of Civil Engineering and Architecture, Guangxi University, Nanning 530004, China
* Correspondence: wjzhang@tju.edu.cn

Abstract: The seismic behavior of the anchor joint in shield-driven tunnel is very difficult to determine with the conventional methods due to the extensive simplifications. This paper proposed an improved approach to investigate the seismic response of the anchor joint, considering both the soil-structure-interaction effect and the actual geometric features. Two three-dimensional numerical models were established, including the soil-tunnel system and the refined model of the anchor joint. A seismic analysis study was first conducted on the soil-tunnel model under different seismic input waves to obtain the responses of the joint opening and offset. Then, these results were imposed on the refined model of anchor joint to further examine its detailed performance under seismic excitations. The joint opening and offset under earthquake excitations from different directions were discussed. The distribution characteristics of the stress of the anchor joint were interpreted. Finally, safety evaluations on the anchor joint were executed based on the overall seismic responses. The results show that the maximum opening and offset of the anchor joint under the two-directional horizontal earthquake are greater than those under the unidirectional conditions, while different deformation trends are observed for the joints at distinct locations. The maximum opening of the anchor joint can reach 0.73 mm, whereas the peak offset is only 0.35 mm. The local plastic strain of the anchor joint increases under the seismic action, but all of the joints are still kept in the safe state under the most unfavorable conditions. The developed method in this paper can also be accessed by the seismic study on other types of joints with complex structural components in shield tunnels.

Keywords: shield tunnel; anchor joint; seismic response; numerical analysis; soil–structure interaction

Citation: Zhang, G.; Zhang, W.; Qi, J.; Niu, R.; Zhang, C. Seismic Response Analysis of Anchor Joint in Shield–Driven Tunnel Considering Soil–Structure Interaction. *Appl. Sci.* **2022**, *12*, 6362. https://doi.org/10.3390/app12136362

Academic Editor: Marco Vona

Received: 1 June 2022
Accepted: 20 June 2022
Published: 22 June 2022

1. Introduction

Shield-driven tunnel has been widely used for the construction of urban underground transportation in the past few decades, and a great number of studies have been conducted to obtain a better knowledge of the mechanical and waterproof performance of segmental lining structures under the static and seismic action [1–5]. According to the published investigations [6,7], tunnels are vulnerable to damage under earthquake motions, especially for the joint area, which has been repeatedly proven to be the weak component in the whole tunnel. It should be noted that if the damage near the joint develops into obvious cracks, more secondary disasters may happen, including water leakage, sand gushing, and even collapse. Thus, it is of great importance to study the seismic response of the segmental joint in shield tunnels.

Generally, the seismic loading should be regarded as a key control factor during the design of a shield tunnel in earthquake-prone areas [8]. To capture the seismic performance of the shield tunnel, quite a few methods have been proposed and adopted. The free-field deformation approach is commonly used in earlier research [9], which neglects the soil-structure interaction (SSI). Using this method, the parameters for executing calculations are easy to gain, whereas inaccurate results may be obtained due to its excessive simplification.

It is well known that the SSI effect has a significant influence on the static and dynamic response of the structures [10–12]. Thus, the SSI effect should be properly taken into account to obtain satisfactory results. In this concern, Brachman et al. [13] conducted full-scale tests to study the effectiveness of conventional orthotropic shell theory in soil-structure interaction analyses. Abdel-Sayed and Salib [14] investigated the possible soil failures caused by centric live loads based on a finite element method considering the SSI effect. Yeau et al. [15] discussed the parameters which may have a great influence on the performance of the in-service culverts, including cover depth, size, and load application. Kasper and Meschke [16] examined the SSI effect on the shield tunneling process from the aspects of grout material properties and soil cover depth. Maleska and Beben [17] established an improved numerical model to explore the impact of soil cover depth on the behavior of soil-steel composite bridges during an earthquake. Maleska et al. [18] carried out a response analysis to reveal the behavior of the soil-steel composite tunnels under seismic excitations. Kuribayashi et al. [19] set up a beam-spring system to determine the longitudinal seismic behavior of the shield tunnel, in which the tunnel can be simulated as the beam, while the SSI effect can be considered through adding springs in different directions. Given that the beam element cannot totally represent the three–dimensional (3D) structural characteristics of the tunnel, more advanced methods were further presented through replacing the beam with cylindrical shells [20–22]. The analytical methods mentioned above provide valuable insights into the seismic response of shield tunnels, while it is insufficient to fully reveal the seismic characteristics of the segmental linings and joints owing to the inevitable extreme assumptions during derivation of the equations. Conducting experiments (e.g., shaking table tests and centrifuge tests) is another effective way [23,24], in which the tunnel model can be manufactured with more local details such as the bolts, gaskets, longitudinal and circumferential joints. Wang et al. [25] carried out shaking table tests to identify the consecutive damage of the tunnel linings, and the test phenomena were compared with those observed in field exploration. Wang et al. [26] examined the dynamic properties of two shallow-buried tunnels. From their findings, there are great differences in the responses of displacement and acceleration of the two different tunnels. Additionally, the numerical method has also gained wide adoption because of its advantages of cost-effectiveness, and flexible behavior in modeling the refined or simplified structures [27–29].

As the shield tunnel is a prefabricated structure consisting of numerous segments connected by joints, its mechanical behavior is highly different from those constructed with in-situ casting. There are many available publications about the weakened effect of the joint under static action [30–32], while that under seismic action is in need of further study. For this purpose, Zhang et al. [33] established a 3D finite element model (FEM) to analyze the dynamic characteristics of different types of bolt joints in shield tunnels. Liu et al. [34] studied the strength degradation, hysteresis behavior, energy dissipation, and the failure modes of the joints between prefabricated segments and cast-in-site structures under conditions of several earthquake excitations. Yang et al. [35] checked the distribution features of the joint opening on the basis of a FEM under multidirectional input waves to grasp the impact of seismic loading. Yan et al. [36] obtained the contact pressure, opening, and offset of the joint in a shield tunnel subjected to seismic loading so that the waterproof behavior of the joint can be considered under such conditions.

Overall, the research mentioned above has made great contributions to the comprehensive understanding of the seismic response of the segmental joint. However, the existing studies only cover the conventional bolt joints, while there are few reports on the seismic performance of the newly developed joints with a more complicated structural system, such as the anchor joint, FRP-Key joint, and CT joint [37–39]. These new joints have the characteristics of great strength, high precision, and quick connection [40], which will be the main trend of shield tunneling. Therefore, it is critical to conduct seismic response analysis of these new types of joints. Nevertheless, there are great challenges to directly reflecting the complex geometric features of the new joints during the dynamic numerical simulation, which may lead to too much increase in degrees of freedom to obtain computational convergence. On the other hand, neglecting the local details of the new joints may result in an unexpected negative influence on the final seismic response. Consequently, it is essential to propose more advanced methods for the seismic analysis of the new joints so that a good balance can be reached between the reliability and implementation difficulty.

This paper aims at investigating the seismic response of the anchor joint in shield tunnels considering soil-structure interaction. An improved method was first introduced to achieve this target, in which a 3D finite element model for the soil-tunnel system should be established to obtain the overall performance response under excitations of different input waves. The circumferential anchor joint was simulated using the multidirectional springs, while the longitudinal bolt joint was replaced by the beam. Based on the joint opening and offset obtained from the integral model, a refined model of the anchor joint was then set up to further examine its mechanical behavior under the seismic action. The responses of deformation and stress were finally discussed to interpret the safety status of the longitudinal and circumferential joints. The method developed in this paper can also be accessed to seismic response analysis of other types of new joints with complicated structures in shield tunnels.

2. Project Overview

A metro line has a total length of approximately 11 km, in which the shield tunneling is adopted to construct the underground section. The design working life of the metro tunnel is 100 years, while the corresponding seismic precautionary intensity is 8-degree. There are nine stations in total. According to the design data, the shield tunnel is approximately 6.7 km in length, and has a maximum buried depth of 28 m from the ground to the roof of the tunnel. Soft soil is widely distributed along the tunnel, while sandy clay, silt, and silty sand can also be found. During the construction, the shield tunnel successively passes through silt sand, sandy clay, muddy clay, fine-silty sand, and clay, whose depths are listed in Section 4.1.2. The ground elevation is 2 m–3 m. The water level is 2 m below the ground surface, implying that the maximum water table can reach 26 m. The metro tunnel is constructed as a new transport line in the suburbs, and there are no other projects in the near vicinity.

The outer diameter, inner diameter, thickness, and width of the segmental lining are 6.6 m, 5.9 m, 0.35 m, and 1.5 m, respectively. Additionally, the universal wedge segment is selected for the assembly of the segmental lining. Each ring has six segments in total, including one K block, two adjacent blocks, and three standard blocks. Different types of joints are adopted for the connections in longitudinal and circumferential directions. M30 bending bolts are selected to connect the adjacent two single segments in the longitudinal joints, while the anchor joint is used for the circumferential joint to make the assembly construction more efficient. There are sixteen anchor joints and twelve bending bolts needed in each ring. The overall view of the cross-section, 3D view, and joints are illustrated in Figures 1–3, respectively.

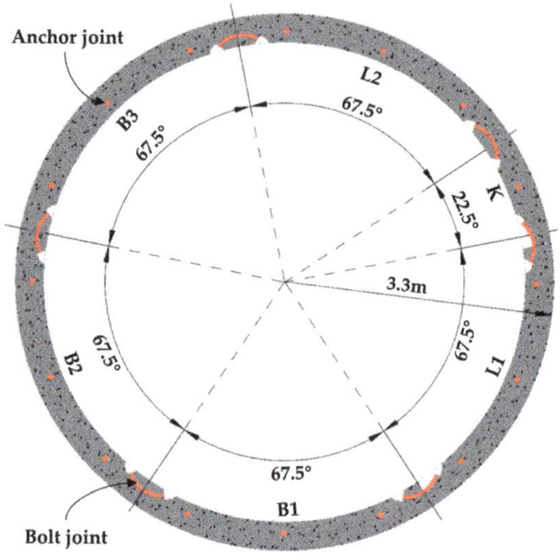

Figure 1. Cross section of the shield tunnel.

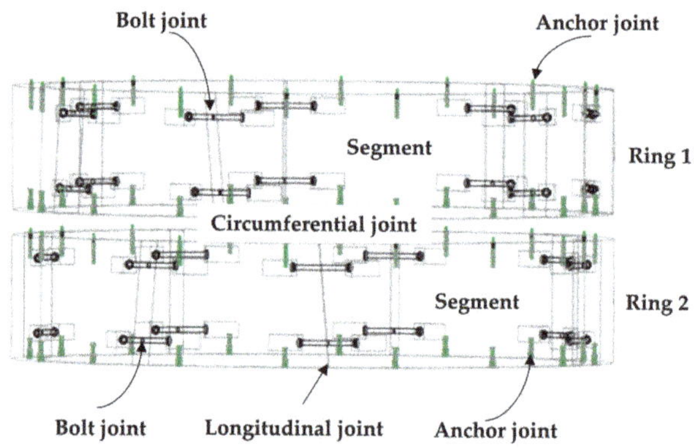

Figure 2. 3D perspective scene of the segmental lining and joints.

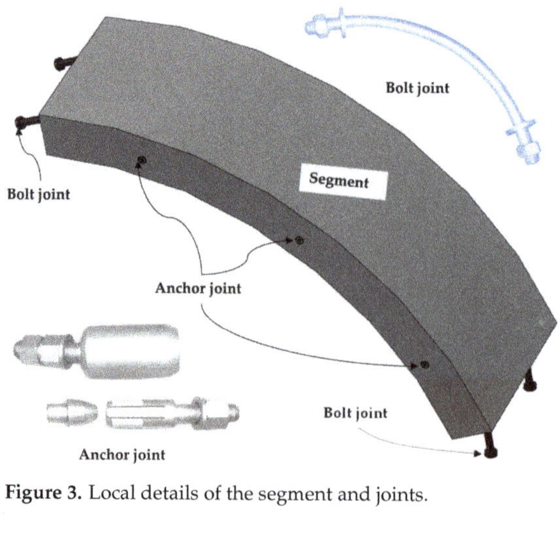

Figure 3. Local details of the segment and joints.

3. Improved Method to Investigate Seismic Response of Anchor Joint

An improved method is proposed and introduced in this section to conduct the seismic response of anchor joint in shield tunnel considering both the SSI effect and the complicated structural features. Using this method, the whole procedure of the seismic analysis is actually divided into two parts. Firstly, a 3D numerical model should be established for the soil-tunnel system, as shown in Figure 4. This soil-tunnel system contains three-dimensional soil and tunnel models as well as joints. Specifically, the anchor joint and bolt joint are simplified as the springs and beam elements. Therefore, both the soil and structure have been included in the established model for considering the SSI effect. In this way, the SSI effect can be taken into account in this step, and a relatively satisfactory result can be obtained with the computation cost in an acceptable range.

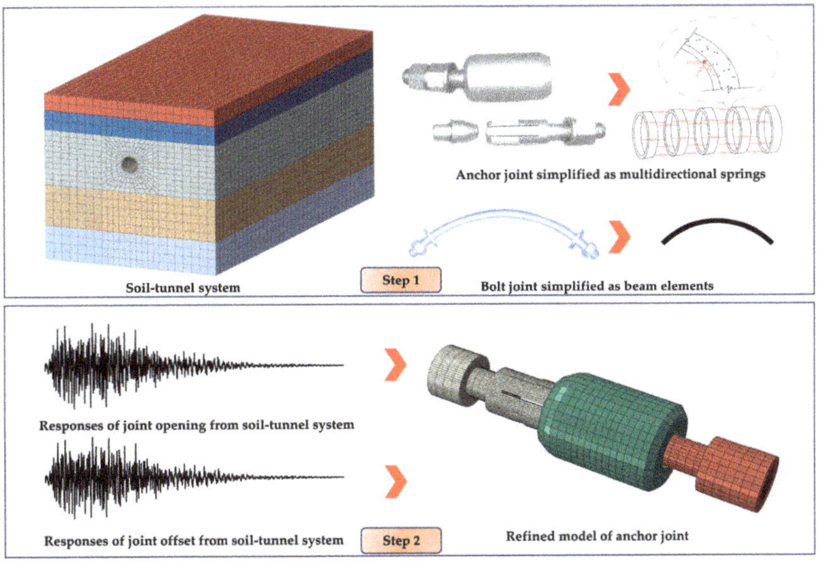

Figure 4. Procedure of the improved method.

Then, the joint opening and offset results can be gained from the integral model mentioned above. After that, a refined numerical model will be set up for the anchor joint, which can totally reproduce its structural characteristics. Combing the refined model of the anchor joint with the obtained joint opening and offset results from the integral model, a more detailed exploration can be conducted on the anchor joint.

4. Numerical Model and Materials

4.1. Soil-Tunnel System

4.1.1. Soil-Tunnel Model

A 3D numerical model was firstly established for the soil-tunnel system to include the dynamic SSI effect under seismic excitations in ABAQUS. The length, depth, and width of the model are 120 m, 60 m, and 100 m, respectively. The soil is regarded as the multilayered media, while the tunnel is composed of equivalent areas without joints and key areas with more details. The numerical model and the corresponding mesh results are shown in Figure 5. In this model, the solid C3D8 element is selected for the segment and soil, while the 3D beam element B31 is used to simulate the bending bolt joint. Shear and normal springs are utilized to replace the connection effect of the anchor joint at the same locations with those in practice. The density of mesh for the whole soil-tunnel system model was determined based on [33].

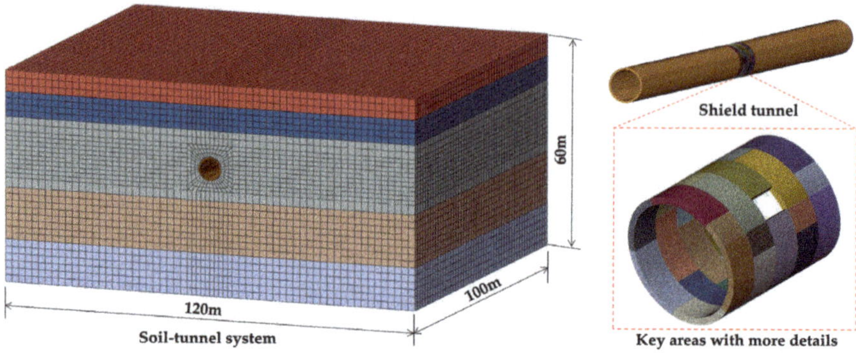

Figure 5. 3D numerical model of the soil-tunnel system and local details.

It should be noted that there are five segmental lining rings in the key areas in total, which are connected by bolt joints and anchor joints in the longitudinal and circumferential directions, respectively. In addition, the reduction effect of the segmental joint on the structural stiffness should be considered for the equivalent areas of the tunnel model. In this regard, the equivalent parameter η is introduced to represent the reduction of the flexural rigidity for the lining structures. In this paper, the η is set to be 0.077 on the basis of the published data from [33].

4.1.2. Material Properties

During the calculations, the concrete damage plasticity (CDP) model is adopted to describe the irreparable damage of the concrete under the seismic action. Based on the CDP model, the damage factor ranges from 0 to 1, representing no damage and complete failure of concrete, respectively. As C60 high performance concrete is used for the segmental lining, the corresponding elasticity modulus, Poisson's ratio, shear expansion angle, flow potential offset value, density, cohesion coefficient, and eccentricity are 3.55×10^4 MPa, 0.2, 0.67, 2450 kg/m^3, 35°, 0.001, and 0.1, respectively. Additionally, the ratio of biaxial ultimate compressive strength to uniaxial ultimate compressive strength is set to be 1.16. All the above parameters were determined as Wang et al. reported [41]. More detailed governing parameters of the CDP model for C60 concrete are listed in Table 1. The damage parameters of the concrete can be obtained by the following equation.

$$d_{\mathrm{k}} = \frac{(1-\beta)\varepsilon_{\mathrm{k}}^{\mathrm{in}} E_0}{\sigma_{\mathrm{k}} + (1-\beta)\varepsilon_{\mathrm{k}}^{\mathrm{in}} E_0}, (\mathrm{k} = \mathrm{t}, \mathrm{c}) \tag{1}$$

in which, d_{k} denotes the damage parameter, while k = t means tension and t = c means compression; E_0 is the initial elasticity modulus of the concrete; $\varepsilon_{\mathrm{k}}^{\mathrm{in}}$ is the strain in the inelastic stage; σ_{k} is the stress; β is the proportional coefficient of plastic strain to inelastic strain.

On the other hand, the elastoplastic model is employed to describe the plastic behavior of the bending bolt joint. For the bending bolt with a strength grade of 8.8, the corresponding elastic modulus, yield stress, and Poisson's ratio are set to be 210 GPa, 640 MPa, and 0.25, respectively [40]. While for the anchor joint, multidirectional springs with bilinear behavior are used, including the tensile stiffness and shear stiffness. This assumption was used in this paper based on the reports published in [42]. The initial tensile stiffness of the anchor joint is set to be 350 MN/m, while that will be decreased to 0 if the joint opening exceeds 1.5 mm. Similarly, the initial shear stiffness of the anchor joint is 80 MN/m, while that will be increased to 230 MN/m if the joint offset exceeds 2 mm.

Table 1. Governing parameters of the CDP model for C60 concrete.

Uniaxial Compressive Stress/(MPa)	Uniaxial Inelastic Compressive Strain	Compressive Damage d_c	Uniaxial Tensile Stress/(MPa)	Uniaxial Cracking Tensile Strain	Tensile Damage d_t
7.239	0	0	2.720	0	0
22.122	6.77×10^{-5}	0.032	2.216	8.28×10^{-5}	0.293
32.149	2.25×10^{-4}	0.112	1.682	1.33×10^{-4}	0.447
34.258	7.68×10^{-4}	0.241	1.426	1.72×10^{-4}	0.579
29.164	1.23×10^{-3}	0.447	1.218	2.26×10^{-4}	0.652
23.297	2.51×10^{-3}	0.622	0.982	2.58×10^{-4}	0.742
19.332	3.56×10^{-3}	0.741	0.872	2.93×10^{-4}	0.791
14.265	4.12×10^{-3}	0.823	0.783	3.24×10^{-4}	0.824
11.173	4.83×10^{-3}	0.872	0.691	3.63×10^{-4}	0.857
10.293	5.63×10^{-3}	0.894	0.635	3.94×10^{-4}	0.872
8.973	6.58×10^{-3}	0.928	0.593	4.32×10^{-4}	0.894
7.254	7.24×10^{-3}	0.933	0.551	4.69×10^{-4}	0.912
4.982	8.73×10^{-3}	0.973	0.514	4.92×10^{-4}	0.939

During the calculations of wave propagation in soft soil and rock, it is important to capture the soil nonlinearity. Consequently, the equivalent linear approach is selected to define the approximate nonlinear behavior of soil through a continuous update of the shear modulus and damping ratio according to the calculated soil strain, which has been widely used in this field. The maximum shear modulus of the soil can be calculated from the following equation. More detailed geological properties of different soil layers can be found in Table 2.

$$G_{\max} = \rho V_s^2 \tag{2}$$

where, $G_{\max}$ is the maximum shear modulus; ρ is the density; V_s is the shear wave velocity.

Table 2. Properties of the soil layers.

Layer	Density/ (kg/m^3)	Shear Wave Velocity/(m/s)	Poisson's Ratio	Friction Angle/(°)	Cohesion/ (kPa)
Slit	1720	180	0.45	36	10
Clay	1780	220	0.32	32	18
Silty clay	1840	270	0.38	30	26
Sand	1950	340	0.25	40	2
Bedrock	1920	600	0.20	-	-

4.1.3. Contact Relationships

As there are lots of different components in the soil-tunnel system, it is of significance to select the correct contact relationships to describe the interaction between them. The surface-to-surface contact in ABAQUS is employed to simulate the interaction between the soil and the segmental lining structures, while the bolt joint is embedded in the segmental linings. Moreover, the hard contact is set for the normal behavior of the surface-to-surface contact, and the penalty friction approach is used with finite sliding formulation in the tangential direction. The friction coefficient between the soil and segments is 0.5, while that between different segments is set to be 0.55.

4.1.4. Boundary Conditions

Since the numerical model of the soil-tunnel system only has a limited range of soil instead of the practical infinite space, it is essential to include an artificial boundary to make sure that the scattering waves can be eliminated on the boundary. In this paper, the viscous-spring boundary proposed by Gu et al. [43] is adopted, whose efficiency has been greatly verified in many publications [44,45]. When using the viscous-spring boundary, a great number of springs and dampers should be added on the boundary nodes. For each boundary node, the viscous-spring boundary is shown as Figure 6.

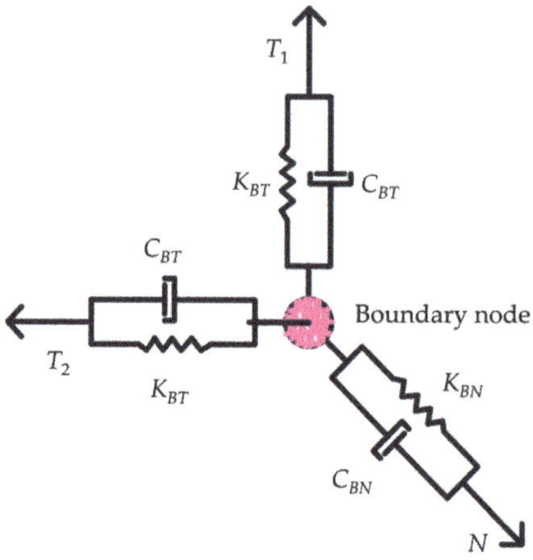

Figure 6. Viscous-spring boundary.

The coefficients of the springs and dampers can be obtained from the following equations:

$$K_{BT} = \alpha_T G/r,\ K_{BN} = \alpha_N G/r \quad (3)$$

$$C_{BT} = \rho c_s,\ C_{BN} = \rho c_p \quad (4)$$

where, K_{BT} and K_{BN} are the spring coefficients in the two different directions, respectively; C_{BT} and C_{BN} are the damping coefficients in the two different directions, respectively; α_T and α_N are modified coefficients for the spring coefficients; r is the distance from calculated nodes to the input wave; G is the soil shear modulus; c_s is the velocity of shear wave, while c_p is the velocity of the P wave.

4.1.5. Earthquake Records

Two seismic waves are selected in this paper to conduct dynamic analysis on soil-tunnel system, defined as wave 1 and wave 2, respectively. The acceleration time histories of the two used seismic waves are displayed in Figure 7. According to the design requirements of the shield tunnel, the amplitude of 0.15 g is set for the two seismic waves with the exceedance probability of 10% in 50 years.

Additionally, as the tunnel has a much greater length than its diameter, the seismic behavior of the tunnel may show obvious difference when subjected to the earthquake from different directions. To take into account the possible influence caused by the incident direction of the seismic waves, three different cases are included, including the earthquake motion along the axial direction of tunnel, along the transverse direction of tunnel, and the two-directional horizontal earthquake. For the two-directional horizontal earthquake conditions, the ratio of transverse acceleration amplitude to axial acceleration amplitude is set to be 1:0.85. Hence, there are six cases in total designed for the seismic responses analysis, which is listed in Table 3.

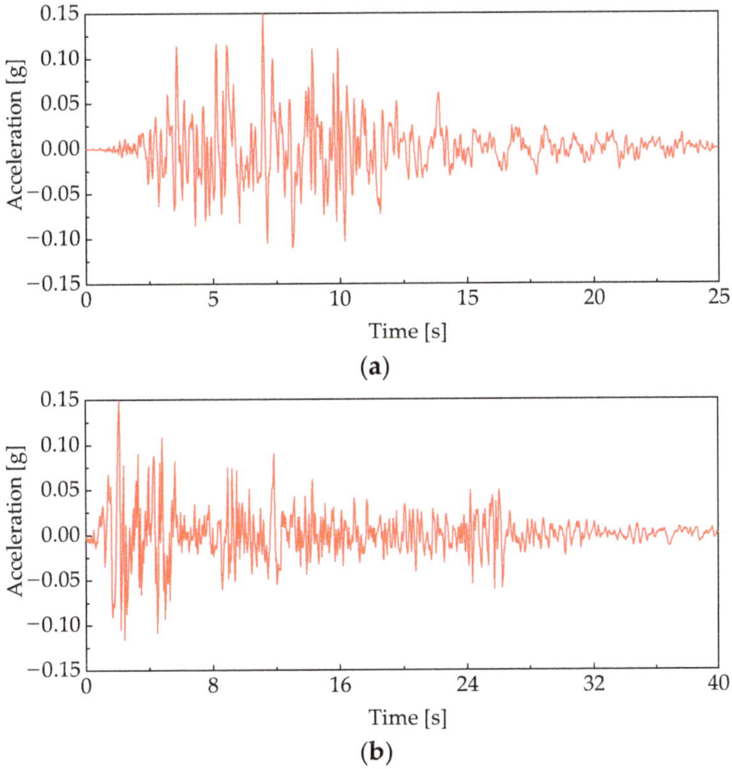

Figure 7. Acceleration time histories of seismic waves. (**a**) Wave 1. (**b**) Wave 2.

Table 3. Cases for seismic response analysis.

Case No.	Incident Direction	Transverse Direction	Peak Acceleration	Axial Direction	Peak Acceleration
1	Transverse	Wave 1	0.15 g	-	-
2	Axial	-	-	Wave 1	0.15 g
3	Two-directional	Wave 1	0.15 g	Wave 1	0.1275 g
4	Transverse	Wave 2	0.15 g	-	-
5	Axial	-	-	Wave 2	0.15 g
6	Two-directional	Wave 2	0.15 g	Wave 2	0.1275 g

4.2. Refined Model of Anchor Joint

4.2.1. Refined Numerical Model of Anchor Joint

Figure 8 shows the 3D refined numerical model and mesh of anchor joint, and two segments are also included in this model. Each segment has a width of 0.75 m, a center angle of 22.5°, and the same thickness with that in the soil-tunnel system. The detailed dimensions of the anchor joint can be found in [40]. The solid C3D8 element is employed to simulate both the anchor joint and segment, and finer mesh is used for the anchor joint as well as the adjacent contact areas for reliable accuracy.

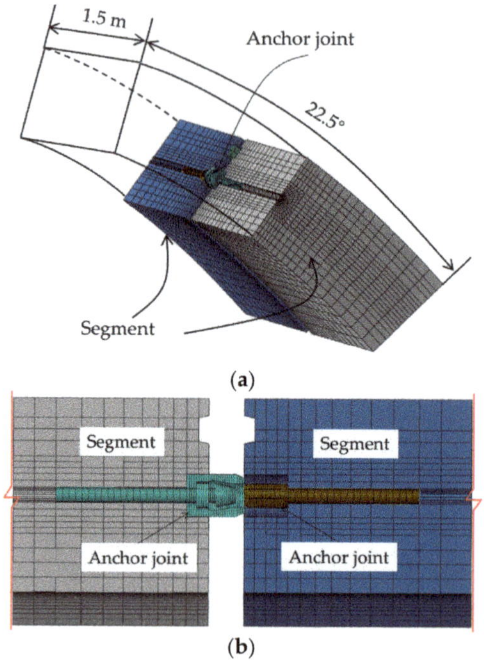

Figure 8. Refined numerical model of anchor joint. (**a**) 3D view. (**b**) Plan view.

From Figure 8, it can be seen that the anchor joint herein consists of three components, i.e., rod, lantern ring, and sleeve. During the construction of the shield tunnel with the anchor joint, the circumferential joint can be directly connected without extra manual operation due to the automatic self-join of the anchor joint. Under the pressure of jack, the rod will be inserted into the sleeve, while the six sleeve walls will further deform to fill the gap between the rod and lantern ring. Based on the above linkage mechanism, the assembly of anchor joint will be completed, which provides high tensile strength.

4.2.2. Material Properties

The CDP constitutive model is used for the two segments, and the corresponding parameters are totally identical to those listed in Section 4.1.2. The elastoplastic constitutive model is adopted for all of the components in the anchor joint. The elasticity modulus of the anchor joint is 181 GPa, while the yield stress is 530 MPa. The Poisson's ratio is 0.26.

4.2.3. Contact Relationships

Given that the anchor joint is actually embedded during the fabrication of the concrete segment in practice, the anchorage zone of the joint is tied with its adjacent concrete, whereas the surface-to-surface contact formulation is employed for the other integrations. The penalty coefficient of interface between anchor joint and concrete is set to be 0.25, while that between different components of the anchor joint is 0.5 to reproduce the fastening effect according to [40].

4.2.4. Boundary and Loading Conditions

Since the refined model is loaded using the obtained joint deformation from the soil-tunnel system, an appropriate boundary condition should be used to simulate the same situation with that in the integral model. For this purpose, the following boundary condition is introduced as shown in Figure 9. That is, the three surfaces of the right segment are set as the constrained boundary, while no constraints exist for the other components.

With such boundary conditions, the loading procedure of the refined model can be divided into two steps. In step 1, the assembly of anchor joint is completed by pushing the left concrete. After that, the joint offset and opening can be loaded on the anchor joint by applying the obtained deformation.

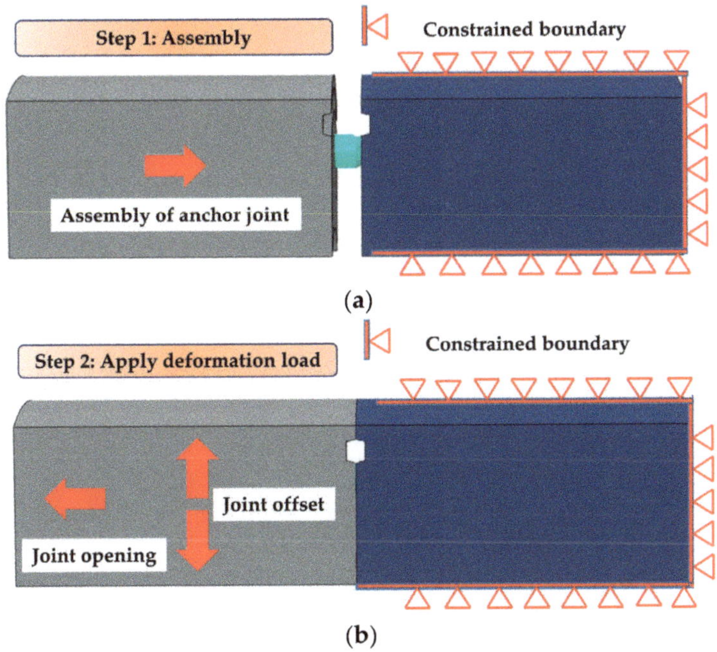

Figure 9. Boundary and loading conditions. (**a**) Step 1. (**b**) Step 2.

4.3. Model Validation and Convergence Study

In this section, to validate the effectiveness and reliability of the refined numerical model of the anchor joint, a comparison study is conducted on the mechanical behavior of the anchor joint during the assembly and tension deformation. Specifically, the results obtained from the model shown in Figure 8 are calibrated against those recorded in the tests conducted by Shirato et al. [38]. In their tests, a single anchor joint was used, and the reinforced concrete segment was simulated by the metal bases. During the tests, the anchor joint was first assembled. After that, the tensile load was imposed on the assembled anchor joint to investigate its tensile performance. For the verification analysis, the anchor joint is first in a separate state as shown in Figure 9a. Then, the anchor joint will be assembled with a thrust action, illustrated in Figure 9b, which are totally the same with those in tests. In the process mentioned above, a displacement-force curve can be drawn. Similarly, another relationship between the displacement and force can be gained during the tension deformation.

Meanwhile, a convergence study is also necessary to determine the reasonable mesh size according to the corresponding accuracy of the models with different mesh results, so that a good balance can be reached between the computational cost and precision. Looking at this target, three models with small mesh size, medium mesh size, and large mesh size are established for comparison. The experimental and numerical results during the assembly and tension deformation of the anchor joint are shown in Figure 10.

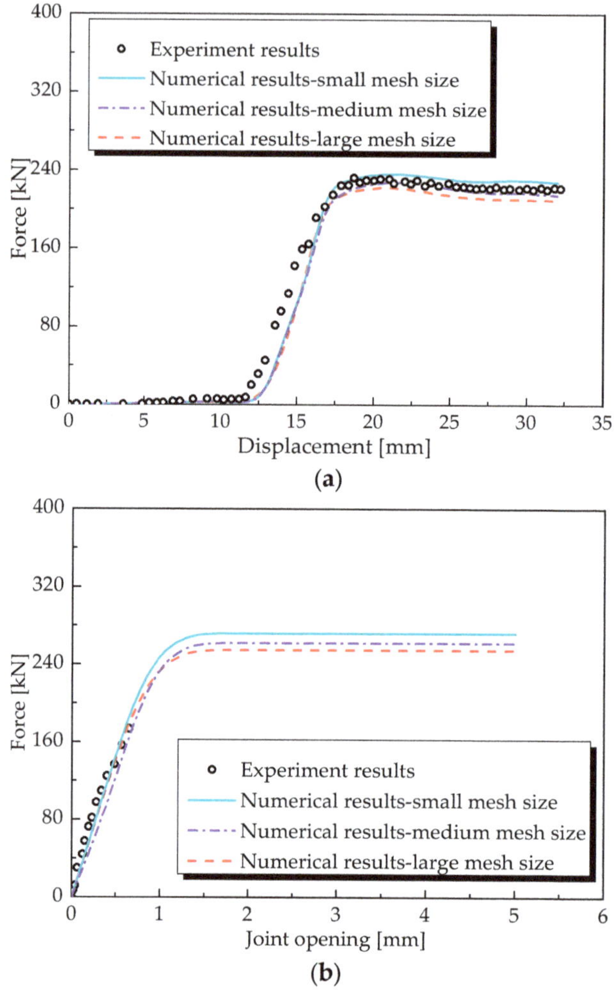

Figure 10. Comparison between experimental and numerical solutions. (**a**) Displacement-force curve during assembly process. (**b**) Joint opening-force curve during tension deformation process.

During the convergence study, the small mesh size has a basic dimension of 1 mm for the rot and sleeve, while that for the lantern ring is 3 mm. The medium mesh size has a basic dimension of 3 mm for the rot and sleeve, while that for the lantern ring is 5 mm. For the large mesh size, the basic dimensions of the rot, sleeve, and lantern ring are 5 mm, 5 mm, and 8 mm. As the concrete segment is not the main component in the model, its mesh size remains unchanged during the convergence study. From Figure 10, it can be found that the numerical results agree well with the experiment results in both the assembly and tension deformation process, whichever mesh size is adopted. The comparison between experimental and numerical solutions provides evidence that the numerical model has the satisfactory reliability. Thus, all of the three examined mesh sizes can be used for further study of the mechanical behavior of the anchor joint. Since using the medium mesh size shows the best computational efficiency according to the calculation process, the medium mesh size is finally employed in the subsequent analysis.

5. Seismic Response of Anchor Joint

5.1. Joint Opening and Offset

The deformation of four circumferential joints in the key areas of tunnel model are shown in Figures 11 and 12. When the seismic wave is input from the transverse directions, the maximum joint opening is found at tunnel arch waist, while the minimum value is at the arch bottom. For the axial seismic conditions, the maximum joint opening is found at arch bottom or the arch vault of the tunnel, while the minimum value is at the arch waist. Besides, it can be seen that these four circumferential joints show similar joint opening distribution behavior, although the detailed values are slightly different. However, this is not true for the two-directional conditions, in which obviously distinct joint deformation can be observed for different joints.

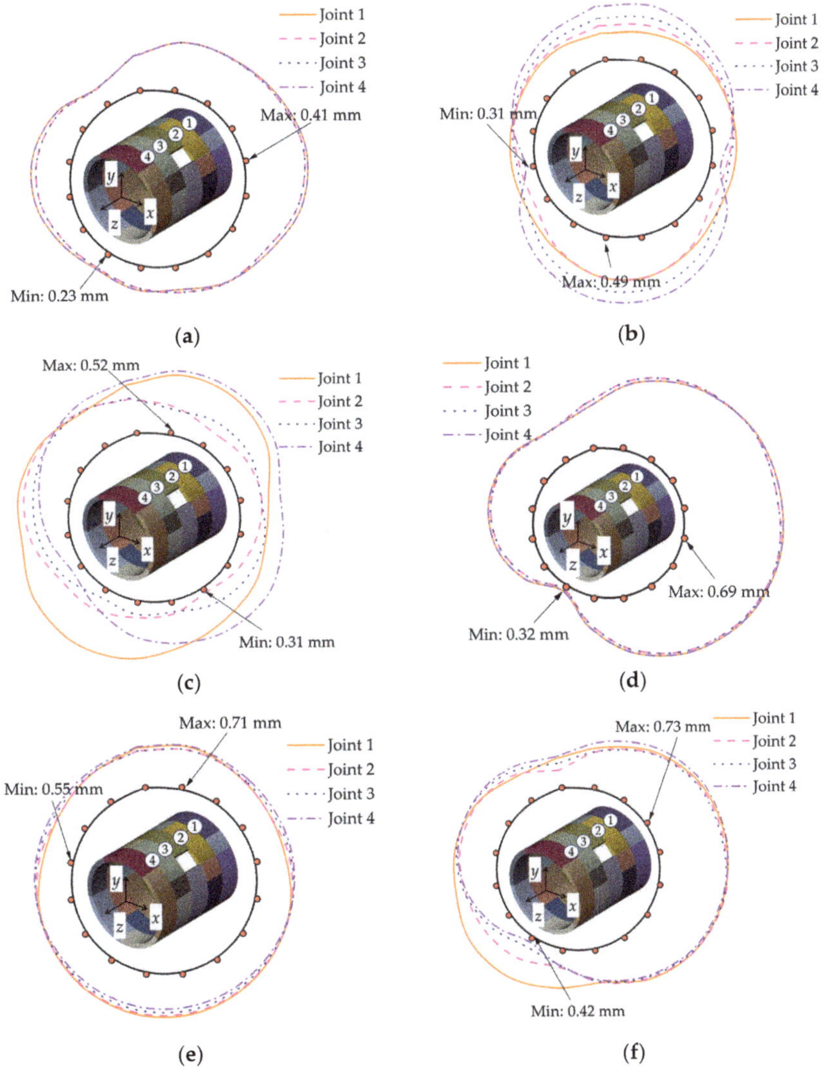

Figure 11. Maximum circumferential joint opening. (**a**) Case1. (**b**) Case2. (**c**) Case3. (**d**) Case4. (**e**) Case5. (**f**) Case6.

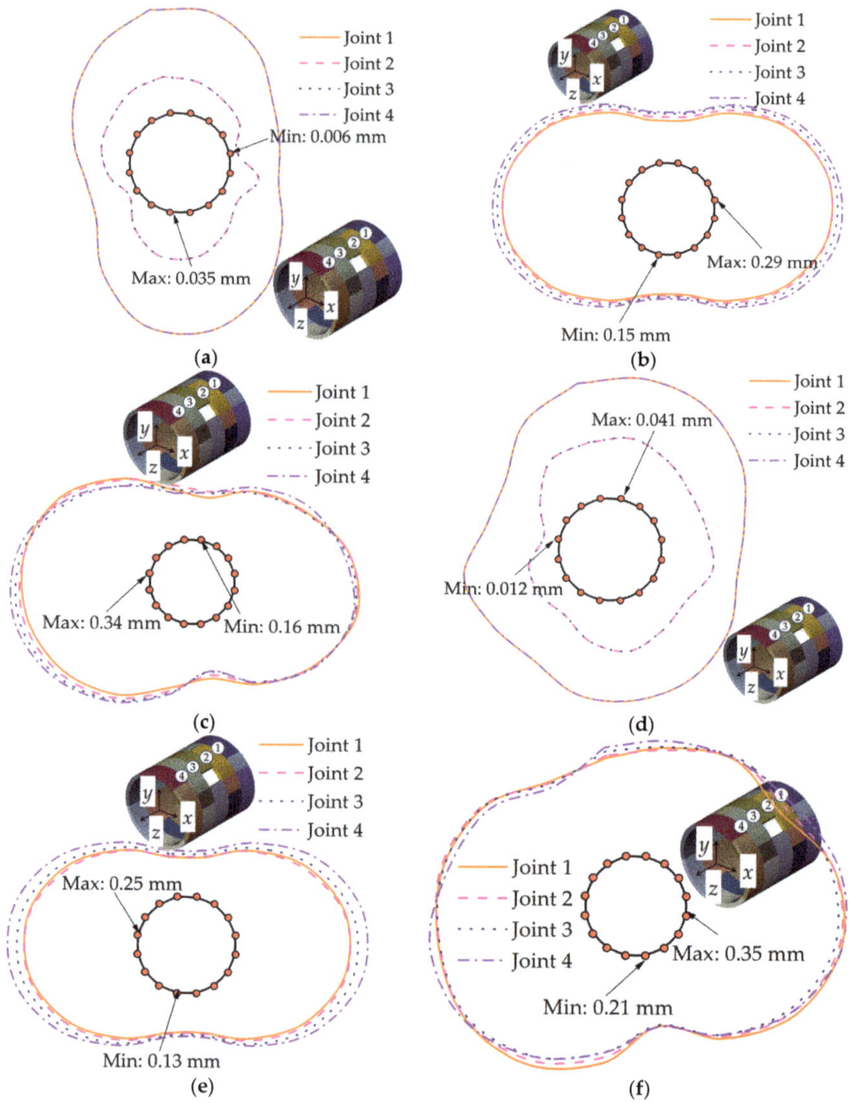

Figure 12. Maximum circumferential joint offset. (**a**) Case1. (**b**) Case2. (**c**) Case3. (**d**) Case4. (**e**) Case5. (**f**) Case6.

Moreover, the maximum circumferential joint opening is 0.41 mm, 0.49 mm, 0.52 mm, 0.69 mm, 0.71 mm, and 0.73 mm in cases 1−6, respectively, which are generally larger than the longitudinal joint opening. According to this finding, the circumferential joint may be vulnerable to larger deformation than the longitudinal joint under the same earthquake motion. Additionally, it is worth noting that the joint opening under the two−directional excitations is a little bit greater than the unidirectional conditions. Though there is only a small increasing level, the amplified effect caused by the two−directional excitations cannot be neglected for the sake of safety evaluations.

Figure 12 shows the envelope curves of the circumferential joint offset results. From Figure 12, apparent distribution characteristics can be found under different seismic conditions. For the transverse excitations, the maximum joint offset locates at the arch bottom and arch vault of the tunnel, while the minimum value is at the tunnel arch waist. For the axial excitations, the maximum joint offset is found at the tunnel arch waist, whereas the minimum one is at the arch bottom and arch vault, which is just opposite to that of the maximum value. Moreover, the joint offset under the transverse excitations is much smaller than that under the axial excitations. Take the wave 1 for example, according to the comparison of case 1 and case 2, the maximum joint offset grows from 0.035 mm to 0.15 mm by increasing nearly 500 percent. In addition, the distribution of the offset under two−directional excitations is similar to that of the axial conditions. This is because the transverse seismic wave has almost no influence on the circumferential joint offset, as the joint offset under the transverse excitations is too small to be considered. The maximum circumferential joint offset is 0.29 mm, 0.34 mm, 0.25 mm, and 0.35 mm in the case 2, case 3, case 5, and case 6, which is nearly half of the corresponding joint opening. Combing the longitudinal joint deformation results, it can be primarily drawn that the seismic action leads to more negative impact on the joint opening than the joint offset.

Zhang et al. [33] conducted seismic response analysis of bolt joints in shield tunnels. According to their findings, the maximum circumferential joint opening and offset can reach 0.7 mm and 0.5 mm under transverse excitations with a peak acceleration of 0.1225 g. While under axial excitation conditions, the maximum circumferential joint opening and offset can reach 0.9 mm and 0.6 mm, respectively. Compared with the bolt joint, the deformation of anchor joint subjected to seismic waves with the same amplitude is smaller. This indicates that the anchor joint has a better bearing capacity. Besides, the deformation of bolt joint under axial excitation is larger than that under transverse excitation, which is similar to that found in this paper.

5.2. Stress Distribution Characteristics of Anchor Joint

The stress distribution results of the anchor joint on the basis of the refined numerical model are displayed in Figure 13. For easy observation, the symmetrical plane passing through the axis of the anchor joint is selected as the sectional view. Figure 13a shows the stress of the anchor joint after finishing the assembly of rot, sleeve, and lantern ring, which is the perfect working state of the joint. In this case, the maximum stress of the joint is 510 MPa, locating at the sleeve walls. It is obvious that most of the stress concentration areas are on the sleeve walls, implying that the three components of the anchor joint have been firmly connected so that the satisfactory tensile property can be obtained. Additionally, the lantern ring is the main component to bear the large deformation and generate the tightening force. The anchor joint has entered the plastic deformation stage under the normal working conditions, which is quite different from the conventional bolt joints.

After loading the deformation responses of the joint, the maximum stress responses of the anchor joint under the six different selected cases are illustrated in Figure 13b–g. It is found that obvious changes happen on both the maximum stress and its distribution features. Specifically, the maximum stress decreases to 464 MPa, 459 MPa, and 462 MPa under the different input conditions of wave 1, while that for the wave 2 is 451 MPa, 449 MPa, and 447 MPa, respectively. Consequently, the maximum stress of the anchor joint decreases in varying degrees after bearing the seismic load, which is mainly due to the change of the relative position of the segments, leading to the change of the relative position between the rod, sleeve walls, and lantern ring. With the joint deformation, especially for the joint opening, a separation trend of the three components will appear, which finally weakens the integrity of the anchor joint. On the other hand, friction is extremely important to keep the tensile bearing capacity of the anchor joint. The change of the relative position of each component may lead to the decrease of the contact pressure between components and further cause the loss of the bearing capacity of the anchor joint.

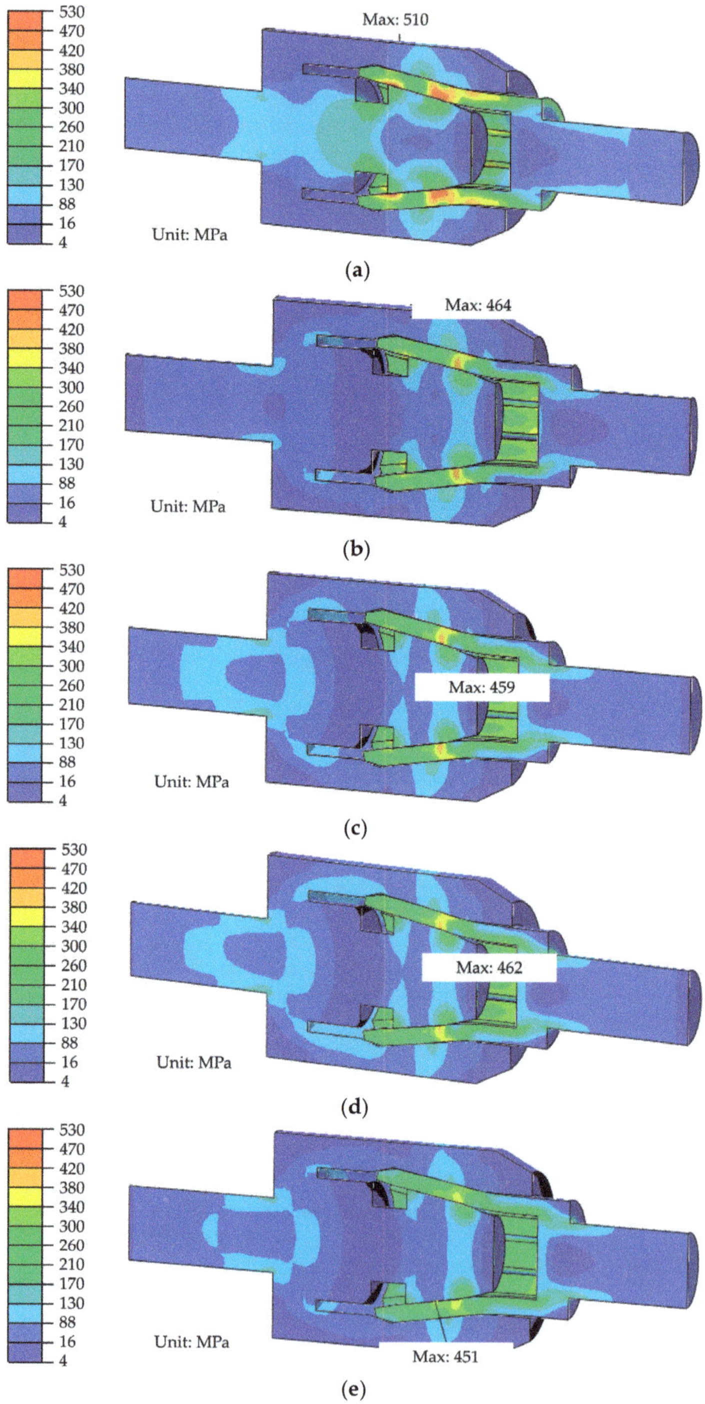

Figure 13. *Cont.*

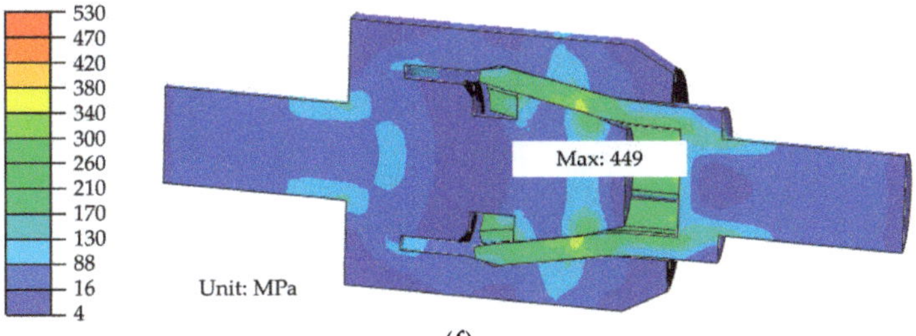

(f)

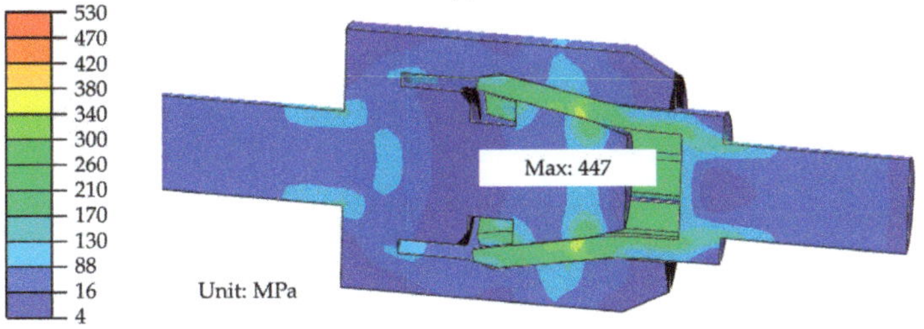

(g)

Figure 13. Stress distribution of anchor joint. (**a**) After assembly. (**b**) Case1. (**c**) Case2. (**d**) Case3. (**e**) Case4. (**f**) Case5. (**g**) Case6.

In addition, the influence of the structural characteristics of the anchor joint on its seismic response must be fully considered. As the anchor joint is a composite structure, if the components are separated only within a certain range under the action of earthquake, the main impact result is the decline of structural integrity. However, the loss of integrity will not cause irreversible structural damage to a single component, which is essentially different from the traditional bolt joint. The recoverability of anchored joints in a certain range of deformation makes them have better resistance to large deformation loads. Therefore, when analyzing the stress characteristics of anchored joints under earthquake, the final deformation of the joints must be taken into account after the earthquake, in addition to the most unfavorable load combination. According to the maximum joint deformation, it can be judged whether the anchor joint is completely damaged during the earthquake. Meanwhile, the final joint deformation can be used to predict the actual state of the anchor joint after the end of the earthquake load.

5.3. Strain Response of Anchor Joint

Based on the stress and deformation results of the anchor joint, the sleeve has the largest deformation among the different components, which play a key role in the bearing capacity of the whole system. Although plastic deformation occurs locally in other components, the value is very small and has little influence on the joint's performance. Therefore, only the main deformation component, namely the sleeve, is analyzed in this section. The plastic strain distribution of the sleeve is shown in Figure 14.

It can be seen from Figure 14a that after the anchor joint is assembled, plastic strain occurs in most areas of the side walls of the sleeve, which is consistent with the stress distribution results. By comparing the plastic strain results at the completion time of assembly and after the application of the most unfavorable seismic displacement responses, it can be found that the plastic strain distribution of the sleeve changes little under the action of different seismic waves, and the maximum plastic strain remains unchanged. It can be inferred that the maximum plastic strain of the side walls of the sleeve occurs at the end of the assembly process. This should happen at the same time with the appearance of the maximum deformation of the sleeve side walls.

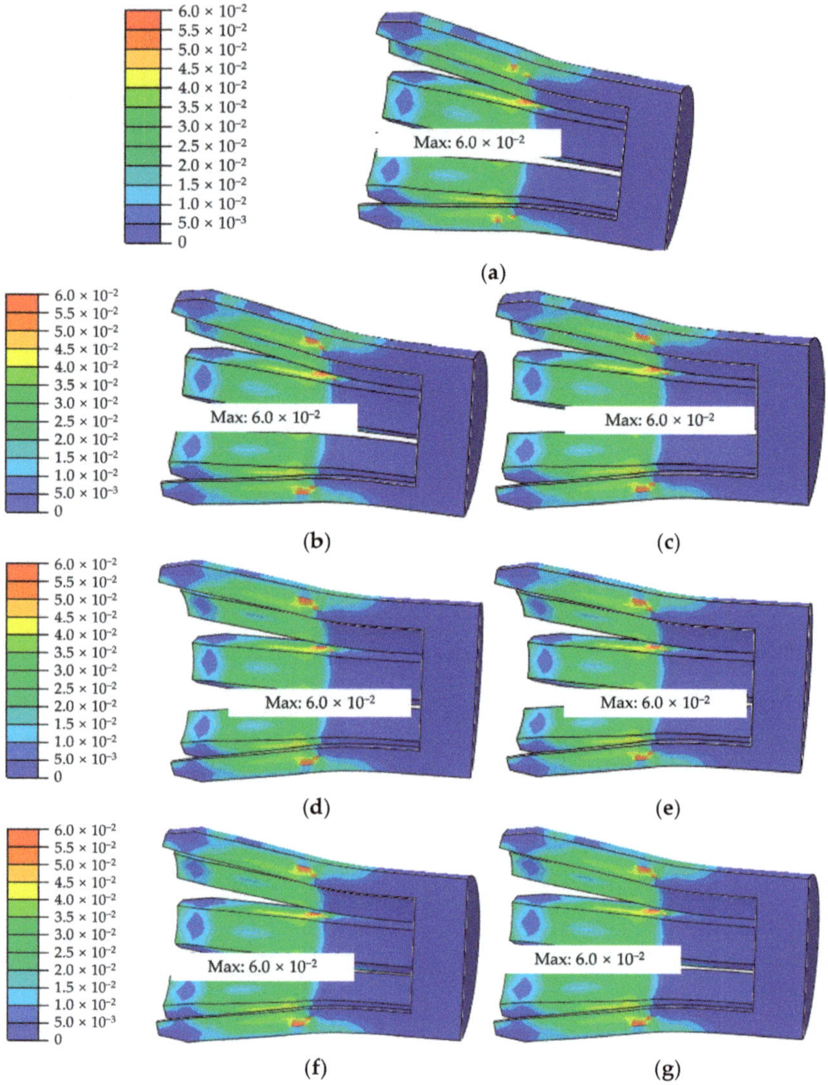

Figure 14. Plastic strain of lantern ring. (**a**) After assembly. (**b**) Case1. (**c**) Case2. (**d**) Case3. (**e**) Case4. (**f**) Case5. (**g**) Case6.

Besides, the mutual position change between the joint members within a certain range will not cause the increase of the maximum plastic strain. However, it is worthy of noting that the distribution range of the maximum plastic strain has expanded in Figure 14b–g, and it changes from the edge of the sleeve side walls to the sleeve center. The above results indicate that although the maximum plastic strain value remains the same, the plastic deformation of the sleeve side walls actually increases after the earthquake. The cumulative effect of plastic strain under seismic action has a negative impact on the mechanical performance of the anchor joint. With the continuous expansion of the distribution range of plastic strain, the recovery level of the bearing capacity of each component in the anchor joint after separation and reassembly is bound to decline. Accordingly, it is of critical importance to conduct a detailed seismic response study on the anchor joint based on the proposed method in this paper. In this way, the deformation characteristics, bearing capacity, and safety state of the anchor joint can be reasonably evaluated in the light of the cumulative plastic deformation of sleeve walls during the earthquake motion.

In fact, most of the existing seismic response studies on shield tunnels have neglected the detailed structural features of the segmental joints. The most important reason is to simplify the model to gain less computational cost. Another critical reason is that it is really difficult to reach convergence during the complicated three-dimensional numerical simulations with too detailed characteristics of the structures. However, in this section, an apparent cumulative effect of plastic strain has been found for the anchor joint under seismic action. This phenomenon demonstrates that it is quite necessary to propose advanced methods, which can reproduce and reflect the real structural features of the segmental joints. Additionally, it can be inferred that for the segmental joints with complex structural components, more detailed investigations are still in need to gain a comprehensive knowledge of their seismic performance.

6. Conclusions

This paper aimed to study the seismic response of the anchor joint in shield tunnel considering the SSI effect. For this purpose, an improved method was first developed and introduced in detail. Then, joint deformation characteristics were discussed based on the results obtained from the soil-tunnel system. Finally, the detailed stress and strain distribution features were interpreted to reveal the seismic performance and evaluate the safety state of the anchor joint. The specific conclusions are as follows:

(1) Combing the dynamic time history method and the response displacement method, an improved approach is proposed for the overall and detailed seismic response study of the anchor joint. Using this method can effectively reduce the modeling difficulty and computing resource consumption caused by the complex structure of the anchor joint.

(2) The maximum opening and offset of the anchor joint under the axial seismic input are greater than those under the transverse seismic input. The distribution features of joint opening and offset are obviously different under different seismic excitations. The seismic action leads to a more negative impact on the joint opening than the joint offset.

(3) Compared with the unidirectional seismic input conditions, the maximum opening and offset of the anchor joint under the bidirectional seismic action are increased. However, the deformation of joints at different locations is rather distinct, which cannot be directly obtained from the sum of results from unidirectional seismic input conditions.

(4) The opening and offset caused by earthquake can cause the separation of the components in the anchor joint, resulting in the decrease of the contact pressure among the components, and further reduce the bearing capacity of the anchor joint. The stress distribution of the sleeve shows obvious changes, while the distribution range of the maximum plastic strain of the sleeve walls is expanded.

(5) For the evaluation on the deformation resistance and post-earthquake performance of the anchor joint, the final deformation of the anchor joint and the cumulative plastic strain of the sleeve can be used as the quantitative indicators.

The seismic response of anchor joint has been preliminarily studied with the proposed improved numerical method. In this paper, only the results of joint opening, offset, stress, and strain were considered, while more detailed investigations are still necessary to gain a comprehensive understanding of the seismic performance of anchor joints. The influence of geology conditions and seismic waves should also be taken into account, which will be the further work.

Author Contributions: Methodology, formal analysis, software, writing—original draft, G.Z.; conceptualization, funding acquisition, writing—review & editing, W.Z.; software, validation, J.Q.; visualization, R.N.; validation, C.Z. All authors have read and agreed to the published version of the manuscript.

Funding: This research was supported by the National Natural Science Foundation of China (Grant No. 51978460), which is gratefully acknowledged.

Conflicts of Interest: The authors declare no conflict of interest.

References

1. Editorial Department of China Journal of Highway and Transport. Review on China's Traffic Tunnel Engineering Research 2022. *China J. Highw. Transp.* **2022**, *35*, 1–40. [CrossRef]
2. Chen, Z.; Liang, S.; Shen, H.; He, C. Dynamic centrifuge tests on effects of isolation layer and cross-section dimensions on shield tunnels. *Soil Dyn. Earthq. Eng.* **2018**, *109*, 173–187. [CrossRef]
3. Gong, C.; Wang, Y.; Ding, W.; Lei, M.; Shi, C. Waterproof Performance of Sealing Gasket in Shield Tunnel: A Review. *Appl. Sci.* **2022**, *12*, 4556. [CrossRef]
4. Zhang, G.; Zhang, W.; Li, H.; Cao, W.; Wang, B.; Guo, W.; Gao, P. Waterproofing behavior of sealing gaskets for circumferential joints in shield tunnels: A full-scale experimental investigation. *Tunn. Undergr. Space Technol.* **2021**, *108*, 103682. [CrossRef]
5. Zhang, G.; Zhang, W.; Cao, W.; Wang, B.; Lai, T.; Guo, W.; Gao, P. A novel test setup for determining waterproof performance of rubber gaskets used in tunnel segmental joints: Development and application. *Tunn. Undergr. Space Technol.* **2021**, *115*, 104079. [CrossRef]
6. Wang, W.L.; Wang, T.T.; Su, J.J.; Lin, C.; Seng, C.R.; Huang, T.H. Assessment of damage in mountain tunnels due to the Taiwan Chi-chi earthquake. *Tunn. Undergr. Space Technol.* **2001**, *16*, 133–150. [CrossRef]
7. Lin, G.; Luo, S.; Ni, J. Damages of Metro Structures Due to Earthquake and Corresponding Treatment Measures. *Mod. Tunn. Technol.* **2009**, *46*, 36–41.
8. Zhang, W.W.; Jin, X.L.; Yang, Z.H. Combined equivalent & multi-scale simulation method for 3-D seismic analysis of large-scale shield tunnel. *Eng. Comput.* **2014**, *31*, 584–620. [CrossRef]
9. Hashash, Y.M.A.; Hook, J.J.; Schmidt, B.; Yao, J.I.C. Seismic design and analysis of underground structures. *Tunn. Undergr. Space Technol.* **2001**, *16*, 247–293. [CrossRef]
10. Wang, X.; Jin, X.; Wang, P.; Yang, Z. Parallel Dynamic Analysis of a Large-Scale Water Conveyance Tunnel under Seismic Excitation Using ALE Finite-Element Method. *Appl. Sci.* **2016**, *6*, 36. [CrossRef]
11. Brandis, A.; Kraus, I.; Petrovčič, S. Simplified Numerical Analysis of Soil–Structure Systems Subjected to Monotonically Increasing Lateral Load. *Appl. Sci.* **2021**, *11*, 4219. [CrossRef]
12. Ma, J.; Han, S.; Gao, X.; Li, D.; Guo, Y.; Liu, Q. Dynamic Lateral Response of the Partially-Embedded Single Piles in Layered Soil. *Appl. Sci.* **2022**, *12*, 1504. [CrossRef]
13. Brachman, R.W.I.; Elshimi, T.M.; Mak, A.C.; Moore, I.D. Testing and Analysis of a Deep-Corrugated Large-Span Box Culvert prior to Burial. *J. Bridge Eng.* **2012**, *17*, 81–88. [CrossRef]
14. Abdel-Sayed, G.; Salib, S.R. Minimum Depth of Soil Cover above Soil–Steel Bridges. *J. Geotech. Geoenviron.* **2002**, *128*, 672–681. [CrossRef]
15. Yeau, K.Y.; Sezen, H.; Fox, P.J. Simulation of Behavior of In-Service Metal Culverts. *J. Pipeline Syst. Eng.* **2014**, *5*, 04013016. [CrossRef]
16. Kasper, T.; Meschke, G. A numerical study of the effect of soil and grout material properties and cover depth in shield tunnelling. *Comput. Geotech.* **2006**, *33*, 234–247. [CrossRef]
17. Maleska, T.; Beben, D. Behaviour of soil-steel composite bridge with various cover depths under seismic excitation. *Steel Compos. Struct.* **2022**, *42*, 747–764. [CrossRef]
18. Maleska, T.; Beben, D.; Nowacka, J. Seismic vulnerability of a soil-steel composite tunnel—Norway Tolpinrud Railway Tunnel Case Study. *Tunn. Undergr. Space Technol.* **2021**, *110*, 103808. [CrossRef]
19. Kuribayashi, E.; Iwasaki, T.; Kawashima, K. Dynamic behaviour of a subsurface tubular structure. *Bull. N. Z. Soc. Earthq.* **1974**, *7*, 200–209. [CrossRef]
20. Luco, J.E.; de Barros, F.C.P. Seismic response of a cylindrical shell embedded in a layered viscoelastic half-space. I: Formulation. *Earthq. Eng. Struct. D* **1994**, *23*, 553–567. [CrossRef]
21. De Barros, F.C.P.; Luco, J.E. Seismic response of a cylindrical shell embedded in a layered viscoelastic half-space. II: Validation and numerical results. *Earthq. Eng. Struct. D.* **1994**, *23*, 569–580. [CrossRef]
22. Kouretzis, G.P.; Bouckovalas, G.D.; Gantes, C.J. 3-D shell analysis of cylindrical underground structures under seismic shear (S) wave action. *Soil Dyn. Earthq. Eng.* **2006**, *26*, 909–921. [CrossRef]
23. Chen, J.; Shi, X.; Li, J. Shaking table test of utility tunnel under non-uniform earthquake wave excitation. *Soil Dyn. Earthq. Eng.* **2010**, *30*, 1400–1416. [CrossRef]
24. An, J.; Tao, L.; Jiang, L.; Yan, H. A shaking table-based experimental study of seismic response of shield-enlarge-dig type's underground subway station in liquefiable ground. *Soil Dyn. Earthq. Eng.* **2021**, *147*, 106621. [CrossRef]

25. Wang, Z.Z.; Jiang, Y.J.; Zhu, C.A.; Sun, T.C. Shaking table tests of tunnel linings in progressive states of damage. *Tunn. Undergr. Space Technol.* **2015**, *50*, 109–117. [CrossRef]
26. Wang, F.; Jiang, X.; Niu, J. The Large-Scale Shaking Table Model Test of the Shallow-Bias Tunnel with a Small Clear Distance. *Geotech. Geol. Eng.* **2017**, *35*, 1093–1110. [CrossRef]
27. Chen, G.; Ruan, B.; Zhao, K.; Chen, W.; Zhuang, H.; Du, X.; Khoshnevisan, S.; Juang, C.H. Nonlinear Response Characteristics of Undersea Shield Tunnel Subjected to Strong Earthquake Motions. *J. Earthq. Eng.* **2020**, *24*, 351–380. [CrossRef]
28. Miao, Y.; Yao, E.; Ruan, B.; Zhuang, H. Seismic response of shield tunnel subjected to spatially varying earthquake ground motions. *Tunn. Undergr. Space Technol.* **2018**, *77*, 216–226. [CrossRef]
29. Amorosi, A.; Boldini, D.; Falcone, G. Numerical prediction of tunnel performance during centrifuge dynamic tests. *Acta Geotech.* **2014**, *9*, 581–586. [CrossRef]
30. Lee, K.M.; Hou, X.Y.; Ge, X.W.; Tang, Y. An analytical solution for a jointed shield-driven tunnel lining. *Int. J. Numer. Anal. Methods Geomech.* **2001**, *25*, 365–390. [CrossRef]
31. Zhao, T.; Liu, W.; Ye, Z.; Mosallam, A.S. Effects of Water Inrush from Tunnel Excavation Face on the Deformation and Mechanical Performance of Shield Tunnel Segment Joints. *Adv. Civ. Eng.* **2017**, *2017*, 5913640. [CrossRef]
32. Jin, H.; Yu, S.; Zhou, S.; Xiao, J. Research on Mechanics of Longitudinal Joint in Shield Tunnel by the Nonlinear Spring Equivalent Method. *KSCE J. Civ. Eng.* **2018**, *23*, 902–913. [CrossRef]
33. Zhang, W.; Zhang, Q.; Cao, W. Study on Stress and Deformation of Bolt Joints of Shield Tunnel under Static and Seismic Action. *KSCE J. Civ. Eng.* **2021**, *25*, 3146–3159. [CrossRef]
34. Liu, Y.; Zhang, M.; Li, P.; Wang, B.; Wang, L.; Yuan, P. Experiments on the seismic performance of Y-shape joints of subway stations built by enlarging two parallel shield tunnels. *Tunn. Undergr. Space Technol.* **2021**, *115*, 104048. [CrossRef]
35. Yang, C.; Mo, H.; Chen, J.; Wang, Y.; Giunta, G.; Guan, Z. Influence of Seismic Loading on Segment Opening of a Shield Tunnel. *Sci. World J.* **2014**, *2014*, 387210. [CrossRef]
36. Yan, Q.X.; Bao, R.; Chen, H.; Li, B.J.; Chen, W.Y.; Dai, Y.W.; Zhou, H.Y. Dynamic response and waterproof property of tunnel segmental lining subjected to earthquake action. *Earthq. Struct.* **2019**, *17*, 411–424. [CrossRef]
37. Zhang, W.; Zhang, G.; Lei, H. Research on Shear Performance of FRP-Key Joint for Shield Tunnel and Rationality of Arrangement Based on Plastic-damage Model. *China J. Highw. Transp.* **2017**, *30*, 38–48. [CrossRef]
38. Shirato, M.; Furuichi, K.; Takimoto, K.; Hara, H.; Mukuno, K.; Yoshida, K. Development of New Composite Segment and Application to the Tunneling Project. *J. JSCE* **2003**, *728*, 157–174. [CrossRef]
39. Nakajima, S.; Akiyama, S.; Yasaka, M.; Sato, H.; Mukuno, K. Development of the Axial-Slide Ductile Segment. *Tunn. Eng. Res. Rep.* **1998**, *8*, 39–46.
40. Zhang, W.; Zhang, G.; Su, R.; Jin, M. Numerical Study on Influence of Manufacturing Process on Mechanical Properties of Longitudinal Anchor Joint for Shield Tunnel. *Tunn. Constr.* **2016**, *36*, 150–157. [CrossRef]
41. Wang, R.; Han, L.; Zhao, X.; Rasmussen, K.J.R. Analytical behavior of concrete filled double steel tubular (CFDST) members under lateral impact. *Thin Wall Struct.* **2016**, *101*, 129140. [CrossRef]
42. Takeuchi, T.; Emori, Y.; Suda, Y.; Koizumi, A. Influence on segments by jack thrust of shield-driven tunneling. *Doboku Gakkai Ronbunshuu F* **2010**, *66*, 599–611. [CrossRef]
43. Gu, Y.; Liu, J.; Du, Y. 3D consistent viscous-spring artificial boundary and viscous-spring boundary element. *Eng. Mech.* **2007**, *24*, 31–37.
44. Xu, C.; Song, J.; Du, X.; Zhong, Z. A completely explicit finite element method for solving dynamic u-p equations of fluid-saturated porous media. *Soil Dyn. Earthq. Eng.* **2017**, *97*, 364–376. [CrossRef]
45. Zhang, X.; Far, H. Effects of dynamic soil-structure interaction on seismic behaviour of high-rise buildings. *Bull. Earthq. Eng.* **2022**, *20*, 3443–3467. [CrossRef]

applied sciences

Review

Waterproof Performance of Sealing Gasket in Shield Tunnel: A Review

Chenjie Gong [1], Yuanye Wang [1,2], Wenqi Ding [2,3,*], Mingfeng Lei [1] and Chenghua Shi [1]

1 Department of Tunnel Engineering, College of Civil Engineering, Central South University, Changsha 410018, China; gongcj@csu.edu.cn (C.G.); yuanye_wang@tongji.edu.cn (Y.W.); mingdfenglei@csu.edu.cn (M.L.); csusch@163.com (C.S.)

2 Department of Geotechnical Engineering, College of Civil Engineering, Tongji University, Shanghai 200092, China

3 Key Laboratory of Geotechnical and Underground Engineering of Ministry of Education, Tongji University, Shanghai 200092, China

* Correspondence: dingwq@tongji.edu.cn; Tel.: +021-6598-1011

Abstract: Rubber gaskets are commonly adopted as the waterproof component in shield tunnels for their outstanding sealing performance. The contact pressure between surfaces generated by the assembly stress ensures that the gaskets resist certain water pressure without leaking. However, with the continuous occurrence of leakage accidents, attention has been drawn to the topic of the waterproof performance of gasketed joint shield tunnels. In this article, prominent contributions to the waterproof performance of sealing gasket in shield tunnels are listed and sorted into four sections: (1) structural behavior of lining and joint; (2) material constitutive model and durability; (3) numerical simulation methods; (4) thermal-mechanical coupling analysis. First, examples of leakage are discussed and tests on gaskets are elucidated, which is followed by a summary of the progress on material mechanical properties and durability. Then, the development of the simulation methods is presented. Finally, the existing research on the thermal-mechanical coupling analysis is summarized. It is found that the contributions to gaskets' waterproof performance are fruitful, however, with stringent construction conditions, such as the material constitutive model and aging mechanism under special conditions, such as high temperature, numerical simulation, and laboratory test methods, which need to be further explored.

Keywords: shield tunnel; sealing gasket; sealing performance; thermal-mechanical coupling

Citation: Gong, C.; Wang, Y.; Ding, W.; Lei, M.; Shi, C. Waterproof Performance of Sealing Gasket in Shield Tunnel: A Review. *Appl. Sci.* **2022**, *12*, 4556. https://doi.org/10.3390/app12094556

Academic Editor: Daniel Dias

Received: 19 March 2022
Accepted: 28 April 2022
Published: 30 April 2022

1. Introduction

With the development of the economy and the improvement of people's living standards, aboveground construction in urban areas is nearly saturated, which accelerates the vigorous development of tunnel construction. At the same time, the highway, railway, and subway tunnels, as the main arteries of transportation, are suffering from leaking accidents one after another due to natural or man-made factors, causing deleterious consequences and huge economic losses [1].

According to China's construction code, the design service life of energy and transportation shield tunnels is usually up to 100 years or even 120 years. Ensuring its safety and functionality during service life under harsh and variable influences [2,3] has become a severe challenge faced by the majority of engineering scientists. The existing construction and operation experiences of cross-river and sea-crossing tunnels present a situation of widespread leakage [4,5]. Table 1 lists the representative tunnel leakage accidents.

Table 1. Shield tunnel leakage accidents.

No.	Time	Accident	Caused by	Results
1	2018	Mud leak in Nanjing Yangtze River North Line Tunnel	The plug of the grouting hole falls off under high water pressure	Traffic stop
2	2018	Hangzhou Wangjiang Tunnel collapse	Frozen soil insulation layer breakdown by groundwater intrusion	Traffic stopped, several pipelines were damaged, and about 150 square meters of pavement collapsed
3	2018	Leakage in Tunnel of Foshan Rail Transit Line 2	Continuous failure caused by Shield tail seal failure	11 dead,1 missing, and 8 severely injured. The direct economic loss is about USD 8.358 million
4	2008	Large deformation in one of the Shanghai Metro Tunnel	large deformation of tunnel cross-section, joint leakage, and bolt breakage caused by the sudden pile load on the surface	Subway outage
5	2003	Leakage in Shanghai Metro Line 4	Insufficient freezing strength	The direct economic loss is about USD 23.55 million

The waterproof performance of shield tunnels follows the principle of "prevention first, rigid and flexible combined, multiple layers of defense, and comprehensive treatment". The key to it lies in the segment joints [6]. The reinforced concrete segment consists of the majority of the shield tunnel, which excels in permeability resistance, as shown in Figure 1a. It is worth mentioning, however, that the segments can be arranged both in alignment and staggered, in practice. Existing engineering examples show that the waterproof performance of tunnels is controlled by joints (Figure 1b). In the construction stage, the gaskets between the longitudinal joints and the circumferential joints are pressed by shielding jacks to form contact stress and resist the intrusion of external water pressure, as presented in Figure 1c. The mainstream strategy for shield tunnel joints in China is to glue EPDM gaskets onto the outer groove of the segment, which is proved reliable under targeted working conditions. However, with hazardous factors (earthquake, fire, etc.) imposed at a certain magnitude, the gaskets are likely to fail.

Waterproof failure is a typical continuous–discontinuous hydraulic expansion failing process [7]. Seepage of the gasket contact surface is a process in which the contact surface of the gasket keeps opening and water keeps penetrating under pressure. When the contact surface of the gasket is completely opened, a water seepage path is formed, resulting in the occurrence of joint leakage (Figure 1d). It can be seen, however, that sealing gaskets play a vital role in the segment joint waterproof process.

Based on the above engineering problems, this paper summarizes the research progress on waterproofing of sealing gaskets, to guide follow-up research and practical referencing for engineering design. Combined with the domestic and foreign scholars' research on the shield tunnel joint sealing performance and analysis methods related topics, in this article, the waterproof performance, material characteristics, numerical simulation, and thermal-mechanical coupling analysis of the shield tunnel joint sealing gasket are discussed.

2. Mechanical Behavior of Lining Structure and Joints

The waterproof performance of gaskets is related to many factors, which can be divided into material factors, structural factors, and external influences. Researchers have conducted in-depth studies on these influencing factors and carried out structural tests. Paul [8] took the lead in the study of the waterproof capacity of gaskets under different joint openings, which was based on the Chicago subway tunnel in the sandy soil, and laid the foundation of the steel-clamp-based joint gasket waterproof test. Table 2 summarizes the related contributions to the indoor waterproof performance test of shield tunnel joints.

Since then, with the large-scale development and utilization of urban underground space and metro tunnel construction, an increasing number of scholars have begun to pay close attention to tunnel waterproof problems. Corresponding tests are conducted to acquire the water pressure admissible value while the gasket joint is in an offset state. Representative cases in recent years include the Singapore metro tunnel [4], the Shanghai

Qingcaosha water tunnel [9], the Su'ai tunnel project [10], the Shanghai deep water storage tunnel [11], and the Changsha metro tunnel [12].

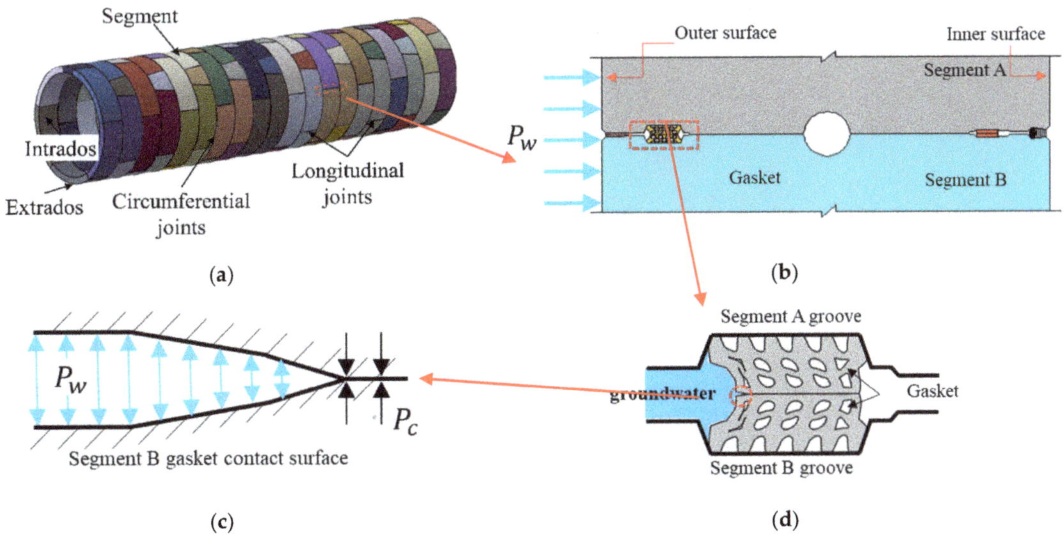

Figure 1. Joint waterproof sealing diagram of shield tunnel: (**a**) lining structure, (**b**) joints [13] (Reprinting with permission from Gong et al., 2019), (**c**) water pressure expansion process, (**d**) joint gaskets [7] (Reprinting with permission from Gong et al., 2018).

2.1. Waterproof Performance Deterioration of Tunnel

With the deepening understanding of the leaking mechanism, scholars devote themselves to the research of tunnel waterproof performance deterioration. In terms of laboratory test technologies and deterioration simulations, Lei et al. [14] carried out investigations and numerical simulations on gasket sealant behavior under high hydrostatic water pressure and a relationship between joint deformation; waterproof performance was obtained and a waterproof performance assessing method was proposed. Liu et al. [15,16] carried out joint waterproof performance tests for full-scale segments and reduced-scale specimens after fire damage. From this, the temperature distributions and the gasket deterioration pattern under high temperatures were acquired. Li et al. [17] and Dong et al. [18] studied the influences of the mechanism of joint opening and offset of subway tunnels on the waterproof performance of gaskets, and corresponding indoor tests were carried out. Aiming at the test characterization of waterproof capacity or seepage water pressure of tunnel joints, the behavior of seepage water under sand leakage conditions is not mentioned by the aforementioned contributions. Zhou et al. [19] established a Coupled Eulerian-Lagrange-based model and the whole leakage process at the segmental joints, considering large deformation and seepage flow in the contact surface, was simulated, and the validity of the hydro-mechanical model was verified. Zheng et al. [20,21] designed a test apparatus to simulate the soil and water losses under the condition of a large joint opening, and an analysis based on the sandy soil of the Tianjin Metro Tunnel was carried out. Given the working conditions of Southern California and Southern China with high seismic intensity, Shalabi et al. [22,23] and Xie et al. [24] conducted joint waterproof performance tests under reciprocating loading conditions, hence, the waterproof capability under vibration was obtained. Zhang et al. [25] investigated the influence mechanism of joint rotation on waterproof capacity; methods were proposed to mitigate the joint leaking situations. Zhang et al. [26,27] developed a waterproof performance testing technology for full-scale segment joints, and impacts of constructing loads on the deformation of gaskets were tested.

With the development of computer sciences and designing protocols, novel technologies and structure forms are introduced for the detection and prevention of the waterproof performance deterioration of joints. Based on a convolutional neural network and EfficientNet, Zhou et al. [28] proposed a defect detecting model, which can be implemented to detect localized leakage in tunnels and monitor the working conditions throughout the design service life (Figure 2). The waterproof performance of a novel structure type with double-layer sealing gaskets at intervals outside the bolt was tested by Xie et al. [29], and the results indicated that the novel structure can substantially enhance the waterproof capacity.

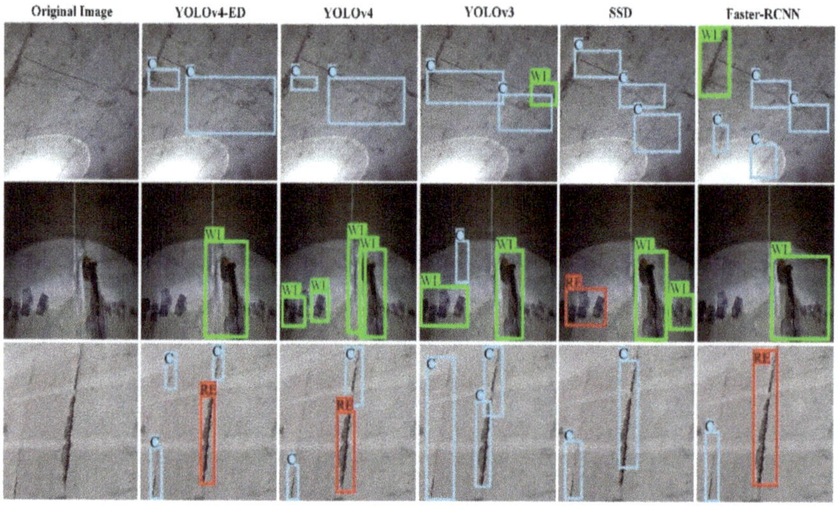

Figure 2. Detection results of different models (Reprinting with permission from Zhou et al., 2022).

Per the literature review, existing research on waterproof performance deterioration captured numerous undermining factors, e.g., high hydrostatic water pressure, high temperature caused by fire, sandy leakage, seismic load, and deformation of the segmental rings. New testing technologies were introduced, and novel structures and waterproof performance evaluation methods were introduced to address the challenges.

2.2. Double-Layer Gasket Configuration and Corresponding Test Technology

Tests and simulations have been conducted for large-diameter shield tunnels with the double-layer gasket configuration. Aforementioned methods are used to determine the functionality under high water pressure. Waterproofing of tunnel joint gaskets depends on the contact condition of the gasket–gasket contact surface and the gasket–segment contact surface, namely, the rubber-rubber contact and the rubber-concrete contact. It is difficult to ensure the smoothness of the in-use segment with reinforced concrete casting and, therefore, it is also hard to determine the contact behavior between the gasket and rubber. Hence, in the tests, the grooves cannot be accurately simulated by a precisely fabricated steel clamp, which makes the results unreliable. Given this, the author participated in the research and development of a novel joint waterproof testing apparatus based on the Nanjing Weisan Road Tunnel [30], and a full-scale concrete segment experiment was performed.

The contributions above on the waterproof testing of tunnel joints are mainly focused on urban railway tunnels and highway tunnels, with a single-layer gasket and the steel clamp specimen. In practice, gaskets are glued in concrete grooves, and the surfaces are rougher than that of steel clamps, so the test accuracy based on steel clamps needs to be further investigated [31]. The most significant difference between the large-diameter deep drainage tunnel and the traditional traffic tunnel is the bi-directional leakage characteristics caused by different drainage and function modes (empty pipe and full pipe).

With the construction of deep drainage tunnels, attention has been drawn to the joint waterproofing of the deep drainage tunnels. Zhang et al. [11] conducted the test of gasket selection according to the Suzhou River test section of the Shanghai Deep Drainage Tunnel. Based on the data collected from the Donghaochong test section of the Guangzhou Deep Drainage Tunnel and the Suzhou River test section of the Shanghai Deep Drainage Tunnel, Gong et al. [32] demonstrated the necessity of adopting a double-layer gasket configuration for joints. However, the above indoor tests of deepwater drainage tunnels adopt the experimental idea of separate testing of inner and outer sealing gaskets or by simply using steel clamps to simulate the contact behavior of sealing gaskets in the segment grooves. Therefore, the test results can not reflect the bi-directional leakage characteristics of the double-layer gaskets in deep drainage shield tunnels under different operating scenarios with high external pressure and variable internal pressure.

Throughout the current studies, the purpose was concentrated on the design and selection of the joint single/double-layer gaskets for urban subway tunnels, urban deep municipal tunnels, and cross-river or cross-sea traffic tunnels, considering the influence of different load conditions (static, earthquake, fire, soil erosion) on the waterproof capacity. To have a scientific and in-depth understanding of the actual leakage failure behavior of the double-layer gaskets in deep drainage shield tunnels under complex internal and external high water pressure, it is urgent to carry out corresponding targeted research and waterproofing performance testing of the double-layer gaskets.

3. Rubber Material Characteristics

The waterproof performance of shield tunnel gaskets is mainly realized by the contact stress between the gasket surfaces [32]. The commonly used materials can be divided into two categories: elastic material, such as ethylene propylene diene monomer (EPDM), chloroprene rubber (CR), etc., and water-swelling rubber material, such as water swelling rubber (WSR). The in-use materials are different, while the overall performance during service is similar. Yang et al. made a detailed review of the material properties, material selection, and chemical mechanism of aging of gasket rubber materials [33]. Test data of water swelling polymer (WSP) and water-swelling rubber polymer (WSRP), under different serving environments, were given by Wang et al. [34], and the performance of the tested materials is compared [35].

Table 2. Summary of joint waterproof performance indoor tests of representative shield tunnel.

Project	Material [1]	Specimen	Purpose	Reference
Chicago Subway Tunnels	NP	Steel clamp	Effects of joint opening	[8]
Singapore Subway Tunnel	EPDM	Steel clamp	Effects of joint opening/offset	[36]
Shanghai Qingcaosha water tunnel	EPDM	Steel clamp	Influence of joint opening/offset/rotation	[9]
Los Angeles Subway Extension	EPDM, NP	Concrete specimen	Influence of seismic load on waterproofing performance	[22]
Beijing Subway Tunnel	COMPOUND	Steel clamp	Effects of high temperatures in fire conditions	[15,16]
Los Angeles Subway Extension	EPDM, NP	Steel clamp	Effects of the opening at the bottom of the sealing gasket	[23]
Nanjing Weisan Road Tunnel	EPDM	Concrete specimen	Summary of test technology	[30]
A Subway Tunnel	EPDM	Steel clamp	Effects of joint opening/offset	[18]
Su'ai Tunnel	EPDM	Steel clamp	Design selection under earthquake	[10]
Nanjing Subway Tunnel	EPDM	Steel clamp	Double-layer waterproof	[17]
Changsha Subway Tunnel	EPDM	Steel clamp	Effects of joint opening/offset	[12]
A Subway Tunnel	EPDM	Steel clamp	Effects of soil erosion	[20,21]
Shanghai deep water storage tunnel	EPDM	Steel clamp	Double-layer waterproof	[11]
Su'ai Tunnel	EPDM	Steel clamp	Comparison of static and dynamic water resistance	[24]

[1] EPDM: ethylene propylene diene monomer; NP: neoprene; COMPOUND: EPDM and water-swelling rubber compound.

3.1. Constitutive Model and Aging Mechanism

The rubber material of gaskets is hyper-elastic, which means that its mechanical constitutive model cannot be described by the typical stress–strain curve. In continuum mechanics, for materials such as hyper-elastic rubber, the phenomenological theory can be used to determine the constitutive model of materials by obtaining the relationship between strain energy density and strain tensor invariants. Commonly used constitutive models include the Arruda–Boyce model, Mooney–Rivlin model (Equation (1)), neo-Hookean model, etc.

$$W = C_{10}(I_1 - 3) + C_{01}(I_2 - 3) \tag{1}$$

where W is the strain energy density, C_{10} and C_{01} are constants, and I_1 and I_2 are the strain invariants.

Due to its simple mathematical form and high simulation accuracy, the Mooney–Rivlin model is often implemented in gasket performance [37]. A constitutive model of hyper-elastic rubber considering thermal-viscoelastic coupling (Equation(2)) and a rheological model based on non-Newtonian fluid (Equation(3)) were established based on different states of the rubber [38]. In terms of the thermal-mechanical aging model, Li [39] has been devoted to exploring the thermal-mechanical aging process of rubber and its relationship with property changes since the 1990s. Based on the Arrhenius formula (Equation (4)) and the Dakin prediction equation (Equation(5)), a P-T-t aging model was proposed, considering the influence of temperature and time on performance deterioration (Equation(6)).

$$\Psi = -\frac{G_s}{2}\frac{T}{T_0}\lambda_m\left[1+b\left(\frac{T-T_0}{T_0}\right)\right]\ln\left\{1-\frac{I_1-3}{\lambda_m[1+b(T-T_0)/T_0]}\right\}+d\left(T-T_0-T\ln\frac{T}{T_0}\right) \tag{2}$$

where G_s is the isothermal infinitesimal shear modulus, λ_m is the isothermal average maximum elongation, d is the material constant, and b is the temperature correlation factor of the maximum elongation.

$$\sigma_{12}= Y+\frac{A\dot{\gamma}}{B\gamma^{1-n}} \tag{3}$$

where Y and B are constant, and γ stands for the shear distance of the non-Newtonian liquid.

$$K = A_0 e^{-\frac{E}{RT}} \tag{4}$$

where K is the chemical reaction rate, A_0 is the pre-exponential factor, E is the activation energy, R is the molar gas constant, and T is the thermodynamic temperature.

$$f_t(P) = kt \tag{5}$$

where $f_t(P)$ is the reaction rate function, *k* is the coefficient, and *t* is the reaction time.

$$\log\left[-\log\frac{y}{B}\right] = B_0+B_1/T + B_2 \times \log t \tag{6}$$

where y = f(P), $B_0 = \log(A_0/2.303)$, $B_1 = -E/2.303R$, B_2 is a constant.

Based on thermodynamics, the mathematical relationship among the prevailing predicting models: the kinetic curve method, linear relation method, variable reduction method, and response function method, were analyzed, and the advantages of the proposed three-element aging mathematical model (Equation (6)) in rubber service life prediction were proved. Shi et al. [40] calculated the activation energy of the thermal-oxygen aging reaction process of phenyl ether silicone rubber by taking the critical value of rubber as the key index and establishing the relationship between the temperature and rubber aging prediction. Wang et al. [41] proposed a time-temperature-strain/stress superposition principle (TTSSP), which is useful for the long-term relaxation and creep prediction of EPDM rubber.

The rubber components in tunnels are facing complex situations with high temperature, offset, mechanical factors, etc. Prominent works have been done given the serviceability of the rubber components under different aging conditions by conducting laboratory tests. Liu et al. [42] carried out compressive stress-hydrothermal aging (CS-HA) tests and the residual performance under different pre-compression forces was obtained, hence, the corresponding institutive model was proposed. Based on undersea shield tunnels, mechanical and microchemical tests (TGA, SEM, FTIR) were conducted on accelerated aged EPDM specimens by Wang et al. [43], and the degradation process of EPDM rubber in seawater was obtained.

3.2. Structural Characteristics and Durability

In the indoor segment joint full-scale test, it was found that with high rubber hardness, the closure stress will exceed the shear strength of the concrete groove and cause shear failure at the joint closure stage, which leads to concrete burst in segments [44,45]. Targeting the engineering problems mentioned above, recent studies have been focusing on optimizing the cross-section of the gasket to control its closure stress. Liu et al. [46] discussed the influences of the Balloon Effect on the compressional performance of the internal closed cavity; the results indicated that due to the presence of pressure in the internal holes, the buckling process of gasket hole wall lags was obvious, and the closure stress increased by over 10%. Lei et al. [47] investigated the effect of the internal hole layout and constitutive model, and corresponding simulation results showed great influences of coefficient C_{10} and the internal hole layout on the average contact stress. In addition to in-depth research on the mechanical properties of joint gaskets at room temperature, some scholars [15,37,48] showed interest in the material property damage of joint gaskets in high-temperature scenarios.

On the other hand, regarding the durability of tunnel joint gaskets during the service period, Sun [49] pointed out the significant effect of the durability of segment joint waterproof materials on the long-term safety and functionality of tunnels. Taking the Shanghai Chongming Yangtze River tunnel as the engineering case, Wu et al. [50] carried out the artificially accelerated aging test of EPDM gaskets and predicted the contact stress of joint gaskets after simulating 100-year aging. Zhong et al. [51] studied the long-term performance of EPDM gaskets under constant compression for the Qiantang River tunnel. Further, Shi et al. [52] proposed a hyper-elastic time-varying constitutive model of EPDM material suitable for the water environment by conducting tensile tests of EPDM dumbbell-type specimens based on the theory of large deformation of elastomer [53] and its extended theory [54]. Gong et al. [55] carried out a full-scale parametric study on the design parameters of the gasket and its influences on the waterproof performance, and the fluid pressure penetration (FPP) module of the FE commercial package ABAQUS/Explicit was implemented to simulate the seepage process of segmental joint leakage. In addition to that, Yang et al. [33] summarized the material characteristics of joint gaskets in shield tunnels.

Scholars have conducted productive studies on the mechanical properties and durability prediction of joint gasket materials, and preliminarily explored the degradation of material properties of gaskets. Existing material characteristic research focused on the constitutive model, time-dependent characteristics, structure response, and gasket layouts while a comprehensive solution is yet to be proposed. The layout study is closely related to the constitutive model, which is highly variable with aging. To begin with, a comprehensive investigation and a universal longevity prediction method should be established.

4. Waterproof State Simulation Methods Development

Numerical simulation methods of the shield tunnel joint waterproof process are twofold [56]: a simplified simulation based on average contact stress and a fine simulation based on the meso-opening process of the contact surface.

4.1. Simplified Method

In terms of the simplified numerical simulation method, the key point is to collect the average contact stress under specific joint deformation conditions, such as the waterproofing threshold due to the material nonlinearity of gaskets, large geometric deformation, and complex contact relations. Zhao et al. [57] proposed the concept of taking average contact stress as a reference for waterproofing ability. Gong et al. [58] proposed empirical formulas for contact stress and waterproofing capacity of seven typical gaskets based on experimental results and numerical analysis data. Sun [59] analyzed the stress deformation of gaskets considering water pressure and studied the corresponding contact stress distribution and waterproof failure mode. Due to the nonlinear nature, large deformation, and complex contact of gaskets, the research of numerical simulation analysis is focused on accurately reproducing the compression deformation process of gaskets in grooves [33,60]. The latest research extends to how to simulate the leaking failure process of joint gaskets: Based on the concept of average contact stress, Gong et al. [13] compared the contact stress on different contact surfaces under the joint opening, offset, and rotation modes to predict the location of leakage. Li Pin et al. [61] proposed the concept of effective contact stress and refined the failure mechanism of different contact surfaces. In essence, the above methods are simplified models that equate the lateral water pressure to uniform pressure and do not consider the mesoscopic opening process of water pressure invading the gasket contact surface. In the preliminary design stage, the simplified numerical simulation method can better predict the deformation performance and contact stress of gaskets.

4.2. Fine Simulation Method

To further investigate the impacts of progressive water pressure intrusion on the waterproof performance of joint gaskets, Yuan et al. [62] introduced the fluid-structure coupling algorithm based on the Euler–Lagrange grid method to perform a fine simulation of the joint seepage process. Zhou et al. [63] further extended the algorithm to the seepage process simulation of t-shaped joints. However, the above calculation method assumes that the fluid has no pressure, and there is a certain deviation from the actual situation. Gong et al. solved the problem while still unable to account for the progressive open leakage process of two-layer gasket waterproofing structures under different water pressure directions [7].

To investigate the tunnel structure and joint deformation under partial joint leakage, a lot of work has been accomplished in recent years. Shin et al. [64] and Zhang et al. [65] established two-dimensional/three-dimensional finite element models based on the formation-structure method and analyzed the evolution of seepage formation and structural deformation under the condition of localized joint seepage. However, the above models adopt the idea of weakening joint stiffness from the Homogeneous Ring method, and the simulation results are not verified by collected data. To tackle this dilemma, Wu et al. [66] proposed a self-defined joint leakage unit, however, the joint opening process in the seepage process is not considered.

The main difficulty of the fine simulation lies in the reappearance of the continuous – discontinuous process of water pressure expansion on the contact surface, and the key to it is overcoming the convergence problem caused by large deformation. Gong et al. [7] adopted the idea of numerical simulation of a large deformation process of static penetration test for a reference [67,68]. Based on the built-in "mesh-to-mesh Solution Mapping" technology of FE commercial package ABAQUS/Explicit, the Mesh Regeneration Method, illustrated in Figure 3, was proposed, and the accurate simulation of the dynamic process of water pressure expansion on the contact surface was realized. A Continuous-Discontinuous Fine Finite Element method for calculating the joint waterproofing capacity of the shield tunnel under normal temperature was established.

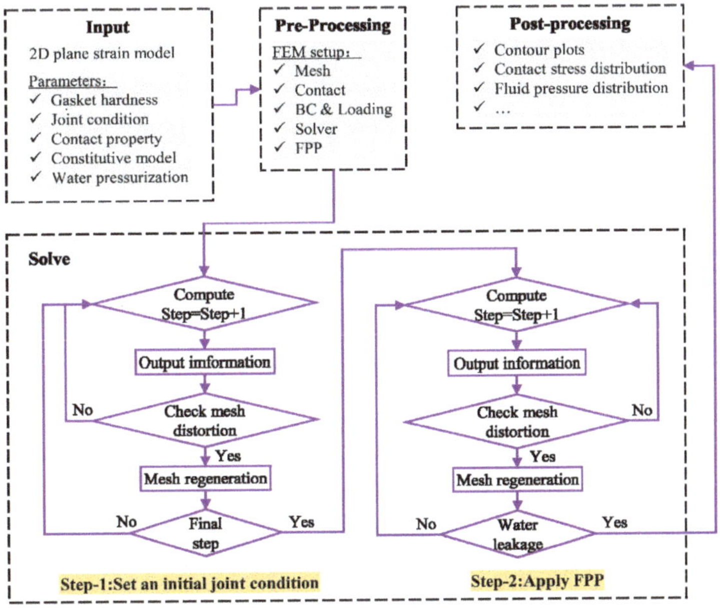

Figure 3. Flowchart of the Mesh Regeneration method [7] (Reprinting with permission from Gong et al., 2018).

The existing fine numerical simulation method of waterproof performance of joint gaskets for shield tunnels and the large numerical simulation model considering the interaction between stratum and tunnel structure did not consider the double-layer gasket configuration, and the adopted model only considers the application of external water pressure, which does not apply to deep drainage shield tunnels.

Therefore, it is necessary to focus on the upcoming requirements of joint bi-directional waterproof sealing of deep drainage shield tunnels under complex internal and external water pressure conditions, and carry out research on the seepage formation mechanism of segment joints and the waterproof failure mechanism of double-layer gaskets.

5. Thermal-Mechanical Coupling Analysis Model

In terms of the thermal-mechanical coupling analysis model, the existing research focuses on mathematical models and corresponding calculation methods for the shield tunnel lining under high temperatures caused by fire.

Zhu et al. [1] made an in-depth and detailed analysis and reviewed the research progress in the theoretical analysis model for segment joints and lining structures under high fire temperatures. Representative research achievements in recent years include Savov et al. [69], who established a layered Beam-Spring model that can characterize the bursting of lining concrete under fire; Yao et al. [70], who discussed the temperature field distribution of lining structure at high temperatures based on the Homogeneous Ring model. Luo et al. [71] used a thermal-mechanical coupling numerical model of tunnel lining structure to analyze the influence of different heating curves on stress. Further, Zhang et al. [72] developed a strong coupling numerical model of heat, water, and steam. Liu et al. [15] constructed a fine thermal-mechanical coupling model for lining structures, considering the cracking and crushing characteristics of concrete at high temperatures. In addition, some scholars have also modeled the mechanical behavior of the structure and waterproofing weak point-segment joints. Shen et al. [73] established the thermal-mechanical coupling finite element model for reinforced concrete and fiber-reinforced segment joints at high temperatures in fire. Liu et al. [15] used the joint thermal-mechanical coupling numerical model to predict the opening amount of joint gaskets and explore the waterproof performance at high temperatures.

In general, compared with the research on mechanical behavior models of shield tunnel lining structures and joints, there are few reports on the sealing capacity and response of lining structures at different joint positions under the effect of temperature.

6. Conclusions

Throughout this article, it can be seen that the research on the waterproof sealing performance of shield tunnel joints is a worthy and in-depth subject. Scholars have carried out a considerable amount of productive research, from full-scale and reduced-scale specimen waterproofing tests to numerical simulations and other aspects. However, the following problems remain to be further probed:

1. Current research mainly focuses on the sealing performance at normal temperature, and the deterioration evolution mechanism of sealing performance considering the variation of temperature is worth exploring.
2. The current aging mechanism of rubber is rather vague, and full-scale, long-term aging tests are not commonly conducted. Starting from the actual working conditions, the relationship between the gasket material aging and the waterproof ability deterioration of the shield tunnel can be obtained through the full-scale test, which will aid engineering practice.
3. The continuous–discontinuous numerical method of gasket waterproof capability is limited to the normal temperature scenarios while failing to cover extreme working conditions, such as fire, etc., which needs to be modified to consider the temperature effect.
4. Research on the thermal-mechanical coupling mechanical performance analysis model of shield tunnel lining structure and the joint is more in-depth and systematic, while the research on the thermal-mechanical coupling sealing performance analysis model is relatively rare.

The waterproof performance of shield tunnel gaskets is affected by materials, structures, and external factors. With the growing number of shield tunnels with large burial depth, high water pressure, and long design life, more attention will be paid to the waterproof performance of the gaskets. The design and inspection of joint gaskets in engineering practice will benefit from research on material durability, structural characteristics, external hazards, and simulation methods.

Author Contributions: Investigation, C.G.; resources, C.G.; data curation, C.G.; writing—original draft preparation, C.G. and Y.W.; writing—review and editing, Y.W.; supervision, W.D., M.L. and C.S.; All authors have read and agreed to the published version of the manuscript.

Funding: This research was funded by the National Natural Science Foundation of China (grant number: 51908557, 52090083, 51978669).

Institutional Review Board Statement: Not applicable.

Informed Consent Statement: Not applicable.

Data Availability Statement: Not applicable.

Conflicts of Interest: The authors declare no conflict of interest.

References

1. Zhu, H.H.; Zhou, L.; Shen, Y.; Yan, Z.G. Review of experiments and mechanical models of shield lining structure exposed to high temperature. *China J. Highw. Transp.* **2017**, *30*, 15–23, 48. (In Chinese) [CrossRef]
2. He, C.; Liu, S.J.; Zhang, Y.C.; Feng, K. Service Security of Underwater Tunnel Lining Structures and Corresponding Guarantee Countermeasures. *Eng. Sci.* **2017**, *19*, 44–51. (In Chinese) [CrossRef]
3. Du, Y.L.; Sun, B.C.; Wu, Z.S.; Qian, Y.J.; Gao, Y. Strategies for Establishing and Perfecting Long-Life Security Strategy of Transportation Infrastructure. *Eng. Sci.* **2017**, *19*, 1–5. (In Chinese) [CrossRef]

4. Howard, A.J. Report on the damaging effects of water on tunnels during their working life. *Tunn. Undergr. Space Technol.* **1991**, *6*, 11–76.
5. Dammyr, O.; Nilsen, B.; Thuro, K.; Grondal, J. Possible Concepts for Waterproofing of Norwegian TBM Railway Tunnels. *Rock Mech. Rock Eng.* **2014**, *47*, 985–1002. [CrossRef]
6. Wu, H.Y.; Dong, M.; Han, T.C.; Xu, R.Q.; Gong, X.N. Riview and prospects of new materials in urban underground development. *Eng. Sci.* **2017**, *19*, 116–123. (In Chinese)
7. Gong, C.J.; Ding, W.Q. A computational framework to predict the water-leakage pressure of segmental joints in underwater shield tunnels using an advanced finite element method. *Int. J. Numer. Anal. Methods Geomech.* **2018**, *42*, 1957–1975. [CrossRef]
8. Paul, S. *Sealability Tests of Gaskets between Precast Concrete Tunnel Lining Segments*; Unpublished Report; Department of Civil Engineering, University of Illinois at Urbana-Champaign: Urbana, IL, USA, 1978.
9. Liu, J.G. Waterproof Mechanism and Application of Shield Segment Gasket. Master's Thesis, Tongji University, Shanghai, China, 2010. (In Chinese).
10. Xie, H.M.; Du, Y.L.; He, C.; Feng, K.; Liu, S.J.; Zhang, J. Waterproof performance of segment joints of large section subsea shield tunnel under strong earthquake. *China J. Highw. Transp.* **2017**, *30*, 201–209. (In Chinese) [CrossRef]
11. Zhang, Z.X.; Sun, J.; Zhu, Y.F.; Huang, X.; Yuan, W.H. Experimental study on waterproof performanceof oint seal for deeply-buried storage and drainage tunnel. *J. Zhejiang Univ.* **2018**, *52*, 431–439. (In Chinese)
12. Shi, C.H.; Cao, C.Y.; Lei, M.F.; Yang, W.C. Sealant Performance Test and Stress-Seepage Coupling Model for Tunnel Segment Joints. *Arab. J. Sci. Eng.* **2019**, *44*, 4201–4212. [CrossRef]
13. Gong, C.J.; Ding, W.Q.; Soga, K.; Mosalam, K.M. Failure mechanism of joint waterproofing in precast segmental tunnel linings. *Tunn. Undergr. Space Technol.* **2019**, *84*, 334–352. [CrossRef]
14. Lei, M.F.; Zhu, B.B.; Gong, C.J.; Ding, W.Q.; Liu, L.H. Sealing performance of a precast tunnel gasketed joint under high hydrostatic pressures: Site investigation and detailed numerical modeling. *Tunn. Undergr. Space Technol.* **2021**, *115*, 104082. [CrossRef]
15. Liu, T.; Yuan, D.J.; Wang, A.H.; Wang, H.; Pan, D. Experimental study on watertight performance of prototype shield tunnel segment joint affected by fire. *China Civ. Eng. J.* **2016**, *49*, 116–122. (In Chinese) [CrossRef]
16. Liu, T.; Yuan, D.J.; Zhang, H. Experimental study on waterproof performance deterioration regulation of shield segment joint rubber in environment of high temperature fire. *China Civ. Eng. J.* **2015**, *48*, 244–249. (In Chinese)
17. Li, X.; Zhou, S.H.; Di, H.G.; Wang, P.X. Evaluation and experimental study on the sealant behaviour of double gaskets for shield tunnel lining. *Tunn. Undergr. Space Technol.* **2018**, *75*, 81–89. [CrossRef]
18. Dong, L.W.; Jiang, Y.S.; Yang, Z.Y.; Cheng, J.G.; Liu, C.Q.; Zhang, J.J. Experimental study and water-resistant mechanism of gaskets in joints of tunnel segments. *Chin. J. Geotech. Eng.* **2017**, *39*, 469–474. (In Chinese)
19. Zhou, W.F.; Liao, S.M.; Men, Y.Q. A fluid-solid coupled modeling on water seepage through gasketed joint of segmented tunnels. *Tunn. Undergr. Space Technol.* **2021**, *114*, 104008. [CrossRef]
20. Zheng, G.; Dai, X.; ZHang, X.S. Experimental study an numerical simulation of leaking process of sand and water in underground engineering. *Chin. J. Rock Mech. Eng.* **2014**, *33*, 2458–2471. (In Chinese) [CrossRef]
21. Zheng, G.; Yao, J.; Dai, X.; YAng, X.Y.; Sun, J.Y. Experimental study on sand inflow under different opening widths of shield tunnel segments. *Chin. J. Geotech. Eng.* **2018**, *40*, 969–977. (In Chinese)
22. Shalabi, F.I.; Cording, E.J.; Paul, S.L. Concrete segment tunnel lining sealant performance under earthquake loading. *Tunn. Undergr. Space Technol.* **2012**, *31*, 51–60. [CrossRef]
23. Shalabi, F.I.; Cording, E.J.; Paul, S.L. Sealant behavior of gasketed segmental tunnel lining—Conceptual model. *Geomech. Tunn.* **2016**, *9*, 345–355. [CrossRef]
24. Xie, H.M.; He, C.; Feng, K.; Liang, M.F.; Li, C.; Maged, A.-S. Unidirectional Vibration Waterproof Test of Ring Joint of Shield Tunnel under the Action of Earthquake. *J. Southwest Jiaotong Univ.* **2020**, *55*, 713–718. (In Chinese)
25. Zhang, D.M.; Liu, J.; Huang, Z.K.; Yang, G.H.; Jiang, Y.; Jia, K. Waterproof performance of tunnel segmental joints under different deformation conditions. *Tunn. Undergr. Space Technol.* **2022**, *123*, 104437. [CrossRef]
26. Zhang, W.J.; Zhang, G.L.; Li, H.L.; Gao, W.Y.; Guo, W.S.; Gao, P. Waterproof performance of sealing gaskets impact of construction loads on segment joints in shield tunnel. *China J. Highw. Transp.* **2020**, *33*, 130–141. (In Chinese) [CrossRef]
27. Zhang, G.L.; Zhang, W.J.; Li, H.L.; Cao, W.Z.; Wang, B.D.; Guo, W.S.; Gao, P. Waterproofing behavior of sealing gaskets for circumferential joints in shield tunnels: A full-scale experimental investigation. *Tunn. Undergr. Space Technol.* **2021**, *108*, 103682. [CrossRef]
28. Zhou, Z.; Zhang, J.; Gong, C. Automatic detection method of tunnel lining multi-defects via an enhanced You Only Look Once network. *Comput. Aided Civ. Infrastruct. Eng.* **2022**, *37*, 762–780. [CrossRef]
29. Xie, H.M.; Wang, S.M.; He, C.; Ma, T.Y.; Peng, X.Y.; Li, P. Performance of a new waterproof system with double sealing gaskets outside bolt hole of segment. *Tunn. Undergr. Space Technol.* **2022**, *119*, 104206. [CrossRef]
30. Ding, W.Q.; Gong, C.J.; Mosalam, K.M.; Soga, K. Development and application of the integrated sealant test apparatus for sealing gaskets in tunnel segmental joints. *Tunn. Undergr. Space Technol.* **2017**, *63*, 54–68. [CrossRef]
31. Wang, S.M.; Xie, H.M. Review and prospect of waterproofing of segment joints of shield tunnel with high hydraulic pressure. *Hazard Control Tunn. Undergr. Eng.* **2020**, *2*, 66–75.
32. Gong, C.J.; Ding, W.Q.; Mosalam, K.M.; Gunay, S.; Soga, K. Comparison of the structural behavior of reinforced concrete and steel fiber reinforced concrete tunnel segmental joints. *Tunn. Undergr. Space Technol.* **2017**, *68*, 38–57. [CrossRef]

33. Yang, C.; Shen, S.L.; Hou, D.W.; Liao, S.M.; Yuan, D.J. Material properties of the seal gasket for shield tunnels: A review. *Constr. Build. Mater.* **2018**, *191*, 877–890. [CrossRef]
34. Wang, Z.N.; Shen, S.L.; Zhou, A.N.; Lyu, H.M. Experimental data of water swelling characteristics of polymer materials for tunnel sealing gasket. *Data Brief* **2020**, *31*, 106021. [CrossRef] [PubMed]
35. Wang, Z.N.; Shen, S.L.; Zhou, A.N.; Lyu, H.M. Experimental investigation of water-swelling characteristics of polymer materials for tunnel sealing gasket. *Constr. Build. Mater.* **2020**, *256*, 119473. [CrossRef]
36. Schurch, M. Small but important–gaskets for tunnel segments. In Proceedings of the International Symposiumon on Underground Excavation and Tunnelling, Bangkok, Thailand, 2–4 February 2006.
37. Tan, W.Y. Research on the Property of Rubber Gaskets in Shield Tunnel Lining Joints after High Temperature. Master's Thesis, South China University of Technology, Guangzhou, China, 2015. (In Chinese).
38. Wei, Y.T.; Fang, Q.H.; Jin, Z.B.; Feng, X.J. Progress in constitutive models of filled rubber. *Polym. Bull.* **2014**, *5*, 15–21. (In Chinese) [CrossRef]
39. Li, Y.J. Study on deterioration of physical and mechanical properties of vulcanized rubber during thermal oxygen aging. *Spec. Rubber Prod.* **1997**, *18*, 44–53. (In Chinese) [CrossRef]
40. Shi, X.B.; Chen, X.Y.; Jiang, L.; Xue, H.H. Activation energy and life analysis of rubber based on accelerated thermal aging test. *Chin. N. Technol. Prod.* **2021**, *11*, 18–20. (In Chinese) [CrossRef]
41. Wang, Z.N.A.; Shen, S.L.; Zhou, A.N.; Lyu, H.M. Investigation of Time-Dependent Characteristics of EPDM Rubber Gasket Used for Shield Tunnels. *J. Mater. Civ. Eng.* **2021**, *33*, 04021251. [CrossRef]
42. Liu, Y.K.; Zhang, Q.S.; Liu, R.T.; Chen, M.J.; Zhang, C.Y.; Li, X.H.; Li, W.H.; Wang, H.B. Compressive stress-hydrothermal aging behavior and constitutive model of shield tunnel EPDM rubber material. *Constr. Build. Mater.* **2022**, *320*, 126298. [CrossRef]
43. Wang, Z.N.; Shen, S.L.; Zhou, A.N.; Xu, Y.S. Experimental Evaluation of Aging Characteristics of EPDM as a Sealant for Undersea Shield Tunnels. *J. Mater. Civ. Eng.* **2020**, *32*, 04020182. [CrossRef]
44. Jin, Y.L.; Ding, W.Q.; Yan, Z.G.; Soga, K.; Li, Z.L. Experimental investigation of the nonlinear behavior of segmental joints in a water-conveyance tunnel. *Tunn. Undergr. Space Technol.* **2017**, *68*, 153–166. [CrossRef]
45. Caratelli, A.; Meda, A.; Rinaldi, Z.; Giuliani-Leonardi, S.; Renault, F. On the behavior of radial joints in segmental tunnel linings. *Tunn. Undergr. Space Technol.* **2018**, *71*, 180–192. [CrossRef]
46. Liu, J.G.; Zhou, S.H. Analysis of the performance of the shield tunnel gasket compression considering the ballon effect. *J. Build. Mater.* **2013**, *16*, 993–997. (In Chinese)
47. Lei, Z.Y.; Liu, M. Sensitivity analysis of water-proof performance of elastic rubber gasket in shield tunnel. *J. Tongji Univ.* **2017**, *45*, 336–341. (In Chinese)
48. Yan, Z.G.; Zhu, H.H.; Ju, J.W.; Ding, W.Q. Full-scale fire tests of RC metro shield TBM tunnel linings. *Constr. Build. Mater.* **2012**, *36*, 484–494. [CrossRef]
49. Sun, J. Durability design of segment lining structure in Chongming Yangtze River shield tunnel. *J. Archit. Civ. Eng.* **2008**, *25*, 1–9. (In Chinese)
50. Wu, Z.Z.; Yang, L.D.; Ji, Q.Q.; Dai, S.; Mo, Y.T. Experimental Study on Stress Relaxation of Waterproof Gasket of River-Crossing Shield Tunnel. *J. Build. Mater.* **2009**, *12*, 539–543. (In Chinese)
51. Zhong, X.C.; Qin, J.S.; Zhu, W.; MIn, F.L. Durability Tests and Analysis on the Waterproof Material for Joint Seam of Shield Tunnel. *Chin. J. Undergr. Space Eng.* **2011**, *7*, 281–285. (In Chinese)
52. Shi, C.H.; Cao, C.Y.; Lei, M.F.; Peng, L.M.; Shen, J.J. Time-dependent performance and constitutive model of EPDM rubber gasket used for tunnel segment joints. *Tunn. Undergr. Space Technol.* **2015**, *50*, 490–498. [CrossRef]
53. Mooney, M. A theory of large elastic deformation. *J. Appl. Phys.* **1940**, *11*, 582–592. [CrossRef]
54. Rivlin, R.S. Large elastic deformations of isotropic materials IV. further developments of the general theory. *Philos. Trans. R. Soc. London Ser. A Math. Phys. Sci.* **1948**, *241*, 379–397.
55. Gong, C.J.; Ding, W.Q.; Xie, D.W. Parametric investigation on the sealant behavior of tunnel segmental joints under water pressurization. *Tunn. Undergr. Space Technol.* **2020**, *97*, 103231. [CrossRef]
56. Gong, C.J.; Ding, W.Q.; Mosalam, K.M. Performance-based design of joint waterproofing of segmental tunnel linings using hybrid computational/experimental procedures. *Tunn. Undergr. Space Technol.* **2020**, *96*, 103172. [CrossRef]
57. Zhao, M.; Ding, W.Q.; Peng, Y.C.; Shen, B.W.; Guo, X.H.; Yang, L.S. Experimental study on joint waterproofing reliability of shield tunnel with high water pressure. *Mod. Tunn. Technol.* **2013**, *50*, 87–93. (In Chinese) [CrossRef]
58. Gong, C.J.; Ding, W.Q.; Soga, K.; Mosalam, K.M.; Tuo, Y.F. Sealant behavior of gasketed segmental joints in shield tunnels: An experimental and numerical study. *Tunn. Undergr. Space Technol.* **2018**, *77*, 127–141. [CrossRef]
59. Sun, L.W. Shield Tunnel Structure Behavior under External Load. Doctoral Dissertation, Zhejiang University, Hangzhou, China, 2016.
60. Zhang, Q.J.; Mo, H.H.; Huang, C.R.; Chen, J.S. Study of Compression Performance and Cross-section Design of Rubber Gasket for Prefabricated Utility Tunnel. *Tunn. Constr.* **2018**, *38*, 399–405. (In Chinese)
61. Li, P.; Xie, H.M.; He, C.; Zhou, Z.Y.; Wang, S.M. Waterproof Performance Analysis of Water Sealing Gasket of Large Open Shield Tunnel Based on Effective Contact Stress. *Tunn. Constr.* **2019**, *39*, 1993–1999. (In Chinese)
62. Yuan, D.J.; Wu, J.; Shen, X.; Jin, D.L.; Li, X.G. Engineering safety of cross-river or cross-sea long-distance large-diameter shield tunneling under superhigh water pressure. *China J. Highw. Transp.* **2020**, *33*, 26–45. (In Chinese) [CrossRef]

63. Zhou, W.F.; Liao, S.M.; Men, Y.Q. Contact stress and waterproof capacity of T-joint in shield tunnel. *Chin. J. Geotech. Eng.* **2020**, *42*, 2264–2270. (In Chinese)
64. Shin, J.H.; Kim, S.H.; Shin, Y.S. Long-term mechanical and hydraulic interaction and leakage evaluation of segmented tunnels. *Soils Found.* **2012**, *52*, 38–48. [CrossRef]
65. Zhang, D.M.; Ma, L.X.; Zhang, J.; Hicher, P.Y.; Juang, C.H. Ground and tunnel responses induced by partial leakage in saturated clay with anisotropic permeability. *Eng. Geol.* **2015**, *189*, 104–115. [CrossRef]
66. Wu, H.N.; Shen, S.L.; Chen, R.P.; Zhou, A. Three-dimensional numerical modelling on localised leakage in segmental lining of shield tunnels. *Comput. Geotech.* **2020**, *122*, 103549. [CrossRef]
67. Tian, Y.H.; Cassidy, M.J.; Randolph, M.F.; Wang, D.; Gaudin, C. A simple implementation of RITSS and its application in large deformation analysis. *Comput. Geotech.* **2014**, *56*, 160–167. [CrossRef]
68. Wang, D.; Bienen, B.; Nazem, M.; Tian, Y.H.; Zheng, J.B.; Pucker, T.; Randolph, M.F. Large deformation finite element analyses in geotechnical engineering. *Comput. Geotech.* **2015**, *65*, 104–114. [CrossRef]
69. Savov, K.; Lackner, R.; Mang, H.A. Stability assessment of shallow tunnels subjected to fire load. *Fire Saf. J.* **2005**, *40*, 745–763. [CrossRef]
70. Yao, J.; Zhu, H.H.; Yan, Z.G.; Zeng, L.J. Experimental research on mechanical property of tunnel lining concrete after high temperature effect. *Chin. J. Undergr. Space Eng.* **2007**, *3*, 66–72, 82. (In Chinese)
71. Luo, J.L.; Xu, Z.S.; Guo, X.J.; Zhang, Y. Analysis of the failure mechanism of a new-type bamboo structure building due to heat and moisture. *J. Saf. Environ.* **2013**, *13*, 159–164. (In Chinese)
72. Zhang, Y.M.; Zeiml, M.; Maier, M.; Yuan, Y.; Lackner, R. Fast assessing spalling risk of tunnel linings under RABT fire: From a coupled thermo-hydro-chemo-mechanical model towards an estimation method. *Eng. Struct.* **2017**, *142*, 1–19. [CrossRef]
73. Shen, Y.; Yan, Z.G.; Zhu, H.H. Full-scale Experimental Research on the Development of the Initial Stage of Fire in a Large Cross-section Road Tunnel. *Chin. J. Undergr. Space Eng.* **2015**, *11*, 1080–1087. (In Chinese)

applied sciences

Article

Experimental Study and Engineering Application of the Spatial Reticulated Grid Bolt-Shotcrete Support Structure for Excavation Tunnels

Mingli Huang [1], Yuan Song [1,2,*], Xudong Zhang [3] and Tong Sun [1]

1 Key Laboratory of Urban Underground Engineering of Ministry of Education, Beijing Jiaotong University, Beijing 100044, China
2 School of Civil Engineering and Architecture, Anhui University of Science and Technology, Huainan 232001, China
3 China Railway 11th Bureau Group Co., Ltd., Wuhan 430061, China
* Correspondence: 18115036@bjtu.edu.cn; Tel.: +86-130-8368-9492

Abstract: In view of the problems of rapid construction and deformation control of large cross-section tunnels in complex urban environments, the concept of "timely high-strength support" was put forward to form strong support over time and bear the load of surrounding rock in real time, and innovatively developed the spatial reticulated grid bolt-shotcrete support structure (SRGB). The research in this paper focuses on the supporting principle, bearing mechanical properties, snatch simulation test, construction process, and engineering application effects. The components of the SRGB and their respective functions were expounded in detail. The failure modes and deformation characteristics of the tested components were revealed through the indoor loading test, and the performance characteristics of the spatial reticulated grid concrete components with high strength and high rigidity were displayed. With the help of the multi-functional operation trolley, the assembly process test of the spatial reticulated grid was carried out, which verified the feasibility of mechanized construction and the rationality of structural design. Relying on the underground tunnel project of Guangzhou Metro, the spatial reticulated grid structure was successfully used as the initial support of the tunnel for the first time, forming a complete set of mechanized rapid construction technology for a large cross-section tunnel, and constructing a timely high-strength support system suitable for the complex urban environment. The research results provide a new idea and new technology for the mechanized rapid construction and efficient support of large cross-section tunnels.

Keywords: excavation tunnel; spatial reticulated grid; bolt-shotcrete support; model test; bearing performance; mechanization; field test

Citation: Huang, M.; Song, Y.; Zhang, X.; Sun, T. Experimental Study and Engineering Application of the Spatial Reticulated Grid Bolt-Shotcrete Support Structure for Excavation Tunnels. *Appl. Sci.* **2022**, *12*, 8506. https://doi.org/10.3390/app12178506

Academic Editors: Mingfeng Lei, Chenjie Gong and Xianda Shen

Received: 3 July 2022
Accepted: 23 August 2022
Published: 25 August 2022

1. Introduction

Since the beginning of the 21st century, the number and the scale of tunnels and underground projects in China have achieved unprecedented development, and the engineering construction conditions and geological environment have become more and more complex. In particular, the large-scale urban underground traffic construction has brought many new challenges to construction safety and operation safety, and has become a scientific problem that needs to be solved urgently in the construction of underground projects. Some scholars pointed out that the safety assessment of tunnel stability is crucial to tunnel construction, and accurate analysis can lead to reliable predictions [1–4]. Under the guidance of the NATM theory, China has formed a variety of tunnel support concepts and technologies. In general, the tunnel support method has gradually changed from single support to combined support. The support concept has also changed from emphasizing the role of secondary lining support to paying attention to the primary support to bear the load of surrounding rock, and the secondary lining is only used as a safety reserve.

The combined support type composed of a grid arch, anchor bolt, and shotcrete is a widely used flexible support type, which can maximize the self-bearing capacity of the surrounding rock and allow appropriate deformation of the surrounding rock. Many scholars at home and abroad have carried out in-depth research on the bearing mechanism and support performance of grid arches, and have reported a series of achievements. Baumann and Betzle [5] took the lead in designing a three-limb grid arch, and verified the rationality of the structural design, combined with field application tests. Nomikos et al. [6] conducted an in-depth analysis of the mechanical characteristics of the three-limb grid arch, and expounded the mechanical response under different parameters. Kim et al. [7] analyzed the differences in ultimate bearing capacity and structural stability between the new four-limb grid arch and the three-limb grid arch through indoor tests and numerical simulations. Lee et al. [8] carried out research on the mechanical properties of a high-strength grid arch, obtained reasonable structural design parameters, and the good stability and bearing performance of this structure are verified. Kim et al. [9] presented an evaluation of the structural performance of tetragonal lattice girders for NATM tunnel construction, and evaluated the ultimate carrying capacity and checked the rationality of the connection method of the girder. Qiu et al. [10] evaluated the relationship between the ultimate bearing capacity of a four-limb grid arch and its component cost, and systematically studied the influence of factors such as web bar diameter and weld length on the mechanical properties of the grid arch. Tan and Han [11] carried out a field test of heat-treated high-strength lattice girders in railway tunnels and highway tunnels, compared the support performance with the original-design I20b steel arch, and achieved good application results. Gong et al. [12] and Ding et al. [13] studied the mechanical properties of concrete segment joints of different materials in shield tunnels, verified the design method, and conducted in-depth research on the waterproof performance of different joints.

Rigid support structures are usually used to bear strong loose ground pressure, such as concrete-filled steel tubulars [14], prefabricated arc plates [15], spatial reticulated grids [16,17], high-strength shotcrete [18], high-strength pre-stressed anchors [19], pretensioned bolt-cable cooperative support [20], etc. It is highly necessary to control the deformation of the deep weak surrounding rock by erecting a supporting structure with high rigidity and high strength. After the excavation of the tunnel, high support strength is applied to the surrounding rock, so that the support object and the surrounding rock within a certain range around the tunnel can work together to improve the bearing capacity of the deep rock mass, preventing the development of rheological failure to ensure long-term stability.

However, in the construction of some high-energy geological environment tunnels, it is not appropriate to adopt the traditional strengthened support [21]. One should gradually release the surrounding rock pressure under certain constraints, allowing large deformation of the surrounding rock and support, and finally achieving a balance between the two. By focusing on the structural characteristics of the support structure, the steel frame can adaptively adjust the deformation of the surrounding rock and release the excess energy accumulated by the support object. The surrounding rock pressure can be released through energy-consuming devices or other measures, allowing large deformation of the surrounding rock and supporting structure, and can provide sufficient supporting resistance to prevent lining cracking, twisting or even breaking of steel arches. In order to realize the effective support of soft rock tunnels, many scholars have also proposed a series of support theories and technical support systems, such as resistance-limiting and energy-dissipating support [22–24], prefabricated arc plate of reinforced concrete [25], stone masonry secondary support [26], concrete-filled steel tubular bolt-shotcrete support [27,28], a π-type steel-concrete composite structure [29], steel grid-concrete core tube [30], semi-rigid wire-shell bolting shotcrete [31,32] and other support forms.

Based on the above analysis of the current situation of tunnel (roadway) support technologies, the research on the use of the spatial reticulated grid is still rarely involved in the tunnel engineering field, and there is a lack of necessary discussion on the mechanism and mechanical characteristics of the SRGB structure. Therefore, this paper designs a high-strength and high-rigidity initial tunnel support structure based on the current situation of the construction of shallow-buried large-section tunnels in complex urban environments, which involve difficult construction processes, high safety risks, and tight construction schedules. The bearing characteristics and failure mechanism are analyzed through indoor model tests, and industrial tests are carried out, relying on the Guangzhou subway tunnel project to provide a reasonable new support type for urban shallow-buried large-section tunnels.

2. The Principle of the SRGB Support Structure

The SRGB support structure is composed of a systematic bolt, shotcrete and spatial reticulated grid, as shown in Figure 1. Among them, the systematic bolt and shotcrete belong to the category of flexible support, allowing moderate deformation of surrounding rock. As a new supporting structure, the spatial reticulated grid is based on high-strength seamless steel tubes as the main force-bearing parts, and threaded steel bars as the connecting webs. The full-section enclosed support structure, connected by bolted end-plates, adopts the continuous erection method, showing the stress characteristics of surface bearing.

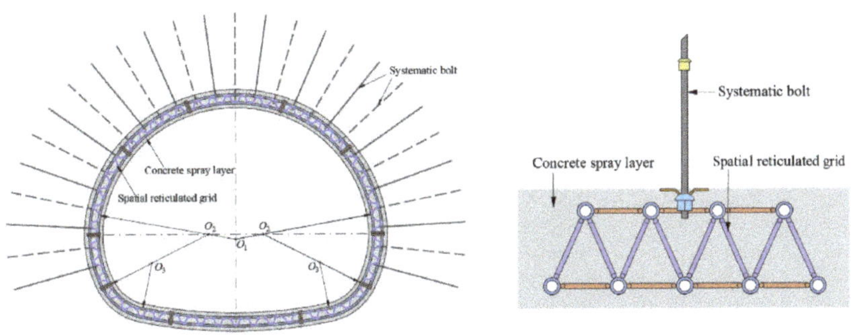

(a) (b)

Figure 1. The SRGB support structure. (**a**) The components of the support system and (**b**) sectional structure feature.

The SRGB support structure takes into account the characteristics of flexible support and rigid support. It gives full play to the advantages of the bolt and shotcrete, and the bolt anchored into the rock mass is integrated with the surrounding rock, so as to improve the stress state and the stability of the surrounding rock. After the tunnel is excavated, the surrounding rock should be sealed by spraying concrete over time, so that the surrounding rock is in a state of three-way stress, forming a combined structure to coordinately bear the load. The spatial reticulated grid draws on the structural form of the large-span grid structure on the ground, and combines the implementation environment of tunnels and underground projects to make a special support structure. This structure has the characteristics of high rigidity, high strength and good stability, and can be used to bear strong loosening pressure. The above-mentioned systematic bolt, spray layer and spatial reticulated grid together constitute the initial support system. The bolt shotcrete support is applied over time in the early stage, which has the effect of quickly sealing the surrounding rock and improving the stress state of the surrounding rock, and then erecting the high-strength arches and re-spraying concrete are carried out to form a permanent lining structure. This structure gives full play to the performance advantages of high-strength steel tubes and concrete materials, and reflects the support concept that the initial support should form strong support in time and bear the surrounding rock pressure in real time.

In tunnel engineering, the support scheme should be reasonably determined according to the different surrounding rock conditions and surrounding environment. For the challenge of strict deformation control in urban complex environment areas, the characteristics of coordinated deformation of bolt shotcrete flexible support and surrounding rock should be brought into play, and the basic load of the surrounding rock could be borne in real time, with the help of a high-strength and high-rigidity spatial reticulated grid structure to meet the bearing capacity and stability requirements of the tunnel.

3. Indoor Tests of the Spatial Reticulated Grid Concrete Components

In order to grasp the synergistic effect of the spatial reticulated grid and shotcrete, the laboratory tests under a pure bending load were carried out for the end-plate joint concrete component and jointless concrete component, the failure mode and stress characteristics of the components were analyzed, and the influence mechanism of the joint structure on the bearing performance of the support structure was also clarified.

3.1. Component Design and Production

3.1.1. Component Design

The tested components include the bolted end-plate joint concrete component (BEJC50 × 6) and the jointless concrete component (JSC50 × 6). The size of the components (length × width × height) is 3200 mm × 1200 mm × 350 mm, and the thickness of upper and lower protective layers is 40 mm. The shotcrete is made of commercial concrete with a strength grade of C25. Combined with the data provided by the laboratory test and the concrete manufacturer, it is concluded that the elastic modulus of the concrete is 26.2 GPa, and the compressive strength is 27.5 MPa. The main structural parameters of the tested components are shown in Table 1. In the components, the main limb steel tube is Q420 high-strength seamless steel, with the diameter of 50 mm and the wall thickness of 6 mm, and the steel tube spacing is 200 mm. π-shaped and U-shaped bars adopt HRB400 threaded bars with the diameter of 14 mm. The end-plate is made of Q235B carbon structural steel, with the thickness of 20 mm. The properties of the steel materials are shown in Table 2. The schematic diagram of spatial reticulated grid concrete components is shown in Figure 2.

Table 1. Main structural parameters of the tested components.

Component Type	Length (mm)	Width (mm)	Height (mm)	Concrete Grade	Steel Tube Size (mm)	End-Plate Thickness (mm)	π-Shaped or U-Shaped Rebar Diameter (mm)	Bolt Size
BEJC50 × 6 component	3200	1200	350	C25	50 × 6	20	14	M24
JSC50 × 6 component	3200	1200	350	C25	50 × 6		14	

Table 2. Mechanical properties of the steel materials.

Material	Elastic Modulus (GPa)	Yield Strength (MPa)	Ultimate Tensile Strength (MPa)	Elongation Ratio (%)
14 mm reinforcement	198	430.34	576.81	23.60
20 mm thickness steel plate	205	254.62	427.25	24.58
50 × 6 mm steel tube	205	457.78	604.61	25.47

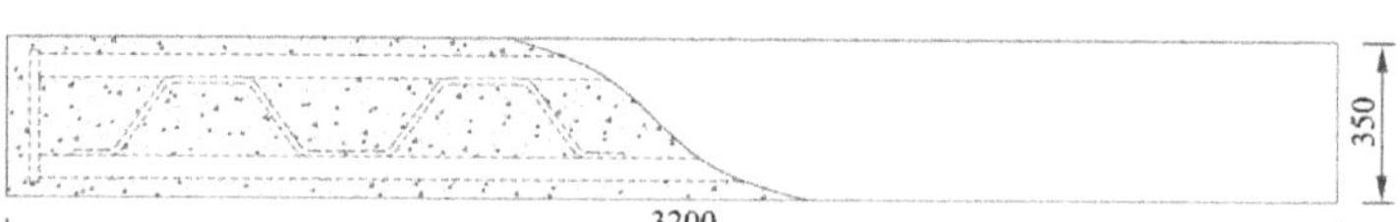

Figure 2. The schematic diagram of spatial reticulated grid concrete components.

3.1.2. Component Production

According to the size of the components, the appropriate pouring formwork is made, and concrete pouring, vibration, troweling, curing and other processes are carried out in the laboratory to ensure suitable curing conditions and effective curing quality control. The specific production process of the components is shown in Figures 3 and 4.

(**a**) (**b**)

Figure 3. Before pouring of the spatial reticulated grid components. (**a**) The bolted end-plate joint component and (**b**) the jointless component.

(**a**) (**b**)

Figure 4. After pouring of the spatial reticulated grid components. (**a**) The BEJC50 × 6 component and (**b**) the JSC50 × 6 component.

3.2. Test Loading and Measurement

The pure bending loading tests were carried out on the Y32-500A four-column hydraulic servo loading system. The two vertical actuators can provide a load pressure with a maximum nominal force of 5000 kN. The four-point loading mode is adopted to ensure that the components are in the pure bending state. The loading is carried out in the following three steps: first, one must preload the components to ensure good contact between the components and the loading device; then, the monotonic step-by-step loading method is adopted, and the load increment of each level is 20 kN; when the load pressure is no longer increased, the displacement loading method is changed until the component is damaged and unstable. The load pressure, mid-span deflection, concrete strain and other data are collected in real time through the JMZX-3006 comprehensive tester and the YBCS-CH60 high-speed static strain test and analysis system. The actual indoor loading is shown in Figure 5.

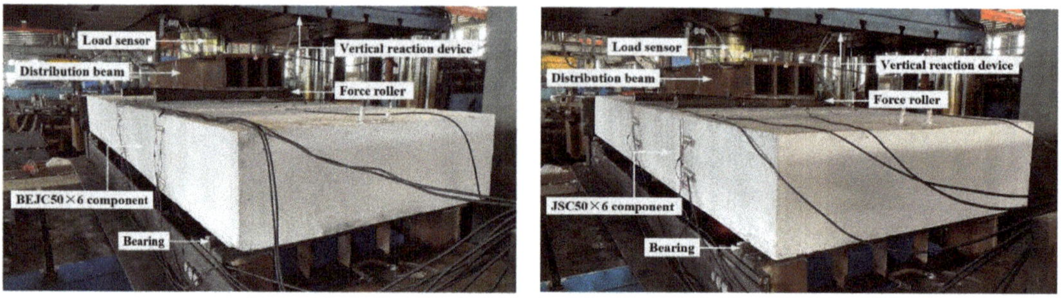

(**a**) (**b**)

Figure 5. Indoor loading tests. (**a**) The BEJC50 × 6 component and (**b**) the JSC50 × 6 component.

By considering the cracking failure of concrete and the bending deformation of components, two concrete strain monitoring sections were arranged at the mid-span and quarter point positions, six sets of strain gauges (vertical and horizontal) were pasted on the top and bottom surfaces, and four sets of strain gauges (vertical and horizontal) were pasted on the front and back surfaces. Two digital displacement meters were installed at the midspan position to monitor the deflection of the component during the loading process. The specific arrangement of strain measuring points is shown in Figure 6.

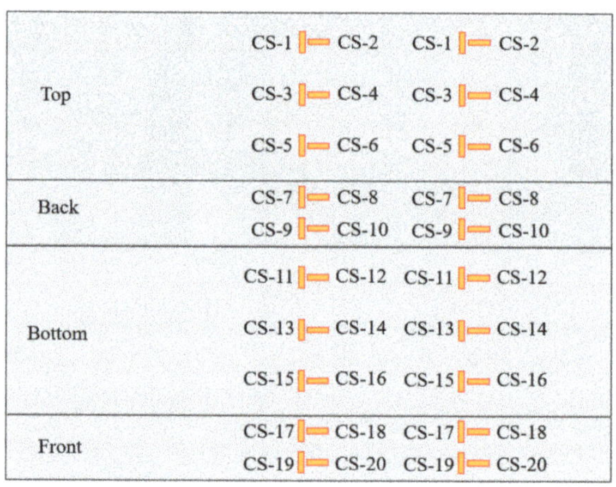

Figure 6. The arrangement of strain measuring points.

3.3. Test Results and Analysis

3.3.1. Experimental Phenomena

(1) The BEJC50 × 6 component

As shown in Figure 7, under the action of the pure bending load, small vertical cracks are generated on the tensile side of the mid-span of the BEJC50 × 6 component, and the cracks continue to expand upward with the continuous application of the load. At the same time, the crack width also increases continuously. When the load increases to a certain extent, the number of vertical cracks increases, and oblique cracks appear. There is a certain regularity in the appearance of the overall cracks, the concrete sprayed layer has good adhesion with the spatial reticulated grid, and no concrete fragments fall off.

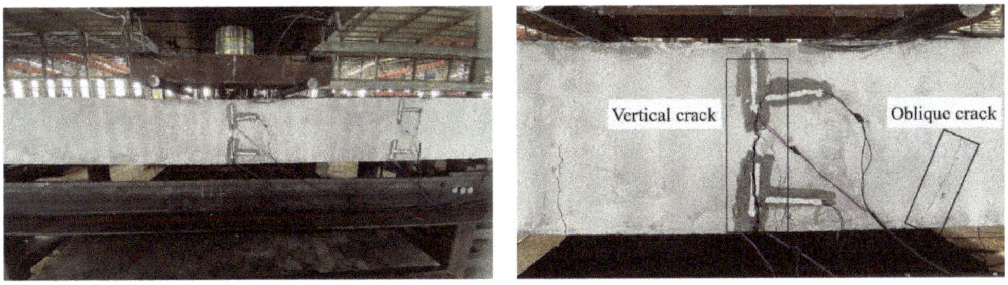

Figure 7. Failure characteristics of the BEJC50 × 6 component.

(2) The JSC50 × 6 component

As shown in Figure 8, during the loading process, a number of vertical cracks occurred on the tensile side of the pure bending section of the JSC50 × 6 component in the middle span. With the continuous increase in the load, the cracks continue to grow upward and generate oblique cracks. When the loading reaches a certain level, small oblique cracks begin to appear between the support and the pure bending section. With the continuous application of the load, the length and width of the cracks gradually increase. The appearance of the overall cracks of the component also has a certain regularity, and the component is unstable and damaged, due to the cracking and crushing of the concrete spray layer at the end of loading.

Figure 8. Failure characteristics of the JSC50 × 6 component.

3.3.2. Deflection Displacement

(1) The BEJC50 × 6 component

Figure 9 shows the relationship between the mid-span deflection displacement and the bending moment of the BEJC50 × 6 component under the pure bending loading. Combined with the experimental phenomenon, it can be observed that the overall deformation process of the component is divided into the following three stages: elastic deformation stage, failure stage, and strength degradation stage. In the initial stage of loading, the bending moment of the component is small, the change rate of deflection displacement is stable, and it simply increases linearly with the load. At this time, it is in the elastic deformation stage. The bending moment continues to be applied, the deflection growth rate gradually increases, and the concrete cracks continue to expand. When the ultimate bearing capacity is reached, the bending moment value of the component suddenly drops sharply, and the component has been damaged. The load continues to be loaded, the bearing capacity of the component gradually decreases, and the deflection changes sharply. This phase is called the strength degradation stage. The maximum bending moment of the component when it reaches the ultimate load-bearing state is 324.5 kN·m, and the mid-span deflection is about

12.0 mm. The loading process is always accompanied by the occurrence and expansion of cracks and the deterioration of bearing performance.

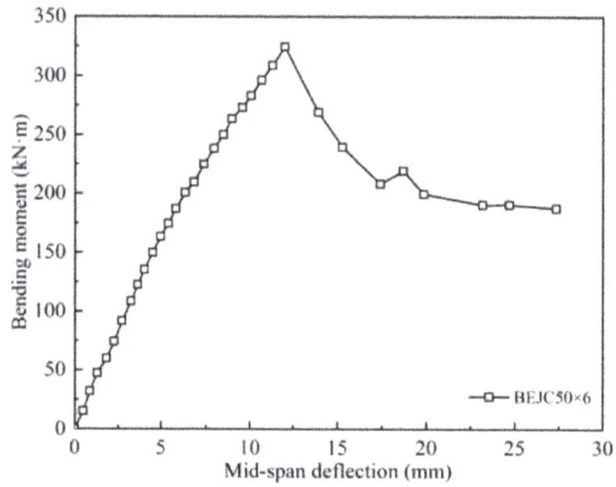

Figure 9. The bending moment–midspan deflection curve of the BEJC50 × 6 component.

(2) The JSC50 × 6 component

Figure 10 shows the relationship between the mid-span deflection displacement and the bending moment of the JSC50 × 6 component under the pure bending loading. Combined with the experimental phenomenon, it can be observed that the overall deformation process can be divided into the following three stages: elastic deformation stage, plastic deformation stage, and failure stage. When the bending moment is small, the deflection displacement increases linearly with the load. When the bending moment increases to a certain level, the bearing capacity remains relatively stable. At this stage, multiple vertical cracks and derivative cracks with uniform intervals appear on the whole component. After approaching the ultimate bearing state, the deflection growth rate is significantly accelerated, and the bending moment value of the component drops significantly, indicating that the component has been damaged. The ultimate bearing capacity of the JSC50 × 6 component is 378.8 kN·m, and the mid-span deflection is about 50.1 mm. Compared with the BEJC50 × 6 component, the allowable deformation and bearing capacity of the JSC50 × 6 component is higher, but the bending stiffness is slightly lower.

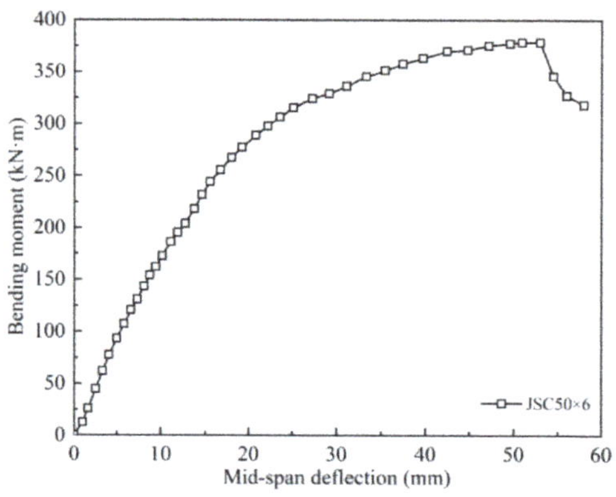

Figure 10. The bending moment–midspan deflection curve of the JSC50 × 6 component.

3.3.3. Concrete Strain

(1) The BEJC50 × 6 component

The relationship curves of the concrete surface strain and bending moment of the BEJC50 × 6 component are shown in Figures 11 and 12, including two monitoring sections at the 1/2 and 1/4 positions.

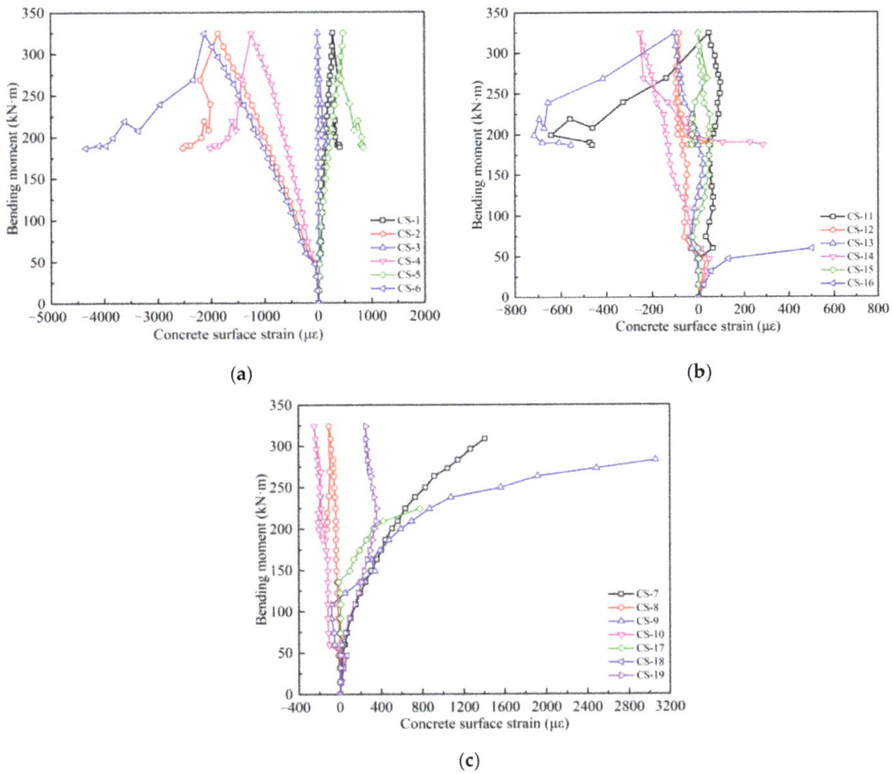

Figure 11. The bending moment–concrete strain relationship curves of 1/2 section: (**a**) monitoring points on top surface, (**b**) monitoring points on the bottom surface, (**c**) monitoring points on front and back surfaces.

From the Figure 11, it can be observed that the measuring points CS-1, CS-3, and CS-5 on the upper surface of the component are in the tension state, the maximum tensile strain occurs at the measuring point CS-5, which is 474 με, and the corresponding stress value is about 10.9 MPa, which is far greater than the ultimate tensile strength of the concrete. The measuring points CS-2, CS-4, and CS-6 show an obvious compressive state. The maximum compressive strain occurs at the measuring point CS-6, which is −2122 με, and the corresponding stress value is about 48.8 MPa, which has exceeded the ultimate compressive strength of the concrete. The measuring points on the lower surface of the component show a certain degree of tension state at the initial stage of loading, and then change to the compressive state. The maximum tensile and compressive strains occur at the measuring points CS-11 and CS-14, which are 46 με and −254 με respectively. In the front and back surfaces of the component, the measuring points CS-7, CS-9, CS-17, and CS-19 show an obvious tension state, and the maximum strain of the measuring point CS-9 reaches 3061 με, which has exceeded the ultimate strength of the concrete. The measuring points CS-8, CS-10, and CS-18 exhibit a compressive state, and the maximum compressive strain occurred at the measuring point CS-10, which is −250 με, and the stress value is about 5.75 MPa.

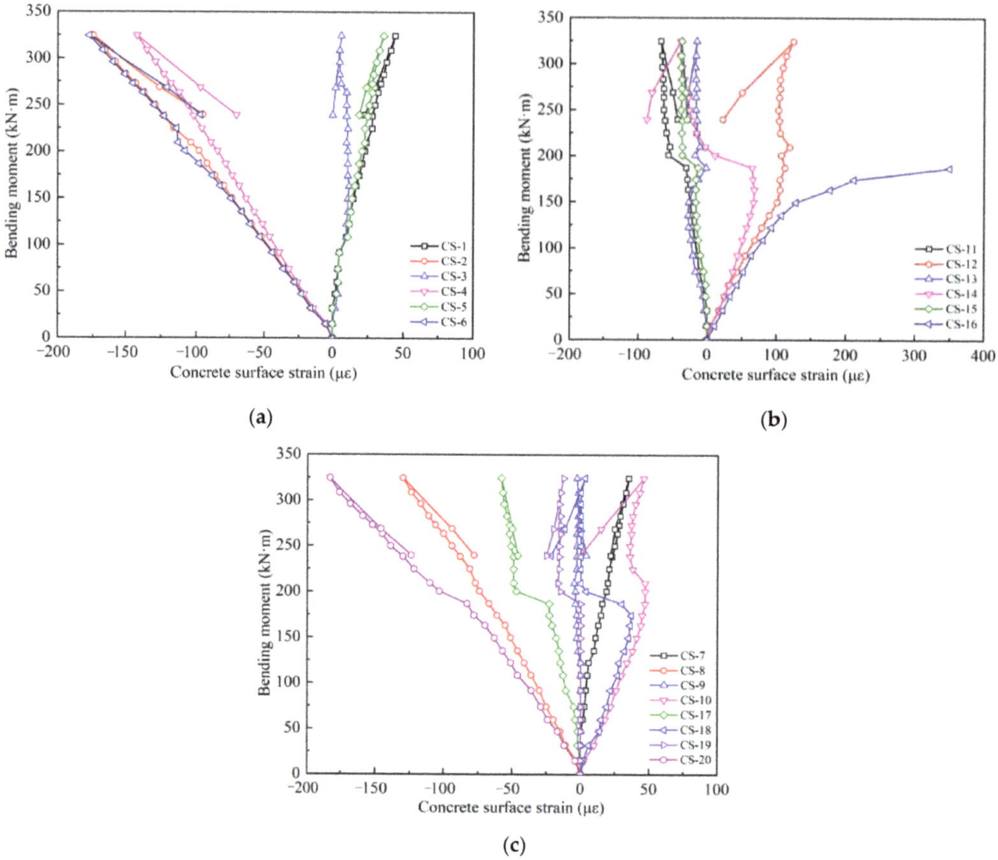

Figure 12. The bending moment–concrete strain relationship curves of 1/4 section: (**a**) monitoring points on top surface, (**b**) monitoring points on the bottom surface, (**c**) monitoring points on front and back surfaces.

From the Figure 12, it can be observed that the measuring points CS-1, CS-3, and CS-5 on the upper surface show a certain degree of the tension state, and the maximum tensile strain occurs at the measuring point CS-1 when the component is fails, and the strain is 44 με. The measuring points CS-2, CS-4 and CS-6 are in the compression state, the maximum compressive strain occurs at the measuring point CS-6, which is −178 με, and the stress value is about 4.1 MPa. The measuring points CS-11, CS-13, and CS-15 on the lower surface show a certain degree of the compression state. The maximum compressive strain occurs at the measuring point CS-11, which is −68 με, and the stress value is about 1.56 MPa. The measuring points CS-12, CS-14, and CS-16 show an obvious tensile state, and the maximum tensile strain occurs at the measuring point CS-16, which is 350 με, and the stress value is about 8.05 MPa, which has exceeded the ultimate tensile strength of the concrete. The measuring points CS-7, CS-10, and CS-18 on the front and back surfaces are in the tension state, the maximum tensile strain occurs at the measuring point CS-10, which is 46 με, and the stress value is about 1.1 MPa. The measuring points CS-8, CS-17, CS-19, and CS-20 are in the compression state. The maximum compressive strain occurs at the measuring point CS-20, which is −183 με, and the stress value is about 4.2 MPa.

(2) The JSC50 × 6 component

The relationship curves of the concrete surface strain and bending moment of the JSC50 × 6 component are shown in Figures 13 and 14, including two monitoring sections at the 1/2 and 1/4 positions.

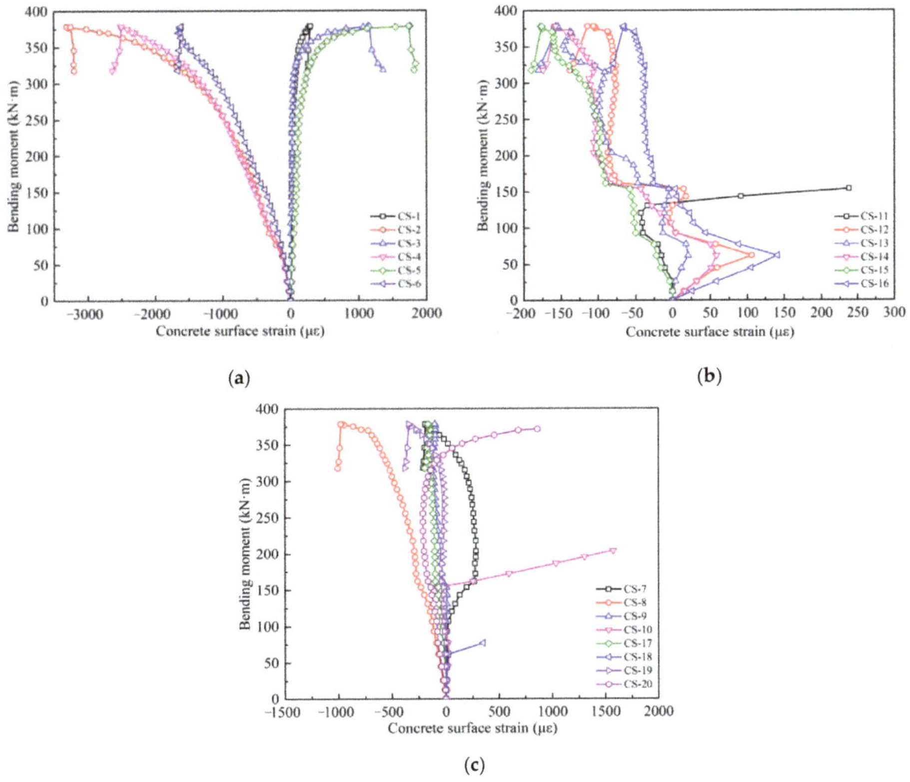

Figure 13. The bending moment–concrete strain relationship curves of 1/2 section: (**a**) monitoring points on top surface, (**b**) monitoring points on the bottom surface, (**c**) monitoring points on front and back surfaces.

From the Figure 13, it can be observed that the strain of each measuring point on the upper surface grows smoothly during the whole loading process. The measuring points CS-1, CS-3, and CS-5 are in the tension state, and the maximum tensile strain occurs at the measuring point CS-5, which is 1728 με, and the stress value is about 39.7 MPa, which is much larger than the ultimate tensile strength of concrete. The measuring points CS-2, CS-4, and CS-6 are in the compressive state. The maximum compressive strain occurs at the measuring point CS-2, which is −3316 με, and the stress value is about 76.3 MPa, which has exceeded the ultimate compressive strength of the concrete. For each measuring point on the lower surface, the strain grows rapidly at the initial stage of loading, and then the strain shows a regular "step" jump, which is related to the regular generation and expansion of concrete cracks. In the later stage, the growth rate of concrete strain is relatively slow. The maximum compressive strain occurs at the measuring point CS-15, which is −177 με, and the stress value is about 4.1 MPa. The measuring points CS-7, CS-10, CS-18, and CS-20 on the front and back surfaces are in the tensile state, and the maximum strain of the measuring point CS-10 reaches 1792 με, and the stress value is about 41.2 MPa. The measuring points CS-8, CS-9, CS-17, and CS-19 are in the compressive state, and the maximum compressive strain occurred at the measuring point CS-8, which is −969 με, and the stress value is about 22.3 MPa.

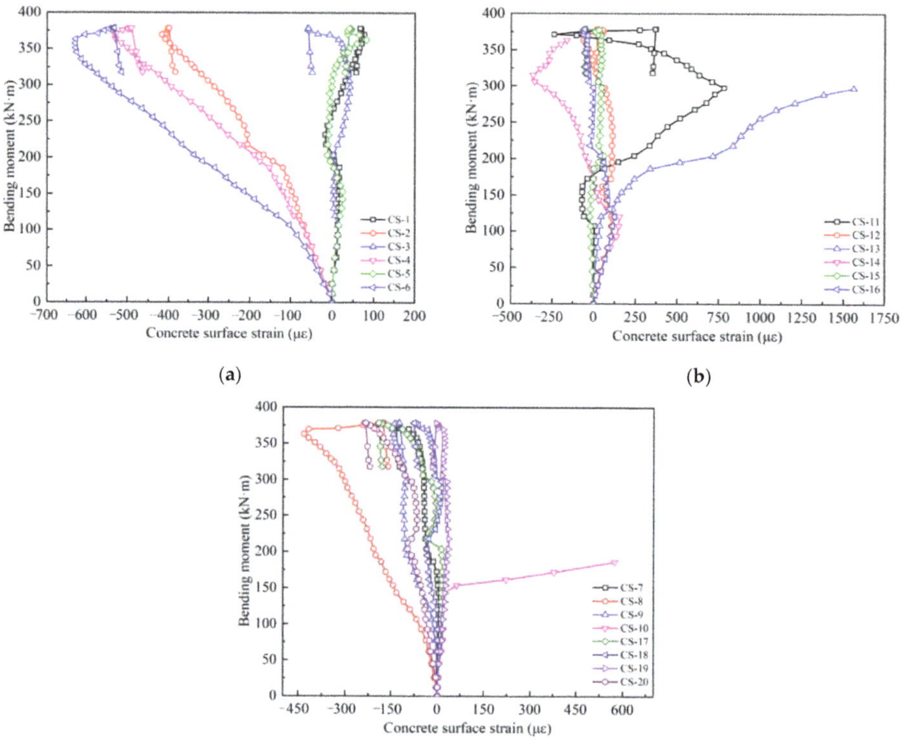

(c)

Figure 14. The bending moment–concrete strain relationship curves of 1/4 section: (**a**) monitoring points on top surface, (**b**) monitoring points on the bottom surface, (**c**) monitoring points on front and back surfaces.

From the Figure 14, it can be observed that the upper surface measuring points CS-1, CS-3, and CS-5 are in the tension state, the maximum tensile strain occurs at the measuring point CS-5, which is 81 με, and the stress value is about 1.9 MPa. The measuring points CS-2, CS-4, and CS-6 are evidently in the compression state, the maximum compressive strain occurs at the measuring point CS-6, which is −628 με, and the stress value is about 14.4 MPa. The measuring points CS-11, CS-13, and CS-15 on the lower surface are in the tension state, and the maximum strain of the measuring point CS-13 reaches 1560 με, which exceeds the ultimate tensile strength of the concrete. The measuring points CS-12, CS-14, and CS-16 are under compression. The maximum compressive strain occurs at the measuring point CS-14, which is −369 με, and the stress value is about 8.5 MPa. The measuring points CS-10 and CS-19 on the front and back surfaces are in the tension state, and the maximum strain of the measuring point CS-10 reaches 574 με, and the stress value is about 13.2 MPa, which exceeds the ultimate tensile strength of concrete; the maximum compression strain occurs at the measuring point CS-8, which is −433 με, and the stress value is about 10.0 MPa.

4. Simulation Analysis of the Spatial Reticulated Grid Structure Segmentation and Snatch Assembly

4.1. Determination of the Joint Position of the Spatial Reticulated Grid

Combined with the underground excavation tunnel project between Hengli and Panyu Square Station of Guangzhou Metro Line 22, research on the subsection scheme of the spatial reticulated grid support structure is carried out. The tunnel passes through moderately weathered granite, the surrounding rock is grade IV, the bulk density is 25,000 N/m^3, the elastic modulus is 5×10^3 MPa, Poisson's ratio is 0.2, the cohesion force is 400 kPa, and the internal friction angle is 35°. The burial depth is 26.5 m, the excavation section span is 14.9 m, the height is 11.95 m, and the spray layer thickness is 0.35 m.

The general finite element software ANSYS is used as the calculation tool to analyze the mechanical characteristics of the tunnel support structure by the load-structure method. According to the "Code for Design of Railway Tunnel" [33], the vertical and horizontal uniform force of the surrounding rock are calculated, and then the equivalent nodal force of each node is obtained, and the formation spring is used to simulate the interaction between the surrounding rock and the structure.

The Link10 element and the Beam3 beam element are used to simulate the interaction between the surrounding rock and the structure, and the lining support structure. A gravity load (g = 9.8 m/s^2) is added to the finite element model through the ANSYS program, the tunnel lining weight is 24,500 N/m^3, Poisson's ratio is 0.28, and the elastic resistance coefficient is 600 MPa/m.

Figure 15 is the internal force distribution diagram of the simulated support structure. From the bending moment diagram, some positions with zero bending moment in the support plane can be obtained. According to this, the joint position of the prefabricated support structure can be preliminarily set at the zero moment.

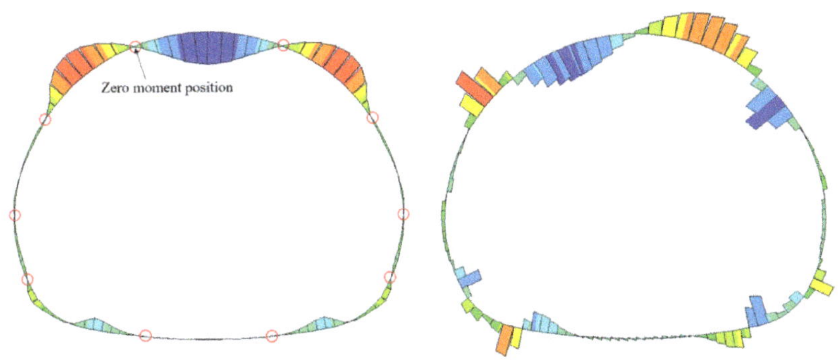

(**a**) (**b**)

Figure 15. Internal force diagram of primary support structure. (**a**) Bending moment diagram and (**b**) shear force diagram.

Combined with the on-site construction method and the load characteristics of the surrounding rock, and taking into account the weight of the components, construction convenience and other factors, the spatial reticulated grid segmentation is systematically analyzed. Each ring of the arch is divided into 10 segments, and the specific joint division is shown in Figure 16a. The height of the upper and lower benches is 7.3 m and 4.65 m, respectively, and the components are divided into five types. The upper bench includes A and B components, and the lower bench includes C, D, and E components. Figure 16b shows the circumferential and longitudinal joint connection types of the spatial reticulated grid. Circumferential joints are connected by the end-plates and high-strength bolts of two adjacent components, and two different U-shaped bolts are used for longitudinal connections to be placed in the reserved holes, respectively. These connection types are easy to install and operate, and reduce the welding workload to a certain extent.

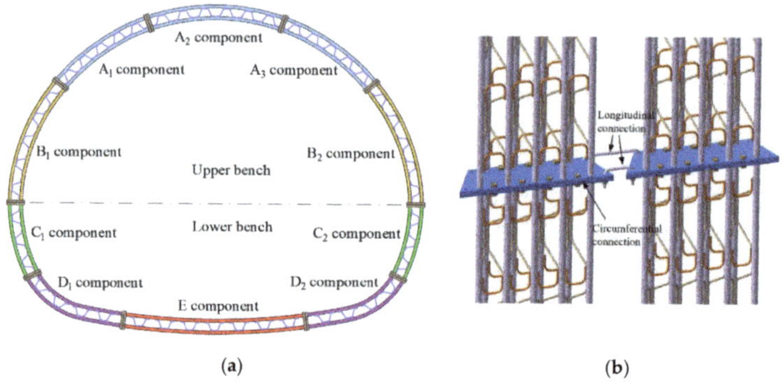

Figure 16. Division scheme and joint connection of the spatial reticulated grid structure. (**a**) Arch joint distribution and (**b**) joint connection types.

4.2. Simulation Analysis of Deformation and Force in the Snatch Process of the Spatial Reticulated Grid

Due to the heavy weight of the spatial reticulated grid components, large deformation and stress concentration may occur in the process of mechanized installation, which will affect the assembly accuracy and installation quality of the arch. Therefore, the finite element software ABAQUS version 6.14 developed by Dassault company in France is used to test the working state of the intermediate boom when lifting the three A components at the arch crown, so as to clarify the stress and deformation laws of the key parts during the lifting process, which provides a reference for on-site mechanized construction.

4.2.1. Numerical Calculation Model

A typical double-lane underground tunnel is selected as the research background, and the spatial beam element model of the spatial reticulated grid is established, in which the number of elements is 5558 and the number of nodes is 10,435. The diameter and wall thickness of the steel tubes are 50 mm and 6 mm, respectively, and the diameter of the connecting rebar is 14 mm, which are simulated by the three-node quadratic space beam element (B32). The mechanical response of the components under the action of gravity is analyzed, the influence of the joint is not considered in the calculation, and the connection between the connecting rebar and the steel tube is simplified, to some extent. The main mechanical parameters of the steel are selected according to Table 2. According to the different snatch positions of the spatial reticulated grid, two comparative working conditions are set, which are the 1/2 position and the 1/3 position. The specific conditions are shown in Figure 17.

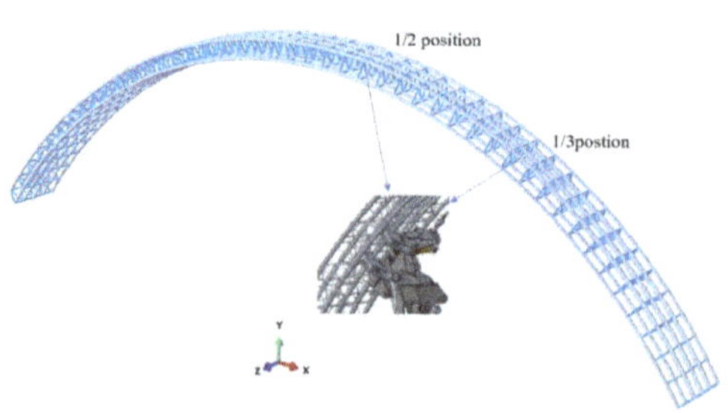

Figure 17. Calculation model of prefabricated spatial reticulated grid.

4.2.2. Analysis of the Simulation Results

(1) 1/2 position

It can be observed from Figure 18 that the maximum stress value of the spatial reticulated grid is 181.8 MPa, which occurs at the position of the vault close to the spandrel, and there is an obvious stress concentration phenomenon at the connection between the rebar and the steel tube. The closer to the two ends of the components, the smaller the stress value, the better the stress state.

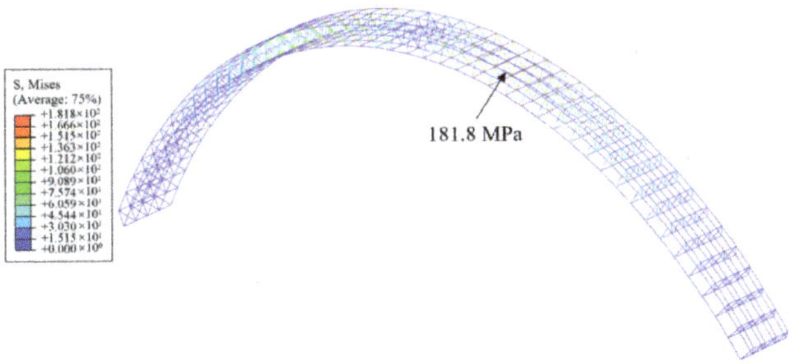

Figure 18. Mises stress nephogram of spatial reticulated grid.

It can be observed from Figure 19 that from the vault to both ends, the vertical and transverse deformation of the spatial reticulated grid gradually increases, the maximum transverse deformation is 14.15 mm, and the maximum vertical deformation is 30.27 mm. In general, all parts of the components are in good working condition, the force is reasonable, and the deformation amplitude is small, which verifies that the designed spatial reticulated grid has sufficient strength and rigidity.

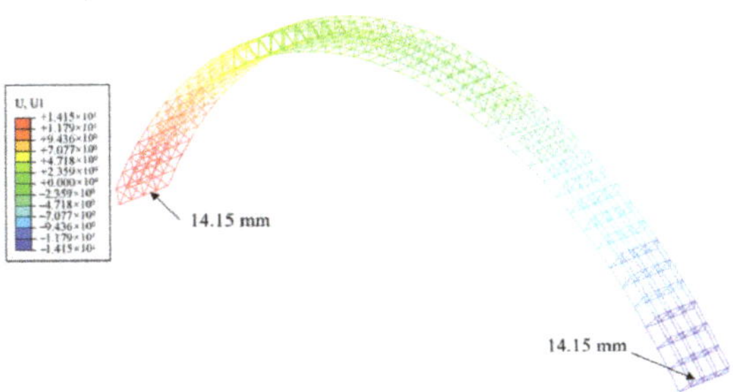

(**a**)

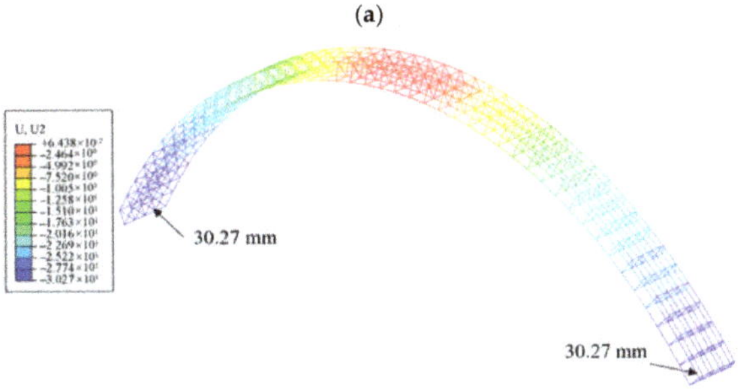

(**b**)

Figure 19. Deformation nephograms of the spatial reticulated grid. (**a**) Transverse deformation and (**b**) vertical deformation.

(2) 1/3 position

It can be observed from Figure 20 that the maximum stress position of the spatial reticulated grid occurs near the snatch point, with a magnitude of 446.7 MPa, which is close to the yield strength of the main limb steel tube, but exceeds the yield strength of the connecting rebar.

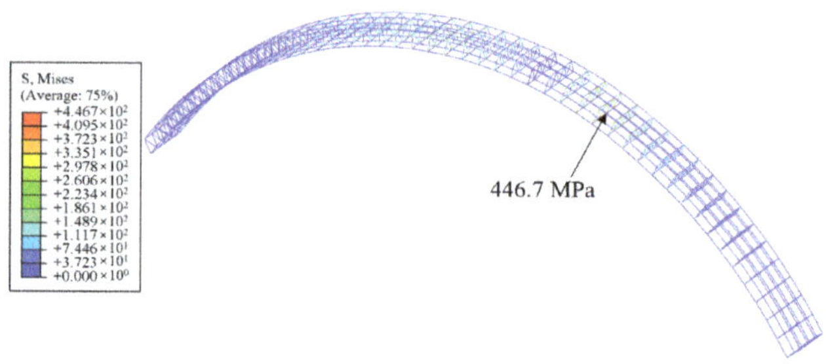

Figure 20. Mises stress nephogram of spatial reticulated grid.

It can be observed from Figure 21 that the maximum transverse and vertical deformations of the spatial reticulated grid both occur at the leftmost end, which are 77.43 mm and 213.6 mm, respectively. In summary, the selection of the snatch position has a significant impact on the deformation and force of the components. When assembling the arch, the middle position should be selected as the snatch point to minimize the adverse effect of the arch installation on the overall structural performance.

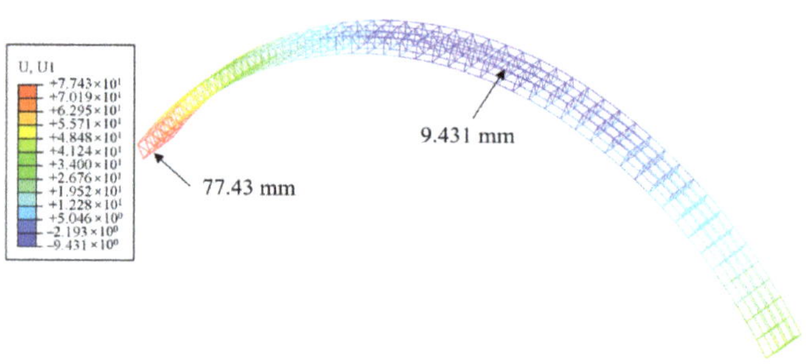

(**a**)

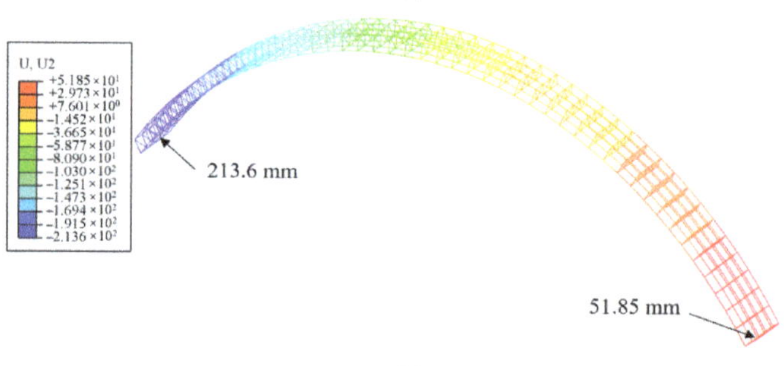

(**b**)

Figure 21. Deformation nephograms of the spatial reticulated grid. (**a**) Transverse deformation and (**b**) vertical deformation.

5. Field Application and Monitoring Analysis

The SRGB support structure belongs to a new type of technology and new process. In order to further clarify the adaptability and feasibility of this support structure in the subway tunnels in bustling urban areas, and to grasp the supporting working status, the field application test of the spatial reticulated grid was carried out. It highlights the performance advantages of the spatial reticulated grid, and provides a safety guarantee for the mechanized and rapid construction of urban underground space engineering.

5.1. Project View

The underground tunnel project of Panyu Square Station of Guangzhou Rail Transit Line 22 has a total length of 262.5 m, and the geographical location of the project is shown in Figure 22. The buried depth of the test tunnel is about 26.5 m, and the size of the excavation section (width × height) is 14.9 m × 11.95 m. The single-hole double-track horseshoe-shaped composite lining structure is adopted, the quality grade of surrounding rock is grade IV, and the bench method is adopted for construction. The minimum horizontal clear distance between the tunnel and Luojiayong bridge piles is only 2.06 m. There are many complex pipelines and major control structures, such as large hotels and libraries, on the ground. The specific location relationship is shown in Figure 23.

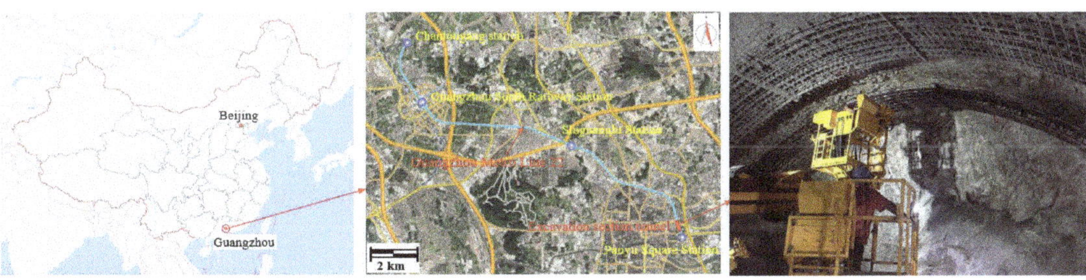

Figure 22. The geographical location of the underground tunnel project in the urban area in Guangzhou, China.

Figure 23. Surrounding environment of section tunnel project.

The design parameters of the primary support of the tunnel are as follows: the longitudinal spacing of the spatial reticulated grid is 1.0 m; and steel mesh diameter is 8 mm, with a spacing of 15 cm × 15 cm; the shotcrete grade is C25, and the thickness of the sprayed layer is 35 cm; early-strength mortar bolts diameter is 22 mm, the length is 3.0 m, and the row spacing is 1.0 m × 1.0 m. The arch foot and side wall are equipped with 42 mm diameter locking anchor tubes, and the single length is 2.0 m.

5.2. On-Site Mechanized Installation Process

(1) The preparation before construction

The staff observes whether the road surface of the tunnel meets the safe passage of the multi-functional operation trolley. If the traffic conditions are not met, the road surface needs to be repaired; one must check the overbreak and underbreak of the tunnel face, if the tunnel shows signs of underbreak, it needs to be dealt with in advance. Field processing is shown in Figure 24.

(**a**) (**b**)

Figure 24. The preparation before construction. (**a**) Road surface treatment and (**b**) the overbreak and underbreak treatment.

(2) Mechanized installation

The erection sequence of the spatial reticulated grid is the arch top A component, the spandrel B component, the arch waist C component, the arch foot D component, and the arch bottom E component. When installing the arches on the upper bench, each of the left, middle, and right arms grab an A component for ground assembly, and at the same time, the reinforcement mesh is bound. After that, the left and right mechanical arms complete the action of releasing the arches, and the middle arm lifts the three A components to the design position. The height and position of the arch installation are based on the previous cycle arch, and there should be no misalignment. Afterwards, the left and right arms respectively grab a B component and assemble it with the A components, tighten the connecting bolts in a reasonable and orderly manner, firmly pad the arch feet, and set up locking foot anchor pipes for fixing. The on-site implementation is shown in Figure 25.

(**a**) (**b**)

Figure 25. Mechanized installation of the arches. (**a**) Assembled arches near the ground and (**b**) the installation of upper bench arches.

When installing the lower bench arches, the left and right arms grab a C component respectively and connect it with the B component in a circumferential direction. Afterwards, the left and right arms grab a D component, respectively, the middle arm grabs the E component, and implement a temporary connection in sequence according to the execution sequence. Then, the arm manipulator is manipulated to adjust the spacing and height of the arches to complete the positioning. The on-site construction is shown in Figure 26.

(**a**) (**b**)

Figure 26. Mechanized installation of arch waist and inverted arch. (**a**) Arch waist and (**b**) inverted arch.

(3) Other processes

When the mechanized installation of the arches is completed, the longitudinal connection was carried out in time, the remaining reinforcement mesh was laid, the system bolts and foot locking anchor tubes were set, and the shotcrete and other processes were also carried out to ensure the construction quality of the initial support.

5.3. *Analysis of Engineering Application Effect*

In order to test the application effect of the SRGB support structure, the stress conditions of the SRGB support structure during the construction process were monitored on site. The main test items include the settlement and convergence of surrounding rock, surrounding rock pressure, steel tube stress and concrete spray layer stress. The layout of measuring points on the test section is shown in Figure 27.

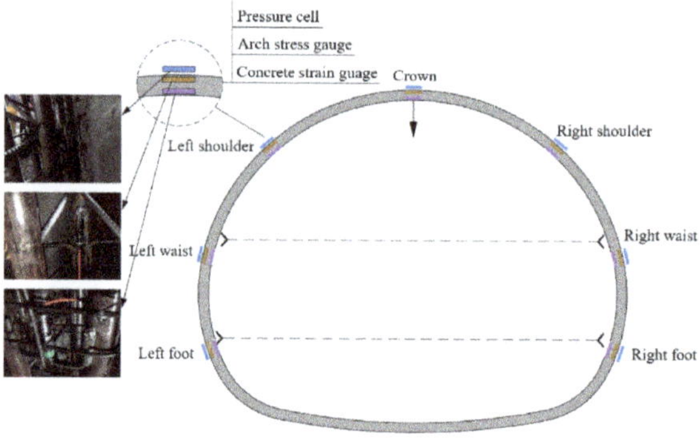

Figure 27. Schematic diagram of the layout of monitoring components.

Figure 28 shows the vault settlement and peripheral convergence duration curves of the SRG arch support section and the grid arch support section. The black line represents the grid arch support, and the red line represents the SRG arch support.

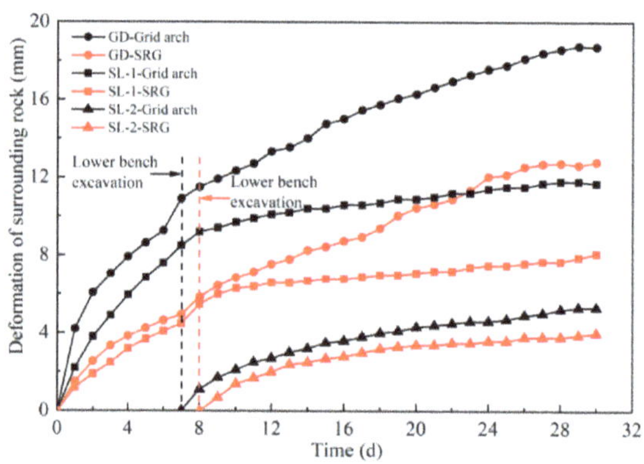

Figure 28. Duration curves of the displacement of the surrounding rock.

After nearly 30 days of real-time monitoring, the maximum settlement of the vault with the SRG arch support is 12.8 mm, which is 5.9 mm less than that with the grid arch support; in addition, the convergence deformation of the upper and lower benches is 8.1 mm and 4.0 mm, accounting for 68.6% and 75.5% of the convergence deformation of the grid arch support, respectively. The main reason is that the SRG arch has high support stiffness and bearing capacity. Under the condition of mechanized construction, it can quickly complete the accurate assembly of a heavy arch frame, close the surrounding rock in time, and effectively control the deformation after tunnel excavation.

Figure 29 shows that the distribution of the surrounding rock pressure in different positions is not the same. The contact pressure in the arch waist and arch shoulder increases more obviously, and the maximum pressure value is significantly higher than that in other positions. The pressure tends to be stable approximately six days after the excavation of the lower bench. The contact pressure of the right arch waist is the largest (244.9 kPa), followed by the left arch shoulder (124.0 kPa), left arch waist (103.5 kPa), right arch foot (70.5 kPa), vault (65.4 kPa) and left arch foot (55.8 kPa). During the whole construction process, the spatial reticulated grid structure was in the elastic deformation stage, and the stress state was good. The stress on the spandrel and the vault is generally larger, followed by the arch waist, and the arch foot is the least stressed. The concrete sprayed layer is dominated by tensile strain, and the local position exceeds the ultimate tensile strength of concrete, but the overall structure still has high safety redundancy.

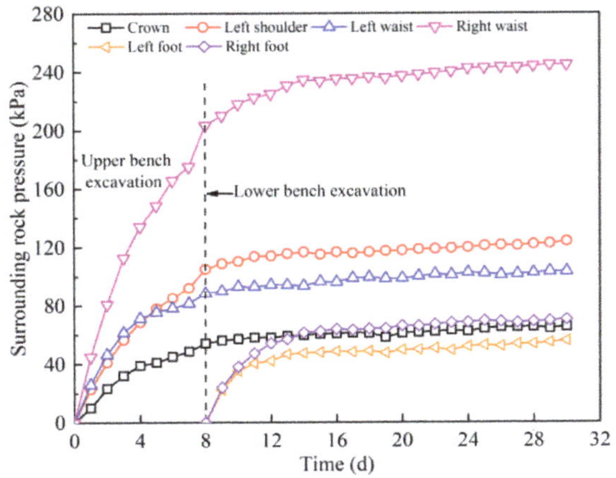

Figure 29. Duration curves of the surrounding rock pressure.

In general, by relying on the field test of the spatial reticulated grid structure carried out by the underground tunnel project of Guangzhou Metro, the first application of this structure in an urban subway tunnel has been realized. The mechanized construction method can effectively shorten the arch erection time. The high-strength, high-rigidity, and high-toughness mechanical properties of the spatial reticulated grid restrain the deformation of the surrounding rock to a great extent, and achieved good application results.

6. Conclusions

(1) In view of the problems of high risk and strict deformation control in the construction of large cross-section underground tunnels in complex urban environments, the concept of "timely high-strength support" was put forward, in which the initial support should form strong support over time and bear the surrounding rock load in real time.

(2) The indoor local full-scale performance tests were carried out around the concrete components with bolted end-plate joints and without joints. The failure modes and bearing characteristics of the two types of flexural components were obtained, and the law of the influence of the joint structure on the bearing performance of the components was revealed. Compared with the jointless concrete component, the bolted end-plate joint concrete component had higher flexural rigidity, and the jointless concrete component had higher ultimate bearing capacity and better deformability.

(3) The simulation analysis of deformation and force in the snatch process of the prefabricated spatial reticulated grid was carried out, and the field test was conducted with the help of the multi-functional operation trolley, which verified the feasibility and superiority of the mechanized construction. The first successful application of the SRGB support structure in the underground tunnel of the Guangzhou metro realized the deformation control and efficient support of the large-section tunnel in the high-density area of the city, enriched the support structure system of the underground tunnel in China, and could provide a reference for similar engineering processes in the future.

Author Contributions: M.H. and Y.S.: conceptualization, methodology, project administration and writing the original draft; X.Z. and T.S.: validation, analysis, and review. All authors have read and agreed to the published version of the manuscript.

Funding: This work was supported by the General Program of National Natural Science Foundation of China (No. 51878037) and the National Key R&D Program of China (No. 2018YFC0808705).

Institutional Review Board Statement: Not applicable.

Informed Consent Statement: Not applicable.

Data Availability Statement: The data presented in this study are available upon request from the corresponding author.

Conflicts of Interest: The authors declare no conflict of interest.

References

1. Zhou, Z.; Zhang, J.J.; Gong, C.J. Automatic detection method of tunnel lining multi-defects via an enhanced You Only Look Once network. *Comput. Aided Civ. Infrastruct. Eng.* **2022**, *37*, 762–780. [CrossRef]
2. Keawsawasvong, S.; Ukritchon, B. Design equation for stability of a circular tunnel in anisotropic and heterogeneous clay. *Undergr. Space* **2022**, *7*, 76–93. [CrossRef]
3. Keawsawasvong, S.; Ukritchon, B. Design equation for stability of shallow unlined circular tunnels in Hoek-Brown rock masses. *Bull. Eng. Geol. Environ.* **2020**, *79*, 1–24. [CrossRef]
4. Sirimontree, S.; Keawsawasvong, S.; Ngamkhanong, C.; Seehavong, S.; Sangjinda, K.; Jearsiripongkul, T.; Thongchom, C.; Nuaklong, P. Neural network-based prediction model for the stability of unlined elliptical tunnels in cohesive-frictional soils. *Buildings* **2022**, *12*, 444. [CrossRef]
5. Baumann, T.; Betzle, M. Investigation of the performance of lattice girders in tunnelling. *Rock Mech. Rock Eng.* **1984**, *17*, 67–81. [CrossRef]
6. Nomikos, P.P.; Sofianos, A.I.; Sakkas, K.M.; Choumanidis, D.; Delendas, S. Nonlinear simulation of lattice girder segment tests. *Tunn. Undergr. Space Technol.* **2013**, *38*, 180–188. [CrossRef]
7. Kim, H.J.; Song, K.I.; Jung, H.S.; Shin, Y.W.; Shin, J.H. Performance evaluation of lattice girder and significance of quality control. *Tunn. Undergr. Space Technol.* **2018**, *82*, 482–492. [CrossRef]
8. Lee, J.W.; Min, K.N.; Jeong, J.W.; Roh, B.K.; Lee, S.J.; Ahn, T.B.; Kang, S.S. Development of high strength lattice girder and evaluation of its performance. *J. Eng. Geol.* **2020**, *30*, 43–57.
9. Kim, S.; Han, T.H.; Baek, J.S.; Kang, Y.J. Evaluation of the Structural Performance of Tetragonal Lattice Girders. *Int. J. Steel Struct.* **2013**, *13*, 31–47. [CrossRef]
10. Qiu, W.G.; Lu, F.; Wang, G.; Huang, G.; Zhang, H.J.; Zhang, Z.Y.; Gong, C. Evaluation of mechanical performance and optimization design for lattice girders. *Tunn. Undergr. Space Technol.* **2019**, *87*, 100–111. [CrossRef]
11. Tan, Z.S.; Han, K.H. Experimental study on the application of heat-treated high-strength lattice girder in tunnel engineering. *Symmetry* **2019**, *11*, 1007. [CrossRef]
12. Gong, C.J.; Ding, W.Q.; Mosalam, K.M.; Günay, S.; Soga, K. Comparison of the structural behavior of reinforced concrete and steel fiber reinforced concrete tunnel segmental joints. *Tunn. Undergr. Space Technol.* **2017**, *68*, 38–57. [CrossRef]
13. Ding, W.Q.; Gong, C.J.; Mosalam, K.M.; Soga, K. Development and application of the integrated sealant test apparatus for sealing gaskets in tunnel segmental joints. *Tunn. Undergr. Space Technol.* **2017**, *63*, 54–68. [CrossRef]
14. Wang, Q.; Luan, Y.C.; Jiang, B.; Li, S.C.; He, M.C.; Sun, H.B.; Qin, Q.; Lu, W. Study on key technology of tunnel fabricated arch and its mechanical mechanism in the mechanized construction. *Tunn. Undergr. Space Technol.* **2019**, *83*, 187–194. [CrossRef]
15. Wang, J.; Hu, C.C.; Zuo, J.P.; Wang, B.; Mao, Q.F.; Ding, H.G.; Zhao, N.N. Mechanism of roadway floor heave and control technology in fault fracture zone. *J. China Coal Soc.* **2019**, *44*, 397–408.
16. Song, Y.; Huang, M.L.; Zhang, X.D.; Li, Z.P.; Peng, X.X. Experimental and numerical investigation on bearing capacity of circumferential joint of new spatial steel tubular grid arch in mined tunnel. *Symmetry* **2020**, *12*, 2065. [CrossRef]
17. Huang, M.L.; Song, Y.; Zhang, X.D. Research and application of a prefabricated spatial reticulated shell support system for large cross-section tunnel in a complex urban environment. *Appl. Sci.* **2022**, *12*, 5058. [CrossRef]
18. Hisatake, M. Effects of steel fiber reinforced high-strength shotcrete in a squeezing tunnel. *Tunn. Undergr. Space Technol.* **2003**, *18*, 197–204. [CrossRef]
19. Li, Z.J.; Guo, X.X.; Ma, Z.W.; Yang, T.L.; Xu, J.Q. Research status and high-strength pre-stressed primary (type) support system for tunnels with large deformation under squeezing conditions. *Tunn. Constr.* **2020**, *40*, 755–782.
20. Luo, J.W.; Zhang, D.L.; Fang, Q.; Li, A.; Sun, Z.Y.; Cao, L.Q. Analytical study on pretensioned bolt-cable combined support of large cross-section tunnel. *Sci. China Technol. Sci.* **2020**, *63*, 1808–1823. [CrossRef]
21. Sun, J.; Jiang, Y.; Wang, B.; Fan, Y. Nonlinear rheological properties and yielding support technology for large squeezing deformation of soft rock tunnel with high ground stress. *Tunn. Constr.* **2021**, *41*, 1627–1633.
22. Wu, K.; Shao, Z.S.; Qin, S.; Wei, W.; Chu, Z.F. A critical review on the performance of yielding supports in squeezing tunnels. *Tunn. Undergr. Space Technol.* **2021**, *115*, 103815. [CrossRef]
23. Kong, X.Y.; Chen, X.; Tang, C.A.; Sun, Z.R.; Hu, E.H. Study on large deformation control technology and engineering application of tunnel with high ground stress and weak broken surrounding rock. *Struct. Eng. Int.* **2020**, *32*, 1–9. [CrossRef]

24. He, M.C.; Gong, W.L.; Wang, J.; Qi, P.; Tao, Z.G.; Du, S.; Peng, Y.Y. Development of a novel energy-absorbing bolt with extraordinarily large elongation and constant resistance. *Int. J. Rock Mech. Min. Sci.* **2014**, *67*, 29–42. [CrossRef]
25. Zheng, Y.T.; Zhu, S.Y.; Li, S.L.; Song, R.C.; Guo, T. Testing research on a composite support system using bolts, shotcrete, mesh and arc plates for a tunnel in soft rock mass. *Chin. J. Rock Mech. Eng.* **1993**, *12*, 3–12.
26. Chang, Q.L.; Zhou, H.Q.; Li, D.W.; Bai, J.B.; Hou, C.J. Stability Principle of Extremely Rigid Secondary Support for Soft and Broken Rock Roadway. *J. Min. Saf. Eng.* **2007**, *24*, 169–172+177.
27. Liu, D.J.; Zuo, J.P.; Wang, J.; Zhang, T.L.; Liu, H.Y. Large deformation mechanism and concrete-filled steel tubular support control technology of soft rock roadway-A case study. *Eng. Fail. Anal.* **2020**, *116*, 104721. [CrossRef]
28. Huang, W.P.; Yuan, Q.; Tan, Y.L.; Wang, J.; Liu, G.L.; Qu, G.L.; Li, C. An innovative support technology employing a concrete-filled steel tubular structure for a 1000-m-deep roadway in a high in situ stress field. *Tunn. Undergr. Space Technol.* **2018**, *73*, 26–36. [CrossRef]
29. Wang, Z.C.; Du, K.; Xie, Y.L.; Su, X.L.; Shi, Y.F.; Li, X.; Liu, T. Buckling analysis of an innovative type of steel-concrete composite support in tunnels. *J. Constr. Steel Res.* **2021**, *179*, 106503. [CrossRef]
30. Xu, F.; Li, S.C.; Zhang, Q.Q.; Li, L.P.; Shi, S.S.; Zhang, Q. A new type support structure introduction and its contrast study with traditional support structure used in tunnel construction. *Tunn. Undergr. Space Technol.* **2016**, *63*, 171–182. [CrossRef]
31. Yao, W.J.; Pang, J.Y.; Zhang, J.S.; Liu, G.C. Key technique study of stability control of surrounding rock in deep chamber with large cross-section: A case study of the Zhangji coal mine in China. *Geotech. Geol. Eng.* **2020**, *39*, 299–316. [CrossRef]
32. Li, C.; Wang, Z.L.; Liu, T. Principle and practice of coupling support of double yielding shell of soft rock roadway under high stress. *Int. J. Min. Sci. Technol.* **2014**, *24*, 513–518. [CrossRef]
33. *TB-10003*; Code for Design of Railway Tunnel. China Railway Publishing House: Beijing, China, 2016.

applied sciences

MDPI

Article

Modeling of Water Losses in Hydraulic Tunnels under Pressure Based on Stepwise Regression Method

Slobodan Radovanović [1,*], Milovan Milivojević [2], Boban Stojanović [3], Srđan Obradović [2], Dejan Divac [1] and Nikola Milivojević [1]

[1] Water Institute "Jaroslav Černi", Jaroslava Černog Street 80, 11000 Belgrade, Serbia
[2] Technical and Business College Užice, Trg. Svetog Save 34, 31000 Užice, Serbia
[3] Faculty of Science, University of Kragujevac, Radoja Domanovića 12, 34000 Kragujevac, Serbia
* Correspondence: slobodan.radovanovic@jcerni.rs

Abstract: In this paper, we presented a methodology for efficient and accurate modeling of water losses in hydraulic tunnels under inside internal water pressure, based on multiple linear regression (MLR). The methodology encompasses all steps needed to obtain an adequate mathematical relation between total water losses and relevant measurements in the tunnel, such as reservoir water level, piezometric levels, concrete and water temperatures, size of cracks, etc. Once the data are preprocessed and input variables were chosen, correlation analysis and PCA (principal component analysis) reduction were performed in order to obtain the pool of regression functions. Through an iterative process, according to stepwise regression principles, the most adequate MLR model in terms of accuracy and complexity was chosen. The methodology presented has been validated in modeling water losses in the hydraulic tunnel under the pressure of PSHPP "Bajina Bašta" in the Republic of Serbia. The obtained results have shown significantly better accuracy compared to the results published by other authors, proving that the developed model can be used as a powerful tool in future analyses of tunnel losses and remediation planning.

Keywords: hydraulic tunnels; water losses; methodology; MLR model; stepwise regression

Citation: Radovanović, S.; Milivojević, M.; Stojanović, B.; Obradović, S.; Divac, D.; Milivojević, N. Modeling of Water Losses in Hydraulic Tunnels under Pressure Based on Stepwise Regression Method. *Appl. Sci.* **2022**, *12*, 9019. https://doi.org/10.3390/app12189019

Academic Editors: Chenjie Gong, Mingfeng Lei and Xianda Shen

Received: 17 August 2022
Accepted: 5 September 2022
Published: 8 September 2022

1. Introduction

Hydraulic tunnels subject to the high internal pressure of water (HT) are mostly used to carry water from the reservoir to the hydropower plant machine building with turbines. The hydraulic fall that exists between the reservoir and the power plant produces the internal water pressure in the hydraulic tunnel. Internal water pressure can cause tensile stresses and cracks in concrete tunnel lining and thus generate static instability in terms of failure and functional instability in terms of water losses from the tunnel. The water loss in the tunnel occurs through various cracks and fissures in the concrete lining. Some cracks are clearly visible, while some smaller cracks are not. On the other hand, the rock mass affects the increase or decrease of total water loss, depending on the porosity and degree of jointing of the rock mass. It is because of this that hydrotechnical tunnels are grouted, so as to fill the cracks around the tunnel and therefore lower the permeability of the rock mass. The water lost from inside the tunnel ends up in the surrounding rock mass, which can have an additional negative impact on the mechanical and deformation properties of the rock mass [1–3]. Besides the grouting measures for filling the rock mass cracks, the elimination of tensile stresses in the concrete lining, and therefore securing functionality and stability of hydraulic tunnels under pressure, is most effectively achieved by stress injection [4]. Other approaches exist as well, such as the use of reinforced concrete tunnel linings in order to prevent the formation of cracks [5].

In HT, appropriate monitoring is carried out during the structure's lifetime. Based on the monitoring, appropriate stability and functionality estimates (SaFE) for the tunnel can be made. The assessment of stability and functionality for the HT is performed based on measurements of significant physical quantities and a variety of physical, mathematical, and other models that describe the processes related to HT. The most frequent measurements are water losses in the tunnel, temperature of concrete, the temperature of water and rock, opening and closing of cracks, groundwater levels, and other relevant indicators. For the purpose of SaFE, two approaches are used. The first approach involves emptying the tunnel and physical inspection, which is occasionally done in the tunnels. However, this approach is expensive, it takes several weeks, and emptying the tunnel can be statically unfavorable for the tunnel structure pre-stressed with injection pressure. The second approach, which is used more often in practice, involves the creation of appropriate analytical, statistical, and numerical models. This approach allows for continuous monitoring and understanding of the structure's behavior. Large deviations between the model and real measurements can be used to make appropriate SaFE indicators.

To the best of the authors' knowledge, there are not many articles dealing with problems of HT safety and functionality, in particular with problems of water losses from the tunnel. Andjelkovic et al. [4] dealt with modeling of total water losses from the HT in the pumped-storage hydroelectric power plant (PSHPP) "Bajina Bašta", located in the Republic of Serbia. The authors established the relationship between water levels in the reservoir, water levels in piezometers, water levels in the surge tank, and water and concrete temperatures as input variables and total water loss from the tunnel as the output variable. Problems related to water loss and loss of other fluids also appear in pressurized pipeline systems that are similar to pressurized hydrotechnical tunnels, from a phenomena point of view. There are several research papers dealing with different detection methods and solutions to intermittent leakage in pressurized pipeline systems (water, gas, fuel, and others) due to cracking [6–9].

Apart from the above, there are papers that tangentially deal with the analysis of water filtration through the rock mass and the tunnel lining. In [10], analytical solutions based on conformal mapping of complex variable methods are derived for two-dimensional, steady seepage into an underwater circular tunnel. In addition to analytical models, a certain number of numerical models have been developed based on methods such as the finite element method. For example, researchers Radovanović et al. [11] analyzed the stress–strain state of the tunnel excavation and stability of the hydraulic tunnel support using the finite element method, taking into consideration the filtration of groundwater into the tunnel. The use of finite element methods is widely used in geotechnical analyses and tunnel design. However, they have their shortcomings—problems with significant deformations, taking into account the discontinuities and joints in rock masses, implementing rock and soil porosity, etc. Because of that, research conducted by [12] analyzes the application of the particle discrete element method taking into account the porosity. This research presents a significant step forward and enables the analysis of hydrotechnical tunnels from the aspect of evaluating the water loss through the geomaterial. The author Rui Vas Rodrigues [5] researched the monitoring of cracks in tunnels pressurized with 1000 kPa of water pressure, taking into account the interaction between the tunnel structure and the rock mass by using analytical equations for stresses, deformations, and reinforcement quantities.

Recently, analytical, statistical, and numerical models have been enriched with various heuristics from the machine learning domain (ML), creating hybrid models that combine their advantages. Javadi [13] analyzed the loss of compressed air in tunnels during construction in situations where there are high pressures from groundwater. In this paper, neural networks were used for estimating air loss in the tunnel, depending on the geological conditions and tunnel dimensions. In [14], phreatic line detection, which is a major challenge in seepage problems, was accomplished with the use of the natural element method (NEM) and genetic algorithm (GA). Zhang et al. [15] analyzed the uncertainty quantification for the mechanical behavior of fully grouted rockbolts using Monte Carlo simulation and the Bayesian method. The use of ML methods in tunnels is present in the paper [16] where the reliability-based design optimization method for rock tunnel support is analyzed.

Although the application of ML models in modern research is often of few benefits, ML models of studied phenomena are characterized by certain disadvantages. The requirement for a large number of measurements, high computational costs, difficulties in assessing individual predictor importance, the difficulty in deriving prediction intervals, and the difficulty of model interpretability are some of the disadvantages. For these reasons, and due to engineering practicality, in this paper, statistical models in the form of multiple linear regression (MLR) were used to model water loss in HT. Ease of understanding, low computational costs, availability of model adequacy tests, and well-founded analysis of the significance of individual regressors represent only some of the advantages of MLR models.

Although MLR represents a classical modeling technique that has been refined for almost two centuries, the problems of creating appropriate regression models in terms of dimensionality and selecting the number and type of regressors are not completely solved. Generally, there are two approaches to solving such problems. One is statistical, based on principal component analysis (PCA) [17,18], partial correlation, and stepwise regression methods [19,20], and the other is a hybrid approach, based on a combination of statistical and ML techniques [21]. The authors have opted for the first approach, which involves PCA dimensionality reduction and the application of the stepwise method in the building of the MLR model.

In accordance with the above, the aim of the paper was to develop a novel methodology for the mathematical modeling of water losses at HT. In addition, the goal was to validate this methodology by building an MLR model of water losses with greater accuracy and reliability, compared to previously developed models of water losses. The case study was carried out for the hydraulic tunnel at PSHPP "Bajina Bašta", located on the Drina River, which represents one of the most significant hydropower structures in the Republic of Serbia for the production of electrical energy.

The different approaches based on statistical and numerical methods that were discussed have their advantages and their shortcomings. Numerical methods are extremely demanding and complex, and they require the determination of several parameters. On the other hand, they enable filtration and stress–strain analyses of tunnels including the interaction with the rock mass. Statistical methods are widely used for the monitoring and assessment of various hydrotechnical structures (majorly dams). They are based on MLR techniques, and there are certain issues in defining the type and number of regressors and their dependencies. In the case of new methodologies that are based on the application of PCA and stepwise regression methods, it is possible to define the most influential regressors for the variable that is modeled (total losses). Besides that, compared to other models, the effect of the crack width is taken into account as it has great influence on the total losses, as does the time component.

2. Theoretical Background

Besides data quantity, the number of variables and potential regressors also affects the complexity of the regression model and the required processing time for its generation. It is advised to consider methods of reducing the potential complexity of the model using techniques of factor analysis. For this purpose, we utilized PCA. The principal components (PC) are the basis for further creation of MLR models of water losses and seepage processes for considered HT. As mentioned earlier, the selection of the number and type of MLR model regressors is entrusted to a stepwise regression technique. In the following sections, the basic elements of the applied methods are given.

2.1. Principal Component Analysis

PCA is a statistical approach that is utilized to analyze inter-relationships among many variables and to describe these variables in terms of their common underlying dimensions (factors) [22]. PCA is concerned either with the co-variances or correlations between a set of observed variables $x_1, x_2, \ldots, x_p$ that can be explained in terms of a smaller number of unobservable latent variables or common factors, $f_1, f_2, \ldots, f_k$, where. The correlation matrix $\boldsymbol{R}$ of observed variables can be transformed according to Equation (1):

$$\boldsymbol{P}^{-1} \cdot \boldsymbol{R} \cdot \boldsymbol{P} = \boldsymbol{D} = \begin{bmatrix} \lambda_1 & 0 & \ldots & 0 \\ 0 & \lambda_2 & \ldots & 0 \\ \ldots & \ldots & \ldots & \ldots \\ 0 & 0 & 0 & \lambda_p \end{bmatrix}, \tag{1}$$

where $\boldsymbol{P}$ is the matrix whose columns are eigenvectors of matrix $\boldsymbol{R}$, and $\boldsymbol{D}$ is the diagonal matrix of $\boldsymbol{R}$. Matrix $\boldsymbol{D}$ keeps the variability of the original matrix $\boldsymbol{R}$, now expressed through eigenvalues $\lambda_1,\ \lambda_2,\ \ldots \lambda_p$. The eigenvalues $(\lambda_i, i = 1, p)$ represent the values given by the root of the characteristic polynomial (2):

$$det(\boldsymbol{R} - \lambda \cdot \boldsymbol{I}) = k(\lambda), \tag{2}$$

where $\boldsymbol{I}$ is the identity matrix.

In order to simplify the system, only some of the principal components, which correspond to the chosen number of first k largest eigenvalues $\lambda_1,\ \lambda_2,\ \ldots, \lambda_k$, are kept without significant loss of variance of the original dataset. Details of PCA are explained in [18,23].

2.2. MLR and Stepwise Regression

The general form of the MLR model can be written as follows (3):

$$y_i = \beta_0 + \beta_1 \cdot x_{1i} + \beta_2 \cdot x_{2i} + \ldots + \beta_j \cdot x_{ji} + \ldots + \varepsilon_i, \tag{3}$$

where y_i $(i = 1, 2, \ldots, n)$ is the ith observation for the response variable y, x_{ji} $(j = 1, 2, \ldots, k)$ are the values of jth predictor variable in ith observation, ε_i are the values of the independent, stochastic, and normally distributed random variable which fulfils the assumption of homoscedascity, n is the number of observations in the dataset, and k is the number of predictor variables. Coefficients $\beta_0 \ldots, \beta_k$ are the unknown parameters of the model, which are usually estimated by using the least squares method. As mentioned in the introduction section, one way of choosing the number and type of predictors in Equation (3) is the stepwise regression method.

The stepwise regression method is a statistical technique designed with the idea of reducing the complexity of the regression model while maintaining satisfactory accuracy. The stepwise method is essentially based on semi-partial correlation, which is expressed through the semi-partial correlation coefficient. The square of the semi-partial correlation coefficient, sr^2, for a specific single variable, indicates by how much the value of the coefficient of determination, R^2, will be reduced if this single variable is removed from the regression equation. In the other words, let χ be the set of all independent variables, expressed through matrix $\boldsymbol{X}$, and ψ_k be the set of all independent variables $\boldsymbol{X}$, excluding x_k. Then the squared semi-partial correlation coefficient is expressed as follows:

$$sr_k^2 = R_\chi^2 - R_{\psi_k}^2, \tag{4}$$

The form of expressing semi-partial correlation coefficients may vary, and one of the most common forms is:

$$sr_k = \frac{t_k \cdot \sqrt{1 - R_\chi^2}}{\sqrt{residualdf}}, \tag{5}$$

where t_k is the Student's t-statistic value for the kth regressor in the MLR model, $residualdf = n - k - 1$ is the number of degrees of freedom for the sum of residuals, nis the number of observations, and k is the number of regressors. Stepwise regression is explained in detail in [19,24].

3. Methodology for Modeling of Total Water Losses in Hydraulic Tunnels

For statistical modeling of water losses and seepage in hydraulic tunnels, we proposed the methodology shown in schematic view in Figure 1. Key features of the proposed methodology are described in the following sections.

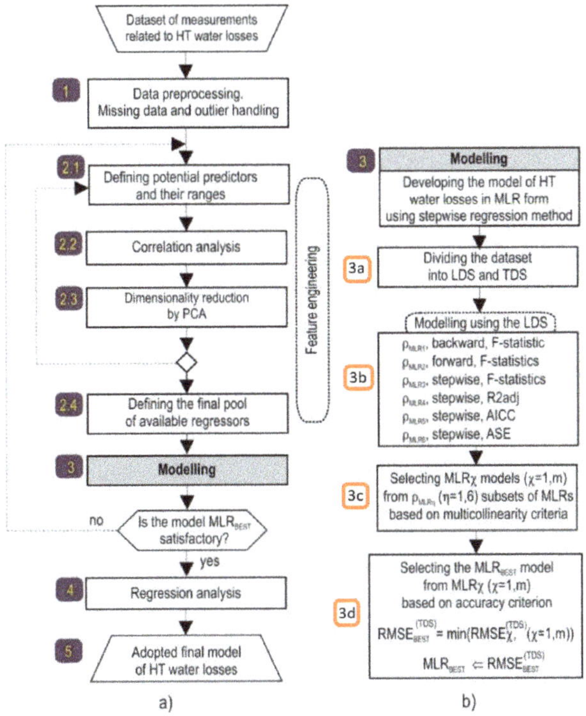

Figure 1. Developed methodology for modeling of total water losses in hydraulic tunnels: (**a**) Schematic overview; and (**b**) stepwise regression modeling method.

3.1. Data Preprocessing and Feature Engineering

The collected dataset consisted of various data related to HT water losses, including measurements of water losses in the tunnel, temperature of concrete, temperature of water and rock, opening and closing of cracks, groundwater levels, etc. Regression models are very sensitive to the occurrence of outliers and missing values in the data. Therefore, the processes of outlier detection and handling, and possible missing data imputation are of crucial importance. This is considered in the first part of the methodology (Figure 1, block 1).

The second part of the methodology relates to feature engineering (Figure 1, block 2). Feature engineering is the act of extracting features from raw data and transforming them into formats that are suitable for predictive modeling [25]. Feature engineering encompasses several stages in our methodology. Building adequate predictive models involves proper selection of predictors and dependent variables and determining valid value intervals for these variables (2.1). Correlation analysis (2.2) is of great importance in this process, because it provides a basis for proper selection of input and output variables, and their potential regressors [19]. Modeling HT water losses involves choosing regressors from a large pool of possible candidate predictors, of which only a few are likely to be important. In our methodology, the dimensionality reduction paradigm is realized through PCA (2.3). The key idea here, as explained in Section 2.1, is to replace redundant features with a few new features that adequately summarize the information contained in the original feature space. In our methodology, PCA consists of three main steps, which are briefly described in the following three paragraphs.

In the first step, data suitability for PCA is assessed in terms of sampling adequacy and strength of the relationship among the variables. Sampling adequacy is assessed based on sample size and Kaiser–Meyer–Olkin (KMO) test of sampling adequacy [26]. The strength of the relationship among the variables is assessed based on examining the correlation matrix and Bartlett's test of sphericity [27].

In the second step, the number of principal components to be extracted is selected. Deciding on the number of principal components is a considerable challenge in using PCA. In our methodology, three well-known methods are available to assist in this decision: Kaiser's criterion [26], Cattell's scree test [28], and Horn's parallel analysis [29].

In the third step, the principal components are subjected to rotation to gain a better understanding and interpretation of the data. Here, rotation refers to the post-processing of PCA results to obtain a so-called 'simple structure' [30], where the division of variables into separate components is more easily interpretable. Two general types of rotation are available: orthogonal, where the new principal axes are also orthogonal to each other, and oblique where the new principal axes are not required to be orthogonal. Our methodology suggests the application of the most frequently used rotation methods: the Varimax method, and the direct Oblimin method, as the most commonly used orthogonal and oblique rotation methods, respectively. A comparison of these rotation methods is given in [18].

Regression models are very sensitive to multicollinearity and singularity, so methodology stages 2.2 and 2.3 are necessary to avoid the pitfalls of model overfitting. Based on previous stages the final pool of candidate regressors is defined in stage 2.4. Candidate regressors are derived by various mathematical transformations of physical quantities that describe the processes related to HT. Furthermore, solving the complex issue of feature engineering can be supported by elements of descriptive statistics and probability distribution laws of physical quantities upon which water losses and seepage depend. The methodological steps in the second stage are often implemented in several iterative steps.

3.2. Modelling

A key issue in creating the model in the form of multiple linear regression (Figure 1, block 3) is the selection of appropriate regressors, considering both the number and type of regressors. To solve this problem, we engaged the method of stepwise regression. In an iterative procedure, the techniques from this method performed automated entering and removing of regressors, thus creating optimal regression models.

At this stage, we defined the following steps:

- Dividing the available data into subgroups: learning dataset (LDS) and test dataset (TDS), according to the given ratio (3a);
- Generating regression models based on LDS during the iterative procedure, using the stepwise regression method with multiple model selection methods: backward, forward, and stepwise (3b). Each of these methods employs additional criteria for entry and removal of regressors. The backward method starts with all regressors in the model, and iteratively performs the removal of regressors based on the criterion of F-statistic and corresponding p-values (probability of F), until no further regressors can be removed without a statistically significant loss of accuracy. This method produced the subset of regression models ρ_{MLR1}. The forward method starts with zero regressors in the model, and iteratively performs entry of regressors, also based on the criterion of F-statistic and their p-values. The procedure stops when there are no variables that meet the entry criterion and generates the subset of regression models ρ_{MLR2}. The stepwise method is a combination of the backward and forward selection methods. This method terminates when no more regressors are eligible for entry or removal based on the chosen criterion. In our methodology, the stepwise method generates four subsets of MLR models (ρ_{MLR3}, ρ_{MLR4}, ρ_{MLR5}, ρ_{MLR6}), based on four different criteria for entry and removal of regressors. These criteria are: F-statistic (F-statistic

maximum for entry, F-statistic minimum for removal), maximum adjusted coefficient of determination R^2, minimum corrected Akaike information criterion (AICC), and minimum average squared error over the overfitting prevention data (ASE). The mentioned criteria are explained in detail in [31];

- Selecting the best MLR model of HT water losses, from MLRχ models $\chi = 1, m$, based on the chosen accuracy criterion (3d). Various indicators can be selected as criteria for assessing model accuracy. One of them is the root mean squared error (RMSE). In our methodology, RMSE for the test dataset, $\text{RMSE}^{(\text{TDS})}$, was chosen as the accuracy criterion, and is calculated according to (6):

$$\text{RMSE}^{(\text{TDS})} = \sqrt{\frac{\sum_{i=1}^{n} (q_i - \hat{q}_i)^2}{n}}, \quad (6)$$

where q_i and $\hat{q}_i$ represent the measured and predicted values of total water losses in HT for the ith record from the TDS, respectively. The model with the $\text{RMSE}^{(\text{TDS})}$, calculated according to (7):

$$\text{RMSE}_{\text{BEST}}^{(\text{TDS})} = min(\text{RMSE}_{\chi}^{(\text{TDS})}, \chi = 1, m), \quad (7)$$

was chosen as the best model, denoted as MLR_{BEST}.

If MLR_{BEST} is satisfactory, in comparison to some previously developed reference model of HT water losses, in terms of complexity and accuracy, or some other criteria arbitrarily chosen by the researcher, our methodology follows up with the regression analysis activity. Otherwise, we should return to block 2, and repeat the methodology stages, starting with (2.1), redefining the pool of potential predictors and their ranges, and/or redefining principal components in (2.3).

3.3. Regression Analysis

Detailed performance indicators of generated models, such as confidence intervals, prediction intervals, significance and confidence intervals of regression coefficients, and other indicators of interest, are obtained by regression analysis for the chosen model, MLR_{BEST} (Figure 1, block 4).

4. Case Study

4.1. Description

In this paper, the hydraulic tunnel of the PSHPP "Bajina Bašta" was analyzed (Figure 2). This tunnel is part of the PSHPP "Bajina Bašta" and was built in the period from 1977 to 1983. During periods of electricity surplus in the energy system of the Republic of Serbia, the supply and drainage hydraulic tunnel serve to pump water from the reservoir of HPP "Bajina Bašta" on the river Drina to the upper reservoir, lake "Lazići", located in the Tara mountain (Figure 3). On the other hand, when there is a need for electricity, the water goes through the HT to the aggregates of the PSHPP "Bajina Bašta". The installed capacity of the PSHPP is 600 MW. The tunnel has a circular cross-section with a diameter of 6.30 m, the tunnel's length is 8030 m, and it has a drop of 4.5‰ (Figure 3). The circular cross-section of the surge tank is 12 m in diameter. Under operating conditions, the maximum hydrostatic pressure in the tunnel can be up to 1.3 MPa.

4.2. Measurements in the Tunnel

In the hydrotechnical tunnel measurements of the temperatures of concrete and water in the tunnel, the water level in the upper reservoir, the water level in the piezometers, and measurements of the width of the crack openings in the tunnel's concrete lining are conducted approximately two times per month. Measurements of the total water losses are conducted 2–4 times per year depending on the weather conditions and the tunnel exploitation regime.

Figure 2. Map of Serbia showing the river Drina and the position of the PSHPP Bajina Bašta.

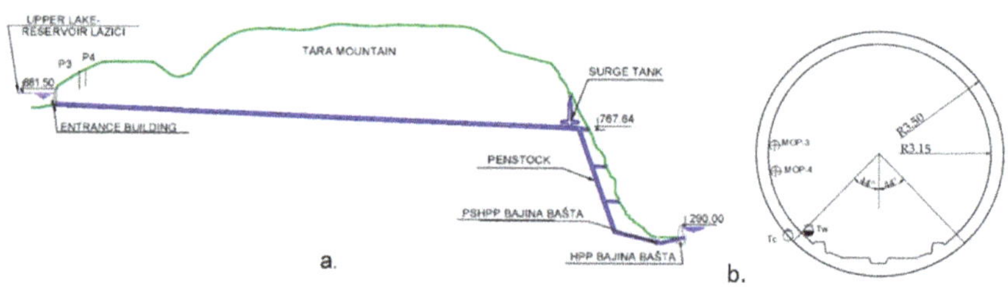

Figure 3. Longitudinal tunnel profile with a schematic view of the upper lake, pipeline, PSHPP, and HPP "Bajina Bašta" (**a**) and cross-section of tunnel (**b**).

Measurement of total water losses is carried out under conditions when the tunnel is closed both at the entrance building at the upper reservoir and at the outlet towards the pipeline. One session of measurements of total water loss is performed in several phases. A single phase includes the rapid reduction of the water level in the surge tank down to some level, H_i, and then, a measurement step. In the measurement step, it is measured how much the water level H_i has lowered in the observed time interval. The decrease in water level is the result of water loss from the tunnel. At the end of the measurement step, the water level is H_{i+1}. Based on the difference, $H_i - H_{i+1}$, the resulting volume of lost water ΔV from the tunnel is obtained according to (8) [4].

$$\Delta V = (H_i - H_{i+1}) \cdot \frac{D^2 \pi}{4}, \tag{8}$$

The total water loss is calculated according to (9):

$$q = \frac{\Delta V}{\Delta t}, \tag{9}$$

where q [L/s], ΔV [l], and Δt [s] are the total water loss, water volume decrease in the surge tank with a circular cross-section (diameter D = 12.0 m), and time interval, during a single measurement step, respectively. The measuring is stopped when the total water loss q becomes negative—ΔV becomes negative, i.e., water volume in the surge tank starts increasing and water loss turns into water inflow. The duration of one measurement step is usually 30 min. According to the same principle, the procedure repeats in the next phases in the session.

Figure 4 shows a diagram of water level change in the surge tank during one session of total losses measurement (in this case, the 75th session, 17 August 2015).

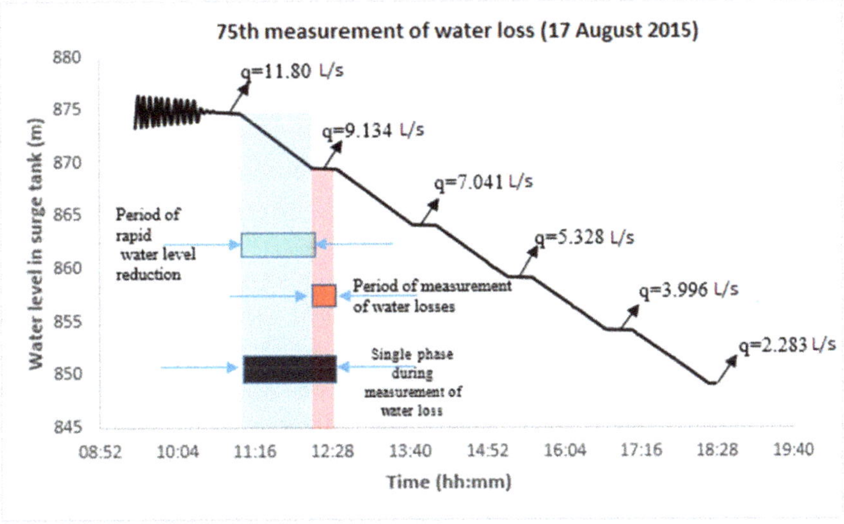

Figure 4. Water level change in surge tank during the measurement of total water losses (75th session).

During the session of measurement of total water losses, measurements of concrete temperature on the extrados of the concrete lining, water temperatures, width of the crack openings, water level in the upper reservoir, and water level in piezometers are also measured at the same time.

For measuring the temperatures of the water and the concrete, thermometers are installed in the tunnel lining (on the extrados and intrados of the tunnel). The instruments are isolated from the effects of water and the measurements are conducted twice a month via a portable measuring station. For measuring the width of the crack openings electroacoustic deformeters are installed in the concrete in the zone of the crack. The instruments are isolated from the effects of water and the measurements are conducted twice a month via a portable measuring station (outside the tunnel).

4.3. Dataset

During the exploitation of the tunnel, for the period from 1983 to 2017, 79 measurement sessions of total water loss were carried out. In this paper, data for the period from 2005 (year of the last tunnel reparation and injection) to 2017 were analyzed—measuring sessions of total water loss from 50 to 79, which generated 180 data measurement records. For the considered input and output variables, the following symbols were introduced: H_r—water level in the upper reservoir, H_{p3} and H_{p4}—water levels in piezometers P_3 and P_4, respectively, T_c —concrete temperature on the extrados of the concrete lining, T_w —temperature of water, H_{st}—average water level in the surge tank for one measurement step, cr_3 and cr_4—width of crack openings at the measuring points M0P-3 and M0P-4, respectively, and q—total water losses from the tunnel. In addition to these variables, variable t, which represents the elapsed time in months from the date of the last tunnel reparation and injection performed in 2005, is introduced. The variable t considers the impact of rheology on the aging and deterioration of concrete as well as possible changes in geology that can lead to irreversible processes and have a lasting effect on total water losses in the tunnel.

The elements of descriptive statistics of measured values are given in Table 1.

Table 1. Descriptive statistics of measured values for measurement sessions 50–79.

	H_r	H_{p3}	H_{p4}	T_c	T_w	H_{st}	cr_3	cr_4	t	q
	[m a.s.l.]	[m a.s.l.]	[m a.s.l.]	[°C]	[°C]	[m a.s.l.]	[mm]	[mm]	[months]	[L/s]
Min.	854.380	862.950	862.780	2.520	3.190	831.825	−0.722	-0.033	4.0	−0.190
Max.	880.260	879.650	879.550	14.560	21.000	879.996	0.129	0.424	139.0	25.511
Average	874.660	874.379	873.697	9.100	11.730	860.617	−0.480	0.103	72.0	9.845
St. dev.	5.116	3.793	3.888	3.105	4.377	10.640	0.224	0.122	41.3	5.176

4.4. Correlation Analysis and PCA of Total Water Losses

4.4.1. Correlation Analysis

A correlation analysis was performed for the available dataset of measuring sessions total water loss of 50–79. The values of Pearson's correlation coefficient are given in Table 2.

Table 2. Correlation matrix for considered input and output variables for the period 2005–2017.

	H_r	H_{p3}	H_{p4}	T_c	T_w	H_{st}	cr_3	cr_4	t	q
	[m a.s.l.]	[m a.s.l.]	[m a.s.l.]	[°C]	[°C]	[m a.s.l.]	[mm]	[mm]	[months]	[L/s]
H_r										
H_{p3}	0.482									
H_{p4}	0.491	0.873								
T_c	0.355	0.334	0.179							
T_w	0.310	0.340	0.148	0.840						
H_{st}	0.441	0.258	0.264	0.129	0.127					
cr_3	−0.384	−0.452	−0.328	−0.903	−0.875	−0.164				
cr_4	−0.384	−0.433	−0.314	−0.898	−0.876	−0.167	0.995			
t	−0.230	0.165	0.128	−0.130	−0.110	−0.072	0.060	0.060		
q	0.164	−0.066	−0.033	−0.257	−0.291	0.758	0.304	0.309	0.015	

The correlation coefficient values in Table 2 indicate that most of the input variables can be considered linearly independent since most correlation coefficient values are less than 0.7 [17].

The correlation coefficient between variables H_{p3} and H_{p4} is 0.873, which indicates a strong correlation. This is to be expected because the piezometers are at a close distance and under the same geological conditions. The concrete temperature on the extrados of the concrete lining T_c is in a strong correlation (0.840) with the water temperature T_w, which is reasonable because the temperature is measured in the interior and exterior of the concrete lining, so these temperatures are interdependent. Variables T_c and T_w are in a very strong negative linear correlation with the sizes of crack openings cr_3 and cr_4. This can be explained by the fact that at high temperatures there is an increase in the expansion of concrete, and consequently the cracks are closed, and in the case of lower temperatures, the concrete contracts and the cracks open up. The average water level during a single measurement step, H_{st} has a strong influence on total water loss, the coefficient of correlation being 0.758. Variables cr_3 and cr_4 are in positive correlation with variable q, because the increase in crack openings leads to an increase in total water losses.

The increase in water and concrete temperature leads to expansion of the tunnel's concrete lining, thereby reducing the size of crack openings which leads to a reduction of water losses from the tunnel. For these reasons, concrete and water temperatures are in negative correlation with the variable q, with the correlation coefficients being −0.257 and −0.291, respectively.

Since the measurement of total losses is carried out at a constant water level in the upper reservoir, in piezometers, and under variable conditions in the surge tank, it is therefore expected that the coefficient of correlation between H_r, H_{p3}, and H_{p4} on q is small. Although these variables individually have little linear effect on the output (the correlation coefficients are significantly lower than 0.3), in order to initially include a more comprehensive set of measured quantities, their interaction effects, and derived variables, the mentioned variables are taken into account for further consideration.

Based on the correlation analysis, it can be concluded that some input variables are strongly linearly dependent and should not be used as such for modeling, because of the problem of multicollinearity [32]. In order to reduce multicollinearity, simplify the model of HT water losses (Ockham's razor) [33], and enable better physical interpretation of the model by dimensionality reduction, PCA was carried out.

4.4.2. PCA Results

A set of 9 variables, described in 4.3, was subjected to PCA using R software [34]. Firstly, the suitability of data for PCA was assessed. The number of samples was 156, which exceeds the minimum recommended sample size of 150 [18]. The value of the KMO index was 0.772, which exceeds the minimum recommended value of 0.6, and indicates sampling is adequate [18]. Most correlation coefficients (23 out of 36) in the correlation matrix were above 0.3 [18] and Bartlett's test of sphericity reached statistical significance ($p < 0.05$) [27], indicating the factorability of the correlation matrix.

In the second step, PCA analysis revealed the presence of 3 components with eigenvalues exceeding 1 (Kaiser's criterion) [26], which explain 49.13%, 20.01%, and 13.58% of the variance respectively, accounting for 82.72% of total variance. Cattell recommends retaining all components above the elbow or break in the plot (Figure 5), as these factors contribute the most to the explanation of the variance in the data set [28]. Based on Cattell's test it has been found that reduction to the 2, 5, or 6 principal components can be considered.

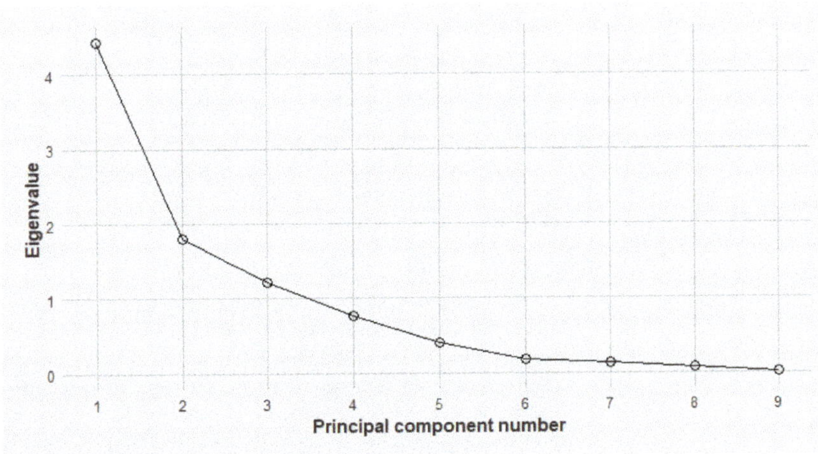

Figure 5. Screen plot of principal components.

Horn's parallel analysis involves comparing the size of the eigenvalues with those obtained from a randomly generated data set of the same size. Only components with eigenvalues that exceed the corresponding values from the random data set are retained [29]. The results of the parallel analysis, shown in Table 3, indicate three components whose eigenvalues exceed the corresponding threshold values obtained using an equally sized matrix of random numbers (9 variables and 162 records). Horn's parallel analysis was carried out using the Monte Carlo PCA for parallel analysis software [35].

Table 3. Results of Horn's parallel analysis.

PC	PC Eigenvalues	Eigenvalues Obtained by Horn's Parallel Analysis	Decision
1.	4.422	1.3549	accept
2.	1.801	1.2326	accept
3.	1.222	1.1374	accept
4.	0.759	1.0594	reject
5.	0.404	0.9872	reject
...	...	...	...
9	0.050	0.6734	reject

Although Kaiser's criterion and Horn's parallel analysis suggest a three-component solution, accounting for 82.72% of total variance, we arbitrarily selected a five-component solution based on Cattell's test, in order to retain more information about the process of generating water losses in the HT and facilitate a more accurate model of water losses. The five-component solution explains a total of 94.61% variance, with component contributions of 49.13%, 20.01%, 13.58%, 8.44%, and 4.45%, respectively.

In the third step, several rotation methods were performed. Finally, the direct oblimin rotation method was chosen as appropriate, enabling the following PC interpretations: T_{wcc}—which includes water temperature, concrete temperature, and cracks opening hP—which includes the influence of water level in the piezometers, $\overline{t}$—which represents time, $\overline{H}_{st}$—which represents water level in the surge tank, and $\overline{H}_r$—which represents water level in the upper reservoir. Table 4 shows correlation coefficients between selected PCs.

Table 4. PC correlation matrix.

	T_{wcc}	hP	$\overline{t}$	$\overline{H}_{st}$	$\overline{H}_r$
T_{wcc}					
hP	0.312				
$\overline{t}$	−0.085	0.161			
$\overline{H}_{st}$	0.147	0.261	−0.067		
$\overline{H}_r$	0.357	0.484	−0.229	0.436	

It can be seen from Table 4 that some components are linearly independent of each other because the correlation coefficients are less than 0.3, and some components have mild linear dependence.

4.4.3. Defining the Pool of Candidate Regressors

In accordance with our methodology, after obtaining the principal components, we defined the pool of candidate regressors. In the modeling of total water losses, PCs can be roughly divided into several groups, based on the physical nature of the input variables: the effects of hydrostatic pressure, the effects of temperature and size of crack openings, and the effects of time (aging, which is considered to be irreversible). The effects of hydrostatic pressure variation are usually represented by polynomials, power, and exponential or logarithmic functions, depending on the water level [21]. In our case study, for hydrostatic principal components ($\overline{H}_{st}$, $\overline{H}_r$, hP) we arbitrarily opted for polynomial terms of the first and second order. As described in the previous section, T_{wcc} includes the water temperature, concrete temperature, and size of crack openings, and the authors have chosen to express its potential predictive effect on total water losses in polynomial and exponential form. It can be assumed that water losses increase with the time elapsed since the last tunnel reparation date. This aging effect can also be interpreted in the following way—the effect of the last tunnel reparation on water losses from the tunnel exponentially decays with time. This long-time behavior is modeled by a negative exponential drift regressor, $-e^{-\overline{t}/k}$, where $\overline{t}$ is time in months elapsed since the last reparation date, and k is a scaling factor, which cannot be determined a priori. Several regressors with different values of k (1, 2, 4, and 10) were placed in the pool of regressors, with an assumption that the stepwise regression method will make an appropriate choice of regressors during the modeling process of water losses. During the application of PCA rotation methods, we did not achieve a simple structure (Table 4). This shortcoming is dealt with in the modeling stage of our methodology (Figure 1, block 3), with the use of interaction terms in forms $T_{wcc}\overline{H}_{st}$, ${T_{wcc}}^2\overline{H}_{st}$, and $T_{wcc}{\overline{H}_{st}}^2$.

Based on previous considerations, an initial set of regressors was created, using the five principal components (T_{wcc}, hP, $\bar{t}$, $\overline{H}_{st}$, $\overline{H}_r$) (Table 5).

Table 5. Potential regressors for the water losses model.

Type	Single Terms					Interaction Terms
Principal components	$\overline{H}_{st}$	$\overline{H}_r$	hP	T_{wcc}	$\bar{t}$	
Regressors	$\overline{H}_{st}{\overline{H}_{st}}^2$	$\overline{H}_r{\overline{H}_r}^2$	$hPhP^2$	T_{wcc}, ${T_{wcc}}^2e^{-T_{wcc}}$	$e^{-\bar{t}}$, $e^{-\bar{t}/2}e^{-\bar{t}/4}$, $e^{-\bar{t}/10}$	$T_{wcc}\overline{H}_{st}$, ${T_{wcc}}^2\overline{H}_{st}T_{wcc}{\overline{H}_{st}}^2$

4.5. *Model of Total Water Losses in the HT Based on Principal Components*

Based on the created pool of regressors, the development of the HT total water losses model was performed in accordance with the strategy described in Section 3 and shown in Figure 1, block 3. The process of modeling was based on the measurements described in Section 4.3. From the total number of records (180), 156 were used for modeling (LDS—learning dataset), and 24 records were used to test the model (TDS—test dataset) (Figure 1(3a)). The TDS, consisted of records from the last four measurement sessions (76–79) of total water losses, in period 2015–2017, i.e., from the last period of HT monitoring.

The automated software procedure, in accordance with the stepwise paradigm and numerous stepwise regression method variations described in Section 3.2, resulted in several MLR models of HTs total water losses which are stored in appropriate subsets of MLR models from ρ_{MLR1} to ρ_{MLR6} (Figure 1(3b)). The numerous models obtained from the iterative procedures are not shown due to their size. The most accurate model, MLR_{Best}, which satisfies the condition of non-multicollinearity, was chosen as appropriate (Figure 1(3d)), from the previous selected set of non-multicollinear regression models ($\text{MLR}\chi$, $\chi = 1$, m) (Figure 1(3c)). The obtained model was the results of the stepwise selection method and the AICC criteria for entry/removal of regressors. AICC is based on the likelihood of the training set given the model and is adjusted to penalize overly complex models [36]. The coefficient of multiple determination (R^2), adjusted coefficient of multiple determination (R^2_{adj}) and RMSE were selected as measures of selected model accuracy. The values of these indicators were $R^2 = 0.94014$, $R^2_{adj} = 0.87386$ and $\text{RMSE}_{\text{LDS}} = 1.8842$ L/s.

The final adopted regression model, with all parameters and results of regression analysis, according to Section 3.3, is shown in Table 6.

Table 6. Parameters of the final adopted MLR model of HTs total water losses.

Regressors	Unstandardized Coefficients		Standard. Coeff.	t	p Value	95% Confidence Interval for B		Correlations		Collinearity Statistics	
	B	Std. Error	β			Lower Bound	Upper Bound	Partial	Part	Toler.	VIF
Constant	10.209	0.321		31.76	0.000	9.575	10.844				
$\overline{H}_{st}$	4.425	0.160	0.855	27.64	0.000	4.109	4.741	0.903	0.814	0.907	1.103
T_{wcc}	−1.787	0.177	−0.345	−10.09	0.000	−2.136	−1.438	−0.610	−0.298	0.743	1.347
hP	−1.151	0.184	−0.222	−6.25	0.000	−1.514	−0.787	−0.430	−0.184	0.686	1.458
$T_{wcc}\overline{H}_{st}$	−1.169	0.155	−0.266	−7.55	0.000	−1.475	−0.863	−0.499	−0.222	0.698	1.433
$e^{-\bar{t}}$	−0.663	0.109	−0.185	−6.10	0.000	−0.878	−0.449	−0.422	−0.180	0.939	1.064
${\overline{H}_{st}}^2$	0.451	0.147	0.100	3.08	0.002	0.162	0.740	0.228	0.091	0.817	1.223
${T_{wcc}}^2$	0.408	0.171	0.094	2.39	0.018	0.071	0.746	0.179	0.070	0.560	1.787

The analytic form of the MLR model of HTs total water losses, based on PCs, is presented in Equation (10).

$$q = 10.20945 + 4.22495 \cdot \overline{H}_{st} - 1.78695 \cdot T_{wcc} - 1.15069 \cdot hP$$
$$-1.16897 \cdot T_{wcc} \cdot \overline{H}_{st} - 0.66310 \cdot e^{-\bar{t}} + 0.45082 \cdot \overline{H}_{st}^{\ 2} + 0.40840 \cdot T_{wcc}^{\ 2}, \quad (10)$$

The measured and predicted values of total water losses in the HT, based on the LDS, for the period 2005–2015 are shown in Figure 6.

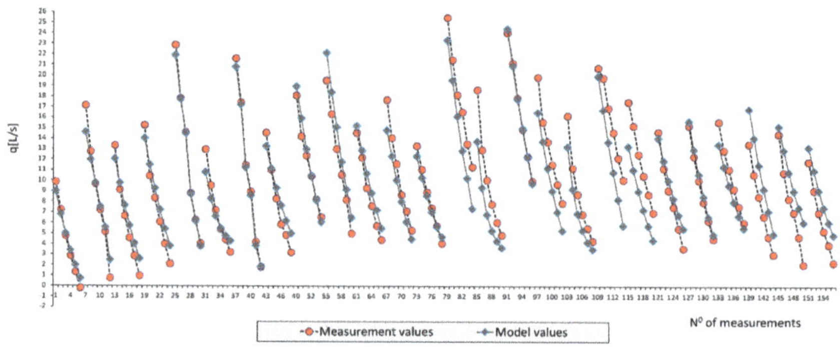

Figure 6. Measured and predicted values of HT total water losses for the LDS, for the period 2005–2015.

Various 3D surface plots can also represent the adopted mathematical model of HT total water losses. Such visualizations (Figure 7) can be useful for the interpretation of physical quantity effects, and the significance of individual predictors. In the diagrams shown, in addition to the predictors shown on the axes, all other predictors are constant and equal to their mean values.

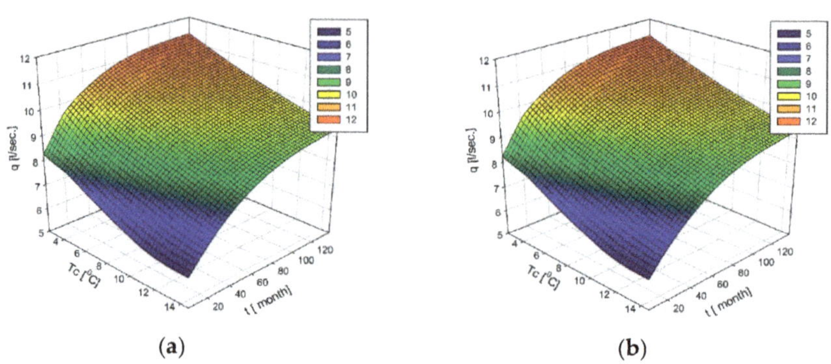

Figure 7. Model of HT total water loss: examples (**a**) q = f(Tc, t) and (**b**) q = f(Tc, Hst).

The diagram in Figure 7a presents a surface showing an increase in total water losses in HT of PSHPP "Bajina Bašta", as a function of time and temperature in the extrados of the concrete lining. It can also be seen that the effect of reparation on water losses from the tunnel is the largest in the first few years after the tunnel reparation was carried out in 2005. After that, this effect slowly dissipates, following an exponential law. The diagram in Figure 7b presents the surface showing that water losses increase with the increasing water level in the surge tank and with the reduction of concrete temperature.

The final adopted model of HT total water losses was verified using the dataset that was not used in the model development process. Figure 8 shows diagrams of predicted and measured values of total water losses during the test period.

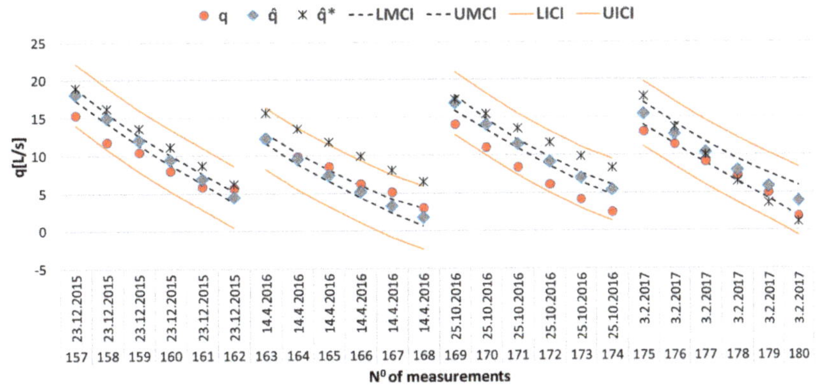

Figure 8. Measured and modeled values of total losses during verification period 2015–2017.

In Figure 8, q represents the measured values, $\hat{q}$ represents the model values according to the model (10) in this paper, and $\hat{q}*$ represents the model values of water losses according to the model given in paper [4]. Test sessions refer to the last four sessions, sessions 76–79, i.e., measured values 157–180. $LMCI$, $UMCI$, $LICI$, and $UICI$ represent the lower (L) and upper (U) limits of the confidence intervals for the expected mean (M) and individual values (I) of the model (10). The accuracy of the model for the test period is $\text{RMSE}_{\text{Test}} = 2.0161$ L/s.

Model (10) is additive, and Figure 9a shows the aggregate effects of regressors grouped by their physical context. The additive component $aHst$ represents the sum of linear and quadratic hydrostatic effects (H_{st} and H_{st}^2), $aTwcc$ represents the sum of linear (T_{wcc}) and quadratic (T_{wcc}^2) thermal and size of crack openings effects, while component ahP is the sum of hydrostatic effects of groundwater measured by piezometers P_3 and P_4. Component $T_{wcc} \cdot H_{st}$ represents the interaction effects of appropriate PCs, and t—encompasses the effect of rheology on aging and deterioration of concrete (e^{-t}).

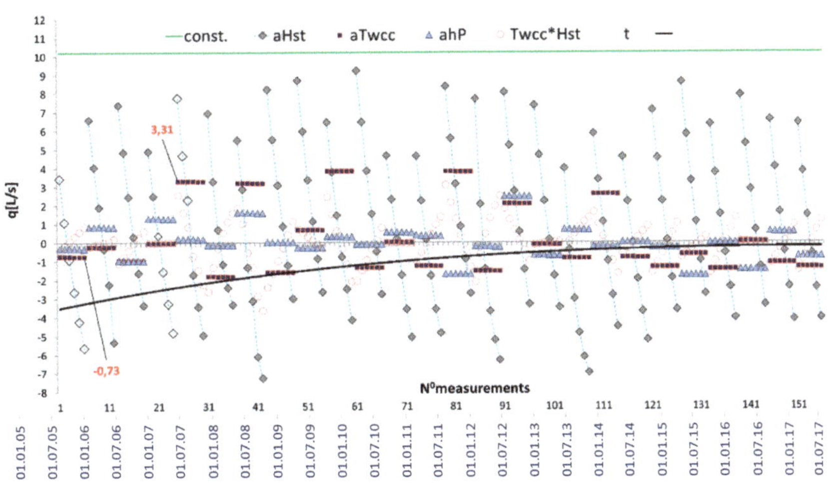

(a)

Figure 9. *Cont.*

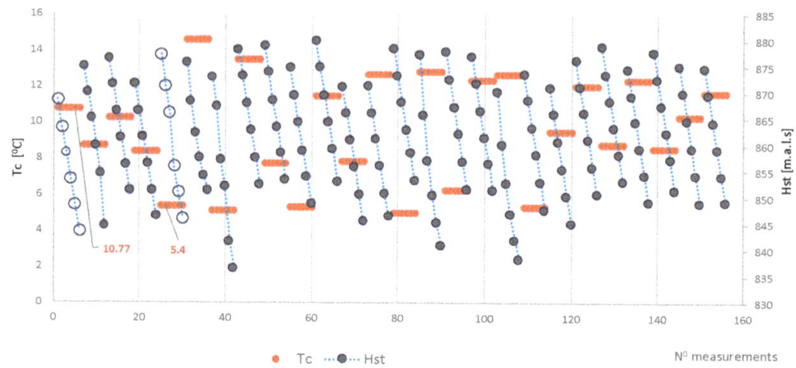

(b)

Figure 9. Aggregate effect of regressors in the model of total water losses in HT at PSHPP "Bajina Bašta" for the period 2005–2015, grouped by their physical context (**a**) and measurement values of predictors Tc and Hst (**b**).

5. Discussion

The adopted model includes 94.61% variance of water losses from the tunnel. Based on the *Tolerance* and *VIF* columns (Table 6), it can be seen that the adopted model has no problems with multicollinearity (for each of the regressors: Tolerance > 0.1 and VIF < 10). The *Standardized Coefficient* column (β) shows the contribution of individual regressors to the entire model. The linear impact of the hydrostatic pressure on the water level in the surge tank ($\overline{H}_{st}$) dominantly affects water losses (0.855). This dominant effect is followed by the thermal effects (temperatures of water and concrete) coupled with cracks (−0.345), water height in piezometers (−0.222), interaction effects of temperature, size of crack openings, the water level in the surge tank ($T_{wcc}\overline{H}_{st}$) (−0.266), and the exponential influence of time since the last repair (−0.185). Standardized beta values indicate the number of standard deviations that scores in the dependent variable would change if there was one standard deviation unit change in the predictor [17]. In model (10), if we could increase the water level in the surge tank ($\overline{H}_{st}$) by one standard deviation (which is 10.64 m a.s.l., from the descriptive statistic in Table 1) the total water loss (q) would be likely to increase by 0.855 standard deviation units (which is 4.42548 L/s, for $\sigma = 5.176$ L/s, according to Table 1). The p-value column shows that each regressor in the model gives a unique contribution to the model (for each regressor, p-value associated with the F-statistic is less than 0.05). The squared values of semi-partial correlation coefficients (*Part*) indicate the values of individual regressor contributions to the total variance of the dependent variable. These unique contributions to the q variable variance for the regressors $\overline{H}_{st}$, T_{wcc}, hP, T_{wcc}, $\overline{H}_{st}$, e^{-t}, $\overline{H}_{st}^2$, and T_{wcc}^2 are 0.6626, 0.0888, 0.0339, 0.0493, 0.0324, 0.0083, and 0.0049, respectively. In other words, the water level in the surge tank contributes 66.26% to the variance of total water losses in HT, temperatures of water. and concrete temperature on the extrados of the concrete lining in the tunnel coupled with cracks, contribute 8.88%, etc. Note that the R^2 value for the model (in this case 0.94014, or 94.014% explained variance) does not equal all the squared part correlation values added up (0.8801). This is because the part correlation values represent only the unique contribution of each variable, with any overlap or shared variance removed or partially taken out.

The quality of the adopted model (10) of total water loss in HT of PSHPP "Bajina Bašta" is visualized by the diagram which shows measured and predicted values of total water losses for the LDS (Figure 6). The 3D diagrams in Figure 7 correspond to the physical nature of the tunnel water loss phenomenon, as described above in Section 4.5.

Accuracy indicators for the adopted model, are given in Table 7. For the purpose of model comparison, RMSE values for the previous reference model of HT water losses [1] are also given in Table 7, for the same observation period.

Table 7. RMSE [L/s] of HTs total water loss.

	LDS	TDS
Model (10) adopted in this paper	1.8842	2.0161
Previous reference model	3.954	3.5983

Using the novel methodology for modeling water losses at HT, we obtained the $\mathrm{RMSE_{LDS}}$ of 1.8842 L/s, which is much better than 3.954 L/s obtained by the previous reference model on the same learning dataset. The model in this paper also performs better on the test dataset (2.0161 < 3.5983). The graphical comparison of the mentioned models, for the last four measurement sessions, together with confidence intervals, is given in Figure 8. In Figure 10, the comparison is made between the measured total water losses, the modeled total water losses using the model suggested in this paper (10), and the modeled total water losses using the model suggested by Andjelkovic et al. [4].

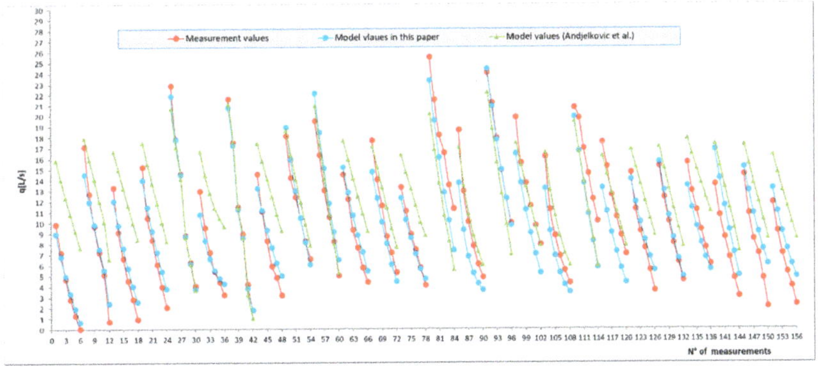

Figure 10. Measured values of total water losses, model values of total losses by the model in this paper and by the model from Andjelkovic et al. [4], for the period 2005–2015.

Further improvements in the quality of the mathematical model of water loss in the HT of PSHPP "Bajina Bašta" are described in the following two paragraphs.

It should be emphasized that the model developed in this paper also takes into account more input variables (size of crack openings and tunnel aging) than the previous reference model. The sizes of crack openings directly affect the increase or decrease in total water losses, which is important for the quality and accuracy of the proposed model. Consideration of the time component as an input variable gives the possibility of taking into account the irreversible process of concrete and structure aging, but also that, which is most important in the SaFE context, the effect of tunnel reparation on the reduction of total water losses in HT.

The advantage of the developed model is also the property of additivity, as there is an implicit assumption that the different model (10) regressors affect water losses in the HT additively. The aggregate and individual effects of PCs and regressors derived from them further shed light upon the process of water losses in the HT (Figure 9). For example, for the measurement sessions in which the temperatures of concrete were 10.7 and 5.4 °C (Figure 9b), the thermal effect is inversely proportional to water losses and amounts to −0.73 and 3.31 L/s (Figure 9a). From the SaFE aspect, the exponential influence of time elapsed from last tunnel reparation date in 2005. (predictor t) is particularly interesting. The diagrams shown in Figures 5 and 9a show that the impact of the last tunnel reparation on water losses decreases by the exponential law and becomes negligible after 8–10 years, i.e., 130 measurements, approximately.

The safety of the tunnel cannot be analyzed using the model of total water losses. For this, other methods based on numerical models are needed (FEM, DEM, PDEM, etc.), which were not the subject of research in this paper. For further research, an integral approach can be applied: the usage of numerical and statistical models. When the tunnel is reopened (upon its inspection), it is necessary to record the condition of the tunnel, conduct mapping of all the cracks in the tunnel, categorize the cracks, and form a three-dimensional numerical model that can analyze the behavior of the tunnel with regards to filtration and stress–strain phenomena. After that, a calibration of the numerical model with the total water loss measurements can be conducted, in order to obtain the best correlation between the calculated losses in the model and the measured total losses. This results in a model that takes into account the realistic field conditions and gives the possibility for a safety analysis of the tunnel. However, the use of statistically based models is an economical method that does not demand complex computing resources and know-how. It can be used daily, and it provides a fast assessment of the conditions in the tunnel and its functionality. For more complex analyses of the conditions in the tunnel and its functionality, more complex numerical models that can take into account significant phenomena of tunnel behavior in interaction with water should be used. Additionally, adequate constitutive material models that take into consideration the joints and porosity of the rock mass can be used to calibrate the numerical model more accurately, and therefore have a more realistic model compared to the field conditions.

6. Conclusions

Most pumped-storage hydroelectric power plants in Europe and the world are 30–50 years old. Within these complex hydro-technical facilities, there are HTs where monitoring and forecast of static instability in terms of failure and functional instability in terms of water losses from the tunnel are of enormous significance. The assessment of SaFE indicators in HT is generally conducted in two ways: (a) by emptying the tunnel and conducting physical inspection (b) by mathematical modeling of significant variables in HT with different types of analytical, numerical, statistical, or ML models. The literature review shows that there are not many articles dealing with problems of HT safety and functionality, in particular with problems of water losses from the tunnel. For these reasons, this paper presents a novel methodology for the mathematical modeling of water losses in HT. In addition, the goal was to validate this methodology by building an MLR model of water losses with greater accuracy and reliability, compared to previously developed models of HT water losses.

The methodology in this paper included the following modeling phases: correlation analysis, dimension reduction by PCA, generation of MLR models based on stepwise method, regression analysis, as well as iterative reduction of model complexity achieved by avoiding multicollinearity of the model, thus improving model stability.

A model of HT water losses developed based on the case study was carried out for the hydraulic tunnel at PSHPP "Bajina Bašta", located on the Drina river, in the Republic of Serbia for the period 2005–2017. The most accurate model, MLR_{Best}, which satisfies the condition of non-multicollinearity, was chosen as appropriate. A detailed regression analysis indicated that the regressors which have the greatest impact on HT water losses are water level in the surge tank, and the regressor that includes the impact of water temperature, concrete temperature and size of crack openings.

By comparing the model adopted in this paper with other literature for the same tunnel, it was concluded that the new model has better accuracy. It should be emphasized that the model developed in this paper also takes into account more input variables (size of crack openings and tunnel aging) than the previous model.

The analyses of water loss in hydrotechnical tunnels are of great significance because the loss of water implies economical losses, and on the other hand can point out that damages or changes in the tunnel have occurred, which can lead to loss of tunnel stability. Based on the proposed model in this paper, it is possible to determine, in a fast, efficient, and economically acceptable manner, any possible problems in the observed tunnel and provide indications for their resolution. Situations where there are significant deviations between the measured and model predicted values of water losses should demand a more detailed analysis, even including tunnel opening and its thorough inspection. Based on this analysis it can be determined if there is a need for new HT reparation.

Due to the immense importance that the SaFE paradigm for HT has in the PSHPP, the authors intend to further improve the model of total water losses. The improvement of the model would be mostly based on the introduction of new modeling methods, such as artificial neural networks, from which higher model accuracy can be expected.

Author Contributions: Data curation, S.R. and S.O.; Formal analysis, S.R. and S.O.; Methodology, S.R., D.D. and N.M.; Supervision, M.M., B.S., D.D. and N.M.; Validation, M.M. All authors have read and agreed to the published version of the manuscript.

Funding: This research received no external funding.

Institutional Review Board Statement: Not applicable.

Informed Consent Statement: Not applicable.

Data Availability Statement: Not applicable.

Conflicts of Interest: The authors declare no conflict of interest.

References

1. Liu, S.; Li, Z. Unloading Behaviors of Shale under the Effects of Water Through Experimental and Numerical Approaches. *Int. J. Geomech.* **2022**, *22*, 04022071. [CrossRef]
2. Li, Z.; Liu, S.; Ren, W.; Fang, J.; Zhu, Q.; Dun, Z. Multiscale Laboratory Study and Numerical Analysis of Water-Weakening Effect on Shale. *Adv. Mater. Sci. Eng.* **2020**, *2020*, 5263431. [CrossRef]
3. Roy, D.G.; Singh, T.N.; Kodikara, J.; Das, R. Effect of Water Saturation on the Fracture and Mechanical Properties of Sedimentary Rocks. *Rock Mech. Rock Eng.* **2017**, *50*, 2585–2600. [CrossRef]
4. Andjelkovic, V.; Lazarevic, Z.; Nedovic, V.; Stojanovic, Z. Application of the pressure grouting in the hydraulic tunnels. *Tunn. Undergr. Space Technol.* **2013**, *37*, 165–179. [CrossRef]
5. Rodrigues, R.V. Crack controlled design of RC pressure tunnels considering rock-structure interaction. In Proceedings of the Fib Symposium Prague 2011: Concrete Engineering for Excellence and Efficieny, Praha, Czech Republic, 8–10 June 2011.
6. Wan, W.; Zhang, B. The intermettent leakage phenomenon of incipient cracks under transient conditions in pipeline systems. *Int. J. Press. Vessel. Pip.* **2020**, *186*, 104138. [CrossRef]
7. Zhang, B.; Wan, W. A transient-features-based diagnostic method of multi incipient cracks in pipeline systems. *Int. J. Press. Vessel. Pip.* **2022**, *199*, 104701. [CrossRef]
8. Elaoud, S.; Hadj-Taieb, L.; Hadj-Taieb, E. Leak detection of hydrogen-natural gas mixtures in pipes using the characteristics method of specified time intervals. *J. Loss Prev. Process Ind.* **2010**, *23*, 637–645. [CrossRef]
9. Roy, U. Leak Detection in Pipe Networks Using Hybrid ANN Method. *Water Conserv. Sci. Eng.* **2017**, *2*, 145–152. [CrossRef]
10. Huang, F.M.; Wang, M.S.; Tan, Z.S.; Wang, X.Y. Analytical solutions for steady seepage into an underwater circular tunnel. *Tunn. Undergr. Space Technol.* **2010**, *25*, 391–396. [CrossRef]
11. Radovanović, S.D.; Rakić, D.M.; Lj Divac, D.; Živković, M.M. Stress-strain analysis and global stability of tunnel excavation. In Proceedings of the 2nd International Conference for PhD students in Civil Engineering and Architecture "CE-PhD 2014", Cluj-Napoca, Romania, 10–13 December 2014.
12. Sun, J.; Huang, Y. Modeling the Simultaneous Effects of Particle Size and Porosity in Simulating Geo-Materials. *Materials* **2022**, *15*, 1576. [CrossRef]
13. Javadi, A.A. Estimation of air losses in compressed air tunneling using neural network. *Tunn. Undergr. Space Technol.* **2006**, *21*, 9–20. [CrossRef]
14. Shahrokhabadi, S.; Toufigh, M.M. The solution of unconfined seepage problem using Natural Element Method (NEM) coupled with Genetic Algorithm (GA). *Appl. Math. Model.* **2013**, *37*, 2775–2786. [CrossRef]
15. Zhang, L.; Wang, M.; Zhao, H.; Chang, H. Uncertainty quantification for the mechanical behavior of fully grouted rockbolts subjected to pull-out tests. *Comput. Geotech.* **2022**, *145*, 104665. [CrossRef]
16. Zhao, H. A practical and efficient reliability-based design optimization method for rock tunnel support. *Tunn. Undergr. Space Technol.* **2022**, *127*, 104587. [CrossRef]
17. Pallant, J. *SPSS Survival Manual*, 3rd ed.; McGrath Hill: New York, NY, USA, 2007.

18. Tabachnick, B.G.; Fidell, L.S. *Using Multivariate Statistics*; Allyn & Bacon/Pearson Education: Boston, MA, USA, 2007.
19. Cohen, J.; Cohen, P.; West, S.G.; Aiken, L.S. *Applied Multiple Correlation/Regression Analysis for the Behavioral Sciences*; UK Taylor Fr.: London, UK, 2003.
20. Milivojevic, M.; Stopić, S.; Stojanovic, B.; Dmdarevic, D.; Friedrich, B. Forward stepwise regression in determining dimensions of forming and sizing tools for self-lubricated bearings. *Metall* **2013**, *67*, 147–153.
21. Stojanovic, B.; Milivojevic, M.; Ivanovic, M.; Milivojevic, N.; Divac, D. Adaptive system for dam behavior modeling based on linear regression and genetic algorithms. *Adv. Eng. Softw.* **2013**, *65*, 182–190. [CrossRef]
22. Jolliffe, I. Principal component analysis. In *International Encyclopedia of Statistical Science*; Springer: Berlin/Heidelberg, Germany, 2011; pp. 1094–1096.
23. Der, G.; Everitt, B.S. *A Handbook of Statistical Analyses Using SAS*; Chapman and Hall/CRC: Boca Raton, FL, USA, 2008.
24. Draper, N.R.; Smith, H. *Applied Regression Analysis*; John Wiley & Sons: Hoboken, NJ, USA, 2014.
25. Zheng, A.; Casari, A. *Feature Engineering for Machine Learning: Principles and Techniques for Data Scientists*; O'Reilly Media, Inc.: Sevastopol, CA, USA, 2018.
26. Kaiser, H.F. An index of factorial simplicity. *Psychometrika* **1974**, *39*, 31–36. [CrossRef]
27. Bartlett, M.S. A note on the multiplying factors for various $\chi 2$ approximations. *J. R. Stat. Soc. Ser. B* **1954**, *16*, 296–298. [CrossRef]
28. Cattell, R.B. The scree test for the number of factors. *Multivar. Behav. Res.* **1966**, *1*, 245–276. [CrossRef]
29. Horn, J.L. A rationale and test for the number of factors in factor analysis. *Psychometrika* **1965**, *30*, 179–185. [CrossRef]
30. Thurstone, L.L. *Multiple Factor Analysis*; Chicago Press: Chicago, IL, USA, 1947.
31. Fox, J. *Applied Regression Analysis and Generalized Linear Models*; Sage Publications: Thousand Oaks, CA, USA, 2015.
32. Montgomery, D.C.; Peck, E.A.; Vining, G.G. *Introduction to Linear Regression Analysis*; John Wiley & Sons: Hoboken, NJ, USA, 2012.
33. Russell, S.J.; Norvig, P. *Artificial Intelligence: A Modern Approach*; Pearson Education Limited: Kuala Lumpur, Malaysia, 2016.
34. R Core Team. *A Language and Environment for Statistical Computing*; R Core Team: Vienna, Austria, 2014.
35. Watkins, M.W. *Monte Carlo PCA for Parallel Analysis [Computer Software]*; Ed & Psych Associates: State College, PA, USA, 2000; pp. 432–442.
36. Hurvich, C.M.; Tsai, C.-L. A corrected Akaike information criterion for vector autoregressive model selection. *J. Time Ser. Anal.* **1993**, *14*, 271–279. [CrossRef]

MDPI
St. Alban-Anlage 66
4052 Basel
Switzerland
Tel. +41 61 683 77 34
Fax +41 61 302 89 18
www.mdpi.com

Applied Sciences Editorial Office
E-mail: applsci@mdpi.com
www.mdpi.com/journal/applsci

www.ingramcontent.com/pod-product-compliance
Lightning Source LLC
Chambersburg PA
CBHW061611220926
51850CB00005B/478
* 9 7 8 3 0 3 6 5 5 5 3 3 1 *